OCEANOGRAP

AN INTRODUCTION TO THE MARINE ENVIRONMENT

OCEANOGRAPHY

AN INTRODUCTION TO THE MARINE ENVIRONMENT

Richard A. Davis, Jr.
University of South Florida

Wm. C. Brown Publishers
Dubuque, Iowa

Book Team
Editor, Edward G. Jaffe
Developmental Editor, Lynne M. Meyers
Designer, Barbara J. Grantham
Permissions Editor, Vicki Krug
Photo Editor, Michelle Oberhoffer
Product Manager, Matt Shaughnessy

wcb group

Wm. C. Brown *Chairman of the Board*
Mark C. Falb *President and Chief Executive Officer*

wcb

G. Franklin Lewis *Executive Vice President, General Manager*
E. F. Jogerst *Vice President, Cost Analyst*
George Wm. Bergquist *Editor in Chief*
Beverly Kolz *Director of Production*
Chris C. Guzzardo *Vice President, Director of Sales and Marketing*
Bob McLaughlin *National Sales Manager*
Colleen A. Yonda *Production Editorial Manager*
Marilyn A. Phelps *Manager of Design*
Faye M. Schilling *Photo Research Manager*

Library of Congress Catalog Card Number: 86-71765

ISBN 0-697-00312-4

Printed in the United States of America
10 9 8 7 6 5 4 3 2 1

Dedication

To Laurie and Lee, they have made us very proud.

Contents

Boxes

PART ONE

PART TWO

PART THREE

PART FOUR

PART FIVE

Preface

Most college textbooks originate from frustration with existing books and the desire to have a volume that rather closely follows one's philosophy of teaching a particular course. I have taught an introductory level course in oceanography for more than twenty years and have tried a variety of texts. I even wrote one several years ago for much the same reason that I have written this book although the two are quite different. This text has been written to provide an alternative approach to that of previously published texts for introducing students to the complexities of the world ocean. This is accomplished by treating each of the major environments in an integrated fashion by incorporating their various attributes under a single chapter.

This book is intended to provide students with an appreciation for and general familiarity with marine environments. The book is written at a level that any undergraduate student, regardless of interest and background, should be able to understand. It is well-suited for an introductory course for nonscience majors wishing to satisfy a college science requirement, although science majors would also benefit greatly from exposure to the breadth of material it contains.

The numerous existing introductory-level texts in oceanography are quite similar to each other in approach although format, organization, and detail vary. The one aspect that does not show marked variation from text to text is the content of the book and the broad way in which it is packaged; that is, the general organization into biological, chemical, geological, and physical aspects of the ocean. Such an approach to learning about the world ocean does not provide the student with an impression of the environment but only of individual and isolated facts; it is a "nuts and bolts" approach. There appear to be two possible reasons as to why this trait has persisted among texts for essentially twenty years: either it is the best way, or it is the most convenient way.

It is my opinion that the reason is the latter one and it is largely for that reason that this textbook is written. Obviously, as in any introductory level book, it is necessary to cover basic principles and those principles are herein organized along the traditional lines. This portion of the book, however, accounts for less than half the total. The main emphasis of the book is to treat each of the major marine environments in terms of the interactions of the biological, chemical, geological, and physical phenomena that are present. It is not possible to convey the real sense of an estuary, a reef, or the deep ocean floor without considering all of its features. This isn't to say that each of the four major aspects of an environment will receive equal coverage in each of the topics discussed. In a delta the physical and geological aspects will be emphasized, in reefs the biological and geological and so on. The intent is to provide the reader with a true feeling for the nature of the marine environments discussed and by this approach provide a much better understanding of the oceans.

The first chapter is an introduction to the subject of oceanography: what it is, why we study it, who studies it, and how they do it, and some aspects of the history of oceanography. The following two chapters provide the physiographic and tectonic setting for the oceans. The next seven chapters are the principles section, containing the usual major topics. The chapters in the first section of the book are organized so that they provide a ready reference in the event that the reader needs to check back on a principle while reading the later chapters. Most of the remainder of the text (chapters 11–19) considers the major marine environments beginning with the coast and proceeding out to the deep ocean basins. The last two chapters deal with some of the economic aspects of the oceans: their resources and the impact of humans on the marine environment.

There are some major additions to topical coverage as compared to other texts. Probably the most obvious is the extensive treatment of coastal environments with four chapters devoted to this topic. These environments—estuaries, beaches, deltas, rocky shores—are the ones with which most people have some familiarity and will have some contact over their lifetime. Reefs are also given an entire chapter for similar reasons. Most existing texts on oceanography have little coverage on these environments.

The book contains some features that are designed to help the student understand the material and to make the subject more interesting. These include box inserts, which are scattered throughout most chapters and which cover such topics as data collection, historical aspects of oceanography, and examples of the subject being covered in the text. There are also brief summaries of the pertinent facts at the end of each chapter along with some suggestions for further reading. All words that appear in bold print in the text are listed in the glossary, with their definitions.

The metric system is employed throughout the book. Because some students may not be familiar with it, a convenient table of conversions is located in the appendixes. It is important for everyone to have a general understanding of what will become the standard system of measurement.

Many people have given valuable assistance throughout the writing and editing process. I am grateful to the many individuals and federal agencies who loaned photographs for use in the text and to the numerous reviewers including:

Stephen G. Lebsack	Linn Benton Community College
Philip A. Meyers	University of Michigan
William F. Kohland	Middle Tennessee State University
Paul R. Pinet	Colgate University
Ronald R. West	Kansas State University
Robert C. King	San Jose City College
Rodney Batiza	Northwestern University
Herbert F. Frolander	Oregon State University

Ed Jaffe, Lynne Meyers, Jan Appleby, and their staff at Wm. C. Brown have been patient and of great assistance.

Richard A. Davis, Jr.
Tampa, Florida
October, 1986

OCEANOGRAPHY

AN INTRODUCTION TO THE MARINE ENVIRONMENT

PART ONE

Introduction to the World Ocean

1
Introduction

Chapter Outline

Oceanography is the study of the oceans: their biology, chemistry, geology, and physics. It is not a science in the usual sense because it is the study of all sciences as they apply to the marine realm. Because oceanography is a young discipline, there has been a tendency to consider it a science in the same sense that we consider the study of outer space a science. This, too, is really the application of the various sciences to space and celestial bodies. An oceanographer is really a biologist, chemist, meteorologist, or any other type of scientist who applies the principles of his or her discipline to the marine world. Someone who is an expert on fishes is called an ichthyologist. This person's interests may be in freshwater fish or in marine fish. If it is the latter, that ichthyologist is considered a marine biologist or a biological oceanographer. Some people in the profession would, however, argue that a biological oceanographer is more than just a marine biologist in that the oceanographer is one who understands and works with all aspects of the ocean environment and the organisms that inhabit it. They consider the marine biologist to be one who studies marine organisms but is concerned primarily with the organisms and not with how those organisms fit into the grand scheme of things in the ocean. This may seem to be a subtle difference; nevertheless those in the profession consider it important.

The Professional Oceanographer

Oceanography is a young discipline, barely a century old. Although persons have been collecting oceanographic data and solving oceanographic problems longer than that, specific training for the profession has been going on for much less than a century: the first specially designed curricula in the marine sciences were developed after World War II. Since that time there has been a great explosion of activity in the education and training of oceanographers. Marine research laboratories began to affiliate with colleges and universities, offered research opportunities for graduate students, and then also began to offer courses. In some instances, marine research laboratories became a part of academic institutions. For example, Scripps Institution of Oceanography at La Jolla, California, became part of the University of California. During the 1950s several coastal universities began their own marine study programs and began to grant degrees in oceanography or marine science, for example, the University of Washington at Seattle and the University of Miami, Florida. Not long after, many land-locked universities, such as Texas A&M University and the University of Michigan, initiated large-scale marine programs, most using classroom and laboratory facilities on the main campus with ship-support facilities and perhaps some laboratories at a coastal facility.

The vast majority of all degree programs in oceanography are at the graduate level, making oceanography a professional curriculum much the same as law, medicine, dentistry, or other post-baccalaureate degree programs. There are a few academic institutions that offer an undergraduate degree in oceanography, but many people feel that such curricula are unnecessary and in some respects are not good preparation for advanced work because of their lack of depth in a particular discipline of science. There are also institutions that offer programs leading to technical preparation for oceanographic employment. These include various aspects of support for oceanographic research, such as ship and equipment maintenance, ship operation, laboratory technology, and other activities that support the research activities of oceanographers.

Nearly all persons who aspire to be oceanographers must first obtain a bachelor's degree in one of the basic disciplines of science or mathematics. Regardless of which discipline one chooses, he or she should take at least one year in each of the other basic sciences, including mathematics through calculus. This provides the best background for graduate study in any of the marine sciences. A basic course in oceanography is also advised, primarily because it provides the student with a taste of what oceanography is all about, not because it is intended as the first course in a professional curriculum.

By the time one has completed such a course of study at the undergraduate level, a decision should have been made not only about continuing on to graduate school but also about whether or not to enter an oceanography curriculum in graduate school.

The student with interests in oceanography should be aware that a graduate degree is mandatory for a professional career in the field. Positions open for bachelor's degree holders are few in number and are restricted to technicians, laboratory assistants, and other research support roles, not professional positions. The majority of research-level positions in this field are restricted to those persons who earn the doctorate. Some desirable positions are available for the master's degree holder, but these are typically restricted to certain areas of specialization and are commonly restricted to governmental agencies or private consulting firms. As is the case in many of the scientific disciplines, employment opportunities vary greatly from time to time due to supply and demand. The student who intends to obtain a degree in oceanography should investigate such opportunities carefully prior to pursuing a degree program if a major goal is employment in the field—and it almost always is!

Because oceanography or marine science is a broad field and covers several traditional scientific disciplines, it is appropriate to consider each major area within the field and briefly describe what a person in the field might do in performing her or his job. Keep in mind that people in this profession are typically trained for specific tasks and the scope of their activities may be narrow.

A physical oceanographer is a person who studies the physics of the ocean and perhaps the atmosphere above it, although the title of marine meteorologist may also be used for the latter. The basic training of such a person is in physics and mathematics as applied to fluids. Such people concern themselves with water density, currents, waves, tides, light, temperature distribution, and many of the other dynamic physical aspects of the oceans. Research of this type not only takes well-trained specialists but requires a great deal of expensive equipment for sampling the phenomena being investigated and for data processing of the measurements. Physical oceanography is probably the area where there is the greatest demand for research personnel, a situation that has persisted continuously.

Chemical oceanographers are chemists who apply their expertise to problems of the ocean. They investigate salinity, concentrations of various elements in seawater, dissolved gases in the oceans, precipitation of materials from seawater, and various cycles of certain elements in the sea. They commonly work closely with biologists on aspects of chemistry that affect organisms in the oceans. A major area of investigation for chemical oceanographers is the pollution of the marine environment. For this reason, as well as a long-term demand for such people, chemical oceanographers enjoy good employment opportunities.

Biological oceanographers study the organisms of the ocean, their interrelationships, and their interactions with their environment. These people tend to specialize in certain groups of organisms; they may concentrate on a broad category like algae (phycologist) or fish (ichthyologist), or perhaps a narrow one, such as euphausiid shrimp or planktonic foraminifera. Some biological oceanographers concentrate on a specific environment and the community that occupies it, such as the intertidal rocky shore or the bottom of the continental shelf. This specialization is by far the largest of the areas of marine science and is also the most crowded, with the poorest employment outlook. Perhaps some of this situation can be attributed to the glamour, as portrayed on television and in the movies, of studying organisms on reefs and other exotic marine environments.

Marine geologists and geophysicists (geological oceanographers) investigate the sediments and rock strata that underlie the oceans as well as the shape, origin, and history of the ocean basins. These scientists tend to specialize as either geologists or geophysicists for specific portions of the ocean basin. For example, there are continental margin geophysicists, geologists who study deltaic sediments and those who are interested in the major fault systems present on the ocean floor. Employment opportunities for geological oceanographers are spotty but overall much better than for biological oceanographers. Marine geophysicists are typically in high demand, almost equal to physical oceanographers. This is largely because of the extensive effort at offshore petroleum exploration and the recent emphasis on understanding the origin and history of ocean basins. Marine geophysicists provide the first line of data gathering and interpretation in such studies. Marine geologists have experienced varied demand for their expertise over the past several decades because much of this employment is tied directly to petroleum exploration, which fluctuates greatly. Both marine geology and marine geophysics are areas in which a person with a master's degree is employable. Positions available to master's degree holders are largely in private industry and are generally not of a basic research nature.

History of Oceanography

In order to consider the history of oceanography it is necessary to look at the history of human involvement with the ocean. The primary uses of the ocean have been as a food source, for transportation, and as a site of conflict. These continue to be among the major uses, although the quest for natural resources is also important. Some of the earliest observations about the ocean include the phenomenon of tides, the fact that sea level stays nearly constant, and the relationship between wind and waves.

The Classical Era

There are indications that as early as 3000 B.C., Indian Ocean traders were aware of the intense storms and rain that came every year during the monsoon season and the effect they had on ocean currents. Ancient Indians and Polynesians had the ability to sail long distances over open water and must have had knowledge of celestial navigation to do so.

The best documentation of ocean activities is in the area of the Mediterranean Sea and adjacent waters, where western civilization began. In fact, many of the ancient Greeks and Phoenicians believed that the world consisted of a landmass surrounding the Mediterranean, which was in turn surrounded by an all-encompassing water mass, Oceanus (fig. 1.1). As early as 1500 B.C., the Phoenicians sailed to the Straits of Gibraltar. The Phoenicians and Greeks sailed throughout the Mediterranean, the Red Sea, and the Persian Gulf.

Although much of the early exploration and trading was conducted by the Phoenicians, it was the Greeks who contributed most of the early scholarly knowledge about the sea. Early maps were produced by Hecataeus (fig. 1.1) and Herodotus in the fifth century B.C. Pytheas was the first to determine latitude, which he did while on a trip to the British Isles in the fourth century B.C., by measuring the angle between the horizon and the North Star, which is directly above the North Pole. This angle is the latitude in degrees north of the equator.

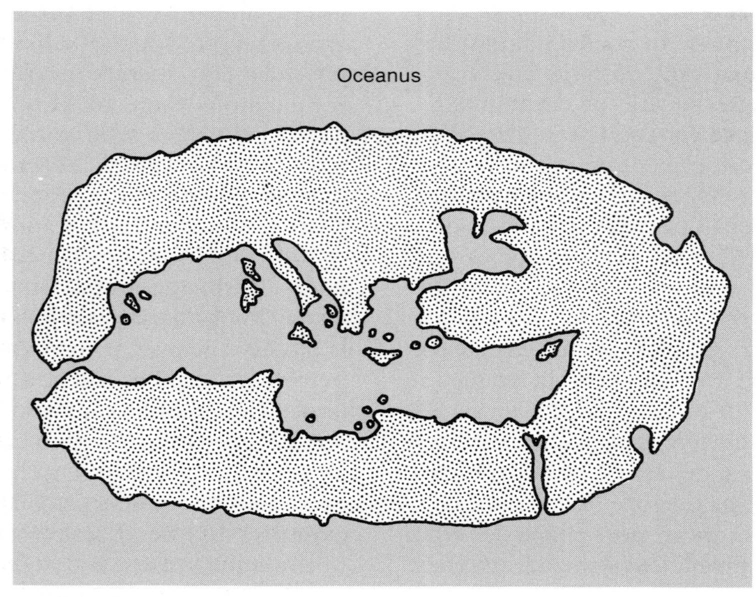

Figure 1.1 Ancient world map according to Hecataeus (500 B.C.). The landmasses surround what we now call the Mediterranean Sea and are surrounded by Oceanus, the unknown world ocean at that time. (After P. Groen, 1967, p. 2.)

About a century later, Eratosthenes, who believed that the earth was spherical, calculated its circumference at 40,000 km, a remarkably accurate value. His method also relied upon a celestial body, the sun. Eratosthenes measured the angle of a shadow of a wall that was a known distance due north of a water well at the same time as the sun was directly over the well thus casting no shadow. Assuming the earth to be a sphere, this angle and the known distance along the earth's surface between these structures (fig. 1.2) permitted the calculation of the circumference of the earth.

$$\frac{\text{angle of shadow } (7.2°)}{\text{distance between structures } (800 \text{ km})} = \frac{360°}{\text{circum. of the earth}} \quad (1.1)$$

The Roman civilization also contributed to our basic knowledge of the oceans, notably through the efforts of Strabo and Seneca. At the time of Christ, Strabo observed that the relative position of land and sea changed after volcanic activity, evidenced by uplift and subsidence of land. Strabo also noted that rivers carried sediment to the sea; however, the full **hydrologic cycle** was first described by Seneca. He recognized that even though rivers flowed into the sea, water level remained constant, which he attributed to the hydrologic cycle: evaporation into the atmosphere, condensation, precipitation, runoff, then evaporation again.

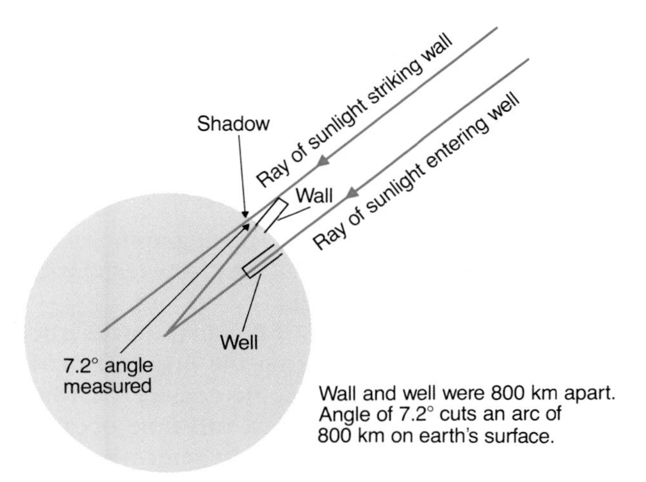

Figure 1.2 Diagram showing how Eratosthenes calculated the circumference of the earth more than 200 years before the birth of Christ. (From H. V. Thurman, 1985, *Introductory Oceanography,* 4th edition, Merrill Publishing Company, Columbus, Ohio. Reproduced by permission of the publisher.)

Age of Discovery

Little significant new knowledge of the oceans was acquired for nearly a thousand years, until the discoveries of the Norsemen about 1000 A.D. Erik the Red and his son, Leif Eriksson, were among the most notable. They discovered Iceland, Greenland, and North America, where they landed on Newfoundland and Labrador.

By far the most concentrated effort in oceanic exploration was spawned by the increasing difficulty in crossing traditional land trade routes to the Far East, the result of extensive and continuous wars in the Middle East. Prince Henry the Navigator of Portugal established a school of navigation in the fifteenth century. This center was responsible for most of the major voyages of discovery during the late fifteenth and early sixteenth centuries, including those of Bartholomeu Dias (1487), Christopher Columbus (1492), Vasco da Gama (1499), Vasco de Balboa (1513), and Ferdinand Magellan, one of whose ships completed the first circumnavigation of the world in 1522 (Magellan himself died before completing the voyage). The prime objective of these voyages was to find new sailing routes to the trading ports of the Far East. Oceanographic data was an important secondary result, including the first depth measurements and the discovery by Ponce de León of the current now called the Gulf Stream. By the end of the sixteenth century, charts of all the major landmasses of the world had been compiled, except those in the high latitudes. Significant additions to these global maps were made by the English explorer James Cook, who charted the South Pacific, including Australia and New Zealand. Captain Cook commanded three voyages between 1768 and 1779 and is credited with being the first to take a scientist on a voyage for the sole purpose of making observations and collecting data on natural phenomena. The first oceanographic publication was Benjamin Franklin's map of the Gulf Stream in 1769 (fig. 1.3). Franklin's map was the result of a synthesis of data from numerous crossings of the North Atlantic.

Figure 1.3 The Gulf Stream as depicted by Benjamin Franklin in 1769. This map is widely credited as the first publication on oceanography.

Science and the Sea

The beginning of the nineteenth century was also the start of widespread scientific curiosity about the oceans. Among the first contributors was William Eaton, who collected marine specimens and made soundings along the coast of England in 1804. The same year, an unknown American sailor tested light penetration in the sea by lowering a dinner plate overboard and noting the depth at which it disappeared from sight. The first federal scientific organization was started in 1807. The Coast Survey, later known as the Coast and Geodetic Survey and now incorporated in the National Oceanographic and Atmospheric Administration (NOAA), was responsible for charting navigable waters.

Numerous noteworthy individuals made their mark on ocean science during the first half of the nineteenth century. Many people consider Matthew F. Maury (1806–1873) the first full-time oceanographer. Lt. Maury (fig. 1.4) is best remembered for his book, *The Physical Geography of the Sea*, the first textbook in oceanography.

British naturalist Edward Forbes was the first person to systematically study life in the sea. In the 1830s he conducted detailed dredging programs for

BOX 1.1

Lieutenant Matthew Fontaine Maury (1806–1873)

M. F. Maury was a United States naval officer who became one of the world's first oceanographers, certainly the first in North America. Maury's shipboard assignments were at a time of peace and were largely concerned with surveying and recording observations on currents, weather, and other oceanic phenomena. During a surveying cruise of southern harbors, he had an accident that injured a leg and rendered him lame. As a consequence of this accident, Lt. Maury was reassigned, in 1841, to land-based duties. Because of his experience and interest in charting and navigation, he was placed in the Navy's Depot of Charts and Instruments (which later became part of the United States Hydrographic Office). During the many years of his association with this agency, Maury devoted most of his efforts to collecting and compiling shipboard observations of wind, waves, currents, and other data. These were compiled and made available to military and merchant ships as navigation aids, a policy that has continued to the present time.

Maury has several "firsts" to his credit in the general area of oceanography. He drew some of the first charts of currents in the North Atlantic, based upon thousands of shipboard observations of the deflection of course and wind. Maury also constructed the first bathymetric chart of the floor of the North Atlantic. His contributions were such that charts currently issued by the Hydrographic Office contain a statement of acknowledgment for the pioneering efforts of Lt. M.F. Maury. He was a primary organizer of the first international conference on marine navigation and meteorology, which was held in Brussels in 1853. Probably the one thing that Maury is best known for is his book *The Physical Geography of the Sea,* which was published in 1855. This book went through eight editions and is considered the first book on oceanography. Lt. Maury was indeed a major contributor to the field of oceanography and was one of the true pioneers in the study of the oceans.

Figure 1.4 Lt. Matthew Fontaine Maury of the U.S. Navy, the man many consider to have been the first full-time oceanographer.

specimens around the British Isles, in the Mediterranean Sea, and in the Aegean Sea. From data collected, Forbes concluded that life was absent below a depth of 600 m, thus he named this the azoic (no life) zone. He was the first to systematically categorize the depths of the sea according to its organisms.

Charles Darwin, a contemporary of Forbes, made important contributions to our knowledge of the oceans, in addition to developing his theory of evolution. Darwin was a keen observer who kept prolific records of his observations. During Darwin's tenure as ship naturalist on the British survey ship HMS *Beagle,* he collected, described, and classified organisms from all places the ship traveled. Some of his most significant observations were on oceanic reefs in the Pacific. Darwin's detailed notes and keen

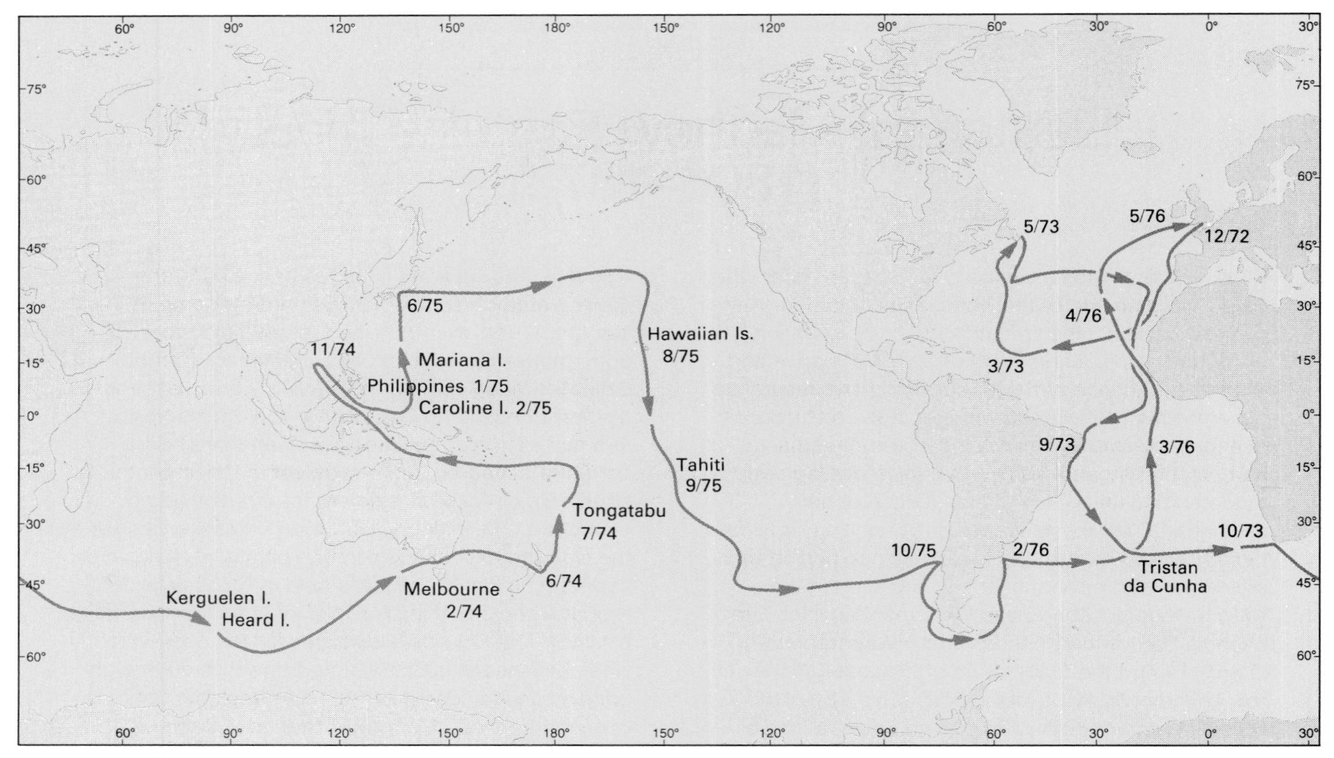

Figure 1.5 World map showing the course taken by the HMS *Challenger* in its around-the-world voyage between 1872 and 1876.

analytical mind led him to propose a theory on the origin and development of circular reefs (atolls) that, although challenged, has withstood both the tests of time and of new techniques for investigation.

Perhaps the major event in the study of the oceans during the nineteenth century was the voyage of the HMS *Challenger,* which took place from December 7, 1872 to May 26, 1876, and was the first extensive voyage for the sole purpose of collecting scientific data on the oceans (fig. 1.5). The expedition was sponsored by the Royal Society and was under the scientific supervision of Sir Charles Wyville Thomson. The scientific party was charged to learn all they could about the oceans. The cruise covered all parts of the world ocean and collected data and specimens from the entire water column. The *Challenger* expedition stands as a major contribution to our knowledge of the ocean environment.

Fridtjof Nansen was a Norwegian oceanographer who explored the high latitudes of the northern hemisphere near the end of the nineteenth century. He was particularly interested in the currents and the drift of ice. In order to test his theories on these topics, Nansen allowed his ship, *Fram,* to be frozen into the ice. He drifted with the ice, making observations for three years (1893–1896). Nansen also made major contributions to physical oceanography by devising water and temperature sampling bottles (Nansen bottles), which enabled accurate measurements of water at depth.

The voyage of the German vessel *Meteor* from 1925 to 1927 was one of the first systematic research cruises designed primarily for nonbiological oceanographic study. It crossed the central and south Atlantic Ocean fourteen times, measured hundreds of temperature-salinity profiles, and conducted a detailed survey of the ocean floor. This was the first comprehensive physical oceanographic study of a major ocean basin.

BOX 1.2

Voyage of HMS *Challenger* (1872–1876)

Charles Wyville Thomson was a professor of natural history at the University of Edinburgh, the same post that Edward Forbes had held. Thomson and a colleague, W. B. Carpenter, were allowed to conduct extensive studies from British naval vessels of marine organisms and water chemistry. Eventually, the British Admiralty supported their efforts with substantial ships, and finally, with a ship and crew for an around-the-world expedition solely for scientific inquiry.

The Royal Society and the Admiralty supported the venture by providing the HMS *Challenger*. Charles W. Thomson was the chief scientist. The scientific party included four naturalists, a chemist, and a secretary. The ship departed port in December 1872 and sailed a total of 127,600 km, establishing many scientific milestones before its return in May 1876. The data and samples collected were voluminous and included 133 dredge hauls of rock and sediment from the ocean floor, 492 soundings to the ocean bottom, including one to a depth of over 8,000 m, 263 water temperature observations, and the discovery of 4,717 new species of organisms. In addition, numerous water samples were collected for later analysis. These analyses became the standard for seawater composition.

This wealth of information was to be compiled by Thomson and colleagues into published form through the *Challenger* Expedition Commission. His sudden death resulted in Sir John Murray, one of Thomson's assistants, replacing him. The result was a fifty-volume set of reports that has become a classic of marine research. The collections of this expedition are still used by researchers for comparison and reference.

As the result of the vast amount of interest in the oceans and data being collected, research laboratories specifically for marine research were established. The first was the Zoological Station of Naples, established in 1872. One of the most famous of all these laboratories is the Marine Biological Association of the United Kingdom at Plymouth, England, established in 1879. The first such facility in the United States was the Marine Biological Laboratory at Woods Hole, Massachusetts. It was later joined by the Woods Hole Oceanographic Institution.

The first facility on the west coast was established in 1892 by Stanford University: the Hopkins Marine Station at Pacific Grove, California. Shortly thereafter (1905) the Scripps Institution of Oceanography at La Jolla was started. During the next several decades there were several private and university-related marine research laboratories established in the United States. The Lamont Geological Observatory (Palisades, New York) was established in 1947 and affiliated with Columbia University. Many people consider Woods Hole, Scripps, and Lamont as the "big three" in oceanographic institutions in this country, although there are several others that may now challenge that designation including the University of Rhode Island, University of Miami, Texas A&M University, Oregon State University, and the University of Washington.

Modern Era of Oceanography

Advances in the study of the oceans made during World War II ushered in oceanography as a sophisticated subject of research, especially because of advancements in data collection devices and methods. Many wartime activities of the Navy have had a profound effect on research efforts in oceanography. One of the most widely used and important devices is the echo-sounder, used in precision depth recorders for bathymetric surveys. Originally developed shortly after World War I, this device was greatly improved and widely used during World War II. Great quantities of bathymetric data provided by this device

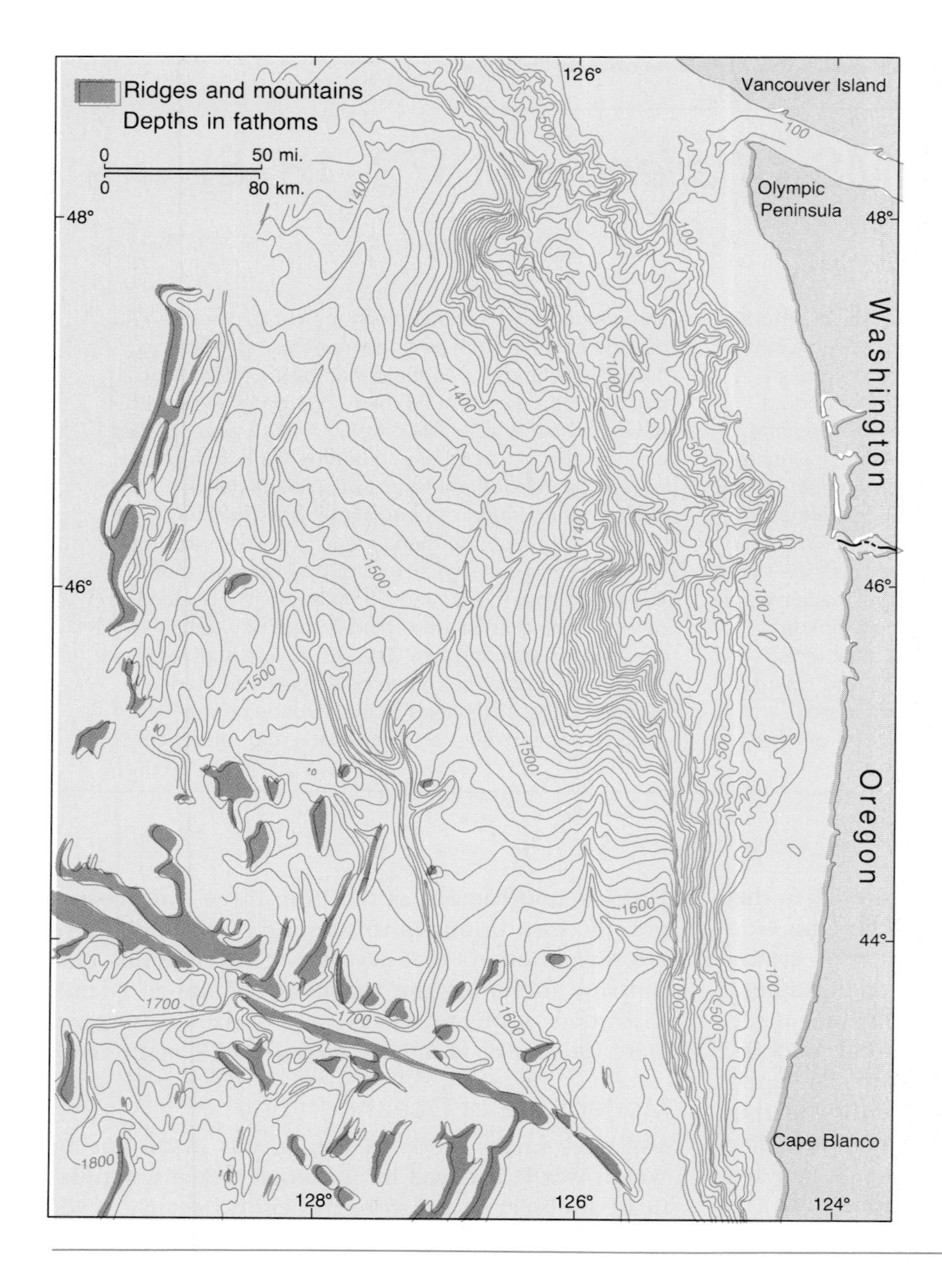

Figure 1.6 An example of a detailed bathymetric map of a portion of the ocean floor off the northwestern United States.

have enabled development of detailed and accurate maps of the earth's ocean floor (fig. 1.6). Another sonic technique developed during World War II is the use of explosives to produce seismic vibrations for study of the earth's crust under the ocean.

One of the most widely used pieces of research and recreational apparatus is the aqualung or SCUBA (Self-Contained Underwater Breathing Apparatus). Although originally invented in the mid-nineteenth century, it was reinvented by Jacques-Yves Cousteau and Emile Gagnan in the Mediterranean Sea in 1943. This apparatus enabled researchers, for the first time, to observe the underwater marine environment for extended periods of time in reasonable comfort. Use of this equipment is limited to reefs, the inner continental shelf, and other shallow environments; however, there are special gas mixtures that enable expert divers to descend to depths of more than 100 m.

(a)

(b)

Figure 1.7 (a) Photograph of the deep-sea submersible *Alvin,* which is owned by NOAA and operated by the Woods Hole Oceanographic Institution. (b) *Alvin* moving toward Lulu, its former tender.

Eventually the desire to descend to greater depths led to a variety of manned deep-water submersibles. The first of these modern vessels, the *Trieste,* was developed in the early 1960s. This began a period of great expansion in building and design of manned submersibles. They varied in size, shape, number of passengers, depth capabilities, and data gathering capabilities. Probably the most famous and most successful of these vessels is the *RV Alvin* (fig. 1.7), which is operated by the Woods Hole Oceanographic Institution with funding support primarily from the National Science Foundation and the Office of Naval Research. In addition to a wide variety of scientific expeditions, the *Alvin* has been used to locate nuclear warheads and sunken ships on the seafloor. Among the most significant recent *Alvin* discoveries is the presence of high-temperature vents, associated sulfer compounds, and large bottom-dwelling fauna on the spreading ridge called the East Pacific Rise in the Pacific Ocean.

BOX 1.3

Deep Sea Drilling Project (DSDP)

The Deep Sea Drilling Project was both a technical marvel and a tremendous scientific success. Some people would say that the engineering and technical accomplishments were as valuable as the scientific data. The *Glomar Challenger* provided a drilling base that allowed accurate and stable positioning, even in fairly rough seas with thousands of meters of drill stem between the ship and the ocean floor. Unlike the nineteenth century *Challenger,* this vessel carried a ship's crew, drilling crew, scientific support party, and a scientific party, totalling nearly one hundred people. The ship was so well equipped that it was nearly like a small town. Not only were the laboratories fully equipped for all types of analytical work, but there was a library, a gymnasium, movie room, and other amenities. The ship made port every two months to take on supplies and to change the scientific crew.

DSDP's expenses were about ten million dollars per year. While this sounds like a great deal of money, there was a very good return on the dollar: there will be nearly 100,000 pages of reports published when all are completed and the average cost of the project will have been about $60,000 per scientist-year. The project drilled a total of over 160 km of holes and recovered more than 50 km of cores. The deepest penetration into the ocean floor was 1.7 km; the deepest water in which drilling took place was 7,000 m. The average depth of the holes was 300 m. The density of drilling sites on the ocean floor is only, however, about one per 800,000 km^2. Therefore, even though this was a tremendously successful program and an enormous data set was collected, it really just scratched the surface.

Probably the biggest single factor in the promotion and perpetuation of oceanographic research in the United States was the establishment of the National Science Foundation in 1950. This federal agency receives its budget directly from Congress and is the major support vehicle for all oceanographic research in this country. The Foundation has built several research vessels and supports the operation of those plus many more. In addition, it supports large numbers of scientific projects in all areas of the marine environment. The other major benefactor of marine research is the Office of Naval Research, which supports applied research projects. Although the goal of ONR research is to enhance activities related to national defense, much basic scientific research has resulted from ONR projects.

The biggest single project sponsored by NSF and one of the biggest research projects of all time was the Deep Sea Drilling Project (DSDP). This was the most significant marine project of the century, perhaps ever. It began in 1963, originally as a project for taking cores of sediments of the ocean basins. The Joint Oceanographic Institutions for Deep Earth Sampling (JOIDES) program included Scripps Institution of Oceanography, Woods Hole Oceanographic Institution, Lamont-Doherty Geological Observatory of Columbia University, and Rosenstiel School of Marine and Atmospheric Sciences of the University of Miami. Later the group also included the University of Washington, Oregon State University, the University of Hawaii, Texas A&M University, the University of Rhode Island, and the University of Texas.

JOIDES expanded to the DSDP when the *Glomar Challenger,* a ship capable of drilling into the ocean floor in thousands of meters of water (fig. 1.8), was developed for this program. The first voyage began in 1968 and continued for the next fifteen years. Over 600 sites were drilled; cores were recovered from all parts of the world ocean. Scientists from all over the world participated in the project and the discoveries they made were both numerous and important. Much of the theory of how the crust of the earth moves (plate tectonics) has been confirmed by the DSDP. Additionally, DSDP provided good data on the

(a)

(b)

Figure 1.8 (a) The *Glomar Challenger,* the drilling ship that was used in the Deep Sea Drilling Project from 1968–1983. (b) Core repository for the DSDP at La Jolla, California.

age of the ocean basins, so that detailed correlations of layers of deep-ocean sediments are now possible. Each two-month cruise of the ship produced a volume of about a thousand pages of data. The number of these volumes has already surpassed the fifty-volume *Challenger Reports* and will total ninety-six volumes when completed.

The DSDP has now been replaced by the Ocean Drilling Program (ODP), which is headquartered at Texas A&M University. The project is currently planned for a duration of ten years and will continue the effort of exploring the world ocean in order to learn more about the composition and history of the earth. Joining the United States as formal cooperating partners are Canada, France, Japan, West Germany, and Great Britain.

The ODP drilling will be from a newer and more sophisticated ship than its predecessor. The *JOIDES Resolution* (fig. 1.9) is 470 feet (146 m) long and is capable of drilling almost anywhere in the oceans. Its positioning equipment can hold the ship steady during drilling and can reenter holes with relative ease. The ship carries a crew of sixty-five with a scientific crew of fifty. The drilling sites for the first two years have been selected and include the Mid-Atlantic Ridge, the Mediterranean Sea, the Peru Trench, the Bahama Banks, and polar waters.

Figure 1.9 The *JOIDES Resolution*, the new drilling ship that has replaced the *Glomar Challenger*. This ship is the next generation in sophisticated deep-sea drilling ships and is the primary vessel used in the current Oceanic Drilling Program (ODP).

Data Collection and Analysis

The nature of the marine environment makes it difficult to collect much of the data required to carry out scientific experiments and research. The vastness of the open ocean environment and the great depths of the sea cause problems for collecting samples. Fortunately, there is a relative homogeneity in the oceans that doesn't occur on land or coastal environments. Other problems that must be dealt with include the weather and sea conditions, the expenses of operating a research vessel, and equipment malfunctions that must be corrected on board the ship. The saltwater environment in which this equipment must operate is also a constant source of problems because of its corrosive effects. As a result of these and other problems associated with oceanographic research, there must be careful planning for any data-gathering expedition. Problems must be anticipated, back-up equipment made available if possible, time must be used efficiently, and participants have to be chosen carefully. In many ways, if an oceanographic research cruise is to be successful it must be like a well-prepared athletic team playing one of its best games.

Research Vessels

Ocean-going vessels designed and built specifically for marine research came into being early in this century. Since that time there have been enormous advances both in the vessels themselves and in the equipment they carry. This development in sophistication and expansion in numbers has taken place primarily since World War II, more specifically, since the early 1960s. There are two primary types of research vessels currently being utilized. One is the all-purpose type, designed to accommodate a wide variety of scientific studies both in terms of data collection and in terms of the preliminary studies that commonly take place on board ship. This is the most numerous type of research vessel currently in operation. The other type is designed primarily for a specific task. Often such ships can accommodate other activities, but typically these vessels are used for a narrow area of research. Some examples are trawlers, designed for collection of various biological specimens, ships with hulls of nonmagnetic material for geophysical surveys, and ships that are specially designed for deep-ocean drilling.

The size and complexity of research vessels ranges widely. In general, the most important thing is that the particular vessel is well-suited for the task at hand. There are many projects that can be carried out with small, unsophisticated vessels, especially in coastal areas such as estuaries, lagoons, and river deltas. These vessels are generally day-boats that go in and out of port on a daily basis. Most "blue-water" research is conducted in larger vessels that can leave port for at least a few days at a time and have a range of at least several hundred kilometers. These larger vessels range in length from about 15 m up to hundreds of meters.

There are hundreds of research vessels in the United States and at least that many more on a global basis. These vessels are operated by various agencies of the federal government, by various academic and other nonprofit research institutions, and also by private industry. The expense of maintaining and operating such research vessels is tremendous. Not only must each ship be kept in good operation, but also a full-time crew must be paid, expensive scientific equipment must be purchased and maintained, and the ship must have a land-based support group as well. As a result, most ocean-going research vessels cost from about two thousand to several thousand dollars per day to operate. The deep drilling ship *JOIDES Resolution* currently costs $75,000 per day to operate.

Most oceanographic research vessels in the United States are funded by various agencies of the federal government. In fact, the National Science Foundation supports the operation of a fleet of several research vessels that are attached to universities. NSF has even funded the cost of construction of some of these vessels. The standard procedure for funding most ship operations is for the scientist to include funds for ship rental in the budget of research proposals that are submitted to funding agencies. Because many of the ships are flexible in their operation, it is sometimes possible for one scientist to share ship time with another scientist working on a different but related project.

The permanent crew of a ship is tied to the size and complexity of the ship and, to some extent, to

BOX 1.4

Anatomy of a Research Vessel—*RV Researcher*

The *RV Researcher* (fig. B1.4a) is one of the largest vessels in the NOAA fleet. It is 84.8 m (278.3 ft) long, 15.5 m wide, and requires 5.6 m of water. This vessel was commissioned in 1970. Based out of Miami, Florida, the *Researcher* carries a crew of fifty, including thirteen officers, and can accommodate fourteen scientists. The cruising speed is 12.5 knots (nautical miles per hour), with a range of 10,800 nautical miles. The ship holds 180,000 gallons of diesel fuel, which will allow its 3,200 horsepower engines to run continuously for thirty-six days. During that period the rate of fuel consumption is 175 gallons per hour.

A research vessel of this type contains numerous laboratories and instrument rooms for data collection and analysis in addition to the usual types of facilities

Figure B1.4a The research vessel *Researcher,* one of the largest of the ships in the NOAA fleet.

the level of funding supporting its operation. In addition to the permanent crew, each scientific cruise or expedition also has a staff unique to the aims of the project. There is a chief scientist or sometimes two co-chief scientists who are in charge of the cruise. In addition, there may be other scientists, technicians, and commonly, students who are also involved in the research project. It is the job of the chief scientist to organize the entire project, including determining where the ship will go, the location of sampling stations, nature of the data to be collected, and the duty roster for the scientific crew. Such research cruises generally operate on a twenty-four-hour basis.

BOX 1.4

on any type of ship (fig. B1.4b). Research facilities include standard wet and dry oceanographic laboratories, a meteorological laboratory, and a darkroom. These are equipped with both fresh and saltwater systems, microscopes, balances, and other sample processing apparatus.

On deck are various types of machinery for sampling organisms, sediments, and water. These include heavy-duty cranes and hoists to lower and lift pieces of equipment that may weigh tons. Cables that can sample the bottom of the ocean are on drums attached to heavy-duty winches.

The *Researcher*, like all oceanographic vessels, has a variety of electronic positioning devices that enable the determination of a location anywhere in the world with great accuracy. In many situations it is necessary to reoccupy a sampling site, thus the need for extreme accuracy in location.

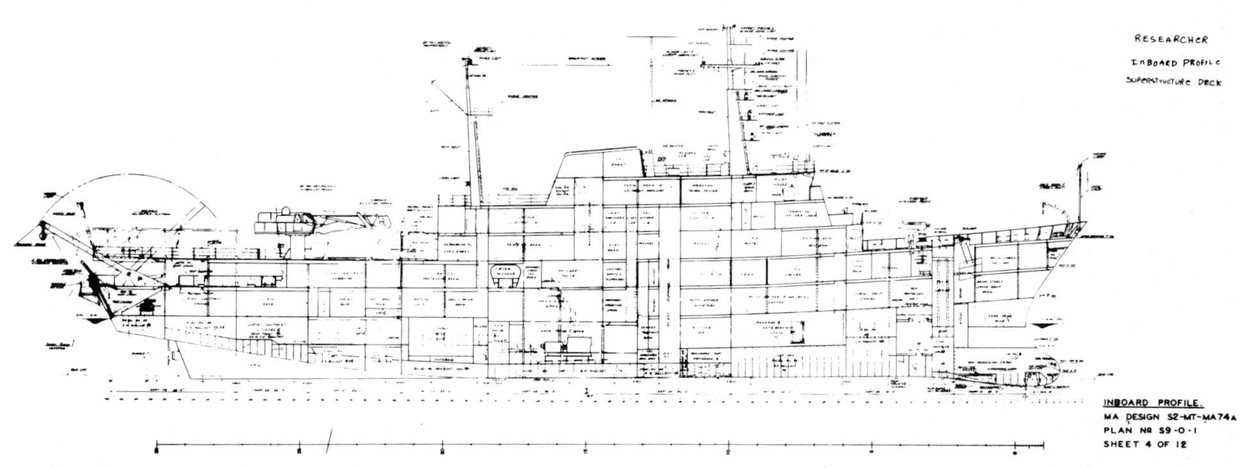

Figure B1.4b Diagram of the interior of the *Researcher* showing the numerous laboratories, storage rooms, and other facilities that are present on a large, sophisticated research vessel. (Photo courtesy of NOAA.)

Sampling and Boat Stations

The data collected on a research vessel vary. Some may be obtained while the ship is under way at cruising speed, whereas some must be collected while the ship is at rest or perhaps even anchored. The general nature of most sampling equipment has remained the same over the years, even though the level of sophistication has increased greatly. This section is not designed to serve as an encyclopedia of oceanographic equipment but to give the reader some feeling for the collection of data over the broad spectrum of oceanography.

While the ship is under way, it is common for a precision depth recorder (PDR) to be operating (fig. 1.10). The PDR charts the configuration of the bottom. It is also possible to collect some biological

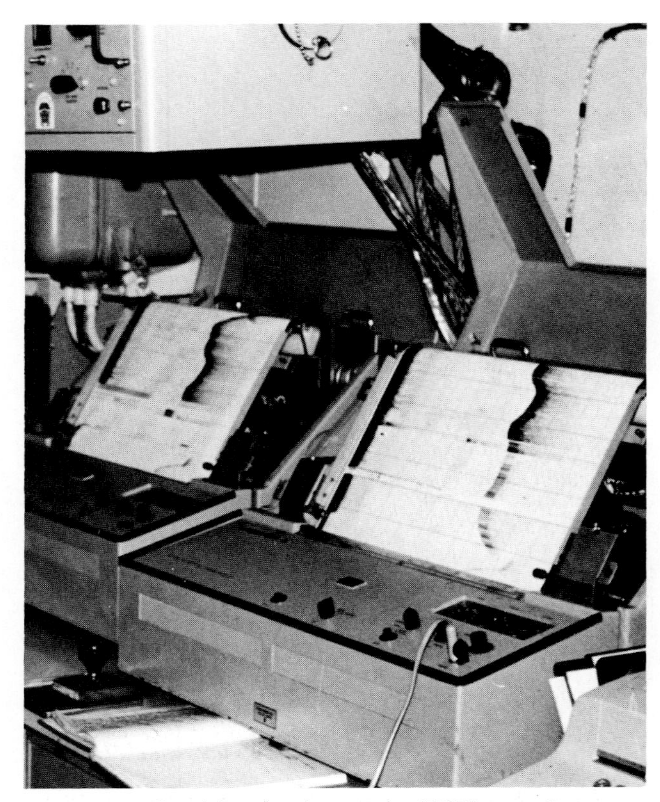

Figure 1.10 Precision depth recorder (PDR) operating on board a ship with the trace of the ocean floor being recorded.

Figure 1.11 Conical-shaped plankton net used to collect samples from research vessels.

samples, particularly small floating organisms that live at or near the surface, with a variety of conically shaped nets (fig. 1.11), which can be towed behind the ship.

While the bottom configuration and small specimen sampling data can be collected at cruising speeds, other types generally require slower speeds. Various types of geophysical data that allow subbottom layers to be perceived and recorded fall into this slow-speed category. Recording can be done by electrical impulse, by explosives, or by pulses of compressed air. A side-scan sonar device is another slow-speed apparatus. It records a picture of the bottom surface for some distance to either side of the sensor. Various types of biological sampling are also carried out while slowly under way, including different varieties of trawling activities for swimming, floating, and bottom-dwelling organisms. Some types of water sampling may also be accomplished while the ship is under power.

The vast majority of shipboard data collecting is done while the ship is at rest. Collecting activities include water sampling, water temperature measurement, salinity and current measurements, various types of biological sampling, bottom sediment sampling, and ocean floor coring. It is not uncommon for all of these activities to go on at a single station. Usually a string of samplers is lowered with various instruments at prescribed depths. Instruments may include current meters, various types of bottles for water samples, and thermometers for temperature readings at depth. Vertically towed nets are used to collect organisms. Various bottom sediment samplers are available including dredges, clamshell type samplers, and box corers. Coring is commonly done with a piston corer (fig. 1.12), which can collect several meters of sediment (fig. 1.13). Special drilling ships, such as the *Glomar Challenger,* have sophisticated positioning devices and other special equipment that allow direct drilling in thousands of meters of water. Cores several hundred meters in length have been collected with such ships.

Figure 1.12 Devices used for coring sediments from the ocean floor. The apparatus on the left (a) is a gravity corer, which takes short cores by penetrating the floor, using a heavy weight. Photo (b) shows a piston corer rigged for descent to the ocean floor. This type of corer is aided by a piston and is capable of collecting several meters of deep-sea sediment.

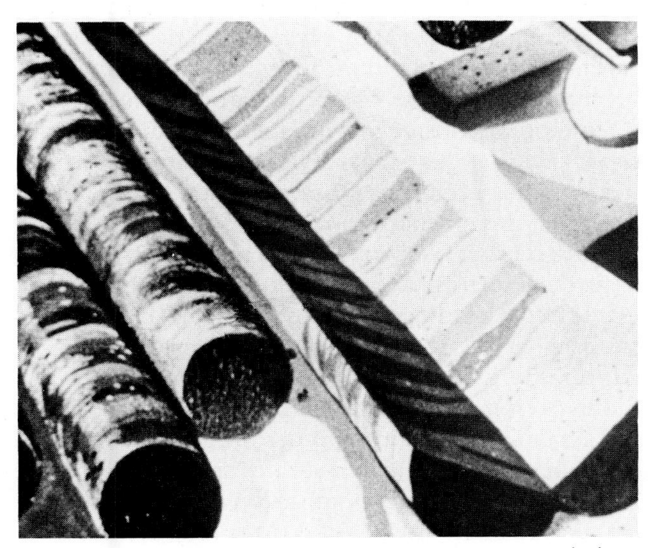

Figure 1.13 Deep-sea sediment that has been extruded from a core barrel.

There are many types of oceanographic data collection that are not conducted from a ship, although in some cases a ship may be involved. Foremost among these activities is instrument mooring to collect samples or record measurements directly. A great variety of instruments can be placed at the surface or at designated depths. These are usually anchored to the bottom, but there may be situations where floating instruments are necessary. These can move with the currents and thereby produce data that will permit plotting of surface or deepwater currents. The instruments used in this general fashion can collect data on temperature, salinity, light penetration, and other parameters from the entire water column of the ocean. The data may be recorded internally within the instrument package or it may be radioed back to a ship or to a land-based laboratory. These instruments are extremely expensive, thus it is important to be able to relocate and recover them. Damage and sometimes loss does occur and further escalates the cost of oceanographic research.

Instruments that actually move with the current are fairly common. These are typically some type of device that floats on the surface or at a predetermined depth. The device produces a radio signal that enables a ship or a shore-based facility to continuously track the device and thereby chart the path of the current moving the device. Crude, inexpensive seabed drifters are sometimes used to study currents. These are typically plastic, in the form of simple cards or small umbrella-shaped drifters, which move with the current and eventually wash up on the beach. The scientist conducting the study must rely on people who find the drifter to send in a card giving the date and location. Obviously this is not a very exact method of charting oceanic currents.

Modern technology has produced some exotic apparatus for sampling and directly observing phenomena in the deep parts of the oceans. Deep research submersibles like the *RV Alvin* (fig. 1.6) have enabled scientists to directly observe the seafloor thousands of meters below sea level. Although remote-controlled deep-sea photography and television have been available for some time, it is now possible for the scientist to be there and aim the camera at the subject being observed. Additionally, many of these submersibles have attached sampling apparatus that can be operated remotely. The samples are typically small and operation is cumbersome but the rewards are often great. There are also independent remote-controlled robots or underwater manipulators. These have somewhat more utility for sampling than those attached to the submersibles.

The general application of remote sensing to oceanographic studies has expanded greatly during the past few decades. At least since World War II, aerial photographs have been used in oceanographic research, especially in studying coastal environments. Aerial photos are also used in studying currents and the movement of plumes of suspended sediments. Now this has been greatly expanded with data from satellites. The various types of imagery that are produced enable study of not only currents, but also heat distribution, pollution of the sea, and marine resources. There are also numerous defense uses.

Helicopters and fixed-wing aircraft are also utilized in oceanographic studies. Helicopters can be used for instrument retrieval, ferrying crews back and forth from a ship, and for a variety of sampling operations. Fixed-wing aircraft are used for various remote sensing activities, including photography, and for airborne magnetic surveys of the earth's crust. By flying a closely spaced grid across the ocean, it is possible to make detailed maps of changes in the earth's magnetic fields as recorded in the rocks of the crust, which may be kilometers below the sea surface.

Data Analysis

There has been a dramatic increase in the sophistication of data analysis since the 1960s. However, even though oceanographers are analyzing data with computers in virtually every aspect of science, there are still many types of data analysis and synthesis that must be done slowly, by a person. This is probably most common in biological and geological oceanography, where there are many descriptive aspects of the research. Biologists must sort out individual organisms and classify them in the same way that they always have, although now they have better microscopes and techniques for preservation of specimens. In much the same way, the geologist must describe core samples, study sediment and rock textures under the microscope, and identify fossils.

Analytical instruments have become extremely accurate, so that nearly any physical or chemical property of an element or compound can now be measured. The only limiting factor is the availability of funds to purchase and operate such machines. Once the organisms or sediments have been identified, the chemical analyses have been determined, or physical properties measured, then the computer becomes the primary tool for further data analysis. Computers are used to rapidly compute statistics, to illustrate relationships between parameters, and to interpret various types of remotely generated signals, such as from satellites or from geophysical studies. Most oceanographic vessels have their own computers on board so that much of this work can be accomplished during the cruise. Once some data have been interpreted, the chief scientist can make intelligent decisions about where to go next or what to sample next.

Summary of Main Points

This chapter covered a broad series of topics to provide a general idea of what oceanography is, how it developed, and how its data are collected from the world ocean. Some of the most significant points of chapter 1 are:

1. Oceanography is not a science per se, but a combination of all sciences as applied to the sea. Those people who make their living as oceanographers or marine scientists are really chemists, biologists, and so on who work on marine problems.
2. Oceanography is a young discipline, barely a century old. It grew from human curiosity about the sea and as a result of the expeditions, as early as the ancient Greeks and Romans, that explored the seas. Much of our modern technology and emphasis on marine research resulted from advances and momentum gained during World War II.
3. The Deep Sea Drilling Project (DSDP), which provided the first extensive data on the nature and history of the strata beneath the ocean floor, was perhaps the most significant single project in the history of oceanography.
4. Oceanographic data collection is expensive and technically sophisticated. There are many types of complicated sampling equipment that are available for specialized collection of virtually any type of information from the oceans.

Suggestions for Further Reading

Deacon, Margaret. 1971. **Scientists and the Sea: 1650–1900.** New York: Academic Press.

Idyll, C.P., ed. 1969. **Exploring the Ocean World, a History of Oceanography.** New York: T. Y. Crowell.

Gordon, D.L., ed. 1970. **Man and the Sea: Classical Accounts of Marine Exploration.** Garden City, New York: Natural History Press.

Schlee, Susan. 1973. **The Edge of the Unfamiliar World: A History of Oceanography.** New York: E.P. Dutton.

Sears, M., and D. Merriam. 1980. **Oceanography and the Past.** New York: Springer-Verlag, Inc.

2
Major Features of the Earth

Chapter Outline

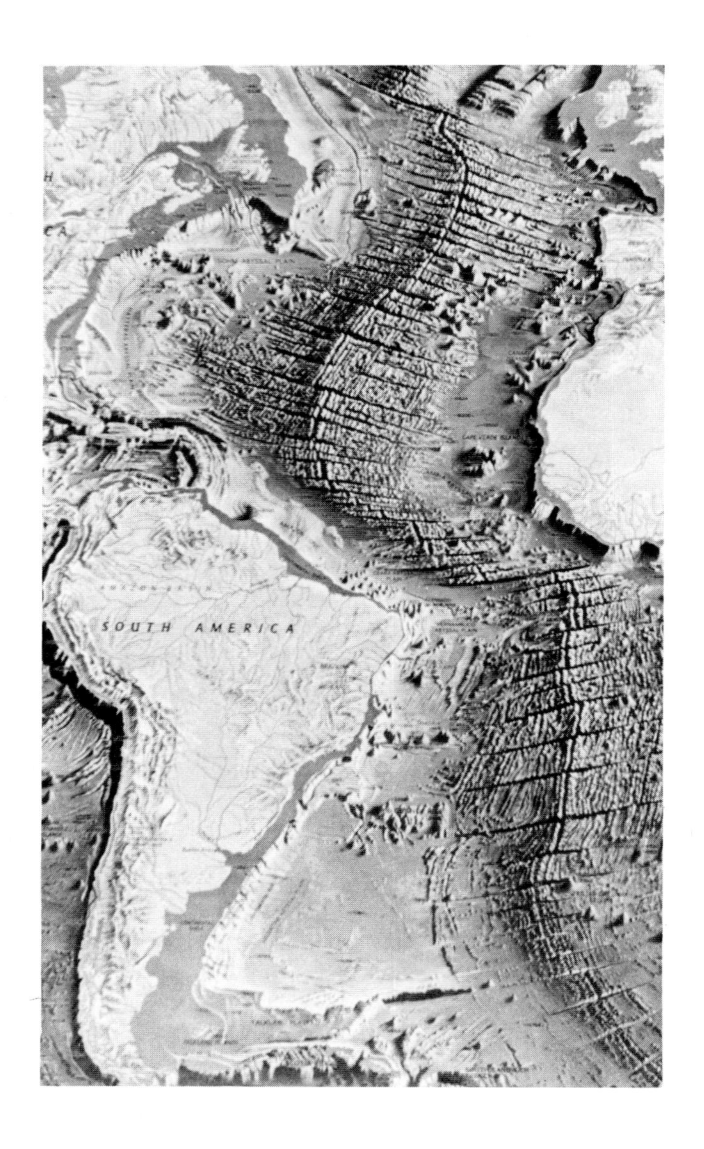

The earth is one of the smallest planets in our solar system, but it has a great deal of variety in its composition, in the configuration of its surface, and in the fluids that envelop it. This chapter will take a general look at the major features of the solid earth and the water envelope that covers most of the planet. The atmosphere will be treated in chapter 5.

Distribution of Land and Water

The earth has a diameter of 12,750 km and a circumference of 40,000 km (table 2.1). Although its shape is essentially spherical, there are deviations, including bulges at the equator and irregularities from surficial features. This planet is unique in that 71% of its surface is covered with water. The oceans cover over 360,000,000 km² with an average sphere depth of nearly 2 km (table 2.1). The distribution of water and land on the earth's surface is not uniform. There is much less land in the southern hemisphere than in the northern hemisphere; in fact, the water-to-land ratio in the southern hemisphere is more than twice that in the northern hemisphere (fig. 2.1). The only place on the globe where there is more land than water is between latitudes 45°N and 70°N; this includes much of North America, Europe, and Asia.

Virtually all the water (98%) in the earth's system is marine or is frozen in sea ice, such as around Antarctica or in the Arctic Sea. Most of the rest is incorporated in glaciers. Liquid fresh water, including

TABLE 2.1
Major Features of the Earth

Feature	Mass ($\times 10^{25}$ kg)	Volume ($\times 10^{6}$ km³)	Thickness or Radius (km)	Area ($\times 10^{6}$ km²)	
Earth	598	1,083,230	6,356 (polar) 6,378 (equatorial)	510	360 water surface 150 land surface
Oceanic crust	0.8	2,660	8		
Continental crust	1.7	6,210	35		
Mantle	407	898,000	2,881		
Core	188	175,000	3,473		

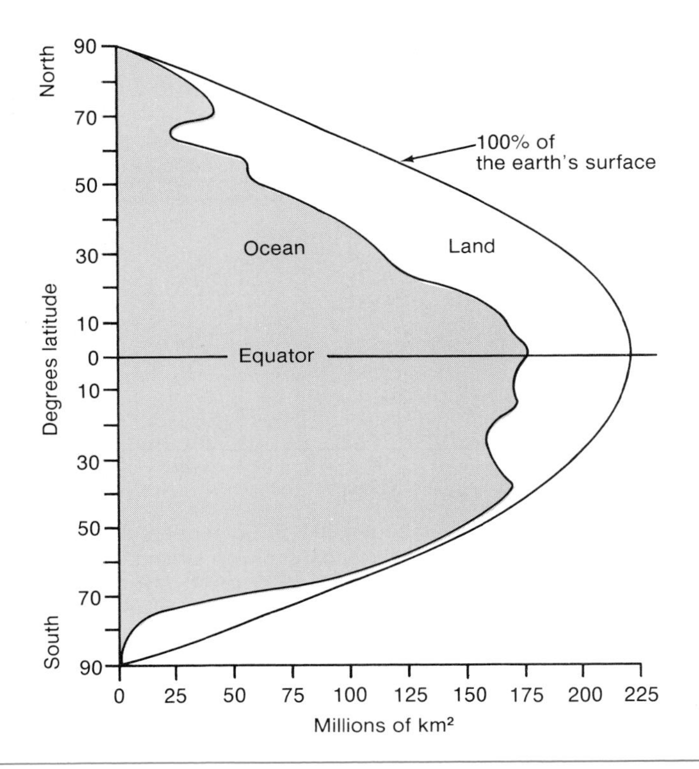

Figure 2.1 Diagram showing the distribution of land and water with latitude. The shaded portion is land and the area to the left of that is ocean. Note the large percentage of land in the northern hemisphere as compared to the southern hemisphere. (After Duxbury and Duxbury, 1984.)

groundwater and all rivers and lakes combined, accounts for only 0.4% (table 2.2). The volume of water in the atmosphere is insignificant.

The average depth of the oceans is 3,750 m, while the mean elevation of the land is 840 m. These numbers are not very informative, however, because the variation in elevation is tremendous, ranging from a peak of 8,850 m on Mount Everest down to a depth of 11,030 m in the Marianas Trench in the southwest Pacific. The **hypsographic curve,** a graphic representation of elevation on the earth's surface, shows a bimodal distribution of the elevations of the earth's surface (fig. 2.2). It shows elevations concentrated just above sea level on the continents and at depths of near 4,000 m in the oceans. Note that there is a distinct break in the distribution at about 2,000 m below sea level (fig. 2.2). The reason for this distribution is that the crust of the earth is comprised of two markedly different materials. These occur at different elevations due to their contrasting composition. More discussion on this topic follows in the section on composition of the earth.

TABLE 2.2
Distribution of Water on the Earth

Environment	Area ($\times 10^6$ km^2)	Volume ($\times 10^6$ km^3)	Relative Abundance (%)
Atmosphere	—	0.015	0.001
Oceans and seas	361	1,370	97.72
Glaciers and ice sheets	15.6	25	1.84
Lakes and rivers	—	0.5	0.04
Groundwater	—	5.1	0.4

Major Water Bodies of the World

There are three major oceans of the world, the Atlantic, Pacific, and Indian. These are well defined by their surrounding landmasses, their own fairly distinct water masses, and their own distinct surface current systems. In contrast, in the Antarctic, the water body surrounds the landmass and there are neither submarine ridges nor distinct water masses

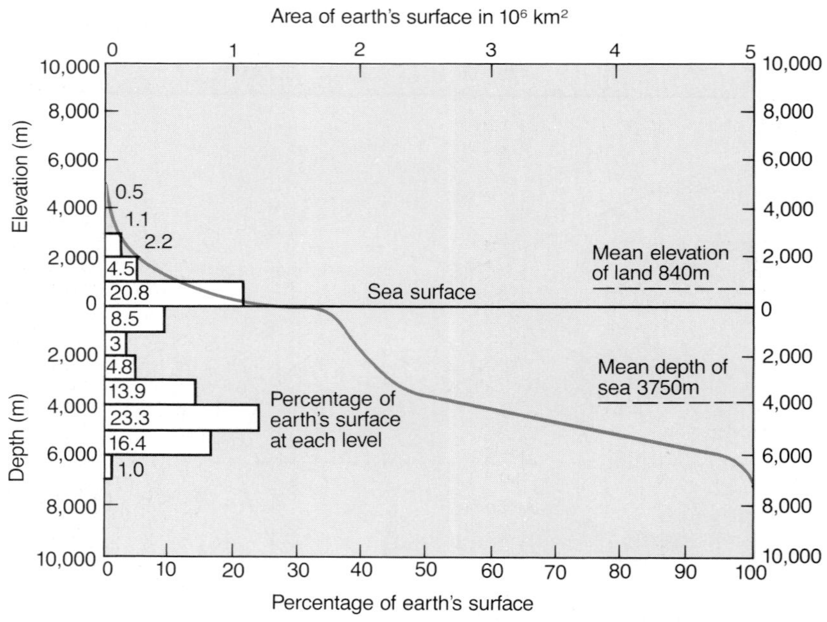

Figure 2.2 A hypsographic curve showing the percentage of the earth's surface at any given elevation or depth. Note that most of the area above sea level is at elevations less than 1,000 m and that most of the area below sea level is about 4,000–5,000 m deep, the depth of the ocean floor. (Sverdrup/Johnson/Fleming, *The Oceans,* © 1942, © renewed 1970, pp. 18, 19, 241, 242, 245. Reprinted by permission of Prentice-Hall, Englewood Cliffs, New Jersey.)

that separate it from the southern portion of the adjacent oceans to the north; consequently, it is not appropriate to consider this an ocean. The Arctic Sea is actually a **mediterranean,** which is a body of water surrounded by land that is relatively small in comparison to an ocean.

The Pacific Ocean is the largest by far, with an area that is a little larger than the other two oceans combined (table 2.3). The Atlantic and Indian Oceans are similar to each other in area but very different in shape and global distribution. The Atlantic is narrow and extends almost the length of the globe, whereas the Indian Ocean is nearly confined to the southern hemisphere. By comparison to the oceans, the Arctic Sea, the Gulf of Mexico, and the Mediterranean Sea are small (table 2.3). Likewise, the Caspian Sea, which is the largest lake in the world, and Lake Superior, which is the largest lake in North America, are insignificant in areal extent compared to the smallest of the three oceans.

It is interesting to note that although there are differences in the shape, size, and bottom features of the oceans, the average depth of the three is reasonably similar (table 2.3). The Atlantic and Indian Oceans are almost the same at just under 4,000 m. The mean depth of the Pacific is approximately 4,300 m, or about 300 m deeper than the other two, because most of the earth's ocean trenches are in the Pacific; when averaged in, their great depths make the Pacific deeper. Other marine basins are much shallower. Mean depths of lakes are even less.

TABLE 2.3
Measurements of the Oceans and Selected Water Bodies

Water Body	Area ($\times 10^6$ km^2)	Volume ($\times 10^6$ km^2)	Mean Depth (m)
Atlantic Ocean	82.4	323.6	3,926
Pacific Ocean	165.3	707.6	4,282
Indian Ocean	73.4	291.0	3,963
Arctic Sea	9.5	9.41	991
Mediterranean Sea	2.9	4.2	1,429
Gulf of Mexico	1.6	2.33	1,512
Caspian Sea	0.44	0.077	180
Gulf of California	0.16	0.13	813
Lake Superior	0.08	0.012	149

General Composition and Structure of the Earth

The earth is not internally homogenous; rather, it is composed of three major concentric layers or zones, the **crust, mantle,** and **core**. The thickness and composition of these layers are known from analysis of the manner in which waves generated by earthquakes travel through the earth. The velocities of the waves over the paths that they take enable scientists to tell the state and density of the material through which they move. This information provides enough data to determine the general composition of the material.

The core of the earth is composed of the densest material and is divided into an inner core and an outer core (fig. 2.3). Although the composition of the entire core is interpreted to be a nickel-iron mixture, the inner portion is solid and the outer part is liquid. The core extends from a depth of 2,900 km and accounts for 31.5% of the mass of the earth. The mantle comprises over 2/3 of the earth's mass and has a general rock composition of iron magnesium silicate.

The **asthenosphere,** a thick zone of the upper mantle, occupies a few hundred kilometers directly below the **lithosphere,** the relatively rigid layers of the earth that include the crust and the uppermost

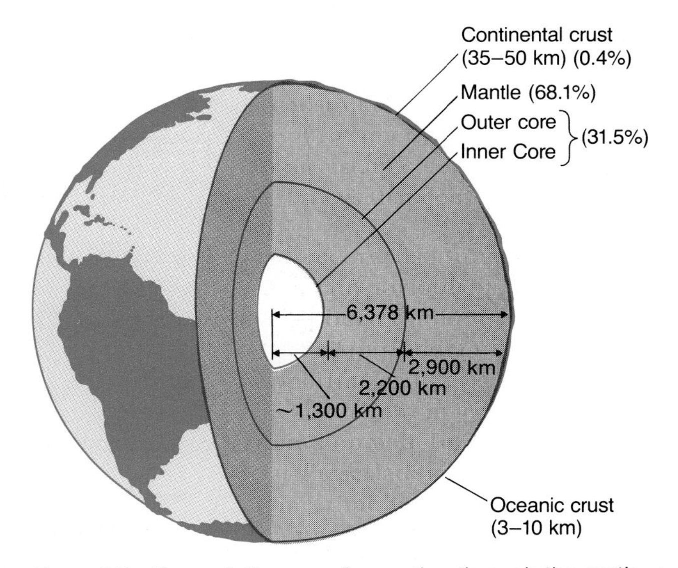

Figure 2.3 General diagram of a section through the earth showing the location and thicknesses of the major layers. The percentages in parentheses are relative to the total volume.

part of the mantle (fig. 2.4). The asthenosphere is somewhat plastic, with an average density of about 3.2–3.3 gm/cm³, whereas the average density of the entire mantle is 4.5 gm/cm³.

The crust of the earth, which is extremely thin relative to the radius of the earth, is divided into two distinct parts based on composition. The **oceanic crust** is 3–8 km thick, has a density of 2.9–3.0 gm/cm³. It has a composition similar to basalt, which is that of silicate minerals rich in magnesium and iron.

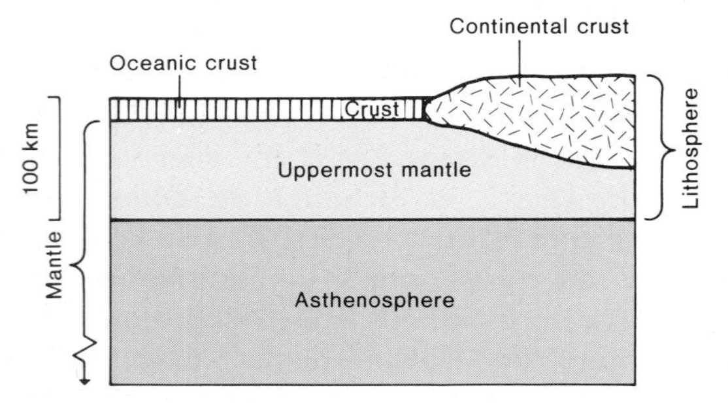

Figure 2.4 Section of the earth's crust showing the relationships among the major crustal types and the upper part of the mantle. (From Plummer and McGeary, 1985.)

Continental crust is up to 40 km thick, has a composition similar to granite, and a density of only 2.7–2.8 gm/cm³. Because of the density difference between these crustal components and the upper mantle (3.2 gm/cm³), the blocks of the crust float on the mantle (fig. 2.4). The surface between the crust and the mantle is called the **Mohorovicic discontinuity (Moho)** after its discoverer, a Yugoslavian geophysicist.

Isostacy

A look at a cross section of the earth's crust and upper mantle shows that there are different thicknesses within continental blocks and that the bases of the continental blocks have irregularities, as does its upper surface (fig. 2.4). This can be explained by the principle of **isostacy,** which is the condition of equilibrium, comparable to floating, of the units of the lithosphere resting on the plastic asthenosphere.

An ice-water analogy can be used to explain this principle. The density of ice is just over 0.9 gm/cm³ and that of water is 1.0 gm/cm³. The ratio of these densities is very close to the ratio of densities between continental crust (granite) and oceanic crust (basalt). If variously sized blocks of ice are floated in a basin of water, the small blocks will have a small

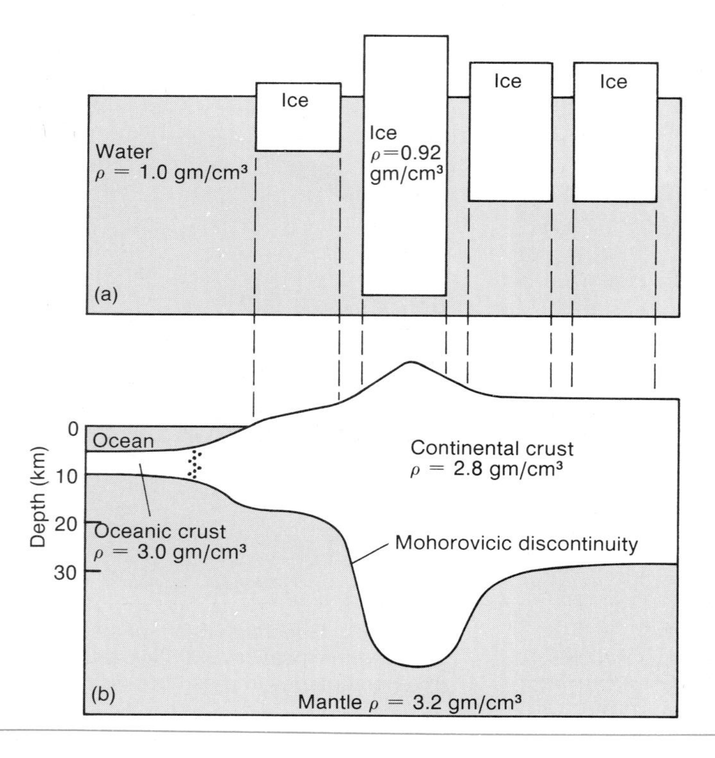

Figure 2.5 Isostacy can be understood by considering the situation in the crust of the earth and comparing it with ice that floats in water. The thick pieces of ice are higher above water level than the thin ones and they also extend further below sea level than the thin pieces. (William A. Anikouchine/Richard W. Sternberg, *The World Ocean: An Introduction to Oceanography,* © 1981, pp. 21, 23. Reprinted by permission of Prentice-Hall, Englewood Cliffs, New Jersey.)

amount of ice above and below the water surface relative to the much larger blocks (fig. 2.5). In effect, the same thing happens in the earth's crust: mountain ranges have deep roots (thick continental crust) relative to places with no mountains (thin continental crust). The material of lower density (continental crust) is floating on the material of higher density (mantle) and the amount of lighter material above the surface is proportional to the amount below the surface.

Major Features of Ocean Basins

The surface of the earth beneath the sea is similar to that above sea level in that there are large-scale geologic and physiographic features that occur on a global basis. Oceanic features are characterized by their depth, bottom configuration, position relative to one another, and composition. They are generally much more predictable in their location than are the major features of the continents. That is, a generalized profile across an ocean basin can be applied nearly universally around the world. The two types of crust also provide a distinct physiographic boundary within the ocean basins. The submerged portion of the continental blocks of the crust contains the **continental margin,** which consists of the continental shelf, continental slope, and continental rise. The remainder of the ocean basin is the **deep sea,** which is more related to the oceanic crust than to the continental crust. It consists of the ocean floor, the oceanic ridge, and the deep-sea trenches (fig. 2.6).

Continental Margin

The continental margin is really just an extension of the continent except that it is under water. The major geologic features, such as stratigraphic units, structural features, and topographic features, continue on both the landward and the seaward sides of the shoreline although there may be some modification. The shelf and slope comprise the **continental terrace.** The **continental rise** is a thick sequence of sediments that was derived from the continental block and deposited at its base overlying the oceanic crust (fig. 2.7).

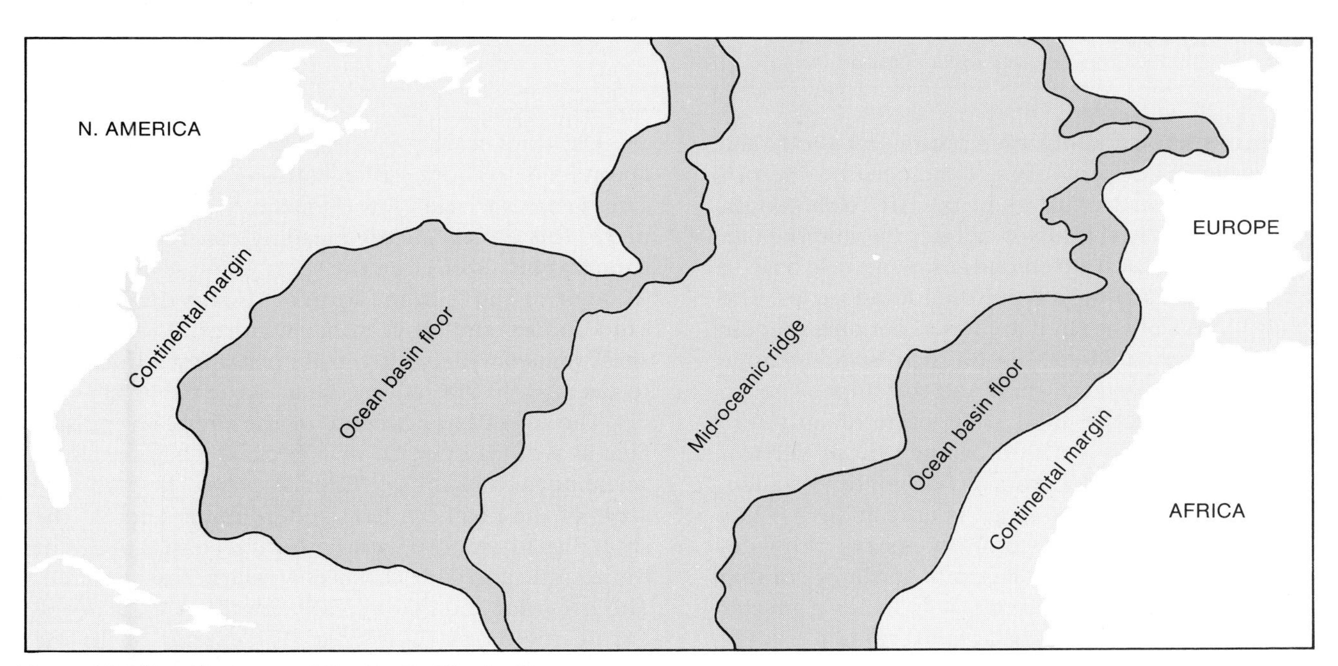

Figure 2.6 Simplified map of the North Atlantic Ocean showing the major provinces. (From Heezen, B. C., et al, 1959, *The Floor of the Ocean–I: The North Atlantic,* Geological Society of America, Special Paper 65.)

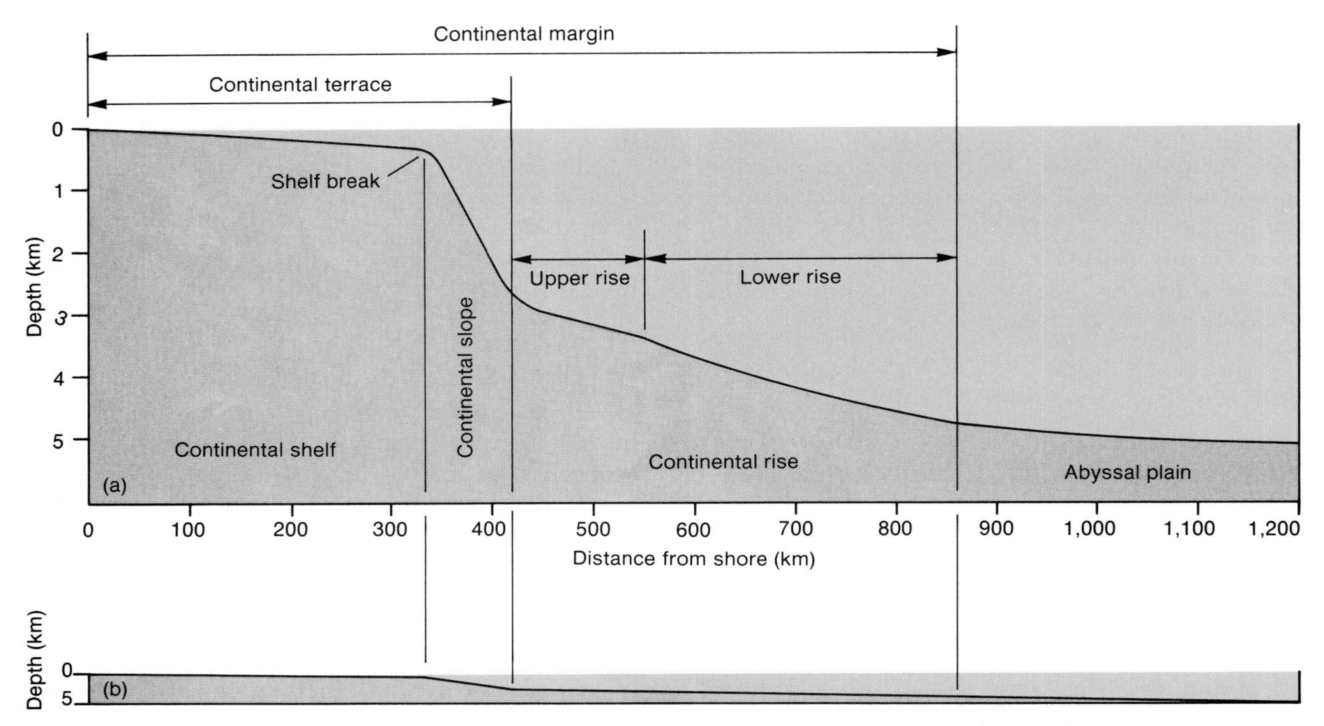

Figure 2.7 Generalized profile across the continental margin showing the relationships between the provinces. Diagram (a) shows great vertical exaggeration, as typically portrayed in textbooks, whereas (b) shows the same profile, but with no vertical exaggeration. (William A. Anikouchine/Richard W. Sternberg, *The World Ocean: An Introduction to Oceanography,* © 1981, pp. 21, 23. Reprinted by permission of Prentice-Hall, Englewood Cliffs, New Jersey.)

More is known about the continental shelf than about other major provinces of the oceans. The primary reasons for this are its proximity to the continent, its relatively shallow water depths, and the fact that it supports such a tremendous economic base in the form of the fishing and petroleum industries. The seaward limits of the shelf are based not on a specific depth of water or width, but on the distinct change in gradient between the shelf and the slope. The average gradient of the shelf is an angle of only 00° 07′, which is nearly 1:1,000 or one unit of vertical change for every 1,000 units of horizontal distance. The average angle of the slope is only about 4° 17′, although there is great variation. As typically depicted in texts, this shows a gentle gradient coming in contact with a steep one (fig. 2.7a), then a gentler gradient at the rise. Such diagrams contain a large vertical exaggeration in order to show differences in gradient and still keep within a single page. A diagram without the exaggeration shows no perceptible difference in gradient, but rather an overall gentle gradient (fig. 2.7b).

Depth of water over the shelf edge ranges from about 35 m to about 350 m with an average of 130 m. The great variation around the world, however, makes this average nearly meaningless. There is some topographical relief on the shelf, but it is only about 20 m or so and is in the form of small valleys, reefs, sand bodies, and other minor features. Width of the shelf ranges widely also, from just a few kilometers to over 1,000 km, with a mean value of 80 km.

The overall appearance of the shelf, on the surface as well as beneath it, reflects the adjacent subaerial portion of the continental mass. The glaciated areas of the northern hemisphere extend across the shelf. Broad, smooth, gently sloping coastal plains of the Atlantic and Gulf coasts continue across the shelf. The irregular and geologically complex areas of the Pacific region of California also extend on to the continental margin.

The continental slope extends from the break in slope to depths of a few thousand meters, where the gradient decreases again (fig. 2.8). The average depth at the base of the slope is 3,600 m, but it may extend to over 8,000 m in areas where there is no rise and

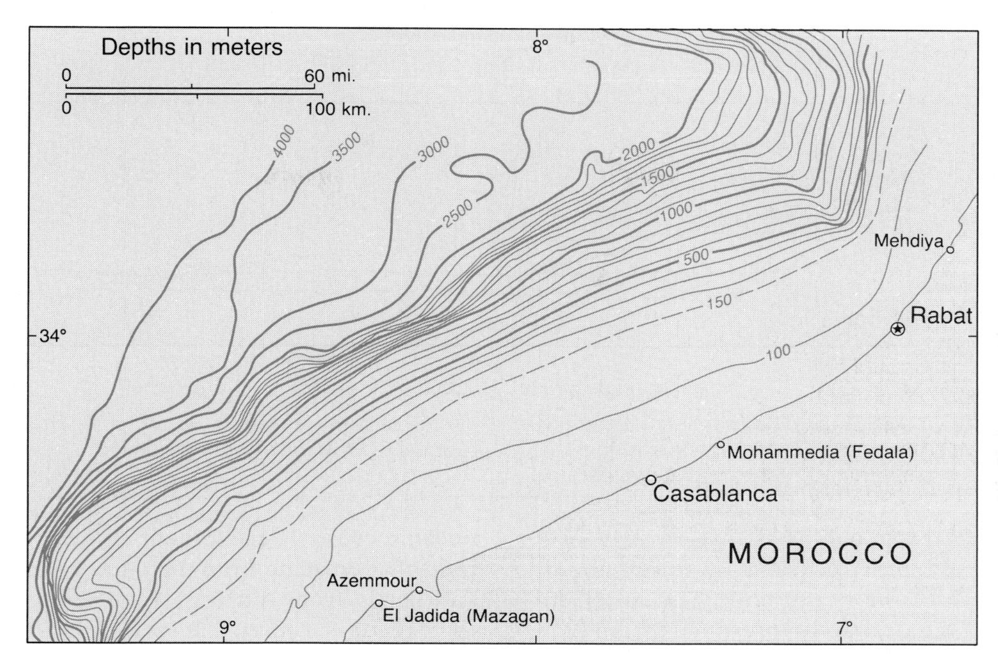

Figure 2.8 Bathymetric map of the continental margin off the coast of Morocco in northwestern Africa showing the continental shelf with widely spaced contour lines, the continental slope with numerous closely spaced contour lines, and the ocean floor with spaced contours. (From Hill, M. N., 1963, *The Sea,* p. 288.)

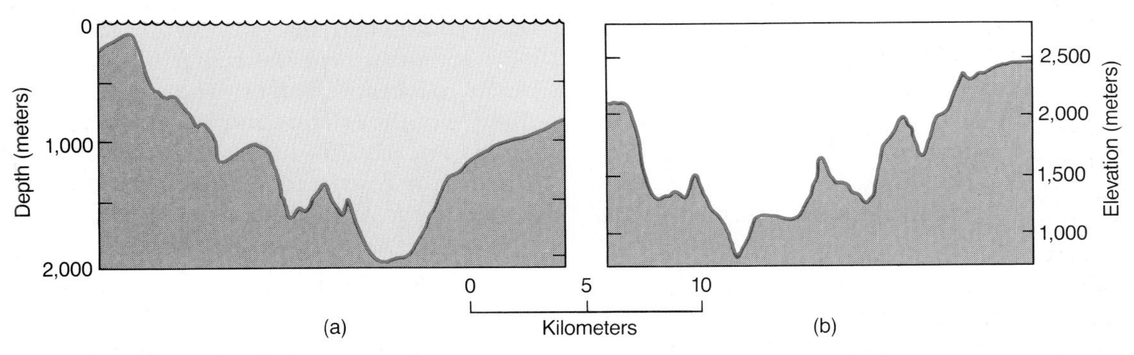

Figure 2.9 A comparison of a profile across (a) a typical submarine canyon, Monterey Canyon, California, and (b) the Grand Canyon, Arizona. Note that the relief on both is similar.

the slope continues into trenches. The steepness of the slope is largely dictated by the structure of the continental terrace. Some terraces have slopes of 45° to 70°. By contrast, there are slopes off deltaic areas where the gradients are only 1° or so.

Submarine canyons are major relief features of the oceans that occur primarily on the continental slope. These features have thousands of meters of relief (fig. 2.9) and may extend from the outer shelf across the slope and onto the rise. They act as sediment-transport pathways for much of the sediment deposited on the continental rise and the ocean floor.

The continental rise is the most seaward and the deepest part of the continental margin. It may be difficult to define physiographically because the break in gradient between the slope and rise may be subtle, similar to that between the rise and the abyssal plain. There are some places in the world where the rise appears to be absent. The gradient of the continental rise ranges widely, from about 1:50 to 1:800 with an average value of 1:150. This is steep relative to the shelf but gentle compared to the slope.

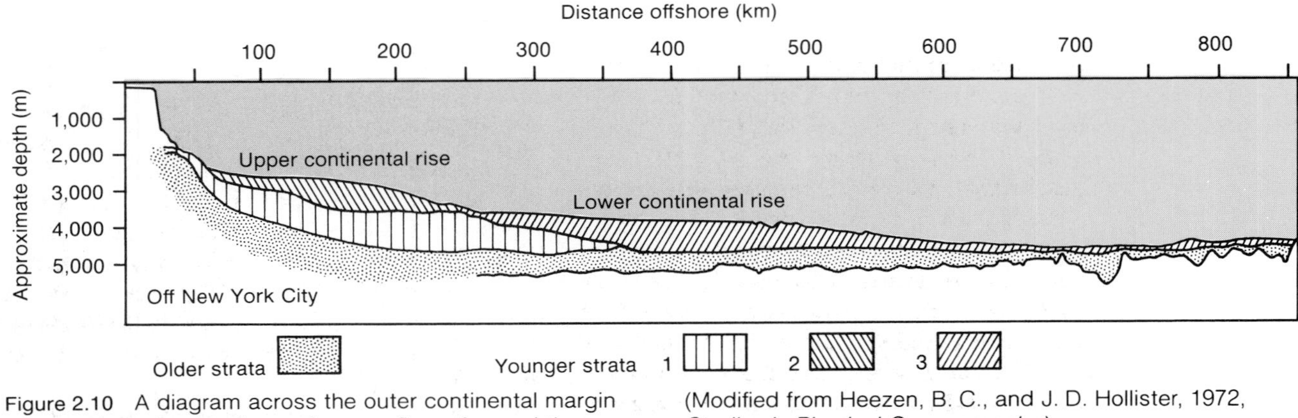

Figure 2.10 A diagram across the outer continental margin off New York showing the surface configuration and the underlying large sediment prisms on the continental rise. (Modified from Heezen, B. C., and J. D. Hollister, 1972, *Studies in Physical Oceanography.*)

The continental rise begins as shallow as 1,400 m and extends to depths of nearly 4,000 m, about the average depth of the ocean floor. The relief on the rise may be high due to the presence of the seaward end of submarine canyons and their related channels, which extend into the continental rise. Except for these features, the relief is low.

The rise is formed by a thick wedge of sediments derived from continents and adjacent continental terraces. Currents and various gravity-generated downslope processes carry the sediment to the base of the slope where it accumulates. The total thickness may be several thousand meters (fig. 2.10). Isostacy may result in some subsidence of this wedge of sediment at the boundary between the continental crustal block and the oceanic crust.

Ocean Basin

A typical ocean basin consists of abyssal floor, oceanic rise, oceanic ridge, and deep-sea trench provinces. Seamounts and guyots may occur throughout.

The abyssal floor is made up of the **abyssal plain** and **abyssal hills**. These are the deepest extensive oceanic environments and at one time were thought to comprise essentially all of the ocean basin. The abyssal plain is defined as that part of the ocean basin where the gradient is less than 1:1,000. It is typically located between the continental rise and the oceanic ridge. The abyssal hills are generally located between the abyssal plain and the oceanic ridge.

The abyssal plain is formed by abundant sediment covering topographic irregularities on the oceanic crust. The sediment comes partly from currents that emanate from the outer continental margin and partly from settling of suspended sediment in the ocean. The related abyssal hills are irregularities of the bedrock oceanic crust and have relief of up to a few hundred meters. Sediment tends to fill in the low areas (fig. 2.11).

There is a pattern to the distribution of the two ocean floor features. Abyssal plains are dominant toward continental margins, while abyssal hills are most abundant near the center of ocean basins, adjacent to oceanic ridges. The abyssal floor covers about ⅓ of the Atlantic and Indian ocean basins compared to nearly 75% of the Pacific Ocean. This province is not restricted to the three oceans: it also occurs in the Arctic Sea, Gulf of Mexico, and other large, deep, mediterranean basins.

The **oceanic rise** is an area of the ocean floor that is elevated above the abyssal floor and is distinctly separate from the continental margin. It typically is about 300 m above the adjacent or surrounding ocean floor and may have a great deal of relief. Rises are aseismic and are thought to result from the upwarping of oceanic crust that is associated with volcanism from **hot spots,** which are sources of magma in the upper mantle. The Bermuda Rise in the North Atlantic and the Chatham Rise of the southwest Pacific are examples.

The **oceanic ridge** system is a continuous, broad, linear belt of irregular and high-relief oceanic crust that extends for 80,000 km around the globe (fig. 2.12). It is also called a mid-oceanic ridge because in the Atlantic Ocean, where it was first described, it is in the middle of the ocean basin. It is

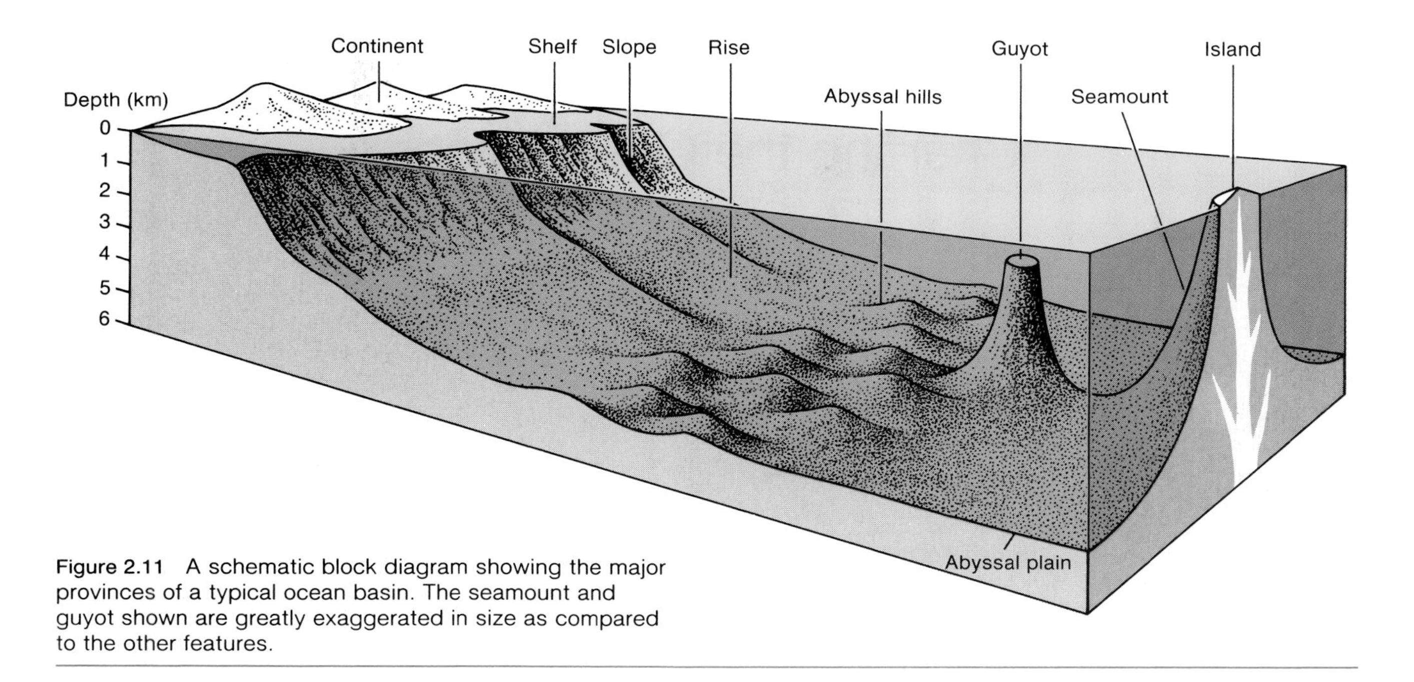

Figure 2.11 A schematic block diagram showing the major provinces of a typical ocean basin. The seamount and guyot shown are greatly exaggerated in size as compared to the other features.

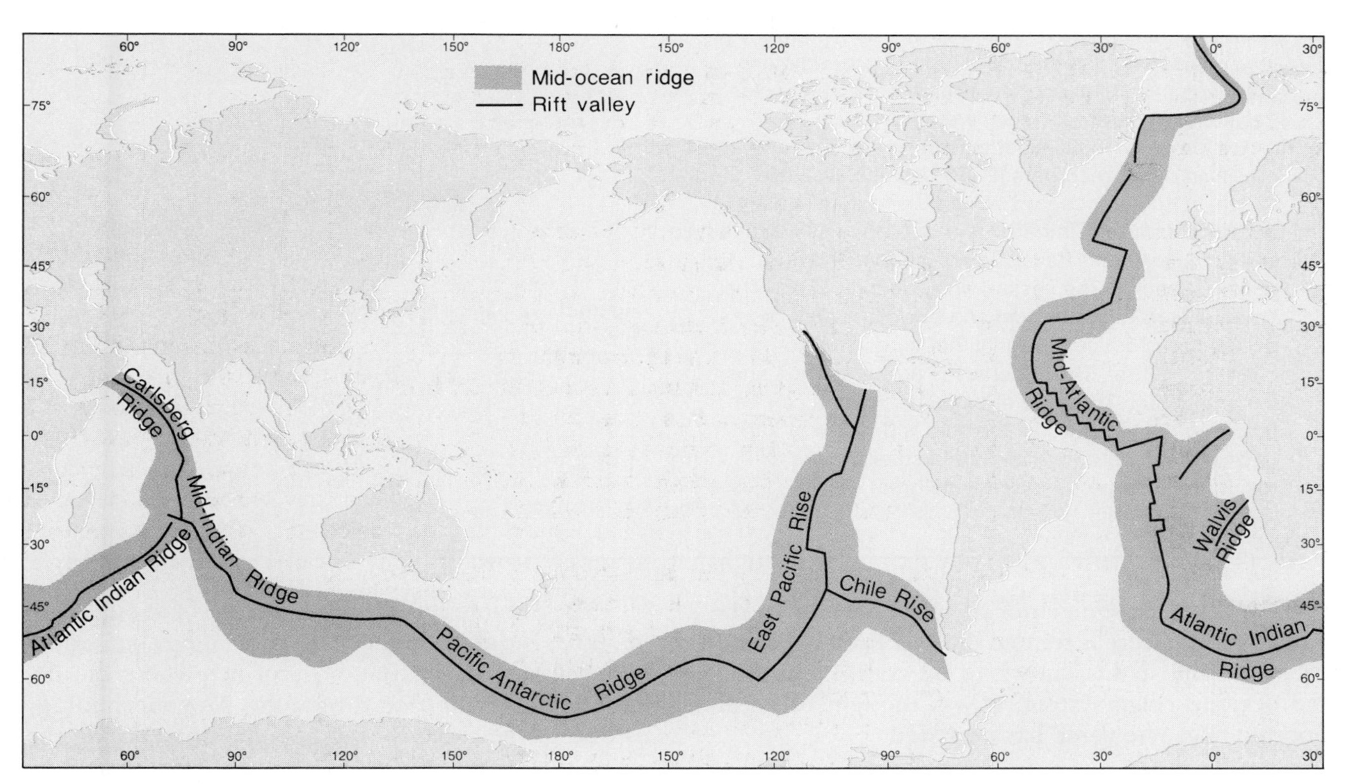

Figure 2.12 World map showing the distribution of the oceanic ridge system, the largest feature on the surface of the earth.

BOX 2.1

Charting the Ocean Floor

Accurate, detailed maps of ocean floor bathymetry require the collection of considerable information and take much time. We presently have moderately good coverage of most of the world ocean, but more detail would improve existing maps and charts, especially in the Arctic and Antarctic where research and ship travel is relatively limited.

Before the availability of electronic instruments, depth readings were obtained by soundings with weighted lines. In order to carry out such readings it was necessary to stop the ship. The ship's location was determined astronomically with a sextant. A weighted, calibrated line was lowered to the bottom and the depth noted. These depth values and locations were then assembled and contoured to provide a bathymetric chart. This was an extremely time-consuming operation and did not provide many data points, considering the size of the ocean basins. Position and depth accuracy were not good.

By the early 1900s electronic devices called precision depth recorders (PDR) became available. These instruments record accurate and continuous depth information. The PDR is based on emission of a sound impulse that travels to the bottom and is reflected back to the vessel, where it is received. The time traveled to the bottom and back is related to depth; this produces a signal that is recorded on a chart (fig. B2.1a). The accuracy is no worse than one meter in 1,500. This type of bathymetric data can be recorded continuously as the ship is under way and locations can be determined from satellites. As a result, the data produced provide great detail, compared to the soundings (fig. B2.1b), and there is much more of it. This has enabled production of accurate bathymetric charts for the entire world ocean. These charts are continuously undergoing revision as new data are collected.

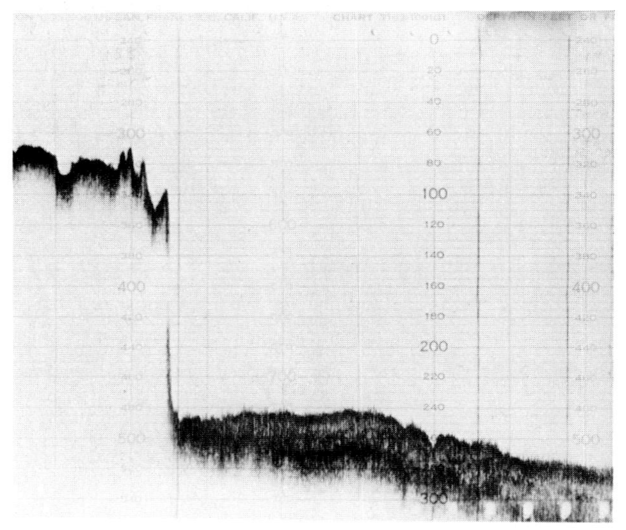

Figure B2.1a Trace of a Fathometer showing the ocean floor.

not, however, in the center of other oceans. In the Pacific, this feature is termed the East Pacific Rise because, before it was known to be continuous with the oceanic ridge system, it was thought to be an oceanic rise. The name has persisted.

The oceanic ridge reaches a depth of about 2,500 m. It has the appearance of a large, high-relief mountain range with maximum relief near the crest, decreasing away in both directions. This morphology is due to the ridges being hot, thereby causing the crust to expand. As you move away from the ridge,

BOX 2.1

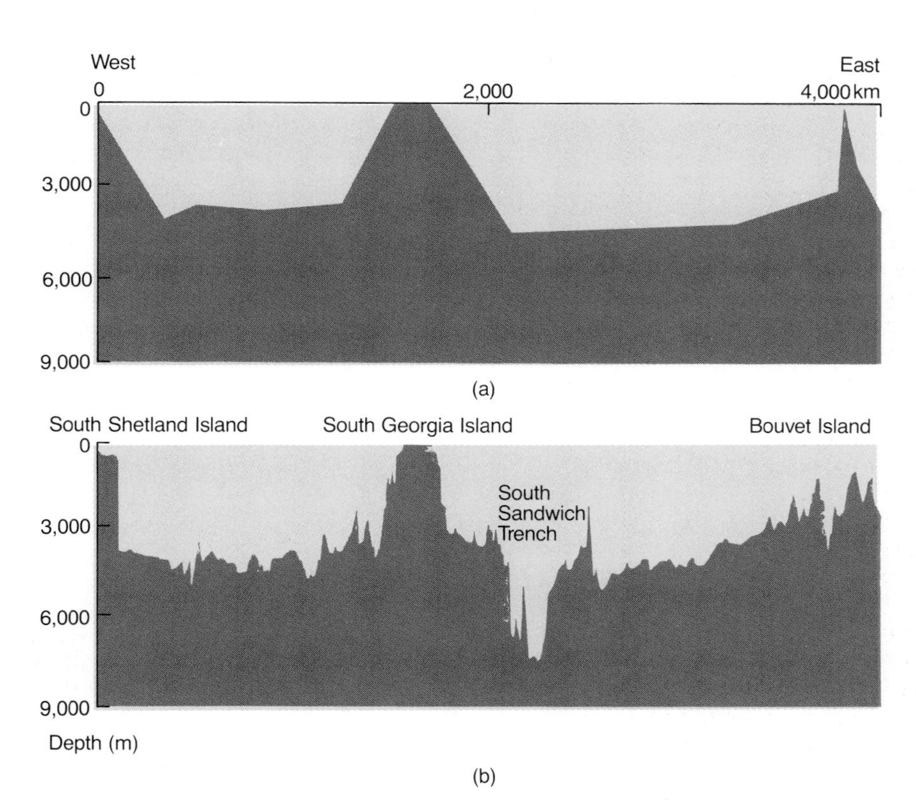

Figure B2.1b Comparison of a profile of a section of ocean floor produced by (a) soundings from a ship and (b) the continuously recording precision depth recorder. (Sverdrup/Johnson/Fleming, *The Oceans,* © 1942, © renewed 1970, pp. 18, 19, 241, 242, 245. Reprinted by permission of Prentice-Hall, Englewood Cliffs, New Jersey.)

the crust cools and contracts thus producing less relief. Width of the oceanic ridge is about 1,000 km. It gradually gives way to the abyssal floor. The oceanic ridge is the largest feature on earth (fig. 2.12). It is characterized by volcanic activity and is seismically active. Numerous faults extend from it at right angles. These **transform faults** denote displacement along the ridge system (fig. 2.13).

The oceanic ridge has some variation throughout its extent. In the Atlantic Ocean, it has very high relief with a central **rift valley** that is 1–2 km deep. By contrast, the ridge in the eastern Pacific, named the East Pacific Rise, has low relief and no deep valley

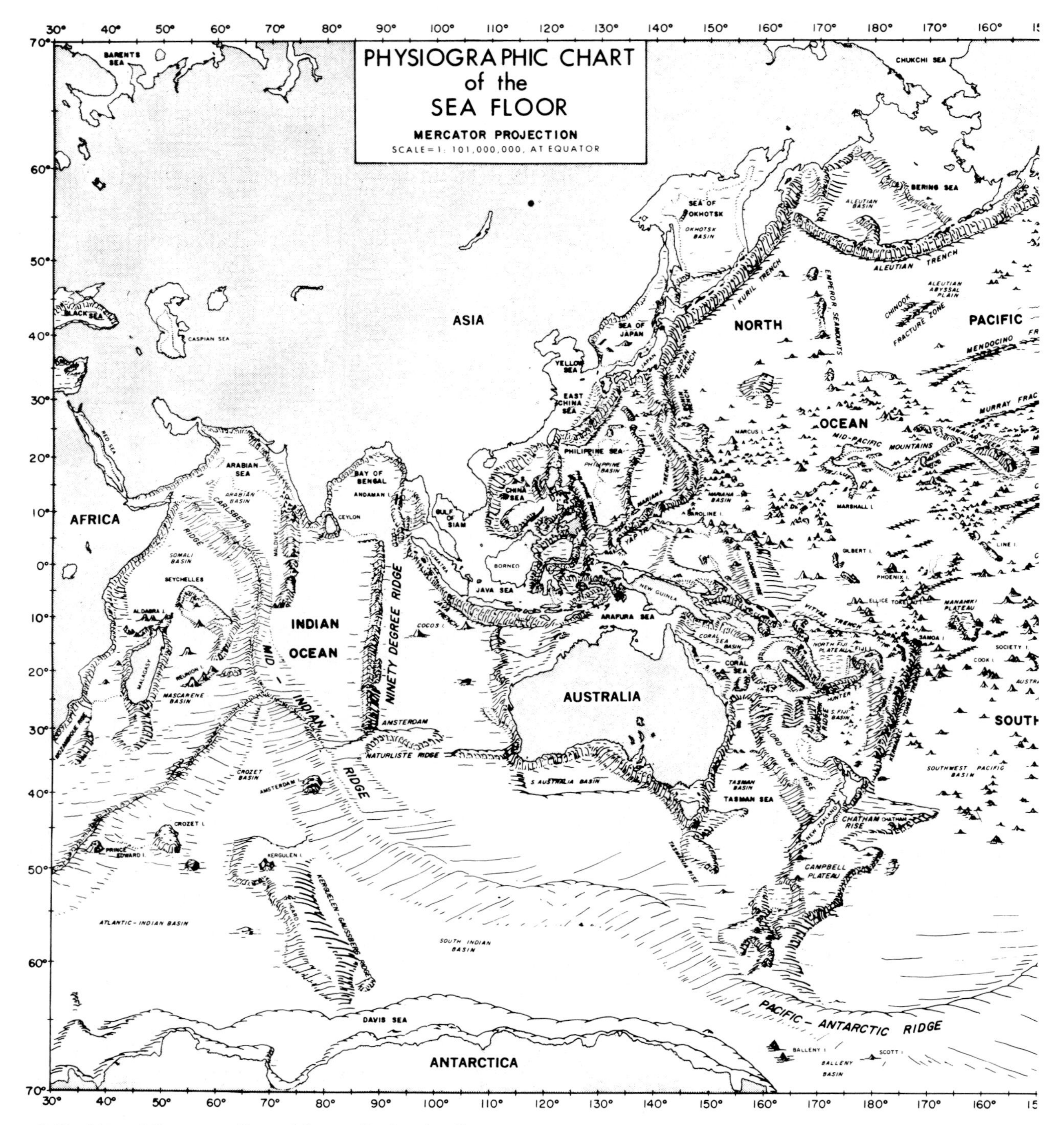

Figure 2.13 Map of the ocean floor of the earth showing its physiographic features. (Courtesy Hubbard Scientific Company, Northbrook, Illinois.)

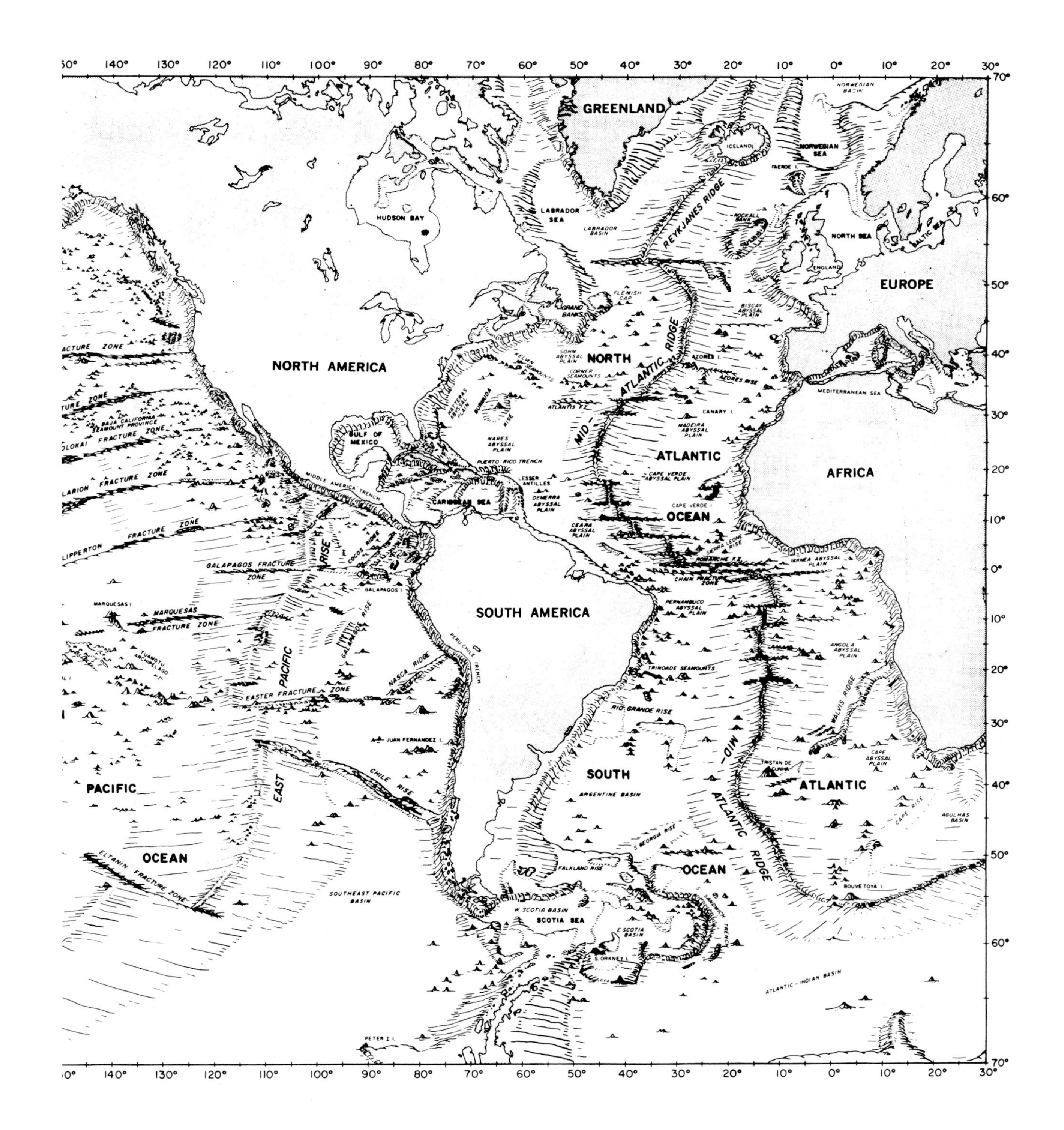

GREENLAND
ICELAND
NORWEGIAN SEA
NORWEGIAN BASIN
HUDSON BAY
LABRADOR SEA
LABRADOR BASIN
REYKJANES RIDGE
ROCKALL BANK
NORTH SEA
BALTIC SEA
ENGLAND
EUROPE
FLEMISH CAP
GRAND BANKS
BISCAY ABYSSAL PLAIN
SOHM ABYSSAL PLAIN
NORTH
AZORES I.
AZORES RISE
CORNER SEAMOUNTS
NORTH AMERICA
MEDITERRANEAN SEA
ATLANTIS F.Z.
CANARY I.
MADEIRA ABYSSAL PLAIN
MID-ATLANTIC RIDGE
GULF OF MEXICO
NARES ABYSSAL PLAIN
PUERTO RICO TRENCH
ATLANTIC
CAPE VERDE ABYSSAL PLAIN
AFRICA
LESSER ANTILLES
DEMERARA ABYSSAL PLAIN
CARIBBEAN SEA
MIDDLE AMERICA TRENCH
CAPE VERDE I.
OCEAN
CEARA ABYSSAL PLAIN
SIERRA LEONE RISE
ROMANCHE F.Z.
CHAIN FRACTURE ZONE
GUINEA ABYSSAL PLAIN
CLIPPERTON FRACTURE ZONE
GALAPAGOS FRACTURE ZONE
COCOS RIDGE
GALAPAGOS I.
MARQUESAS I.
MARQUESAS FRACTURE ZONE
SOUTH AMERICA
PERNAMBUCO ABYSSAL PLAIN
GALAPAGOS RISE
PERU-CHILE TRENCH
TUAMOTU ARCHIPELAGO
NASCA RIDGE
ANGOLA ABYSSAL PLAIN
TRINDADE SEAMOUNTS
EASTER FRACTURE ZONE
EAST PACIFIC RISE
RIO GRANDE RISE
WALVIS RIDGE
JUAN FERNANDEZ I.
SOUTH
TRISTAN DE CUNHA
CAPE ABYSSAL PLAIN
PACIFIC
CHILE RISE
ARGENTINE BASIN
ATLANTIC
CAPE RISE
AGULHAS BASIN
S. GEORGIA RISE
OCEAN
ELTANIN FRACTURE ZONE
FALKLAND RISE
OCEAN
BOUVET TOYA I.
SOUTHEAST PACIFIC BASIN
W SCOTIA BASIN
SCOTIA SEA
E SCOTIA BASIN
S. ORKNEY I.
ATLANTIC-INDIAN BASIN
PETER I. I.
140° 130° 120° 110° 100° 90° 80° 70° 60° 50° 40° 30° 20° 10° 0° 10° 20° 30°
70° 60° 50° 40° 30° 20° 10° 0° 10° 20° 30° 40° 50° 60° 70°

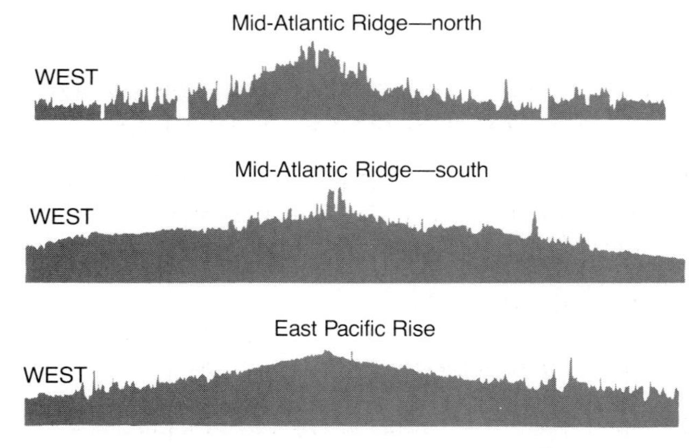

Figure 2.14 Selected profiles across the oceanic ridge system showing the range in relief from the high-relief, rugged profile of the North Atlantic Ocean to the low relief of the East Pacific Rise. (After Heezen, B. C. and M. Ewing, 1963.)

at its crest (fig. 2.14). It is broad, relative to the narrower ridge in other areas. The reason for this difference is believed to be at least in part related to the difference in spreading rates of the East Pacific Rise as compared to the Mid-Atlantic Ridge. This topic will be explained in detail in the following chapter.

The **oceanic trenches** or **deep-sea trenches** are the deepest parts of the oceans. They are elongate and narrow, somewhat arcuate in shape with the convex side facing oceanward, and they occur along the outer margins of the ocean basins. Although there are trenches in all three oceans, they are most common in the Pacific (fig. 2.15). The trench floors are below depths of 6,000 m. Trenches are generally less than 200 km wide and up to 6,000 km long. The deepest trench is the Mariana Trench in the western Pacific, which plunges 11,030 m below sea level.

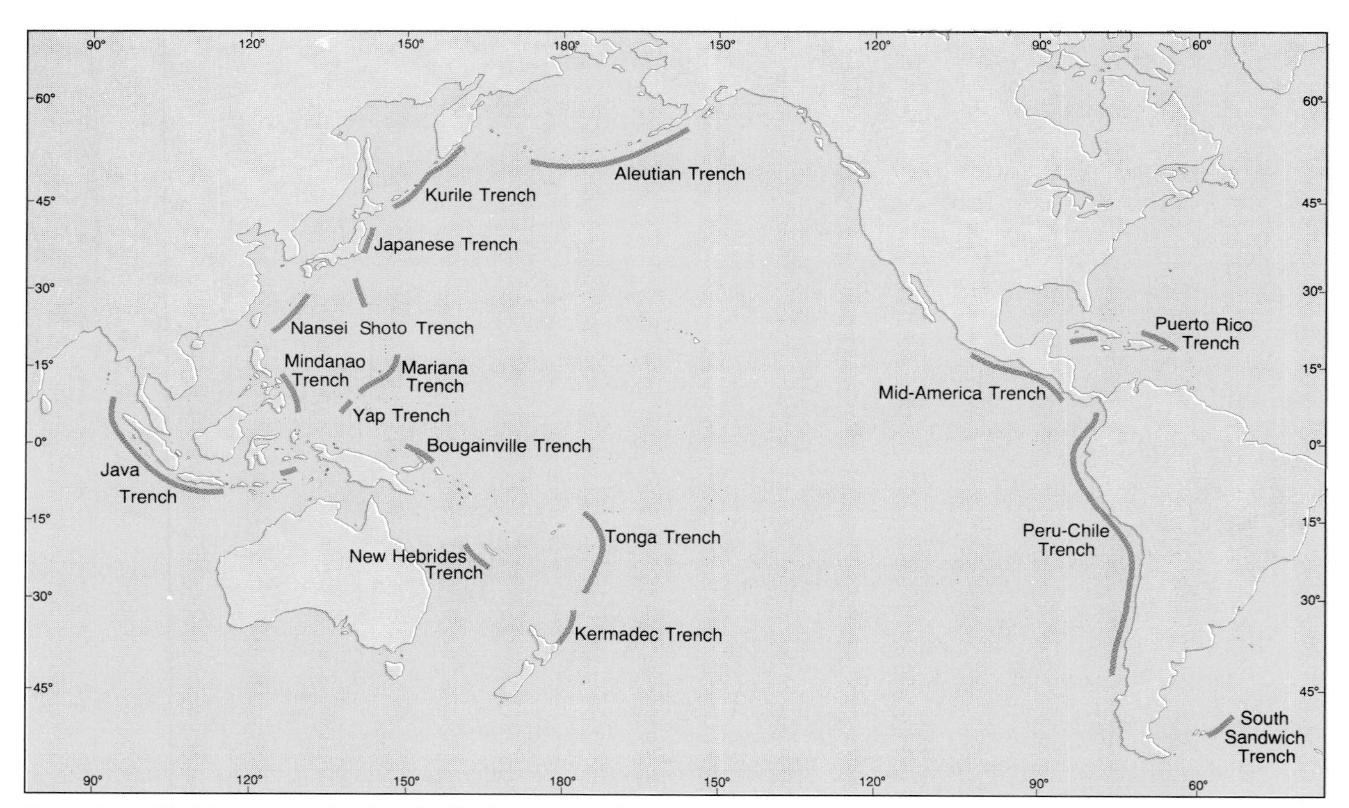

Figure 2.15 World map showing the distribution of oceanic trenches. Most of them are associated with the many island arc systems that surround the Pacific Ocean.

Most trenches are associated with an **island arc** system or with a volcanic range on the immediately adjacent continental land mass. Island arcs are volcanic belts that parallel the trench and are located adjacent to them on the landward side. The profile of a trench is asymmetrical, with the steep side toward the island arc. High earthquake activity, low pull of gravity, and low flow of heat from the earth also characterize trenches.

Local, high relief features of volcanic origin are scattered throughout the ocean floor. **Seamounts** are more or less circular in plan view and rise at least 1,000 m above the seafloor. They have a peaked profile. Seamounts are most abundant in the Pacific Ocean and are concentrated near the transform fault zones. Some occur in linear groups and are thought to indicate hot spots. **Guyots,** also called tablemounts, are similar to seamounts in all respects except that they have a flat-topped profile. These features apparently extended above sea level at one time and were eroded to their present configuration. Their present depth is far in excess of estimated sea level changes, but this can be explained by subsidence of these large masses on the thin oceanic crust. Tilted upper surfaces and areas of depression at the base support the subsidence theory.

Summary of Main Points

The earth contains a number of important major provinces both within it and on the surface. These range from features that are global in extent to those that may only cover a square kilometer or so. Together they form a complex planet.

The earth's interior has only three layers: core, mantle, and crust. The rigid upper layer is called the lithosphere and contains the crust and part of the upper mantle. The crust itself contains a relatively light granitic portion, which is represented by the continental land masses, and a denser basaltic portion, which is beneath the ocean basins.

The major surface features of the oceans are the continental margins and the ocean basins. The continental margins are actually part of the granitic crust and contain the continental shelf, continental slope, and continental rise provinces. The ocean basins include the abyssal floor, oceanic ridge, and oceanic trench provinces.

Suggestions for Further Reading

Burke, C.A., and C.L. Drake (eds.). 1974. **The Geology of the Continental Margins.** New York: Springer-Verlag, Inc.

Heezen, B.C., et al. 1959. **The Floor of the Ocean—I. The North Atlantic.** Geological Society of America, Special Paper 65.

Shepard, F.A. 1973. **Submarine Geology,** 3d ed. New York: Harper and Row.

Seibold, E., and W.H. Berger. 1982. **The Seafloor: An Introduction to Marine Geology.** New York: Springer-Verlag, Inc.

Turekian, K.K. 1976. **Oceans.** Englewood Cliffs, New Jersey: Prentice-Hall.

3
Plate Tectonics

Chapter Outline

Probably the most significant advance in the earth sciences during the last half of the twentieth century is the establishment of plate tectonics theory along with the associated phenomena of **continental drift** and **seafloor spreading.** It is now known that the oceans have changed their size and configuration through time and that landmasses have come together and broken up in the past. The crust of the earth is dynamic, not only in the vertical sense as evidenced by mountain building, but also through the extensive, essentially horizontal movements that take place in both the basaltic (oceanic) and granitic (continental) components of the crust.

The great mountain ranges of the world are belts of large-scale crustal movement that may extend for many thousands of kilometers and are relatively narrow compared to their length. Mountain belts are characterized by folding and fracturing, along with the introduction of new crustal material and changes to that material already present. All of these phenomena tend to suggest or to be associated with compression, which led many people to theorize that the earth was shrinking or contracting and that the relatively brittle outer shell of the earth, the crust, was being deformed similar to the way that an orange wrinkles after many weeks of being at room temperature.

Another school of thought believed that the opposite situation was taking place. The recognition and delineation of the oceanic ridge system, along with the association of normal faulting with ridges in the oceans and on land, such as the rift valley in Africa, led to the concept of an expanding earth. Earthquake activity around the trench-island arc systems and the patterns of heat flow from the earth's crust gave support to the idea.

Data were available to support each of these hypotheses, however, the earth could not be experiencing both contraction and expansion simultaneously. Another explanation was necessary.

History of Continental Drift Theory

Benjamin Franklin is credited with many scientific discoveries and numerous inventions, but is not usually mentioned in discussions of continental drift. He was, however, one of the first people to recognize that the earth's crust was mobile and subject to considerable change. Franklin also recognized that this intense change in the crust was restricted to specific areas.

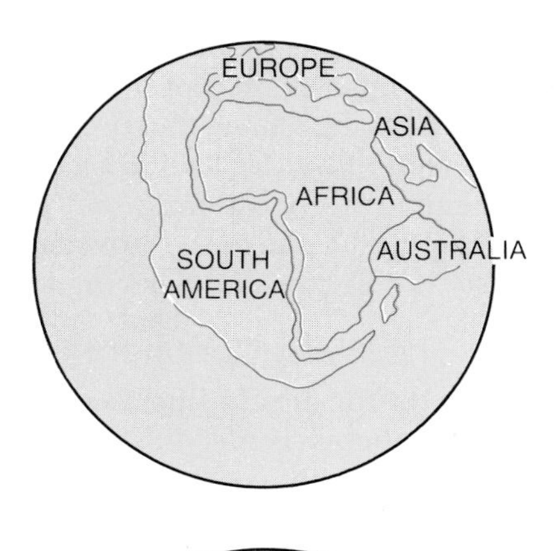

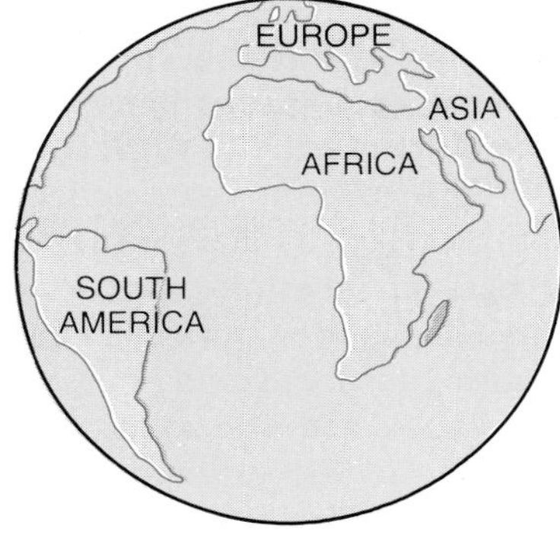

Figure 3.1 This diagram by Snider was the first to explain geological similarities between Europe and North America using continental drift. (From Tarling, D. and M. Tarling, *Continental Drift.*)

Even before Franklin, Sir Francis Bacon (1561–1626) recognized the geometric compatibility of the eastern side of South America with the western side of Africa. Subsequent naturalists and explorers noted the same relationship, but it was not until 1858, when Antonio Snider suggested that the Atlantic Ocean was formed by the breaking apart of Africa and South America, that the idea of continental drift was first suggested. Snider produced maps showing the nature of the landmass prior to the breakup and after the formation of the Atlantic Ocean (fig. 3.1).

Various ideas and speculations about continental movement and ocean basin changes were proposed in the early twentieth century. One, proposed by Howard Baker, an American paleontologist, stated that the close passage of a celestial body generated enough gravitational attraction between it and the earth to cause a large piece of the crust to be pulled away, resulting in the Pacific Ocean. The crust that was removed became the moon! Frank B. Taylor, an American glaciologist, came closer to the theory of continental drift by suggesting that the young mountain ranges surrounding the Pacific Ocean resulted from the opening of the Atlantic and subsequent shrinking of the Pacific due to the movement of landmasses, but he had no plausible driving mechanism.

Wegener's Theory of Continental Drift

Most of the credit for developing a comprehensive continental drift theory is given to Alfred Wegener, who conceived his hypothesis in 1910, but did not make it public until his presentation to the Geological Society in Frankfurt in1912. Although Wegener was a meteorologist by training, he had remarkable insight into global geological phenomena. He presented a supercontinent called Pangaea, which broke up about 200–250 million years ago (Late Paleozoic) with the pieces moving about the earth's surface to their present locations (fig. 3.2). Wegener diligently pursued his idea until his death in 1913 without gaining widespread acceptance.

Wegener believed that the rotation of the earth caused the breakup of the supercontinent and subsequent drift. The breakup began about 200 million years ago along two main fractures: one between the Americas and Europe-Africa and the other between Africa-India and Australia-Antarctica. As separation proceeded along these fractures, the Atlantic and Indian oceans developed. The predrift ocean was the ancestor to the present Pacific Ocean.

Pangaea contained a shallow sea, Tethys (fig. 3.2), which separated India from the rest of present-day Asia until about 75 million years ago (Cenozoic). Eventually India collided with Asia, causing the development of the Himalayan Mountains. Wegener believed that all the major folded mountain ranges formed as a result of collisions or as the continental crust dragged along over the underlying lithosphere. The drifting of the continents also resulted in detached pieces being left behind in the form of islands, such as the island of Madagascar, east of Africa, and Tasmania, off the southeast coast of Australia.

Wegener's substantiation of drift was based on circumstantial information. Physicists did not accept the forces of rotation as a mechanism for drift and the idea of moving continents was implausible to geologists. Consequently, the theory of continental drift was discounted by scientists as unfounded and undemonstrable.

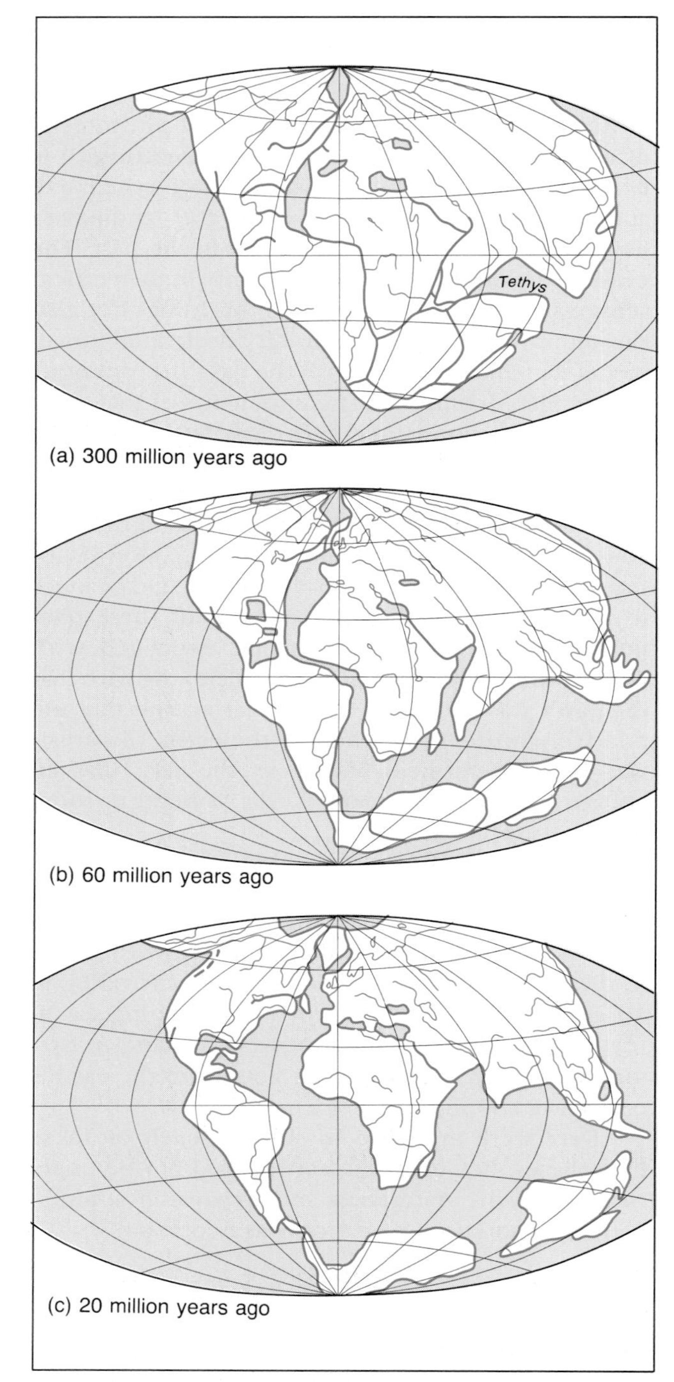

Figure 3.2 Three stages in the continental drift model as proposed by Alfred Wegener in 1912.

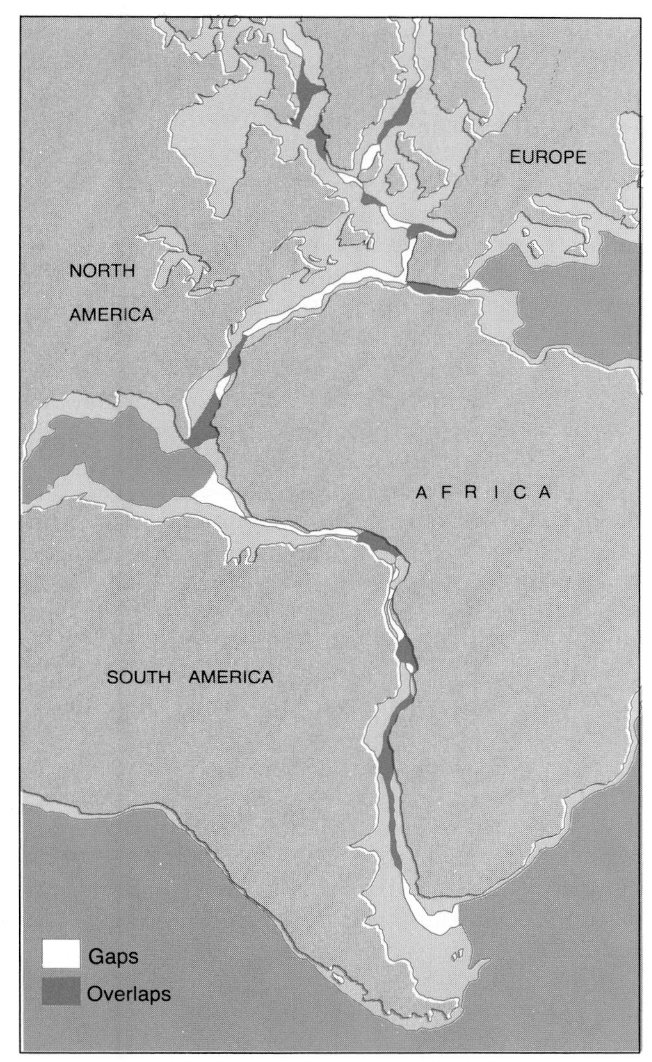

Figure 3.3 The fit of North and South America with Africa and Europe as taken at the present 6,000 ft (1,835 m) contour. The fit is reasonably good, although both overlap and gaps exist. The continental margin has changed greatly over the past 190 million years.

New Data and Rebirth of the Drift Theory

The end of World War II brought a return to pure research in the earth sciences and a great expansion of activity in the ocean basins parallel to that on the continents. Even though the number of supporters of the continental drift theory was small, it included some vociferous scientists. During the 1950s there were several symposia on the topic. Most proponents of the theory came from Tasmania and South Africa; only a few were from North America. The data discussed in these forums ranged widely in nature but can be categorized into that from the landmasses and that from the ocean basins.

Data from Continents

The apparent fit of the various pieces of the Pangaea puzzle has already been mentioned. In an attempt to provide credibility to such reconstructions, maps of the various landmasses were prepared, based on the continental margin dimensions rather than the usual sea level ones (fig. 3.3). A contour of several hundred meters depth was used to insure that all of the shelf was included. This approach caused some problems, because by using the edge of the modern continental mass, investigators were using modern pieces for a puzzle that was over 100 million years old.

Better-substantiated matches were demonstrated for the ocean basins, generally by comparing structural similarities for such features as mountain ranges, faults, and large intrusive bodies. The Great Glen fault complex in Scotland was matched with the Cabot fault system of northeastern North America. The Hebrides Islands in Scotland were connected with related landforms in Newfoundland (Labrador). A similar connection between folded mountains in Argentina and mountains on the Cape of Good Hope in South Africa could be made. Similarly, the ancient cores of continents, called shield areas, of South America and Africa can be matched on the basis of their rock types and age (fig. 3.4). Similar connections can be postulated across the Indian Ocean and between South America and Antarctica. In the Indian Ocean, for instance, the island of Madagascar contains folded rocks that compare well with folded rocks in India.

Data from the fossil record also support the idea of continental drift. Remains of organisms that appear to be identical have been found in widely separated landmasses. Land plants and animals are the best candidates for supporting drift because an ocean would prevent or impede them from migrating. Thus if the same fossil is found on widely separated landmasses, it implies that these landmasses must have been contiguous at one time. As an example, the *Glossopteris* flora is an assemblage of plant fossils from coal beds formed between the Early Carboniferous and the Middle Cretaceous (300–100 million years before present). These fern-like plant fossils are found in all landmasses of the southern hemisphere and in India. Similarly, fossils of a certain type of

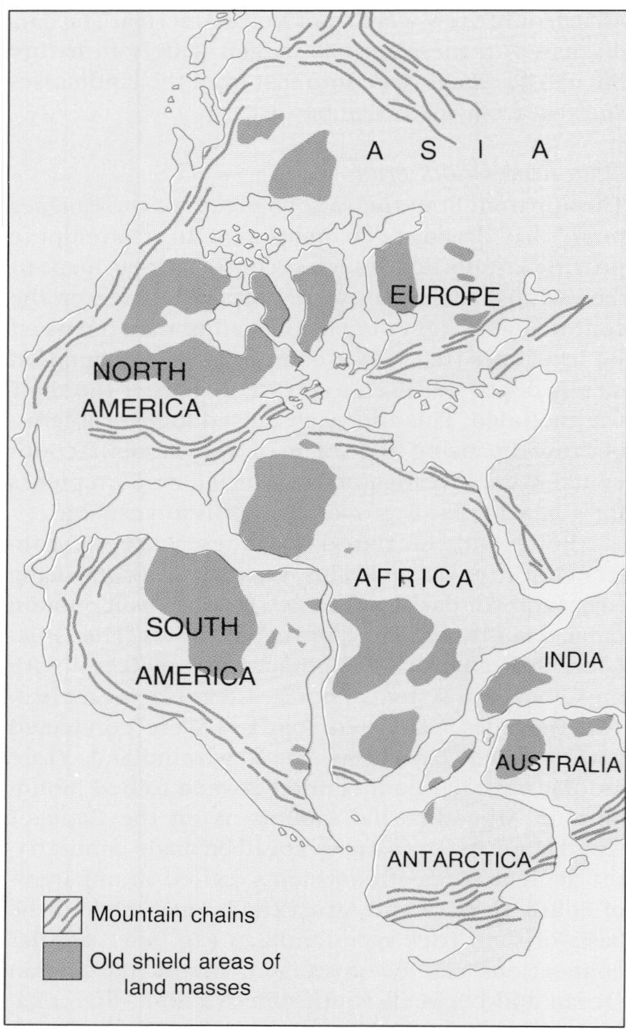

Figure 3.4 Large-scale geological connections across the Atlantic Ocean. Many of the old shield areas and mountain ranges match well if the Atlantic is closed and the continents moved together.

lemur are found in Africa, India, and southeast Asia, areas separated by large water masses or environments that would not permit migration. Fossils of numerous varieties of shallow marine invertebrates that lived about 450 million years ago in the Appalachians of North America and the British Isles show great similarity but are different from varieties of similar age in the Rocky Mountains. In all of these cases, unless these landmasses were once connected it would not be possible for the same organisms to exist in such geographically distinct areas.

In addition to matching structural features, rock types, and fossils, there are also many absolute age dates for correlating comparable rocks by their composition. This lends support to the idea that such correlative but currently separated rock bodies may have at one time been joined together. Absolute age dates are **radiometric age dates,** based on the decay rates of certain unstable isotopes of various elements. The ratio of a stable isotope to an unstable isotope changes through time as the unstable one decays. Measurement of this ratio permits determination of the absolute age of the rock.

If, for example, a granitic body in South America has the same chemical composition as a granitic body in west Africa and if the granitic bodies are also a geometric fit, then there is evidence to suggest that the now-separated rock bodies might once have been one. However, it is also critical to the argument to demonstrate that the granitic body in South America is the same age as its counterpart in Africa. Radiometric age dating has demonstrated that those suggested correlative rock bodies on separated landmasses are in fact compatible as far as their age is concerned.

Rocks of the continental crust that are magnetic provide another line of evidence for continental drift. These rocks contain records of the earth's magnetic field at their time of formation. When magnetic minerals crystalize, they align with the earth's magnetic field, thus creating a **paleomagnetism** or ancient record of the magnetic field. The magnetic field of the earth reverses, causing the north and south poles to flop back and forth, as they have done frequently in the past. The magnetic poles of the earth also tend to wander slowly. As a consequence of this wandering, if one were to collect data from North America on the location of the magnetic pole 250 million years ago, there would be a marked migration from the north Pacific Ocean across to western Canada. This in itself is not evidence for continental drift; however, if we compare the paths of the wandering magnetic pole from North America with that of a landmass in the eastern hemisphere, for example Europe, a very different pattern would be displayed (fig. 3.5). The map shows that a path for the paleomagnetic pole of North America is distinctly separate from that of Europe. Note also that the two paths converge. This suggests that the two landmasses in question were not always in their present positions. The distance separating the paths of these poles is about the same as the width of the

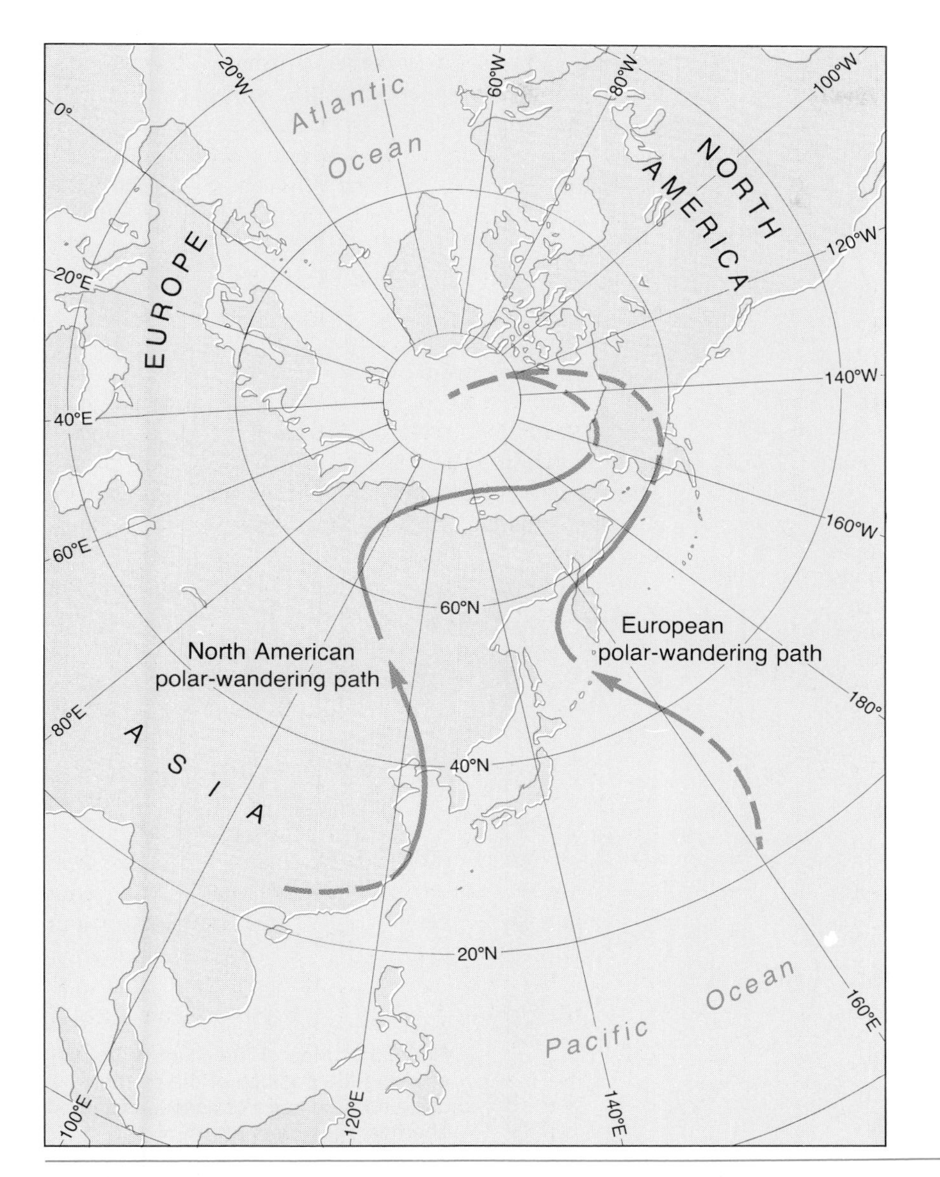

Figure 3.5 Map of the North Pacific showing the paths of the north magnetic pole from both North America and Europe. If the Atlantic Ocean were closed and the continents moved together, the paths would be essentially the same.

North Atlantic. Moving these landmasses together causes the polar paths to merge. This is strong support for the concept of continental drift.

Data from the Ocean Basins

Ocean floor charting showed a huge mountain chain that was continuous on a global scale. This oceanic ridge system and its origin became the subject of much discussion and served as a point of argument for the supporters of continental drift. In the Atlantic and Indian oceans, the ridge system essentially bisects the basins. Coupled with this enormous feature is the age distribution of rocks in the oceanic crust. Those at or near the ridge system are very young and there is a regular pattern of increasing age away from the ridge (fig. 3.6). This is displayed in all ocean basins. Coupled with these data is the fact that no rocks older than about 180 million years (Triassic Period) are present in the oceanic crust. These data suggest that the floors of the oceans are moving away from the ridges and that no oceanic crust exists that is older than about 180 million years. Related evidence is the age of volcanic islands (seamounts and guyots) along the large fracture zones that run perpendicular to the ridge. These volcanic islands also display an increasing age away from the ridge.

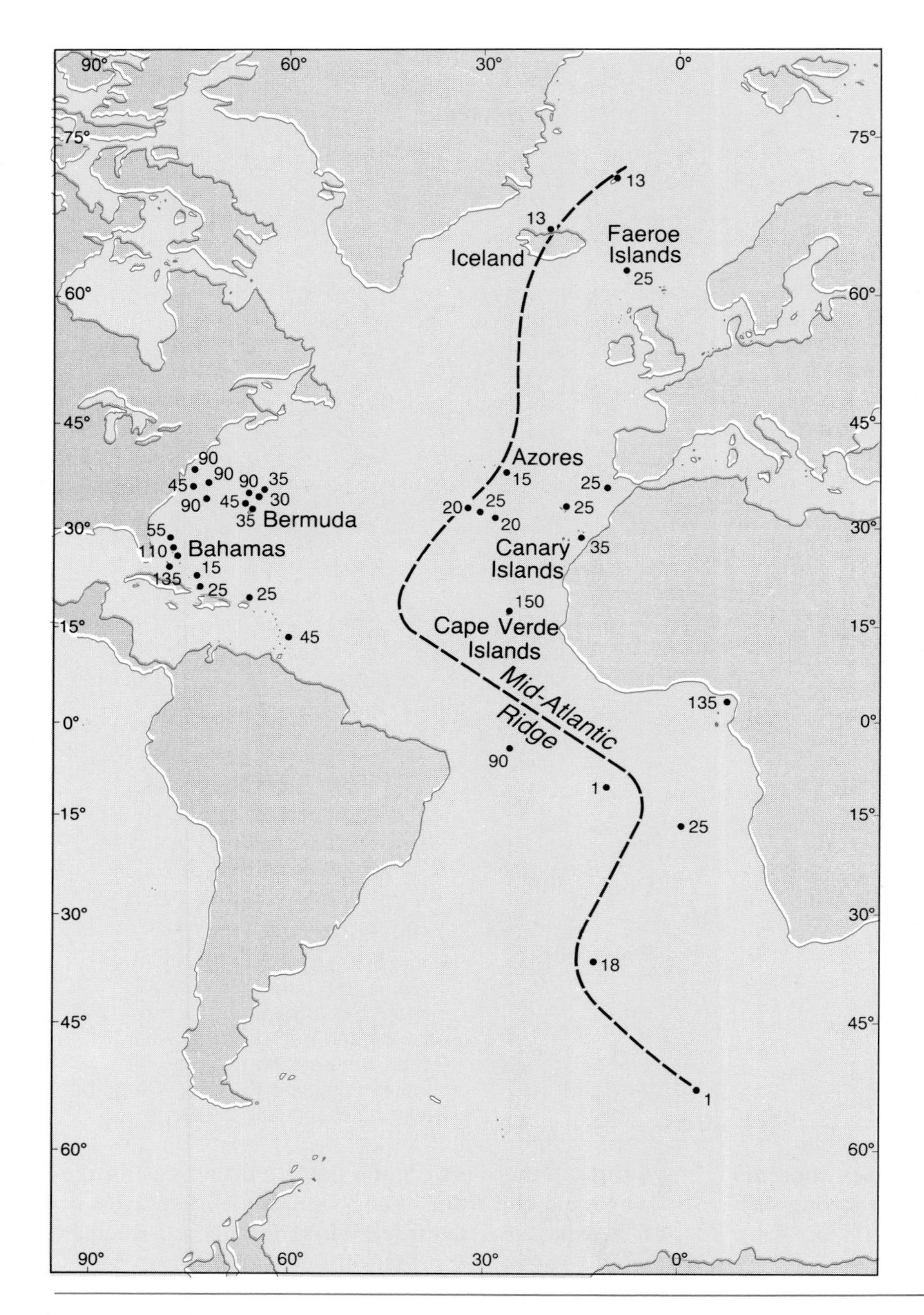

Figure 3.6 Map of the Atlantic Ocean showing the position of the oceanic ridge and the ages of various islands. There is a general pattern of increasing age of these islands going away from the oceanic ridge. (Adapted from "Continental Drift," by J. Tuzo Wilson. Copyright © 1963 by *Scientific American,* Inc. All rights reserved.)

Paleomagnetism in the oceanic crust can also be used to support the idea of continental drift. Basaltic rocks contain abundant magnetic minerals and display strong, easily measured magnetic fields. As mentioned previously, the earth's magnetic field reverses, generally once or twice each million years. Magnetic reversals produce a striped pattern in rocks of the oceanic crust, with normal and reverse magnetism forming the stripes.

Even a casual glance at the pattern of magnetic reversal stripes shows that there is a bilateral symmetry to it with the oceanic ridge in the center. A plot of the magnetic profile with time as the vertical scale shows that the pattern on one side of the ridge is a mirror image of the other side (fig. 3.7). Additionally, the ages of the reversals increase, without exception, away from the ridge.

BOX 3.1

Paleomagnetism

It has long been known that certain minerals are magnetic; in fact, the Norwegian explorers used magnetite, an iron oxide, to aid in navigation. It was also recognized that the earth itself behaves like a giant magnet. It is the magnetic field of the earth that causes the preferential orientation of magnetic materials. This field is thought to result from electrical charges generated by differential movement of the earth's core and mantle.

The drift of the magnetic poles of the earth's field has been substantiated by noting the movement of magnetic north over several centuries. Over a period of 250 years, the north magnetic pole wandered 32° as measured from London. From this information, it has been estimated that the magnetic poles wander around the geographic poles in a cyclic period of 500 years.

Basaltic magma has a high iron content. The basaltic or oceanic crust that forms from this magma contains much magnetite, the most abundant magnetic mineral in the earth's crust. These magnetic crystals act like small magnets floating in a soup of viscous material: as the molten magma cools and the magnetite crystals begin to form, they align with the magnetic field of the earth. When the entire magmatic mass is cooled and lithified, these magnetite crystals are frozen in the crust with this preferential alignment.

The magnetic field of the earth reverses polarity and has done so many times in the past. This means that at certain times the north-seeking end of the compass points to the magnetic pole that is in the geographic north portion of the earth but during a magnetic reversal this north-seeking end of the compass points toward the south magnetic pole. Data show that each period of reversal lasts about half a million years and that it takes a few thousand years for the reversal to occur.

Although the magnetic orientation of a rock sample is easily determined in the laboratory, it is necessary to look at the global scale to determine the nature of paleomagnetism of the rocks of the basaltic crust. In order to accomplish this, a magnetometer is used. The magnetometer can be towed by a ship or an airplane and can determine the magnetic orientation of the crust below.

Surveys of this type have shown that the magnetic minerals in the oceanic crust display stripes showing normal and reversed magnetic fields of the earth. These stripes parallel the oceanic ridge systems and are mirror images on opposite sides of the ridge crest. This type of information has been used to help confirm seafloor spreading and also to calculate the rates of spreading.

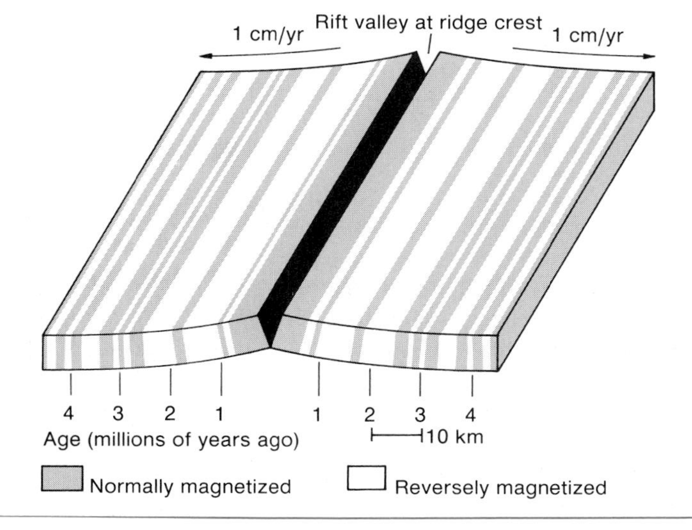

Figure 3.7 Stripes on the oceanic crust are developed by normal and reversed magnetism of the earth's magnetic field. These magnetic reversals make it possible to determine the age of the seafloor and measure the rate of spreading from the oceanic ridge.

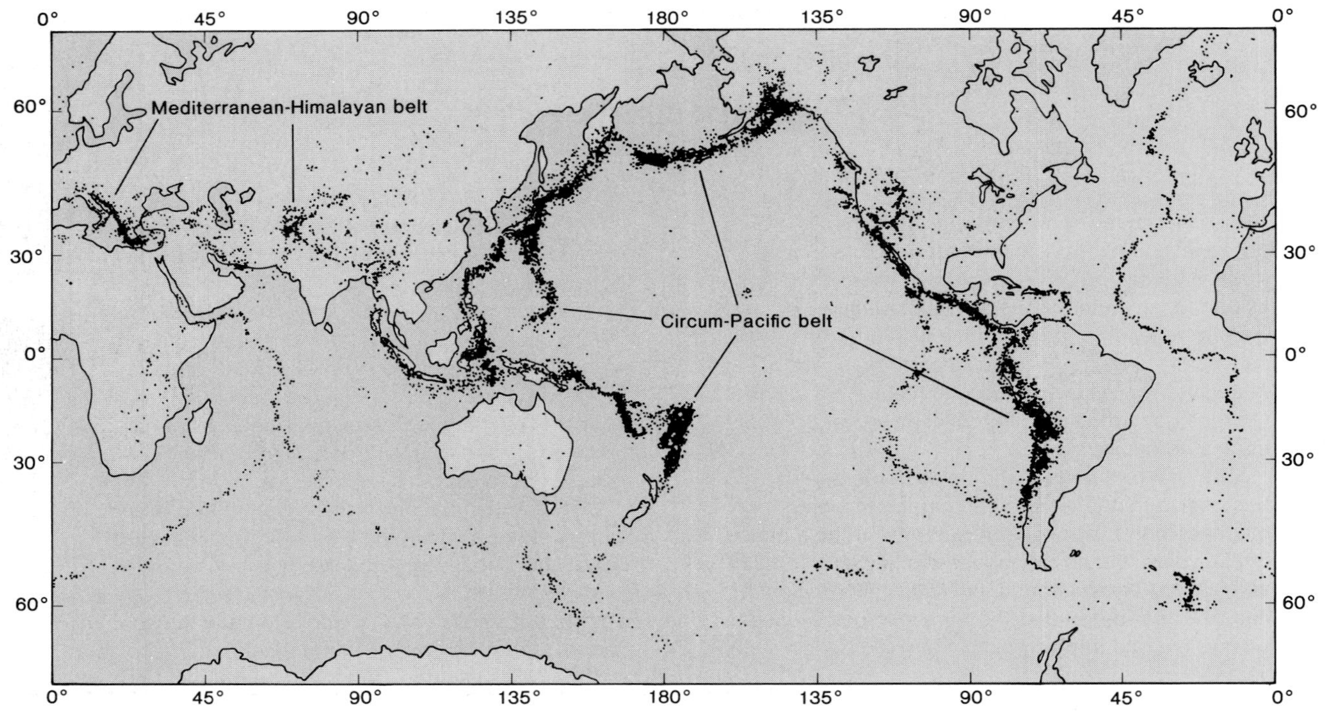

Figure 3.8 Map of the world showing the earthquake epicenters. There are two zones of concentration: the "Ring of Fire" around the Pacific Ocean, where oceanic trenches are located, and the oceanic ridge system.

The locations of earthquake epicenters around the world provide other evidence for continental drift. Although they are widely scattered, there are places where significant concentrations occur. The circum-Pacific belt of oceanic trenches is the greatest concentration of earthquake activity. A second but much less intense concentration is associated with the oceanic ridge system (fig. 3.8). At this stage in the discussion it is premature to consider the reason for this distribution of earthquakes, but there is a distinct pattern.

All of the previous data lend support to the idea of continental drift, but the evidence is circumstantial. There is no serious contradiction within the data set. There is no significant evidence that indicates that drift did not happen but the preceding data offer no proof nor has there been, to this point in the discussion, a reasonable mechanism proposed that could produce the preceding information.

Theory of Plate Tectonics

As information in support of continental drift began to accumulate, it became necessary to find a mechanism for continental drift or find some alternative to drift that could explain all of these data. It was through this information and the ideas of men like Arthur Holmes of the University of Edinburgh, H.H. Hess of Princeton University, and Robert Dietz of NOAA, that the idea of **plate tectonics** was born. This concept said that the earth's lithosphere is composed of several plates of differing sizes and shapes (fig. 3.9), which move over the plastic asthenosphere in the upper mantle. Convection cells caused by heat from radioactive decay in the mantle are the driving mechanism. Convection cells circulate the heat upward, causing upwelling in the mantle. As the magma in the convention cell cools, downwelling occurs. Such a mechanism was suggested by Holmes in the 1920s (fig. 3.10) and then proposed again by Hess in the early 1960s.

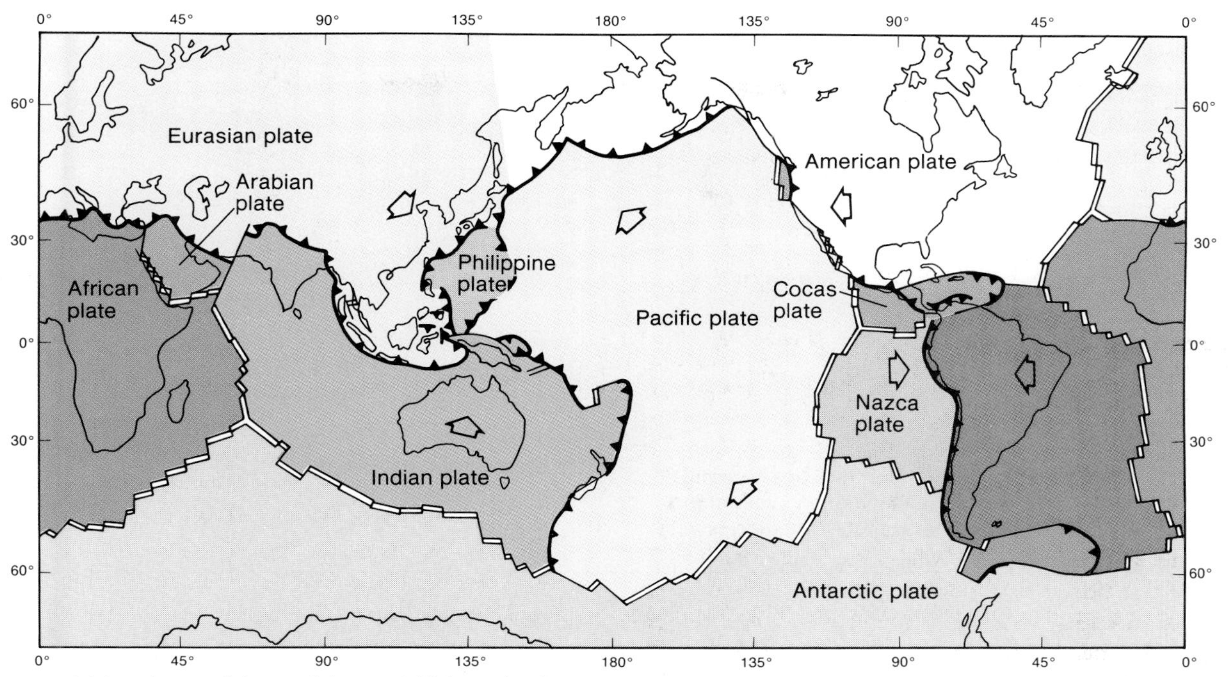

Figure 3.9 Major plates of the earth's crust. This projection makes the areas near the poles appear much larger than they really are. (After W. Hamilton, U.S. Geological Survey.)

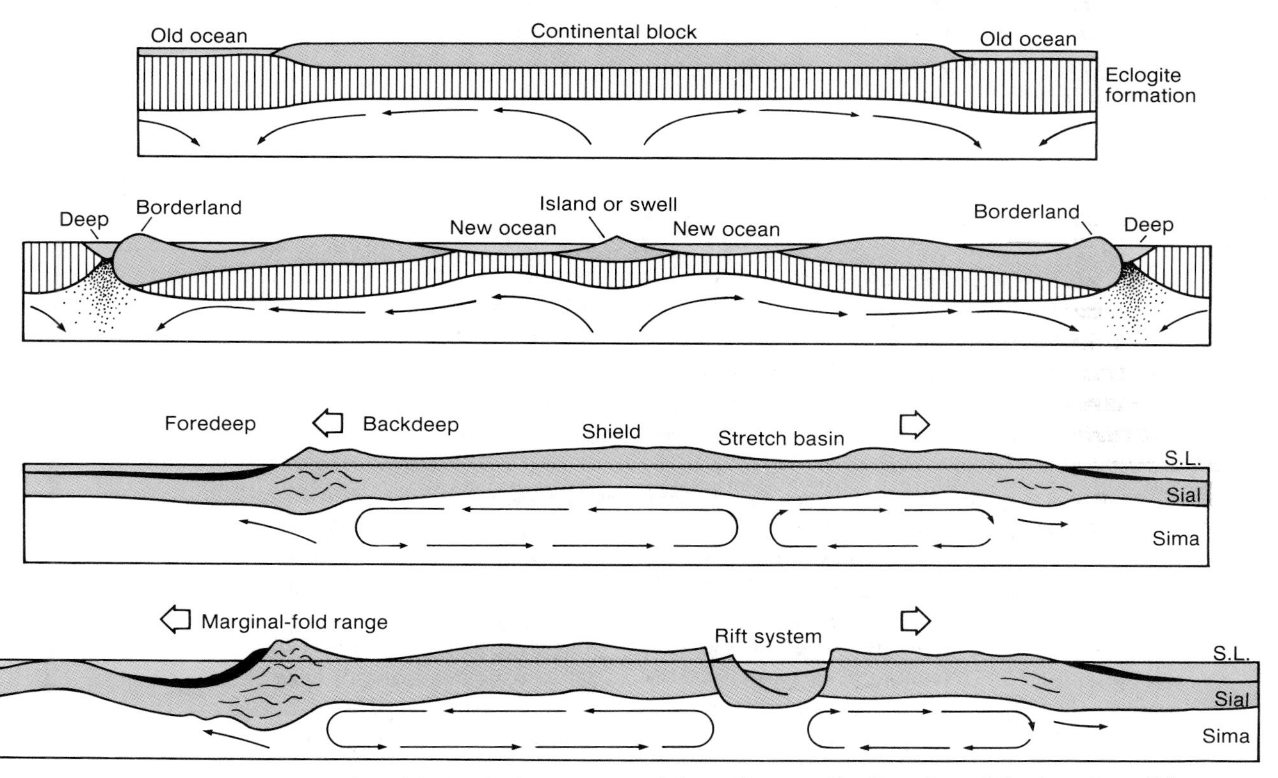

Figure 3.10 A schematic representation of Arthur Holmes's idea for the mechanism of seafloor spreading. Heat rises and produces volcanism along the ridges, and the crust sinks and causes earthquakes at the trenches. (After Holmes, A.)

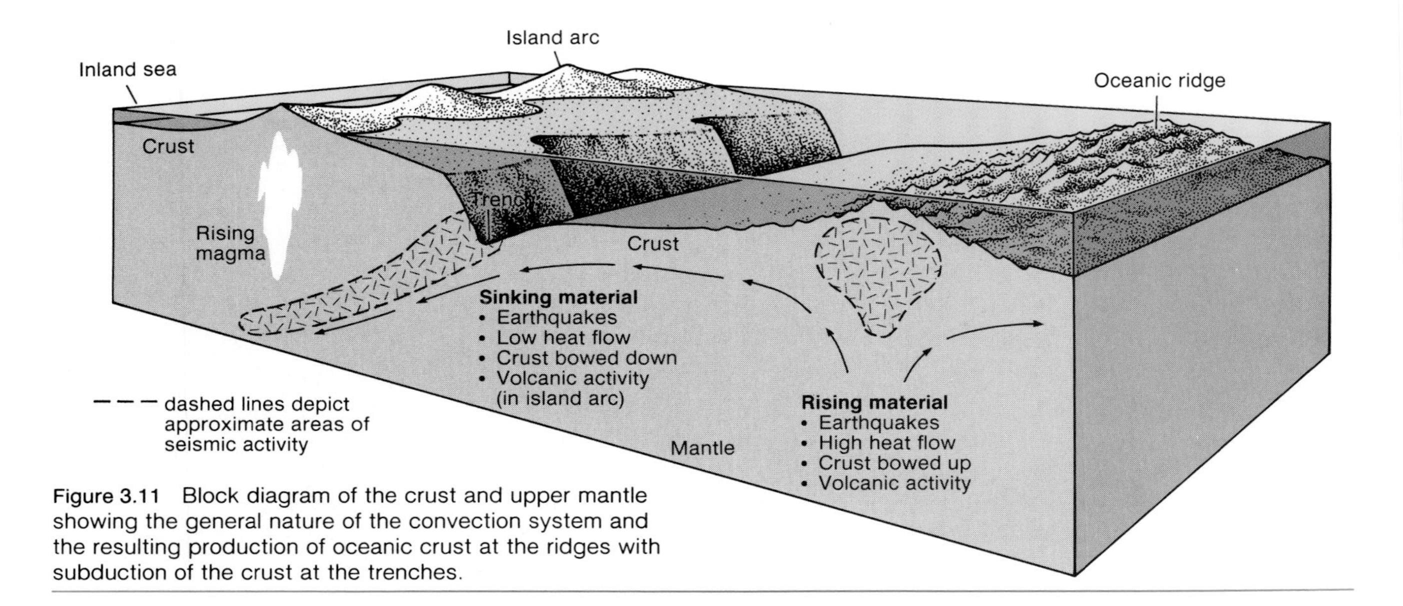

Figure 3.11 Block diagram of the crust and upper mantle showing the general nature of the convection system and the resulting production of oceanic crust at the ridges with subduction of the crust at the trenches.

The movement of the plates results in areas of spreading or plate separation explained by the oceanic ridge system. Where plates collide earthquakes and volcanism prevail, and the deep-sea trenches and their associated island arc systems develop. Heat flow data show high values over the oceanic ridges and low values over the trenches, which fits the convection cell model.

The distribution of earthquakes supports the concept of plate tectonics. In addition, data from deep earthquakes show that their epicenters descend at an angle in the direction of theorized plate movement. This happens because when plates collide, one is forced under the other, forming a **subduction zone.** As the edge of the lower plate descends, heat and pressure absorb the plate into the mantle and it loses its identity (fig. 3.11). Continental plate collisions cause folded mountain belts to form, such as the Appalachian Mountains.

Several specific features must be considered to give the theory of plate tectonics credibility. The first consideration is the question of how the various rigid, oddly shaped plates move over the earth's sphere. The data indicate that there are three types of plate boundaries: (1) oceanic ridges, where plates separate or spread; (2) subduction zones, where plates collide; and (3) **transform faults,** where plates slip with lateral movement. Plates may move in opposite directions or at different rates along transform faults. Associated with the transform faults are the numerous fracture zones that allow for dislocations due to variation in spreading rates within the curved plates (fig. 3.12). The confluence of three plate boundaries, called **triple junctions,** may consist of ridges, subduction zones, or transform faults. Some combination will allow for plate motion over the spherical earth.

Mechanism for Drift

Geological and geophysical information show that the oceanic ridges are spreading centers where new oceanic crust and lithosphere are formed (fig. 3.13). All data, including ages of the ocean floor, physiography, heat flow, paleomagnetic data, and structure of the ocean floor substantiate this phenomenon. Correlating the age of the oceanic crust with its position relative to the ridge at which it formed gives seafloor spreading rates ranging from about 1 cm to 10 cm per year. Both biostratigraphy and paleomagnetism data show increasing age with increasing distance from the ridges, although they vary somewhat with time (fig. 3.13).

The rate of spreading apparently affects the physiography of the ridge system. Rapid spreading produces a broad, relatively low ridge without a deep central valley, such as in the Pacific, west of South America. Slow spreading results in a high-relief ridge with a deep central rift valley, such as in the central Atlantic Ocean.

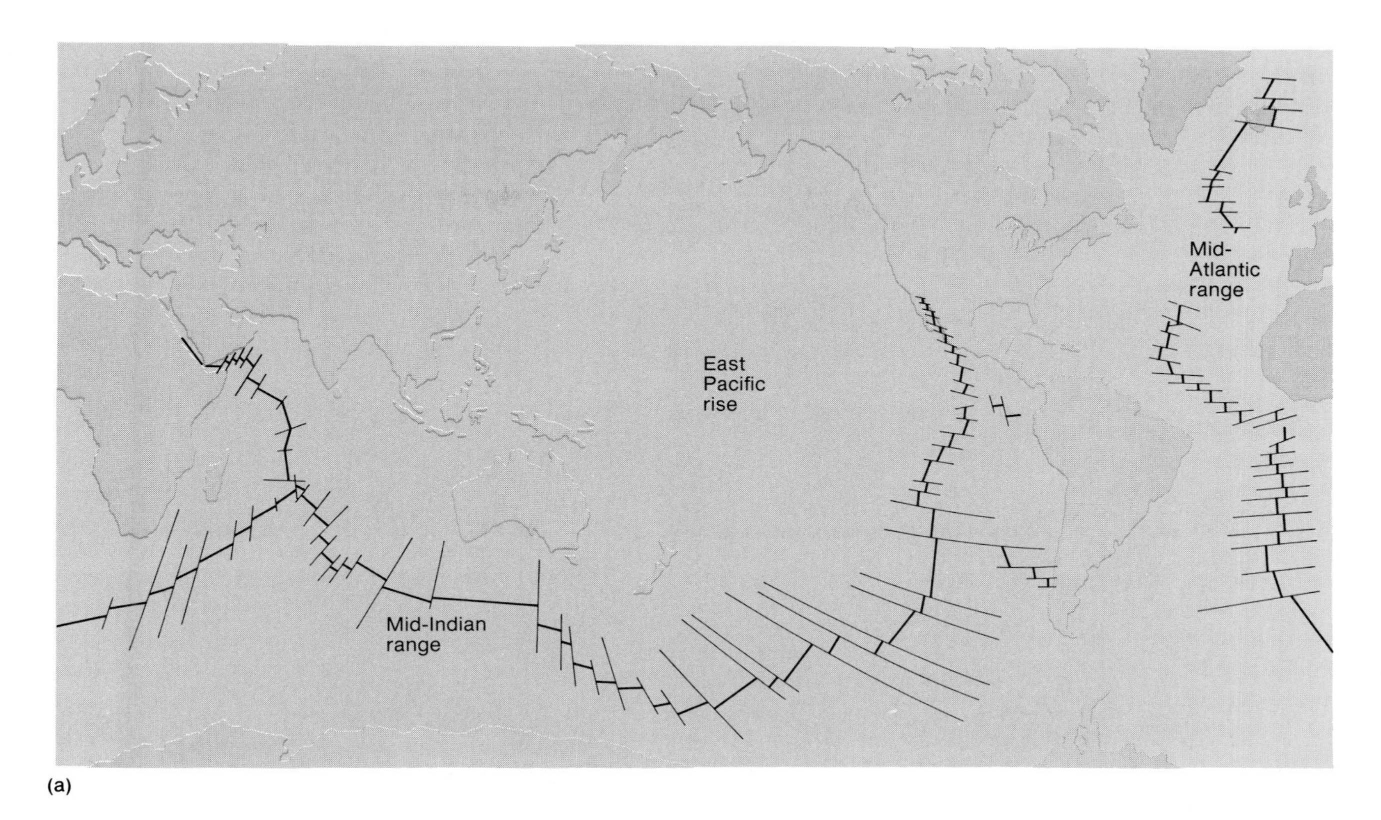

(a)

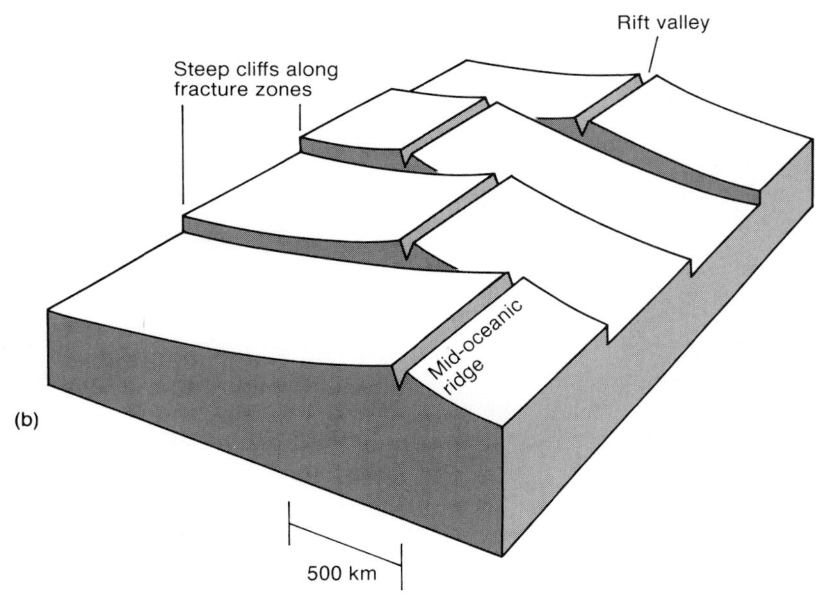

(b)

Figure 3.12 (a) Map showing numerous dislocations along the oceanic ridge system where transform faults occur and (b) a close-up diagram of these transform fault zones.

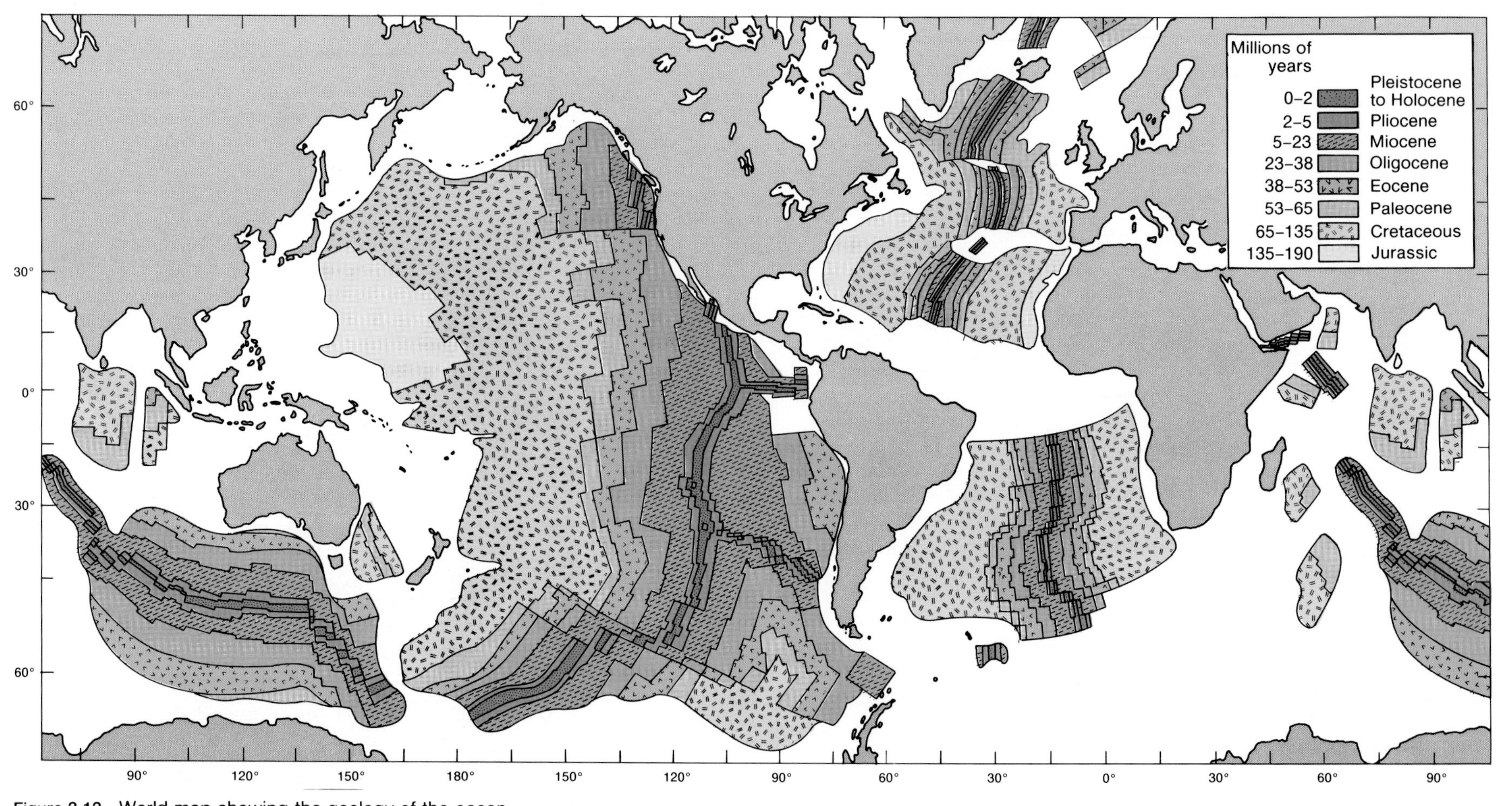

Figure 3.13 World map showing the geology of the ocean floor with the ages of areas for which data have been collected. There are still many places where drilling must be done, especially in the Indian Ocean.

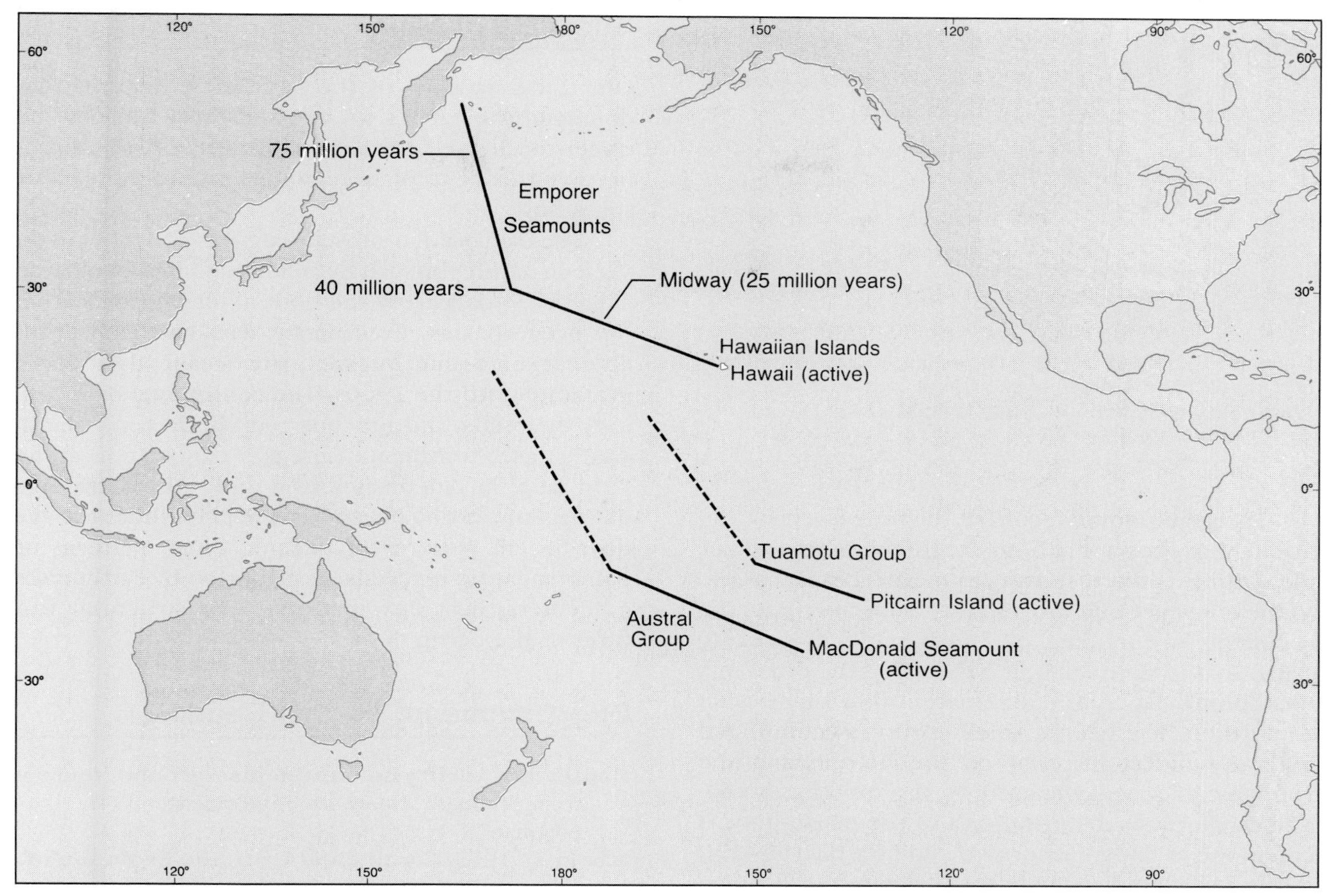

Figure 3.14 Each island chain in the ocean basins can be attributed to the passage of an oceanic plate over a stationary hot spot in the upper mantle. Examples shown here include the Hawaiian Islands, which are older toward the northwest.

The earth's internal convective motion is produced by a combination of heat flow in the asthenosphere and increasing density of oceanic crust away from the ridge. Heat and magma surface at the ridges. As the crust solidifies, it spreads away from the ridge crest and down the gradual slope of the ocean floor. As it cools and ages, the crust becomes denser, thus facilitating its subduction when collisions between plates occur. The system acts like a conveyer belt, rising under the ridges, spreading, and then converging and descending at the trenches and volcanic arcs. The crustal plates are in a sense riding on this conveyer belt and become partially consumed as descent occurs. The belt turns very slowly, taking over 100 million years to cross an ocean.

There are some local areas where **hot spots** exist under the lithosphere. These are places where extreme heat and magmatic conditions are close to the upper surface of the mantle. As the crustal plate passes over this hot spot, volcanoes may form on the ocean floor. Over many millions of years, a line of several volcanic islands may result from this type of activity as a plate passes over the asthenosphere and its contained hot spot (fig. 3.14). The Hawaiian Islands represent such a situation.

Confirmation of the Theory

To date, all information from the continents and from the seafloor supports the theory of plate tectonics and continental drift. Essentially all of the previously mentioned data were collected prior to the initiation of the Deep Sea Drilling Project or were being collected during the early stages of the project. The data provided by this project, in the form of numerous deep cores from all parts of the world ocean basins, confirm the theory.

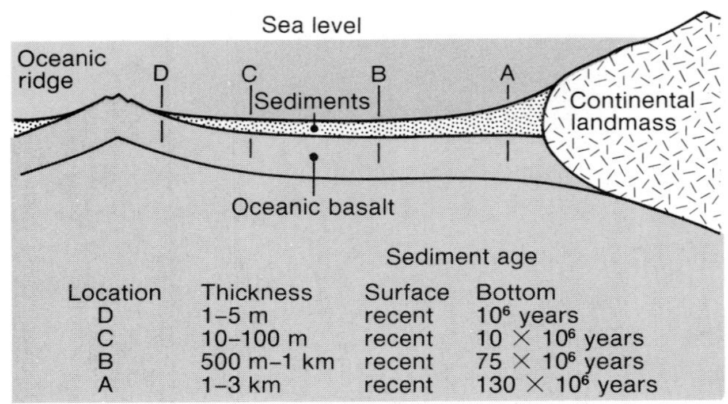

		Sediment age	
Location	Thickness	Surface	Bottom
D	1–5 m	recent	10^6 years
C	10–100 m	recent	10×10^6 years
B	500 m–1 km	recent	75×10^6 years
A	1–3 km	recent	130×10^6 years

Figure 3.15 General diagram showing the increasing age and increasing thickness of sediments over the oceanic crest going away from the oceanic ridge. (After Duxbury and Duxbury, 1984.)

The stratigraphic record of the rock layers on the ocean floor shows that both the thickness and age of the sediments increase away from the oceanic ridges. As the conveyer belt moves away from the ridge, the sediments accumulate on the basaltic or oceanic crust. By the time the belt has carried the crust to a location near a continental margin or a subduction zone, there has been a thick section accumulated with the oldest material on the bottom and the youngest or recent material on the top (fig. 3.15). The thickness at the distal part of the ocean floor is commonly a kilometer or two and may be 100–130 million years old at the base. Adjacent to the ridge, there is little sediment accumulated and it is quite young. Therefore the combination of the age and stratigraphy of the oceanic sediments and the patterns and ages of magnetic reversals in the basaltic crust provide good documentation of the rates, direction, and conditions during spreading of the seafloor.

Summary of Main Points

Our understanding of the large-scale features that dominate the earth's surface and crust can now be understood due to the development of a comprehensive theory of plate tectonics including seafloor spreading.

The data used to substantiate this concept are diverse in nature, but can be subdivided into two broad categories: data from the continents and data from the ocean basins. Continental data has been available for some time, but adequate oceanic data wasn't available until the 1960s. The continental data consists of fossils, old igneous rock bodies, mountain chains, and continental margins that match, along with data on paleomagnetism and radiometric age dating. Supporting ocean basin data includes the age distribution patterns of oceanic crust, patterns of paleomagnetic reversals, distribution of earthquake epicenters, heat flow patterns, and structural features of the ocean floor.

Suggestions for Further Reading

Continents Adrift and Continents Aground, readings from *Scientific American.* San Francisco: W.H. Freeman.

Marvin, U. 1973. **Continental Drift, the Evolution of a Concept.** Washington, D.C.: Smithsonian Institution.

McKenzie, D.P. 1972. "Plate tectonics and seafloor spreading." *American Scientist,* 60:425–35.

Press, F., and R. Seiver. 1974. **The Earth.** San Francisco: W.H. Freeman.

Wegener, A. 1966. **The Origin of Continents and Oceans.** New York: Dover Publications (paperback reprint of original translation from 1915 original).

PART TWO

Basic Principles of the Marine Sciences

4

Physical and Chemical Properties of Seawater

Chapter Outline

Water is truly an amazing substance! It is the only naturally occuring substance on earth that is present in all three states, gas, liquid, and solid. It is the dominant compound in all life forms and it is necessary to sustain life. Not only does water cover 71% of the earth's surface, but it is also the most abundant compound in the outer few kilometers of the crust: there is six times more water than feldspar, which is the most abundant mineral in the crust. Water is one of the few inorganic, naturally occurring liquids. Its great transparency allows penetration of light for photosynthesis and visual observations. Unlike most compounds, as water changes from liquid to solid it increases in volume. Some other unusual properties of water are its great heat capacity, its high boiling and melting points relative to its low molecular weight, its great dissolving powers, and its high surface tension.

Much of the unusual nature of water is due to the structure of the molecule. It is comprised of one oxygen atom and two hydrogen atoms, thus giving a molecular weight of 18. The arrangement of these atoms is such that the hydrogens are at an angle of 105° with each other as they attach to the oxygen atom (fig. 4.1). This angle increases about 4° in ice. Because of the asymmetrical arrangement of the hydrogen atoms with respect to the oxygen atom in this structure, there is a dipolar nature to the molecule. The oxygen end of the molecule is negatively charged and the hydrogen end is positively charged. As a result, water acts somewhat like a magnet, in that water molecules tend to be preferentially oriented in an electrical field. They also tend to be attracted to one another because of this polar nature.

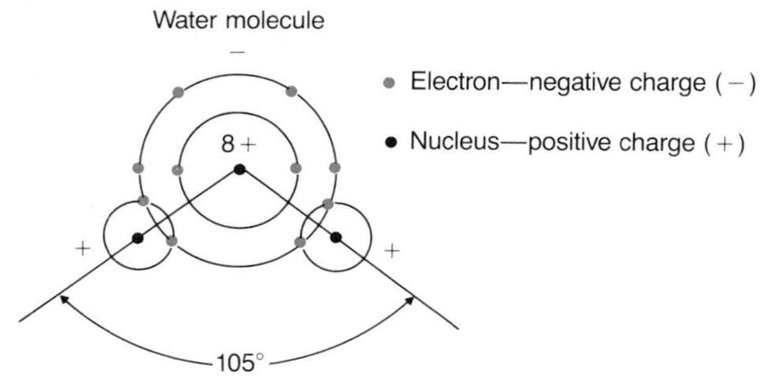

Figure 4.1 Schematic molecular diagram of a water molecule, which contains one atom of oxygen and two atoms of hydrogen. The structure of the molecule is such that it is dipolar; that is, one end is negatively charged and the other end is positively charged. (From H. V. Thurman, 1985, *Introductory Oceanography,* 4th edition, Merrill Publishing Company, Columbus, Ohio. Reproduced by permission of the publisher.)

Physical Properties of Water

Water's unique chemical makeup and its widespread distribution in all three states make its physical properties and their distribution critical to an understanding of the ocean environment.

Temperature

At sea level, pure water exists in the liquid state from a maximum of 100° C to a minimum of 0° C. Because of the effects of salinity and pressure, seawater may still be liquid at temperatures below zero. The range of natural temperatures in the ocean is from that minimum value to about 33° C. Seawater freezes at −1.9° C at sea level. Changes in temperature can have a great affect on other properties of seawater and can effect life in the sea. In general, there is an indirect relationship between temperature and density of the water due to the excitation of atoms in the water molecules. As the temperature is raised the density decreases. This also tends to result in increased dissolving power as temperature increases.

The combination of the great capacity of water to retain heat and the circulation of oceanic water masses results in the ocean acting as a giant heat

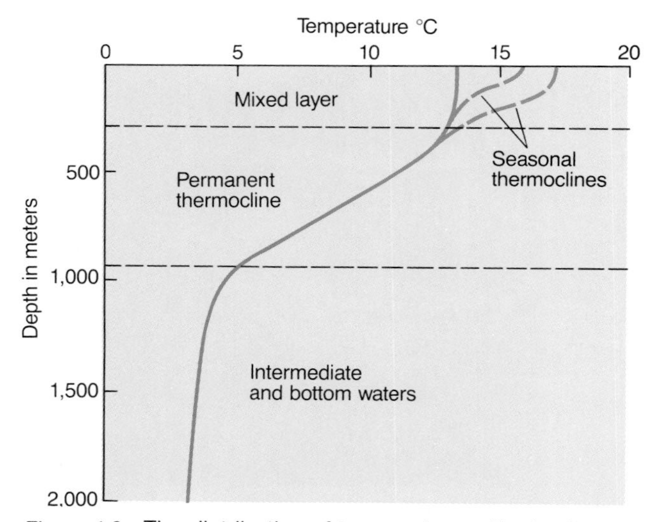

Figure 4.2 The distribution of temperature with depth shows the development of both seasonal and permanent thermoclines. Variation in the development of a thermocline also occurs with latitude.

pump. Most heat from the sun reaches the earth in the lower latitudes, where it warms the water. The circulation of surface waters transfers this heat to the relatively high latitudes. This results in some high-latitude coastal areas having a milder climate than the adjacent inland areas, such as Nova Scotia in Canada and the north coast of Europe.

Overall, however, the vast majority of the ocean is cold. Less than 10% of the ocean is warmer than 10° C and more than 75% is colder than 4° C. One of the prime reasons for this is the fact that the sun's energy can only warm the upper few hundred meters of the water column. The seasonal effects of the sun's radiation only extend down a hundred meters or so. As a result, there tends to be a relatively warm surface layer of water with an abrupt transition to the cold water that comprises the remainder of the water column. This zone of rapid change in temperature with depth is called the **thermocline** (fig. 4.2). Sometimes the overlying layer is called the mixed layer because the temperature ranges widely, both seasonally and geographically. The thermocline has a great influence on many phenomena in the sea, such as oceanic circulation, distribution of organisms, chemical cycles, and distribution of related physical properties. As might be expected, the thermocline is most pronounced in the low latitudes because of high surface water temperatures, but the seasonal variation tends to be greatest in the mid- to high-latitudes where climatic changes are greatest from one part of the year to another.

The geographic distribution of ocean surface water temperature tends to generally correspond with latitude, much like the weather; however, the influence of major circulation patterns alters this pattern (fig. 4.3). For example, the Gulf Stream, which flows northerly along the eastern margin of North America, carries warm surface water far to the north of where it would otherwise be expected without this current. The same phenomenon exists on the west side of all major ocean basins. On the east side of the basins there is a tendency for the temperature to be abnormally low, due to the upwelling of deep, cold water (see later section of this chapter) such as on the coast of California.

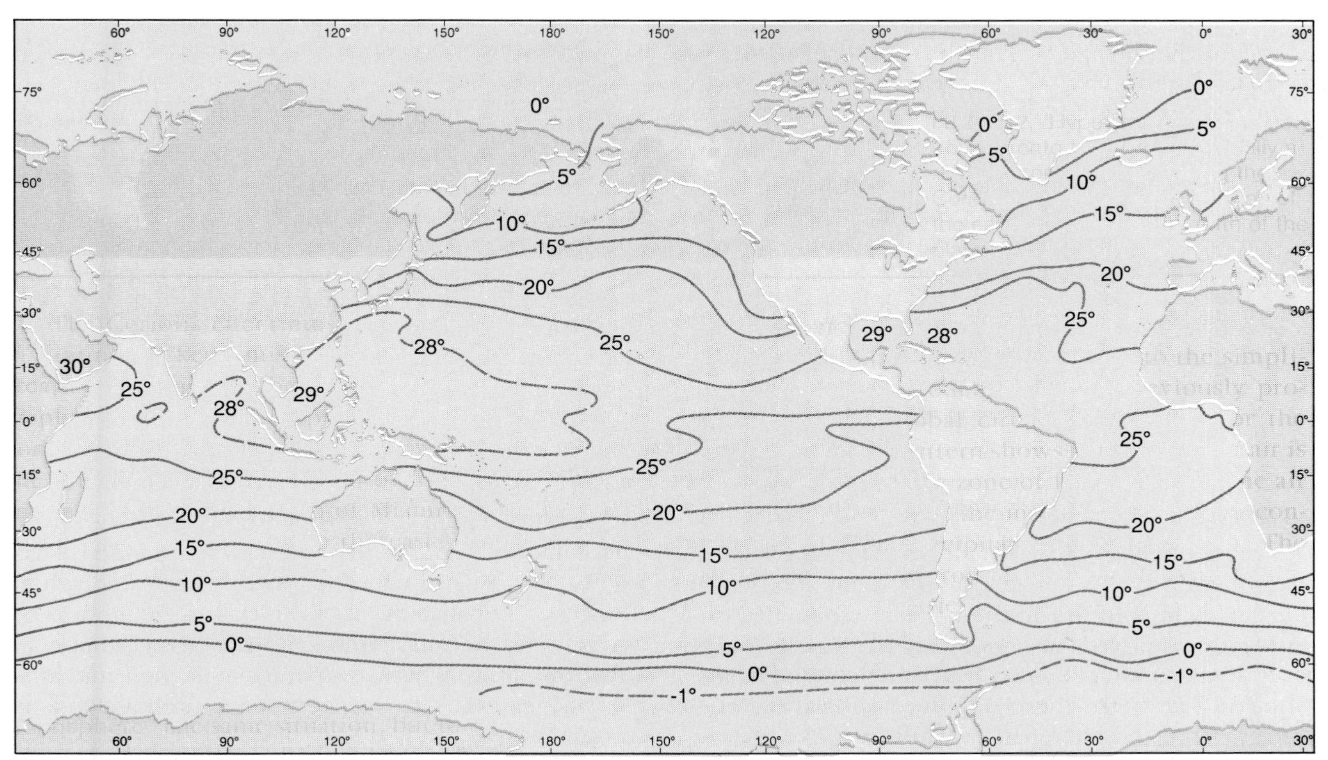

Figure 4.3 Global distribution of surface temperatures during the month of August. Highest values are in the center of each major ocean basin. (After Pickard, G. L., 1963.)

BOX 4.1

Measuring Water Temperature at Depth

Measuring the temperature of seawater can be a difficult, complicated exercise. Surface waters are not a problem because there is little time involved with collecting a water sample and measuring its temperature on the deck of the ship. The temperature of water that is hundreds or thousands of meters below the ocean surface is more difficult to sample. Collecting a sample and bringing it to the surface is not satisfactory because of the time involved and the changes that can occur during that period. Water temperature must be measured to at least tenths of a degree; therefore, it must be measured *in situ,* that is, it must be measured in place.

The **bathythermograph** is an instrument that continuously records temperature with depth. It contains a thermometer and a pressure-sensitive bellows that moves a small stylus to draw a temperature-depth profile as the torpedo-shaped instrument is towed through the water. This apparatus is not accurate enough for detailed temperature research and is only useful to a depth of about 300 m.

For many years, the standard temperature-measuring device has been the **reversing thermometer.** This instrument is accurate to 0.02° C and can be used to measure temperature at great depths. It is a mercury thermometer that is attached to a water sampling bottle. When the desired depth is reached, the bottle is released and pivots on a swivel, collecting a sample and at the same time reversing the thermometer. The thermometer's orientation changes 180°, which fixes the temperature so that it will not change upon ascent to the ship.

This thermometer consists of two units: one protected by an outer glass jacket that is also filled with mercury to protect the instrument from pressure and false readings, and another that is unprotected. Both thermometers have a thin constriction in the mercury tube that causes the mercury column to separate upon reversal (fig. B4.1). This fixes the appropriate amount of mercury for that sampling site in the temperature column and prevents the thermometer reading from changing during ascent to the ship.

The reversing thermometer has two units so that the readings on the two can be compared to ensure an accurate calculation of the depth at which the temperature measurement was obtained. Because pressure is directly related to depth (1 atmosphere per 10 m), it is possible to compare the temperature readings on the protected thermometer with those on the unprotected one and determine the depth.

It is now common to use temperature-sensitive electrical sensors called **thermistors** for temperature determinations. These are placed on a cable and lowered to the desired depth. Temperature values are then transmitted continuously to the ship. The thermistors are arranged so that they can be towed behind the ship while under way. This tends to restrict their use to moderately shallow water. Some thermistor devices are dropped from the ship or from airplanes and transmit temperature information by radio signal to the ship as the instrument falls through the column of water.

Salinity

The most obvious property of seawater is its salty taste or salinity. This is due to the many dissolved substances in solution in the oceans, especially sodium chloride (NaCl or common salt). Nearly every naturally occurring element has been detected in solution in seawater. These dissolved substances are derived from the chemical erosion of rocks of the earth's crust. All natural waters have some dissolved materials. Their concentration is expressed in parts per million (ppm) or parts per thousand (‰). Rivers have about 0.1‰ salinity and seawater generally has 35‰. There is great range in this seawater value, however, because of low salinities at the mouths of rivers, and because high salinities are common in shallow areas where evaporation is high and dissolved materials are concentrated, such as lagoons or places like the Red Sea.

The distribution of salinity is such that only along the coasts or in isolated water bodies does it deviate markedly from the average of nearly 35‰. This uniformity is due to continual mixing. The open ocean

BOX 4.1

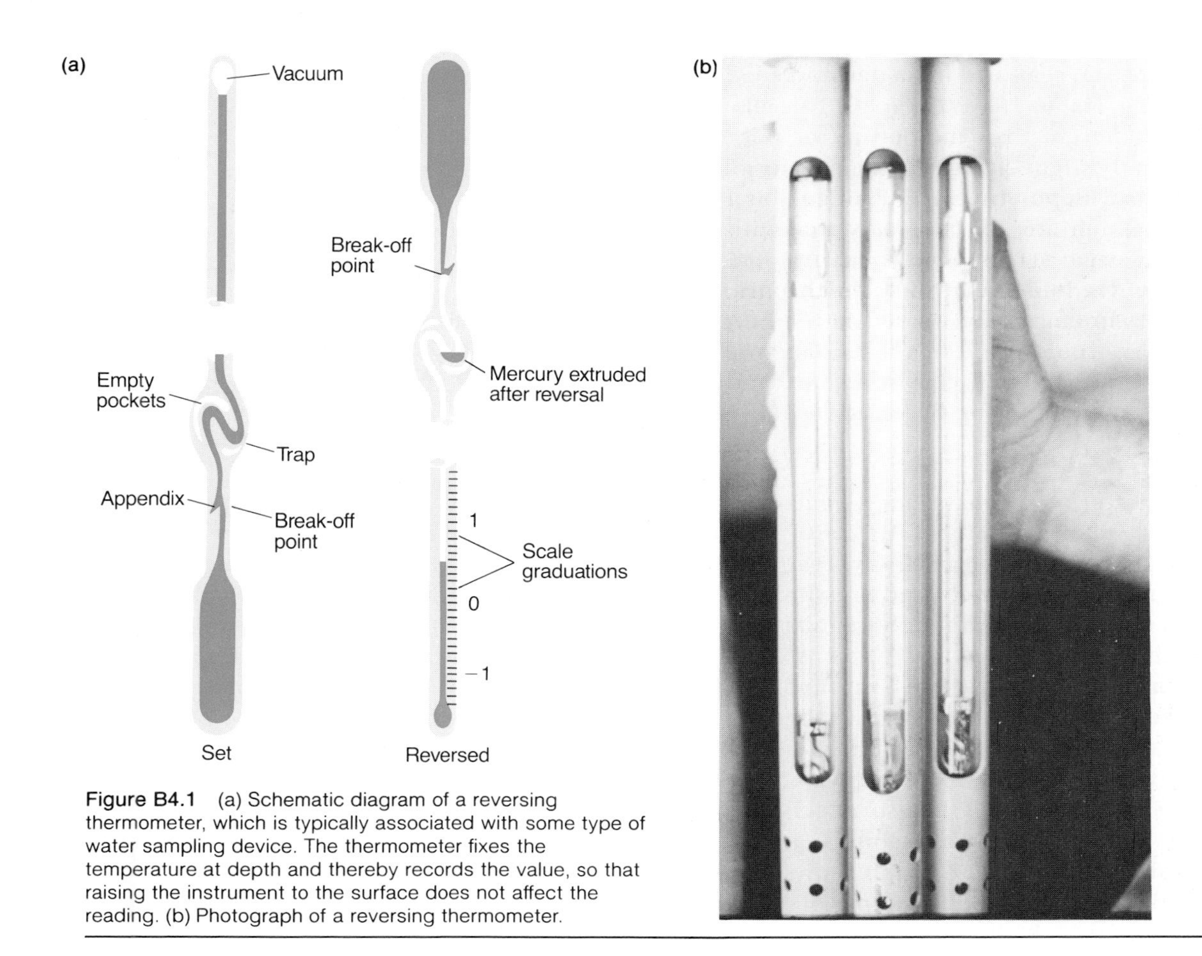

Figure B4.1 (a) Schematic diagram of a reversing thermometer, which is typically associated with some type of water sampling device. The thermometer fixes the temperature at depth and thereby records the value, so that raising the instrument to the surface does not affect the reading. (b) Photograph of a reversing thermometer.

surface waters typically range from about 33‰ to 37‰, with the higher values concentrated in the centers of each of the major ocean basins (fig. 4.3) and the lower values in the high and low latitudes. Primary reasons for this pattern are (1) the relative lack of circulation in the center of the ocean as compared to the outer portion and (2) the diluting effect of precipitation and runoff in low latitudes combined with dilution by meltwater from ice in the high latitudes.

Vertical distribution of salinity shows even less variation. Most of the 33‰ to 37‰ range for open ocean salinity is restricted to the upper 100 m. Below this level, salinity tends to be uniform with a range of less than 1‰. The only significant exceptions to this generalization are the presence of slightly higher salinity water at great depths in the Antarctic and in areas adjacent to seas like the Mediterranean.

Density

Pure water has a density of 1.0 gm/cm³ at 4° C, the temperature of maximum density. There are several factors that cause variation in the density of water, such as temperature, salinity, and pressure, all of which can cause substantial changes. The spatial and temporal distribution of water density is of great importance in understanding the ocean environment.

In the open ocean, seawater has a density range of about 1.02400 g/cm³ to 1.03000 g/cm³. This may not seem much different than the density of pure water; however, only slight changes in the density of water are of great importance in distinguishing **water masses,** which are volumes of water with distinct properties, such as temperature, salinity, and density. Because of the importance of slight changes, water density measurements are carried to the fifth decimal place. This would result in a redundancy of density values if expressed in g/cm³, so a shorthand notation has been adopted using the relationship

$$\sigma_t = (\text{density} - 1) \times 10^3 \qquad (4.1)$$

Thus normal seawater would have a σ_t (sigma-tee) value of 26.8 instead of a density of 1.02680.

Density changes greatly with salinity, temperature, and depth as a result of increased pressure. Of these, the only one that is completely regular and predictable is the effect of pressure. Water is compressed uniformly as one descends through the column with each 10 m equal to one atmosphere of pressure. Increased pressure also lowers the temperature of maximum density for fresh water. Salinity has the same effect and it also affects the freezing point. These interrelationships can be shown graphically (fig. 4.4). This graph shows that as the salinity increases the temperature of maximum density decreases. It should be kept in mind that the density of the water is increasing as this temperature decreases. Note also that the freezing point of water decreases as salinity increases. This is the reason that some deep water of the ocean has a temperature below zero. This graph shows that the temperature of maximum density and the freezing point converge until salinity reaches 24.7‰. At this point they are equal and above this value, the density of water increases directly with decreasing temperature until freezing.

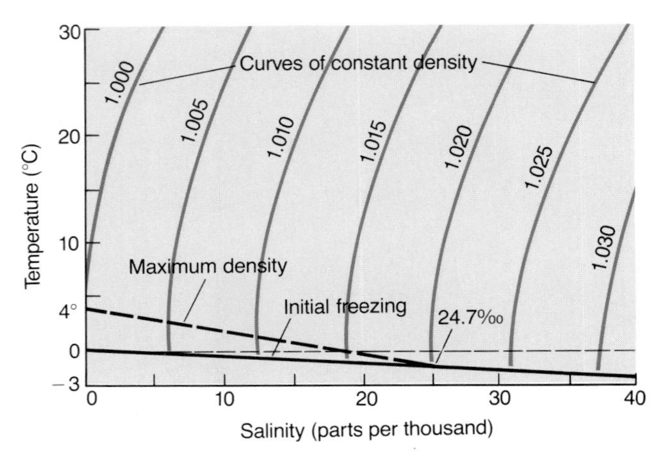

Figure 4.4 Seawater density as affected by both temperature and salinity. Note that the temperature of maximum density also varies with salinity and that at a salinity of 24.7‰, the temperature of maximum density and the freezing temperature are the same. (From M. Grant Gross, *Oceanograpy: A View of the Earth,* © 1982, pp. 88, 156. Reprinted by permission of Prentice-Hall, Englewood Cliffs, New Jersey.)

Thus an increase in both salinity and pressure causes a decrease in the temperature of maximum density. Open ocean water has a high enough salinity to eliminate the somewhat anomalous temperature of maximum density that exists for fresh water or low salinities. The density of seawater along the coastal zone is controlled primarily by salinity, whereas the density of water in the open ocean is primarily dictated by temperature, because salinity is relatively constant.

Light

The capacity of water to transmit light may be its most important physical attribute because without it aquatic photosynthesis could not take place. In addition, water reflects and absorbs light. These are the properties that help to give water its apparent color.

Incident light is part of the radiation that the sun supplies to the surface of the earth. Much of this energy is in the form of heat, which warms the atmosphere and the surface waters of the ocean. This radiant energy is filtered twice, first by the atmosphere and then again as it passes through the surface layers of the ocean. The vast majority of the light spectrum is absorbed within the upper 10 m of water and almost none penetrates below 150 m, even in the clearest water. The presence of microorganisms

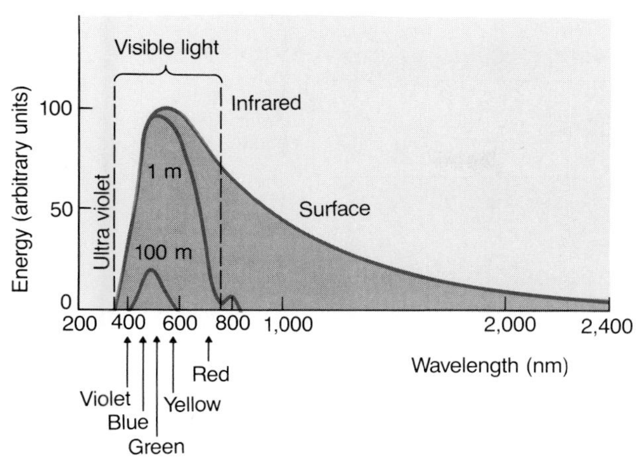

Figure 4.5 The spectrum of light that can reach various ocean depths. The wavelengths are presented in nanometers (nm), which are billionths of a meter. (After Sverdrup, et al., 1942.)

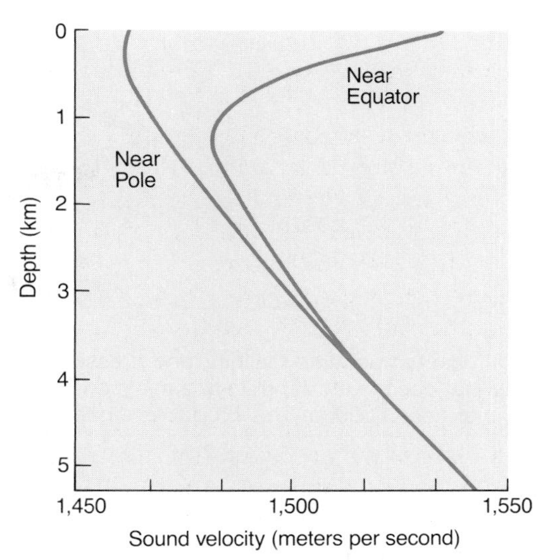

Figure 4.6 Changes in sound velocity with depth of water at both high and low latitudes. Note that the minimum velocity is near the surface in high latitudes, but is at >1,000 m in low latitudes, due to temperature effects. (Bell Laboratories *Record*.)

or suspended sediment particles in the water also greatly reduces the penetration of light or the transmissivity of water. Reflection of light from the water surface is also an important phenomenon affecting penetration. When the sun is directly over a smooth water surface, there is only about 2% reflection; however, when the sun drops to 60°, then this value increases to about 6%. From 60° to 90° the reflection increases sharply, to essentially 100%. Any type of wave action on the water surface will greatly increase the amount of reflectance of the incident light.

There are many forms of radiant energy from the sun. One form is the visible spectrum of light, which contains wavelengths of about 0.4 to 0.8 microns. The shortest of these are on the violet end of the spectrum and the longest are the reds. The various portions of this spectrum are absorbed differently within water and this is what gives clear water its apparent color. The red and orange part of the spectrum is absorbed most readily and penetrates only to depths of about 15 m and 50 m, respectively (fig. 4.5). On the other hand, the blues and greens are absorbed the least and penetrate to the greatest depths. This is the reason that the open ocean water appears to be blue or green. There are also organic phenomena, such as great concentrations of floating microorganisms, that may contribute to a greenish color.

Sound

Sound travels much faster in water than in air: the velocity of sound in air is 334 m/sec, whereas in the ocean it averages 1,445 m/sec. Almost everyone has had the experience of being underwater and hearing an approaching boat that sounds very close, only to look up and find the boat some distance away. This is evidence of the facility with which sound is transmitted in the water. There is a range in velocity of about 100 m/sec in the marine environment due to variations in temperature, salinity, and depth. All of these variables increase sound velocity as each variable is correspondingly increased. In the open ocean, salinity is nearly constant, so variations in sound velocity are due to temperature or depth (pressure) changes.

Because the velocity of sound changes with temperature, the thermocline is an important factor in determining the velocity of sound in the upper part of the water column. Velocity increases slightly with depth to the level of the thermocline, where there is a decrease to the minimum value (fig. 4.6). Below the thermocline there is a general increase in sound velocity. The presence of the thermocline results in

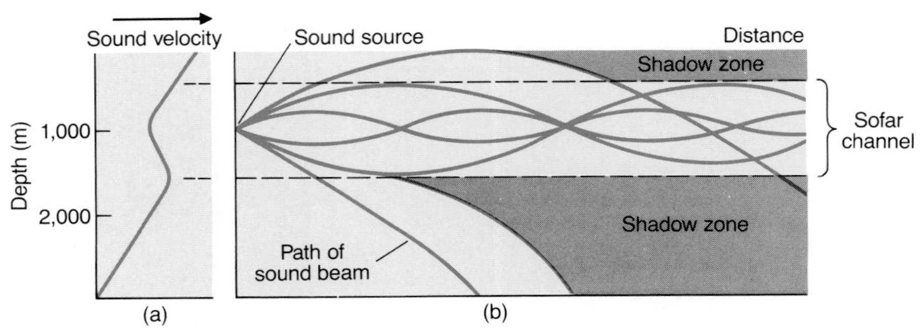

Figure 4.7 The temperature, salinity, and pressure variations that occur with depth give a minimum sound velocity layer near 1,000 m. This produces a phenomenon known as the Sofar channel, where sound tends to be trapped between layers of higher velocity. (From Duxbury and Duxbury, 1984.)

a sound channel associated with the low velocity zone. Sound waves are bent or refracted into low velocity areas due to variation in the speed of sound relative to temperature and pressure. Consequently, there are shadow zones associated with high velocity areas and caused by the refraction of sound waves away from high velocity areas toward low velocity areas (fig. 4.7).

These phenomena are important in the use of SONAR (sound navigation and ranging), which is the detection of objects by reflection of sound waves. It is employed as the basis for precision depth determinations and location of submerged vessels such as submarines.

Chemical Composition of Seawater

The chemistry of the marine environment is a complex aspect of the world ocean. Not only is it important to understand the chemical composition of seawater, but it is also necessary to consider the origin of the constituents that are present in seawater, how long they reside in the oceans, and where they go if they leave. Seawater chemically interacts with the atmosphere, the organisms in the marine environment, and the sediments on the bottom.

The increased sophistication of research in marine chemistry over the past couple of decades has provided many insights and answers about the history of the oceans and the earth as a whole. This is especially true for ancient climatic conditions, the early composition of the oceans and atmosphere, and for the generation of minerals on the seafloor.

On the one hand, the composition of seawater is simple in that only a few ions comprise nearly all of the dissolved constituents. On the other hand, it is complex because virtually every element known is present in seawater; however, the vast majority are present only in trace quantities. The bulk of the ocean environment is uniform in its chemistry, although there are marked changes along the coasts and subtle but important variations within the open ocean.

Origin and Development of the Hydrosphere

The hydrosphere of the earth must have had an origin similar to that of the atmosphere and the lithosphere. Most scientists believe that the hydrosphere, along with the other major components of the earth system, attained its character early in the earth's history. The atmosphere and the hydrosphere could not, however, have originated from the weathering of the earth's crust, because the composition of the granitic and basaltic crust could not have yielded water vapor, carbon dioxide, nitrogen, and other volatile compounds and elements.

The critical elements and compounds that are present in earth's atmosphere and hydrosphere do occur in certain meteorites. Some scientists believe that carbonaceous chondrite meteorites provided these necessary materials in the form of a thin shell around the earth which was derived from early fractionation within the earth. This took place quickly and the shell released volatiles by melting and recrystallizing. Proponents of this theory believe that the atmosphere and hydrosphere formed quickly after the earth formed and that the ocean had its present composition almost immediately, but that the atmosphere of that time was quite different than the present. The early atmosphere was largely carbon dioxide, nitrogen, and water vapor. The latter began to dissociate to produce oxygen, which then reacted to weather rocks of the earth's crust. The carbon dioxide also reacted with rocks and diminished greatly. The presence and activities of bacteria and photosynthetic organisms eventually brought oxygen and carbon dioxide to their present levels of concentration in the atmosphere.

Another view on the origin of the atmosphere and hydrosphere is that the necessary volatiles were produced slowly, via volcanic activity and hot springs. In this concept, the ocean was initially acidic. From about 3.5 billion years to about 1.5 billion years before present, the ocean gradually changed to its present basic condition. There is support for this idea in the rock record in the form of the abundance and distribution of iron, limestone, and fossil algae. According to this scenario, the composition of the ocean has remained essentially constant for the past 1.5 billion years.

Regardless of which argument is correct, both hold that the nature of the hydrosphere has not changed substantially for at least the past billion or so years. This includes the time during which all complex life evolved.

Constancy of Composition

The amount of dissolved solids in seawater is termed the **salinity.** More specifically, salinity is the total dissolved solids by weight when the carbonate is converted to oxide, the bromide and iodide are converted to chloride, the organic matter is oxidized, and the remainder is dried to 480° C. Salinity is measured in percent, or more commonly, in parts per thousand. Values in the open ocean are typically between 33‰ and 39‰, with normal salinity considered 35‰.

One of the amazing aspects of seawater is that, regardless of the overall concentration of dissolved solids, the relative concentration of the major constituents is the same. This was determined by Dittmar in 1884 when he analyzed 77 samples collected on the voyage of HMS *Challenger* from various depths and geographic areas of the world. This series of chemical analyses stands as one of the landmark contributions to oceanography in that it demonstrated the constancy of composition of major components in seawater. These salinity data have proven very accurate in comparison with current analyses conducted with modern analytical techniques and instrumentation. Although there was some concern about the geographic distribution of the samples and the fact that they were stored in glass containers for more than two years prior to study, the results have stood the test of time.

TABLE 4.1
Major Dissolved Constituents in Seawater

Ion	Chlorinity Ratio g/unit Cl	g/kg of H_2O of 19‰ Chlorinity
Chloride (Cl^-)	0.99840	19.3530
Sodium (Na^+)	0.55610	10.7620
Sulfate (SO_4^{2-})	0.13940	2.7090
Magnesium (Mg^{2+})	0.06680	1.2940
Calcium (Ca^{2+})	0.02125	0.4130
Potassium (K^+)	0.02060	0.3870
Bicarbonate (HCO_3^-)	0.00735	0.1420
Bromide (Br^-)	0.00348	0.0673
Strontium (Sr^{2+})	0.00041	0.0079
Boric acid (H_3BO_3)	0.00023	0.0044
Fluoride (F^-)	0.00007	0.0013

(From Horne, R. A., *Marine Chemistry,* 1969, Elsevier Science Publishing, THE NETHERLANDS.)

Major Elements

Common salt or sodium chloride (NaCl) is a prominent constituent of seawater. In fact, just over 30‰ of the 34.9‰ dissolved solids in normal seawater, or 85%, is sodium and chlorine (table 4.1). These and four other ions comprise over 99% of the total dissolved solids in seawater and are termed **major constituents.** Others include sulfate, magnesium, calcium, and potassium. Each has a concentration of at least one percent (table 4.1).

These constituents are derived from a variety of sources. The chlorine (Cl^-) is produced largely through volcanic activity, with gases being dissolved in surface waters and carried to the ocean. Sodium (Na^+) comes from the weathering of sodium-rich feldspar. The sulfate (SO_4^{2-}) also comes from volcanic activity. Magnesium (Mg^{2+}) and calcium (Ca^{2+}) are primarily derived from weathering of iron-magnesium minerals and calcium-rich feldspar, respectively. Potassium (K^+) also comes from feldspar. Thus, the major constituents dissolved in seawater initially come from volcanic activity and/or the weathering of minerals in igneous rocks.

TABLE 4.2 List of All Elements in Seawater, Their Concentration, and Residence Time					
Element	**Concentration in mg/l (ppm)**	**Residence Time in Years**	**Element**	**Concentration in mg/l (ppm)**	**Residence Time in Years**
Hydrogen (H)	108,000		Rubidium (Rb)	0.12	2.7×10^{5}
Helium (He)	0.000007		Strontium (Sr)	8.1	1.9×10^{7}
Lithium (Li)	0.18	2.0×10^{7}	Yttrium (Y)	0.0003	7.5×10^{3}
Beryllium (Be)	0.0000006	1.5×10^{2}	Zirconium (Zr)	0.000025	
Boron (B)	4.5		Niobium (Nb)	0.00001	3.0×10^{2}
Carbon (C)	28		Molybdenum (Mo)	0.01	5.0×10^{5}
Nitrogen (N)	0.5		Technetium-Palladium (Tc-Pd)	0.0003	2.1×10^{6}
Oxygen (O)	857,000		Silver (Ag)	0.00011	5.0×10^{5}
Fluorine (F)	1.4		Cadmium (Cd)	0.00001	
Neon (Ne)	0.0001		Indium (In)	0.00001	5.0×10^{5}
Sodium (Na)	10,800	2.6×10^{8}	Tin (Sn)	0.005	3.5×10^{5}
Magnesium (Mg)	1,350	4.5×10^{7}	Antimony (Sb)		
Aluminum (Al)	0.01	1.0×10^{2}	Tellurium (Te)	0.06	
Silicon (Si)	3	8.0×10^{3}	Iodine (I)	0.0001	
Phosphorus (P)	0.07		Xenon (Xe)	0.0005	4.0×10^{4}
Sulfur (S)	885		Cesium (Cs)	0.03	8.4×10^{4}
Chlorine (Cl)	19,400		Barium (Ba)	0.000012	1.1×10^{4}
Argon (Ar)	0.6		Lanthanum (La)	0.0004	6.1×10^{3}
Potassium (K)	387	1.1×10^{7}	Cerium (Ce)		
Calcium (Ca)	413	8.0×10^{6}	Praeseodymium (Pr)	<0.0000025	
Scandium (Sc)	0.000004	5.6×10^{3}	Tantalum (Ta)	0.0001	1.0×10^{3}
Titanium (Ti)	0.001	1.6×10^{2}	Tungsten (W)		
Vanadium (V)	0.002	1.0×10^{4}	Rhenium-Platinum (Re-Pt)	0.000011	5.6×10^{5}
Chromium (Cr)	0.0005	3.5×10^{2}	Gold (Au)	0.00005	4.2×10^{4}
Manganese (Mn)	0.002	1.4×10^{3}	Mercury (Hg)	<0.00001	
Iron (Fe)	0.01	1.4×10^{2}	Thallium (Tl)	0.00003	2.0×10^{3}
Cobalt (Co)	0.0008	1.8×10^{4}	Lead (Pb)	0.00002	4.5×10^{5}
Nickel (Ni)	0.0054	1.8×10^{4}	Bismuth (Bi)		
Copper (Cu)	0.003	5.0×10^{4}	Polonium-Astatine (Po-At)	0.6×10^{-15}	
Zinc (Zn)	0.01	1.8×10^{5}	Radon (Rn)		
Gallium (Ga)	0.00003	1.4×10^{3}	Francium (Fr)	6.0×10^{-11}	
Germanium (Ge)	0.00006	7.0×10^{3}	Radium (Ra)		
Arsenic (As)	0.003		Actinium (Ac)	0.00005	3.5×10^{2}
Selenium (Se)	0.0001		Thorium (Th)	2.0×10^{-9}	
Bromine (Br)	68		Protactinium (Pa)	0.003	5.0×10^{5}
Krypton (Kr)	0.0025		Uranium (U)		

(From Horne, R. A., *Marine Chemistry,* 1969, Elsevier Science Publishing, THE NETHERLANDS.)

Minor Elements

The minor constituents of seawater vary in abundance depending on locality and therefore are not reliable for determining salinity by the constancy of composition relationship. They are, however, important to the overall chemical scheme in the oceans. Some are of economic importance, others are critical to various organisms in the sea.

The vast majority of elements in seawater occur in concentrations of less than one part per million (table 4.2). Gold, zinc, aluminum, and iron all have economic value, but not in the dispersed state in which they occur in the sea. Gold, for example, occurs at about one part per trillion; however, even at this low concentration, there is about a million dollars worth of gold in every cubic kilometer of seawater.

Some of the minor elements in seawater play an important role in the precipitation of minerals on the seafloor. For example, iron and manganese are primary constituents of manganese nodules, which occur on the deep ocean floor around the globe. Some of the dissolved gases, such as oxygen and nitrogen, are critical to the life cycles of large groups of organisms. Silicon is incorporated into the skeletons of various organisms, both plants and animals.

Residence Time

One of the earliest attempts at calculating the age of the oceans was by dividing the total mass of sodium currently in the ocean by the rate of influx of sodium into the ocean to obtain the age. In theory, this would give the age of the ocean. What was not taken into account was the fact that there is considerable removal of sodium through precipitation of various minerals. Consequently, the figure produced in the calculation was very low, only about 100 million years. What this value represents, if the concentrations and rate of influx are correct, is the **residence time** of that element, that is, the average length of time that the element, in this case sodium, stays in solution in the ocean.

Seawater has maintained a fairly constant salinity throughout most of the history of the earth. This is due to the dynamic equilibrium that exists between the amount of dissolved material contributed by the landmasses and the removal rate of these materials from seawater. Removal is accomplished by various reactions with sediment particles, or by organisms, or by precipitation of new compounds directly from seawater. Residence time is the length of time an element remains in the ocean before it is removed. Residence time varies greatly for each dissolved element and is defined by the total amount that is present in the ocean and the rate that it is carried into the marine environment.

$$\text{RT} = \text{Amount in oceans/Rate added to oceans} \qquad (4.2)$$

Each element has its own residence time depending on how reactive the element is in the marine environment. Some, such as sodium, magnesium, and potassium, have long residence times. On the other hand, there are elements with very short residence times. Among these are iron and aluminum with times of only about 100 years. This is one of the reasons that these elements are only present in trace concentrations in the oceans; they don't stay around very long!

Salinity

The strict definition of salinity is the total dissolved solids in one kilogram of seawater when all carbonate is converted to oxide, bromine and iodine are replaced by chlorine, and organic matter is oxidized to 480° C. This is a cumbersome operation that can now, because of the constancy of composition, be stated that

$$\text{Salinity} = 1.80655 \times \text{chlorinity} \qquad (4.3)$$

Distribution

Geographical variation in salinity on a global scale shows two major trends: one, an overall decrease near the margins of the oceans due to runoff and two, variations related to latitude and climate.

The term **brackish** is used to designate salinity levels considerably below normal marine concentrations. This term is typically applied to water of less than 17‰ salinity. The Baltic Sea is a brackish water body with salinities generally about 10‰ or less throughout. By contrast, the term **hypersaline** is used to designate higher than normal salinity level. Concentrations must be above about 38‰ for this designation to apply. Places where there is little fresh water influx and considerable evaporation typically display such salinity levels. Examples are the Persian Gulf and the Red Sea.

The mouths of rivers and their nearby areas may have very low salinities. In many cases fresh water extends beyond the river mouth, with large rivers, such as the Mississippi and the Amazon, having essentially fresh water many kilometers beyond their mouths. Away from the landmasses, generally near the outer portion of the continental shelf, there is little variation in salinity. Ranges vary by only a few parts per thousand and the changes are gradual.

Surface salinity is largely controlled by the meteorological and climatic conditions of the ocean basins. The central areas of the major circulation cells of each ocean basin are regions of relatively high salinity, with values typically about 2‰ above normal

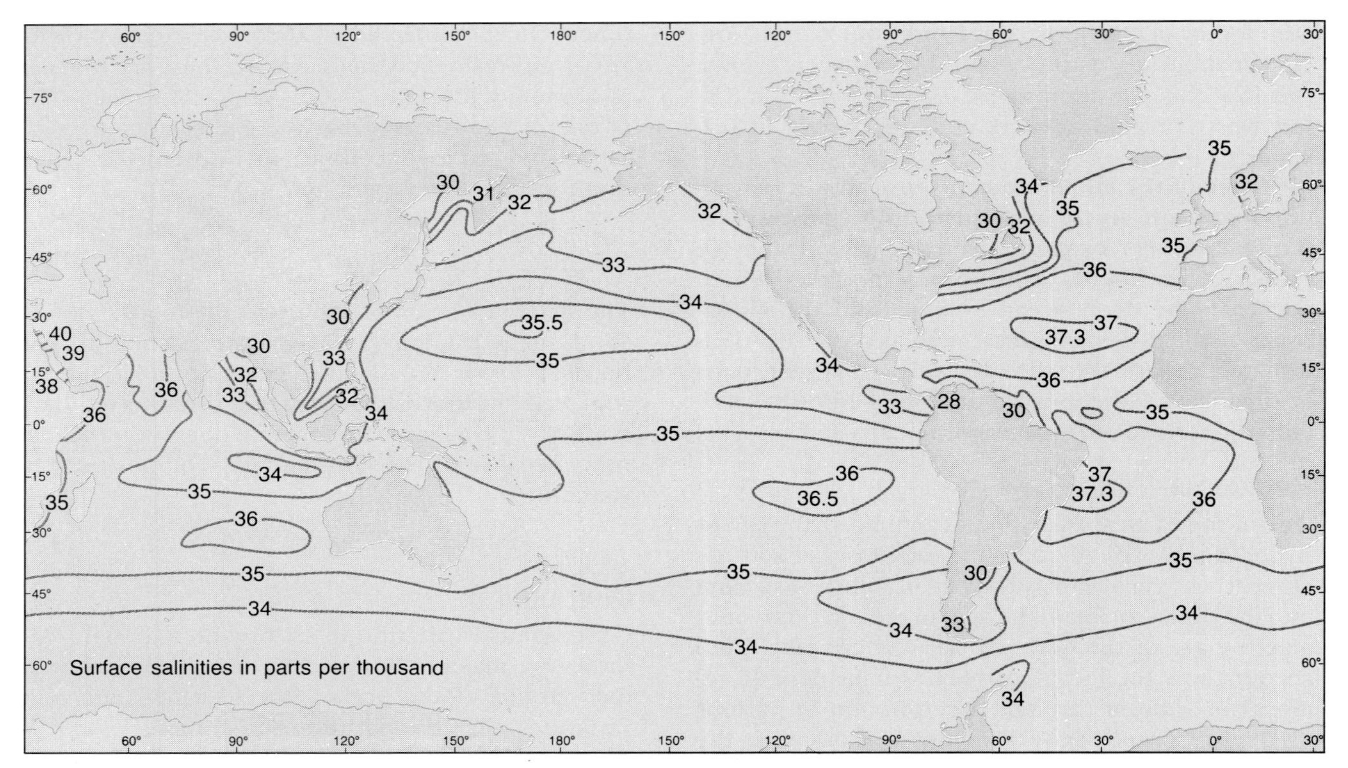

Figure 4.8 Surface salinity distribution over the world ocean. Highest values in each ocean basin are in the center due to the relative lack of circulation and high evaporation there.

(fig. 4.8). The increased rate of evaporation and diminished circulation in these areas is the cause of this phenomenon. There is also an overall lower salinity in the northern hemisphere than in the southern hemisphere. This is not surprising because of the much greater landmass in the northern hemisphere, which produces extensive drainage basins and therefore more freshwater runoff relative to the southern hemisphere. It is interesting to note that the North Atlantic Ocean has a higher overall salinity than does the North Pacific (fig. 4.8). Because of the many large rivers entering the Atlantic and relatively few entering the Pacific, the opposite relationship would be expected. As yet, the reason for this odd salinity distribution is not known, although evaporation seems a likely contributor.

A general latitudinal distribution of salinity shows that lowest values are near the equator and in the high latitudes (fig. 4.9). High rainfall in the low latitudes and dilution from meltwater of glaciers and ice sheets in high latitudes account for these low trends. This distribution closely parallels the curve for evaporation minus precipitation.

These geographic patterns for salinity distribution hold only for the surface waters of the oceans. Below the thermocline, a few hundred meters deep, seawater is essentially uniform in its salinity. This is illustrated by a histogram of salinity for the entire volume of the world ocean (fig. 4.10). This shows that the vast majority of the water in the oceans has a salinity of between 34.3‰ and 35.0‰.

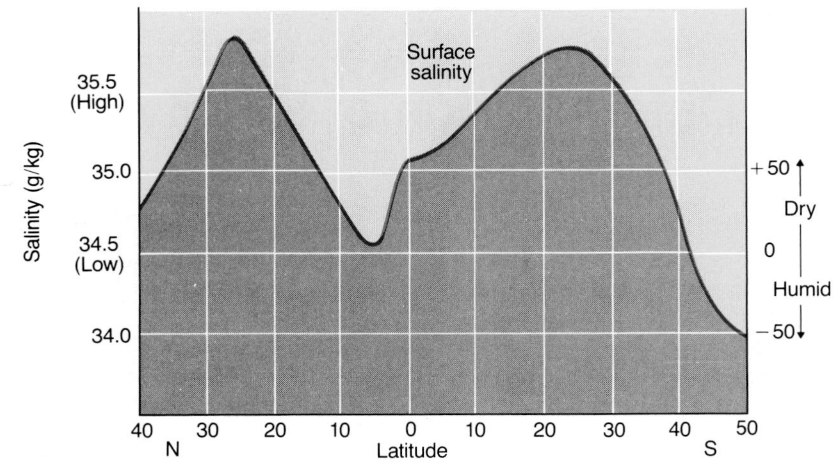

Figure 4.9 Average distribution of salinity over the world ocean based on latitude. Note that low values are near the equator and high values are in the mid-latitudes. Compare with figure 4.8.

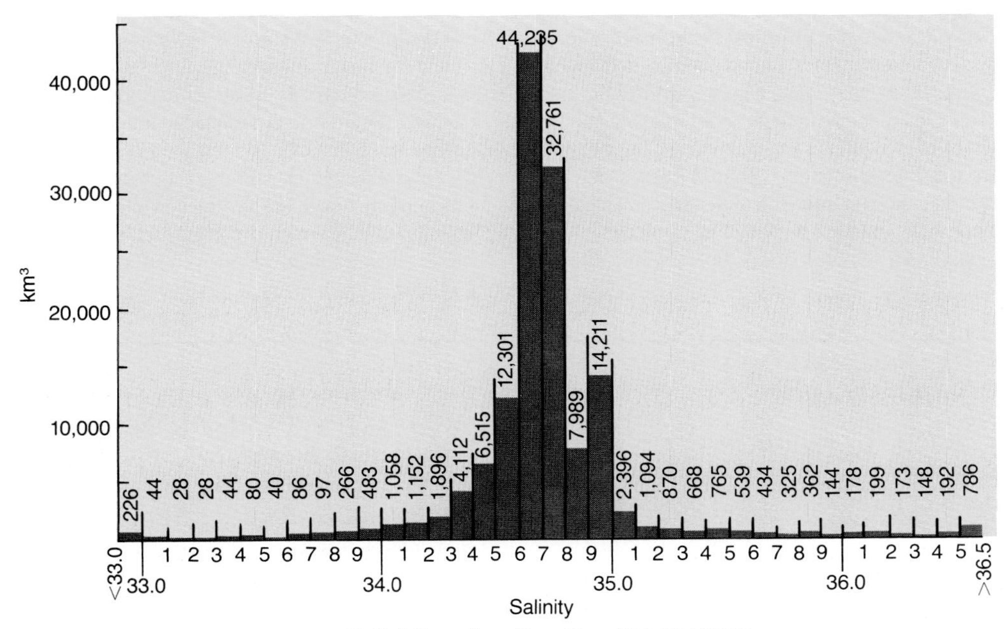

Figure 4.10 Histogram showing the global distribution of salinity, with each tenth of a part per thousand being expressed in area. The vast majority of the ocean surface shows a range of less than one part per thousand. (After Montgomery, 1958.)

BOX 4.2

Determination of Salinity

There are various ways for determining salinity, some tedious and time-consuming, and others relatively easy. Accuracy is an important consideration in all salinity determinations.

One of the most straightforward approaches is to take a known amount of seawater and evaporate all of the liquid. The residue that remains is the material that was dissolved in the seawater. Its mass, relative to that of the volume of water at the start of the experiment, gives the concentration of dissolved material in the seawater, that is, the salinity. This method, however, takes a lot of time and is not very accurate.

A second simple technique is to place a hydrometer in a vessel of seawater and read the density of the water directly from the hydrometer. This value is used to calculate the total mass of dissolved solids in the volume of seawater. Simple arithmetic, as in the first example, is then used to yield the concentration of dissolved solids. For example, let us say that the experiment started with 100 grams of seawater and that the density of the seawater as read from the hydrometer indicated that 3.5 grams of dissolved solids were present; then the concentration is expressed as 35‰, the concentration of normal seawater. This technique is also not very accurate but it is fast.

A third technique is based on the observation that the way that light passes through water is directly related to the salinity of the water. An optical refractometer is commonly used to make such measurements, especially in the field, where other instrumentation is not available. The optical refractometer requires only that a drop of water be placed on a glass surface and then held up to the sunlight. A scale calibrated to the relationship between salinity and the index of refraction of the water is seen through the eyepiece on the instrument and the salinity is read directly. This technique is also not very accurate, but it is rapid and inexpensive.

By far the most widely used technique for determining salinity is by measuring electrical conductivity using a **salinometer**. The conductivity of seawater is directly related to the concentration of dissolved solids or salinity. It is therefore possible to place electrodes in a water sample and read the salinity directly from a calibrated dial. This method is rapid and accurate with reproducibility at better than 0.005‰.

Acidity and Alkalinity (pH)

The acidity or basicity of natural waters is an important consideration of the overall environment. It has a profound effect on the presence or absence of a given species of organism and on growth of individuals. The nature and rate of chemical reactions that take place in the marine environment are also affected by pH.

An **acid** is a substance that has a sour taste, contains hydrogen, can dissolve many other substances, and strongly dissociates when combined with water. A **base** is a substance that has a bitter taste, can neutralize an acid, and removes hydrogen ions from water solutions. Bases contain the hydroxide ion (OH^-), which is made available when the compound dissociates. Hydrochloric acid (HCl) and sulfuric acid (H_2SO_4) are strong acids, which means that they dissociate readily, whereas acetic acid ($HC_2H_3O_2$) and carbonic acid (H_2CO_3) are weak acids, which means they only partially dissociate.

The statements above indicate that the key to acidity and basicity is hydrogen, with the hydrogen ion (H^+) meaning acidic and the hydroxide ion (OH^-) meaning basic conditions. The shorthand term **pH** is used to express the acidity-basicity relationship. What it really stands for is the negative log of the hydrogen ion activity (log /[H^+]). This is closely related to the concentration of the hydrogen ion and therefore it is commonly stated that pH is the concentration of the hydrogen ion (H^+). The values of pH range from 1 to 14, with 7 being neutral. Values of less than 7 are acidic and those above 7 are basic.

Examples of pH of some common substances will help to clarify the concept of acidity and basicity. Pure water and human blood have pH values that are

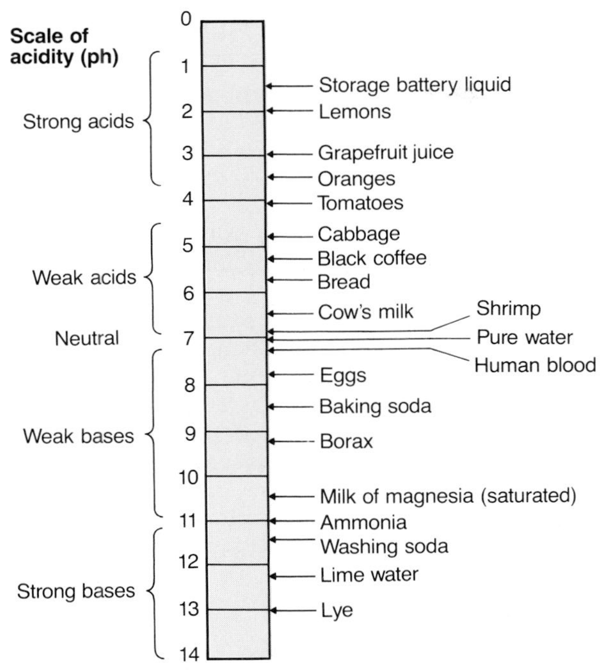

Figure 4.11 A scale of possible pH values and corresponding common liquids.

essentially neutral (fig. 4.11). Milk is only slightly acidic and black coffee a bit more so. Lime or lemon juice has quite a low pH because of the high content of citric acid. On the basic side are such things as baking soda ($NaHCO_3$) and milk of magnesia ($Mg[OH]_2$). Household ammonia (NH_4OH) used for cleaning is about 12.

Natural environments also have waters with a wide range of pH values. Many volcanic lakes are very acidic, about 2.5–3.0. Near the other end of the scale are alkaline lakes in desert regions like southern California and Nevada where pH can be over 10.

Distribution of pH in the Ocean

The average pH of seawater is 7.8, which is slightly basic. There is some variation in open water but it is rather narrow; generally from 7.5 to 8.4. These subtle but important variations in pH occur primarily through the water column and are not of a geographic nature. They can be shown graphically (fig. 4.12) with minimum values just below the zone of photosynthesis and near the seafloor. Highest values are at the surface and at about mid-depth (2.0 km–2.5 km). The high surface values are caused by a high rate of photosynthesis. This phenomenon takes

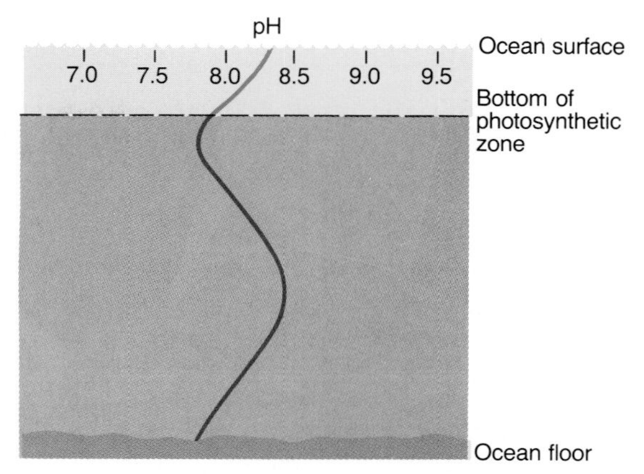

Figure 4.12 The vertical distribution of pH through the water column in the ocean. The range is less than one pH unit.

up carbon dioxide (CO_2), thus preventing it from combining with water (H_2O) to form carbonic acid (H_2CO_3). The result is an elevated pH. In places where photosynthesis is absent or depressed, it may be possible to have lower than normal pH values because of the formation of carbonic acid. This entire scenario is restricted to near-surface waters because of the equilibrium that exists between the carbon dioxide in the atmosphere and that in the oceans, plus the photosynthetic process.

Just below the zone of photosynthesis, animals are taking up oxygen and producing carbon dioxide. This allows for increased production of carbonic acid and a lowering of the pH (fig. 4.12). The effect is diminished with depth because of the decrease in organisms. Near the bottom, the pH decreases and may even drop below 7 due to the absence of oxygen and the production of hydrogen sulfide (H_2S).

There are a few other factors that have an effect on hydrogen-ion concentration, including temperature and pressure. There is a direct relationship between temperature and hydrogen-ion concentration; thus a lower temperature means a lower pH. A change in temperature, however, causes only a very small change in hydrogen-ion concentration (0.0003 pH unit per 10° C), which is insignificant through the entire range of temperature in the ocean. Pressure, on the other hand, does have a measurable effect in that there is a direct relationship between it and the rate of carbonic acid dissociation that amounts to about 0.02 pH unit per 1,000 m depth. The result is a bit of a decrease in pH with great depth.

Summary of Main Points

Water is probably the most unusual compound on the earth. It has a variety of properties that are not found in any other liquid: it occurs naturally in all three states, liquid, solid, and gas; it is transparent; it expands upon freezing; and it has a high heat capacity.

The temperature of maximum density of pure water is 4° C. The density of water is dependent upon temperature, pressure, and salinity. Only pressure shows a uniform trend in that it increases with depth. Temperature and salinity variations range widely and thereby cause a wide range in distribution patterns of water density.

The temperature of seawater ranges widely at and near the surface, generally reflecting the climate and latitude. At depths of about 200 m and below there is only subtle variation in temperature with the range only a few degrees.

Salinity is similar to temperature in that there is a modest variation at the surface, due primarily to climate, but at depth, salinity shows little range. The only major exception is in the North Atlantic where the relatively high-salinity water of the Mediterranean Sea maintains its character for a few thousand kilometers into the ocean basin.

The composition of seawater is reasonably constant in the open ocean, but shows great variety near the coast, where influx from rivers and human activities causes major fluctuations. The relative abundance of major constituents dissolved in seawater is the same, regardless of the absolute concentration. Trace elements and minor elements do show some range, however.

The ocean as a whole is slightly basic, although there are naturally occurring aquatic environments that are quite acidic, such as volcanic lakes, and some that are quite basic, such as saline lakes.

Suggestions for Further Reading

Groen, P. 1957. **The Waters of the Sea.** London: Van Nostrand.

Harvey, H.W. 1957. **The Chemistry and Fertility of Sea Water.** 2d ed. Cambridge: Cambridge University.

Knauss, J.A. 1978. **Introduction to Physical Oceanography.** Englewood Cliffs, New Jersey: Prentice-Hall.

McClellan, H.J. 1965. **Elements of Physical Oceanography.** New York: Pergamon Press.

Pickard, G.L., and W.L. Emery. 1982. **Descriptive Physical Oceanography.** Oxford: Pergamon Press.

Sverdrup, H.W., M.W. Johnson, and R.H. Fleming. 1942. **The Oceans.** Englewood Cliffs, New Jersey: Prentice-Hall.

Von Arx, W.S. 1962. **An Introduction to Physical Oceanography.** Reading, Massachusetts: Addison-Wesley.

5
Circulation Systems of the World

Chapter Outline

The major circulation patterns of both the atmosphere and the hydrosphere are closely integrated. Patterns of atmospheric circulation have a profound influence on major oceanic circulation patterns. Landmasses also greatly affect the circulation of the atmosphere and in a different, but related way, they influence oceanic circulation. This section will consider some of the fundamental principles of atmospheric circulation and the next section will integrate these principles with oceanic circulation.

Atmospheric Circulation

Atmospheric temperatures experience a significant diurnal change at most latitudes during nearly all times of the year. An ocean or other large body of water, on the other hand, does not undergo much change in temperature on a daily basis. As a result of these two phenomena, diurnal or daily atmospheric circulation patterns occur along most coastal areas.

Land Breeze and Sea Breeze

During the daytime, the land heats up, relative to the sea, because land has a lower heat capacity. This causes the air above the ground to warm, causing it to expand and rise, forming a zone of low atmospheric pressure over the land. The adjacent air over the sea is at a higher atmospheric pressure, which generates circulation and produces a breeze that flows from the sea toward the land—the sea breeze (fig. 5.1a). This commonly takes place about nine or ten o'clock in the morning and is the reason that sailboats rarely are out in the early morning hours.

In the evening, the reverse situation prevails. Air over the land cools rapidly, relative to the sea, because water has a high heat capacity. This causes higher atmospheric pressure over the land than over the water. The result is a circulation on the surface that moves air from the land toward the water—a land breeze (fig. 5.1b). Because the difference in temperature between land and sea is greatest during the day, the sea breeze is typically brisker than the land breeze.

Global Circulation

In order to understand the basic principles of fluid motion about the earth it is best to consider the earth as a uniform sphere surrounded by a thin envelope of water. The overlying atmosphere circulates as the result of temperature changes. Heat expands the atmosphere and cooling contracts it. Because these phenomena occur in different places at different times, there are gradients in between warm, expanding air, and cool, contracting air masses.

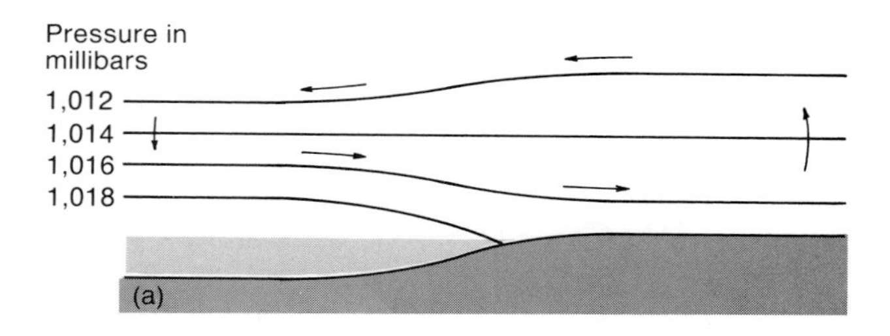

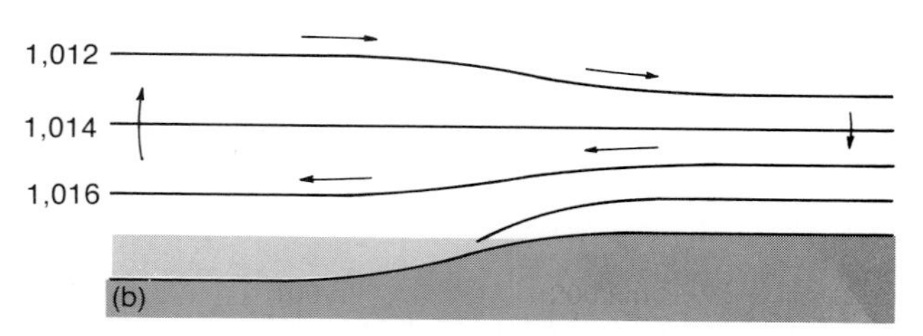

Figure 5.1 Diagrams showing the coast with barometric pressure contours (isobars) expressed in millibars. The situation in (a) represents the sea breeze that develops in the morning, and (b) shows the land breeze that develops during early evening.

In a nonrotating earth the radiation per unit of surface area from the sun is greatest near the equator and least at the poles. This causes the air near the equator to heat and rise, whereas the contracting, cold air at the poles is descending. The result is a global-scale convectional circulation with low pressure conditions near the equator and high pressure conditions near the poles.

Coriolis Effect

The rotation of the earth complicates this simplified circulation. The movement of an object over a rotating sphere is influenced by the earth's gravity field, so the object appears to move in a curved path rather than a straight line. This phenomenon is called the **Coriolis effect.** From the observer's point of view, an object moving over the earth's surface in the northern hemisphere appears to be deflected to the right and in the southern hemisphere it appears to be deflected to the left. The deflection is actually an apparent change in the path of the moving object as viewed from the observer's position on the rotating earth.

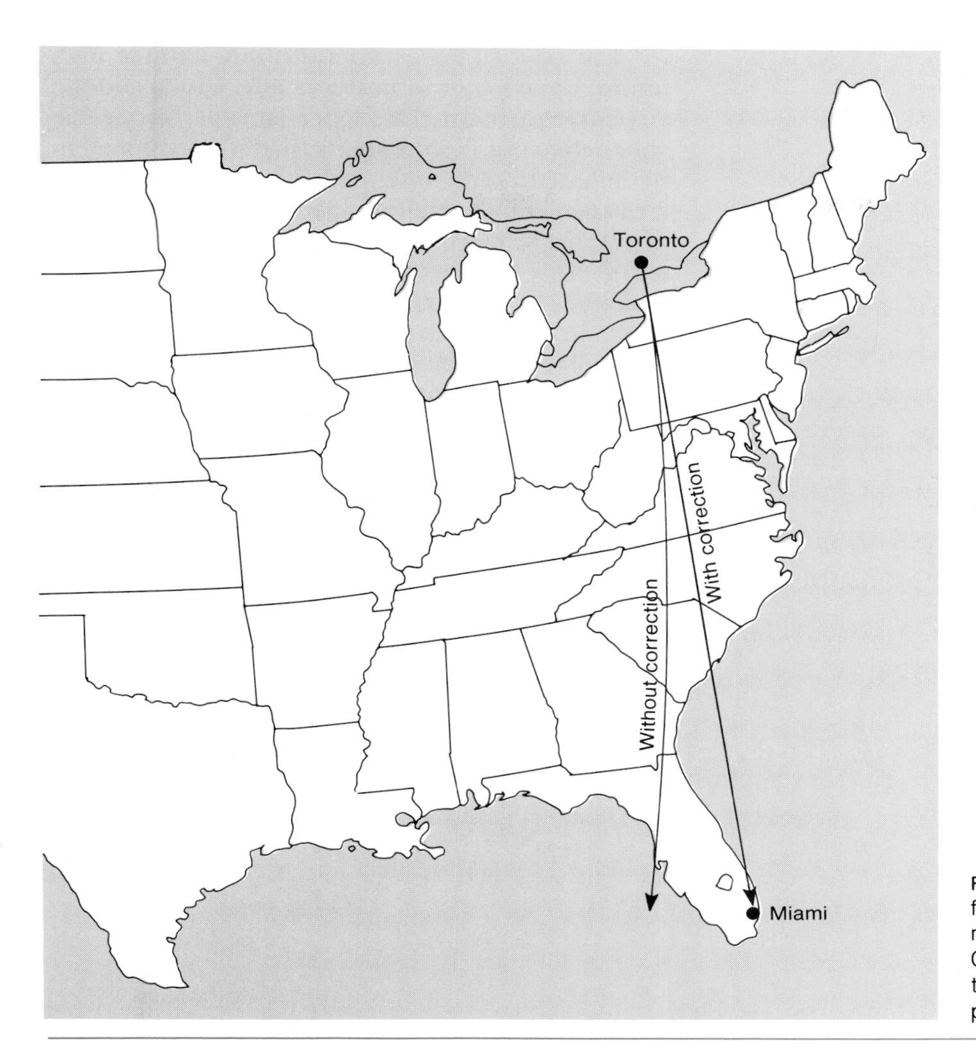

Figure 5.2 Hypothetical plane flight from Toronto to Miami, essentially a north to south path, showing the Coriolis effect due to the rotation of the earth under the flight path of the plane.

The Coriolis effect must be taken into account in aiming rockets, missiles, and other long-range weapons. It must also be considered when planning airplane flights. For example, a plane flying from Toronto to Miami would take about two hours. During the time the plane is in the air traveling over the rotating sphere, the position of Miami would move a few hundred kilometers to the east. If the Coriolis effect is not considered, the plane will head into the Gulf of Mexico (fig. 5.2). Because velocities on the earth's surface vary with latitude (due to its spherical shape), this phenomenon of apparent right deflection takes place in all directions in the northern hemisphere. The same situation, but to the left, takes place in the southern hemisphere. There is no deflection for an object traveling directly above the equator.

Atmospheric Circulation Patterns

The application of the Coriolis effect to the simplified hypothetical model described previously provides a general global circulation pattern for the atmosphere. This pattern shows that equatorial air is warmed and rises in a zone of low pressure. The air cools upon rising and the moisture it contains condenses and falls as precipitation near the equator. The dry air moves away from the equator and descends near 30° latitude, due to the high pressure created by the compression of this air with the southerly counterpart from the high latitudes (fig. 5.3). Air in this latitude sinks and spreads both to the north and to the south.

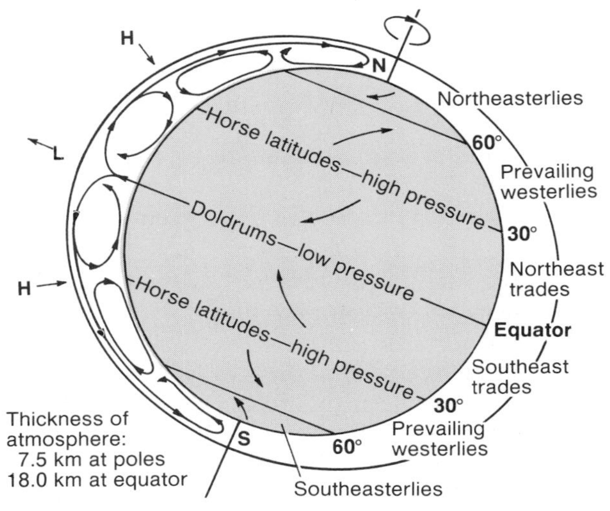

Figure 5.3 Generalized diagram of atmospheric circulation over the globe. These patterns are affected by the Coriolis effect and by prevailing pressure systems. (After Bowditch, 1977.)

The air that moves from this mid-latitude high toward the equator forms the **trade winds** and the air moving toward the higher latitudes forms the **westerlies** (fig. 5.3). A similar pattern developed in the high latitudes, where the **polar easterlies** are formed. Some generalities can be made about each of the major circulation belts. In belts of low atmospheric pressure where the air is rising (fig. 5.3), rainfall is abundant, and cloudy conditions with light, variable winds prevail. Conversely, the high pressure regions where air sinks are characterized by cool, dry air with little rainfall.

In should be noted that the Coriolis effect plays an important role in the patterns displayed by these global wind cells. In the northern hemisphere, the wind is deflected to the right and in the southern hemisphere, it is deflected to the left.

Now that the rotating earth with its envelope of atmosphere has been considered, it is appropriate to add the landmasses. The presence of these terrestrial bodies results in surface winds tending to form cir-

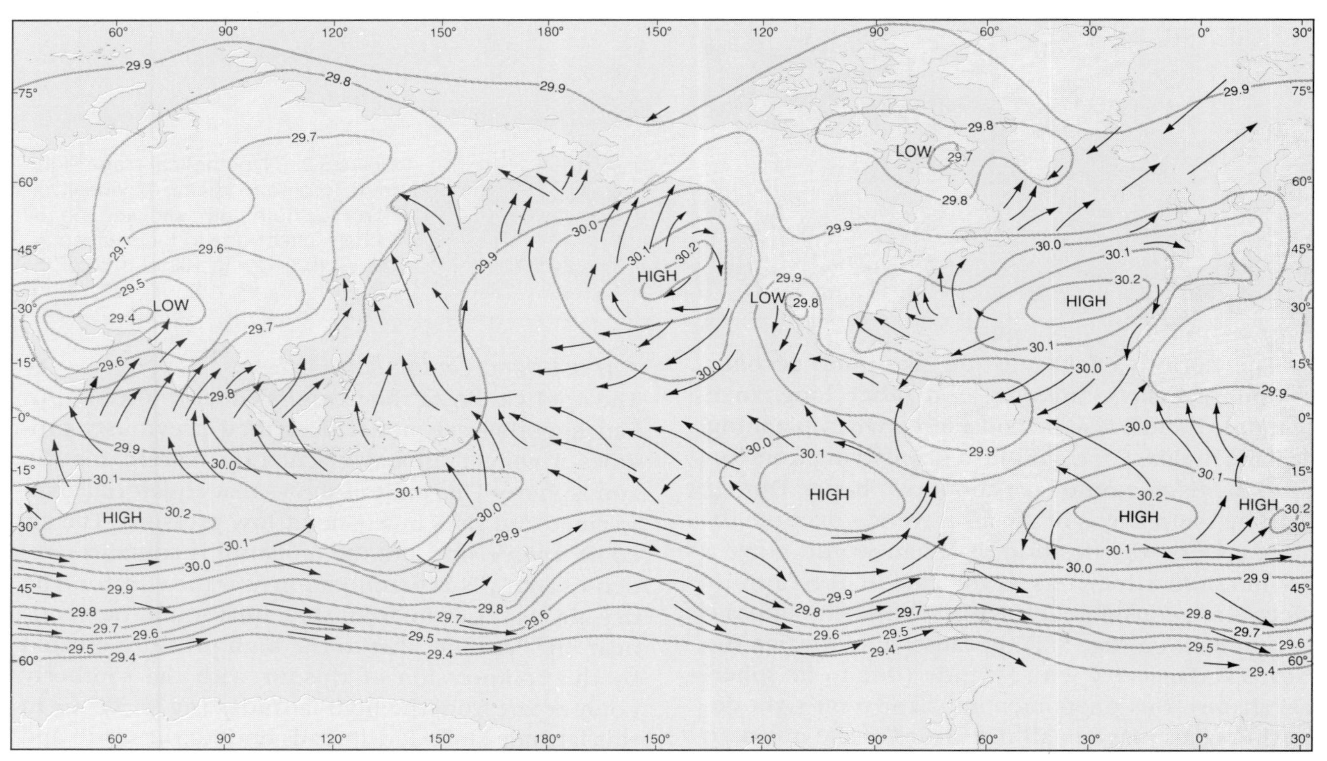

Figure 5.4 General diagram showing the influence of landmasses on global wind patterns and related barometric pressure.

circular or elliptical cells of circulation over the oceans (fig. 5.4). The fact that land masses are not uniformly distributed over the globe further complicates the situation. This is especially true in the high latitudes where there is no land mass in the northern hemisphere but where one is present in the southern hemisphere.

Another important influence of the landmasses is in the seasonal temperature variation. The warm temperatures of the continents in the summer cause low pressure conditions to prevail over the landmasses due to the heating of the air, which causes it to rise. When a high-pressure system moves toward land, it is weakened or broken up because of this heating effect, but if a low-pressure system moves from the ocean over land, it intensifies. During the winter the opposite situation exists with cold, high pressure dominating the landmasses.

Weather Fronts

Moving now from the global scale to the scale of continents, it is important to consider some of the basic weather systems that affect us on a regular, continual basis. These systems are pertinent to our consideration of the oceans because they have a profound impact on coastal and shelf environments.

The term **front** is applied to the boundary between warm and cold air masses. When the leading edge of a denser cold air mass forces its way beneath a lighter, warm and moist air mass, a cold front is formed (fig. 5.5). This is typically a steep front, accompanied by unsettled weather such as thunderstorms or squalls. Under extreme conditions, tornadoes may be generated. The passage of a cold front brings cool, dry air, and a shift in wind direction. Warm fronts exist when a warm, moist air mass moves into an area after departure of a cold air mass. The

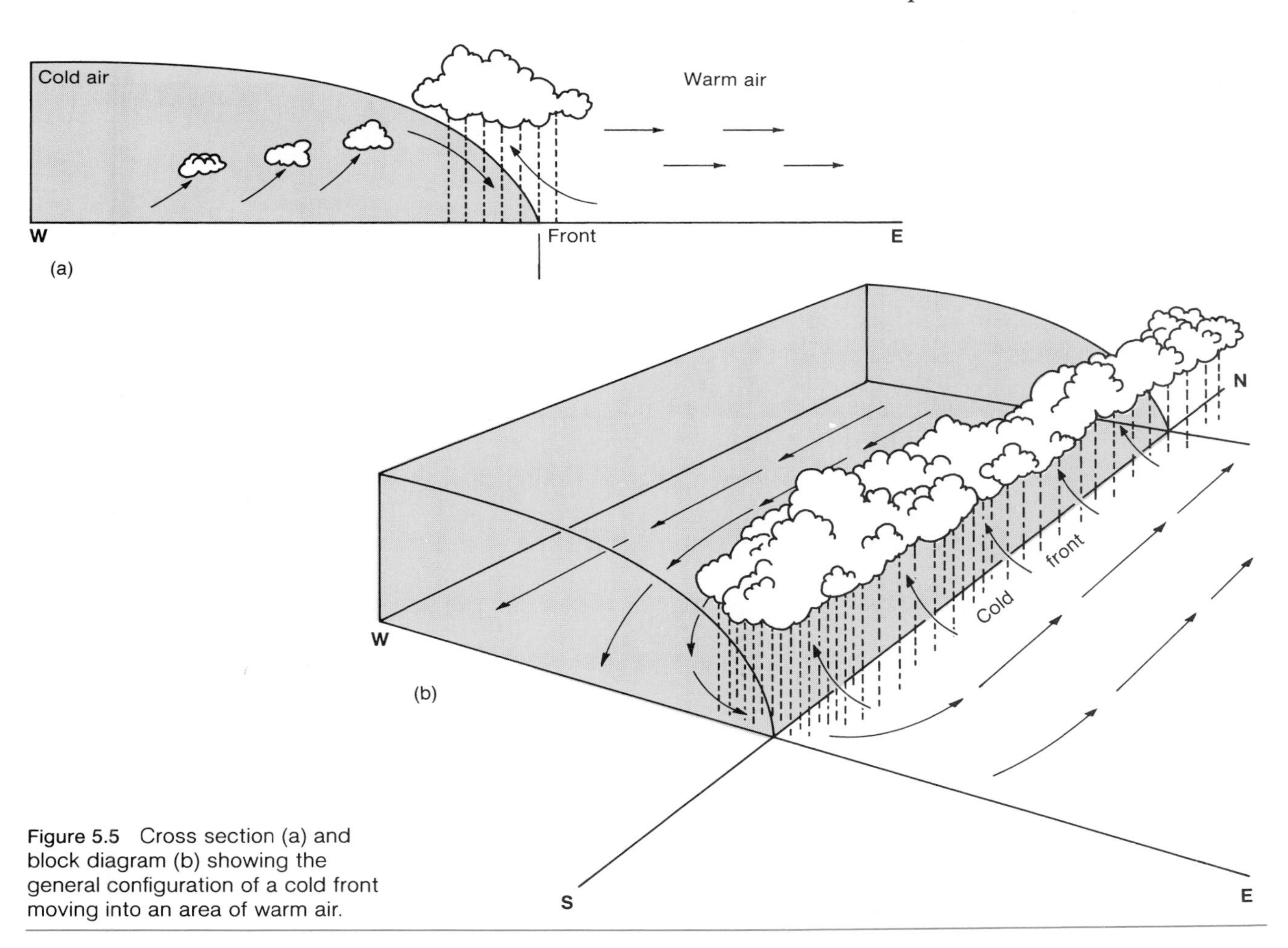

Figure 5.5 Cross section (a) and block diagram (b) showing the general configuration of a cold front moving into an area of warm air.

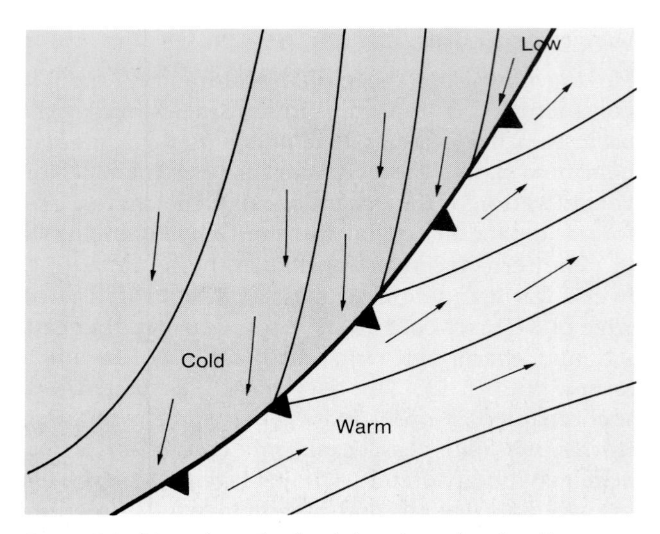

Figure 5.6 Map view of a frontal system showing the general wind circulation pattern associated with a cold front.

warm air moves over the gradual slope of the cold air and precipitation in the form of drizzle or steady rain may persist for a day or more.

The wind circulation associated with frontal systems is essentially one of a shearing relationship (fig. 5.6). Circulation in the cold air mass moves parallel to the front and that in the warm air mass does likewise but in the opposite direction.

Cyclones

Large waves may develop along a front as the result of this shear between air masses. Eventually such a wave produces a gyre-type circulation pattern with counterclockwise circulation around a central low pressure area in the northern hemisphere. These low pressure cells are among the most prevalent types of weather systems that characterize the mid-latitudes of the globe. They move in a generally west to east direction as the result of the prevailing westerly

Figure 5.7 Satellite image of Hurricane Diana on the North Carolina coast in September, 1984.

winds. These cyclonic systems are extremely important to coastal processes as they move across the coastal zone because their winds generate large waves and strong currents.

Tropical cyclones are formed over the open ocean as the result of energy provided by warm, moist air and the circulation provided by the Coriolis effect. These cyclones begin as low pressure waves and eventually develop a closed counterclockwise circulation if in the northern hemisphere. They are called tropical depressions until they have good circular motion and wind speeds of at least 60 km/hr, at which stage they become tropical storms. When wind speed reaches 120 km/hr these cyclones are classed as hurricanes (fig. 5.7). These storms are very powerful and are commonly destructive. Hurricanes occur only during the summer and early fall. They often have landfall with high wind speeds accompanied by large waves and elevated water level. Loss of life and property (fig. 5.8) is common, with the latter regularly in the hundreds of millions of dollars.

Figure 5.8 Structural damage to residences along the North Carolina coast after the passage of Hurricane Diana in 1984.

BOX 5.1

Reading a Weather Map

Everyone has seen the weather maps that are displayed on the television during the evening news or in the newspaper but probably few of us really have an understanding of what information is contained in these maps. Here we will examine an example of the weather maps that are produced by the United States Weather Bureau, a branch of NOAA. It is this type of map from which the maps used in newspapers and by television weatherpeople are derived.

Basic elements in each weather map include contours of barometric pressure called **isobars,** data on wind at several locations throughout the map area, and location and configuration of frontal systems and defined cyclonic systems.

The isobars connect points of equal barometric pressure and define the low-pressure and high-pressure systems. Low pressure cells commonly will have central values of less than 1,000 millibars, whereas high pressure cells may have pressures greater than 1,020 millibars (fig. B5.1). Atmospheric circulation flows from high pressure to low pressure with a circular component of motion. Thus, the wind blows counterclockwise and toward the center in a low-pressure system, and clockwise and outward in a high-pressure system.

Frontal systems are also shown on weather maps by heavy lines with symbols indicating the nature of the front. Half circles on the frontal line indicate a warm front and triangles indicate a cold front. These frontal designations on the map can be related to the preceding diagrams shown for frontal systems (fig. 5.6).

Information on winds at specific locations is also commonly given on a weather map. This is provided at first-order weather stations, which are comprehensive weather observation and recording facilities operated by NOAA. Wind direction and speed is shown for each such station for the period of the map. The direction is indicated by a line oriented in the direction from which the wind blows. The speed in knots (nautical miles per hour) is indicated by the bars on the line. Each bar represents 10 knots and each half-bar, 5 knots. Most TV and newspaper weather maps omit this type of detail.

Major Circulation Patterns in the Ocean

Large-scale circulation systems in the oceans are of two different types caused by two markedly different driving mechanisms. The circulation of the atmosphere and the earth's rotation interact with the surface waters of the ocean to produce **wind-driven circulation,** also called surface circulation. These oceanic currents are restricted to the upper few hundred meters, or that part of the ocean that is essentially above the thermocline. The deep-water portion of the ocean, which is the bulk of the volume, has a circulation system that is termed **thermohaline** because it is driven by the differences in salinity and temperature of the water masses of the world. Thus, the deep-water circulation patterns are density currents because they are driven by gradients in water density. Surface currents or wind-driven currents move rapidly relative to thermohaline or density currents. In addition, they primarily move horizontally whereas thermohaline currents have a distinct vertical component to their movement.

BOX 5.1

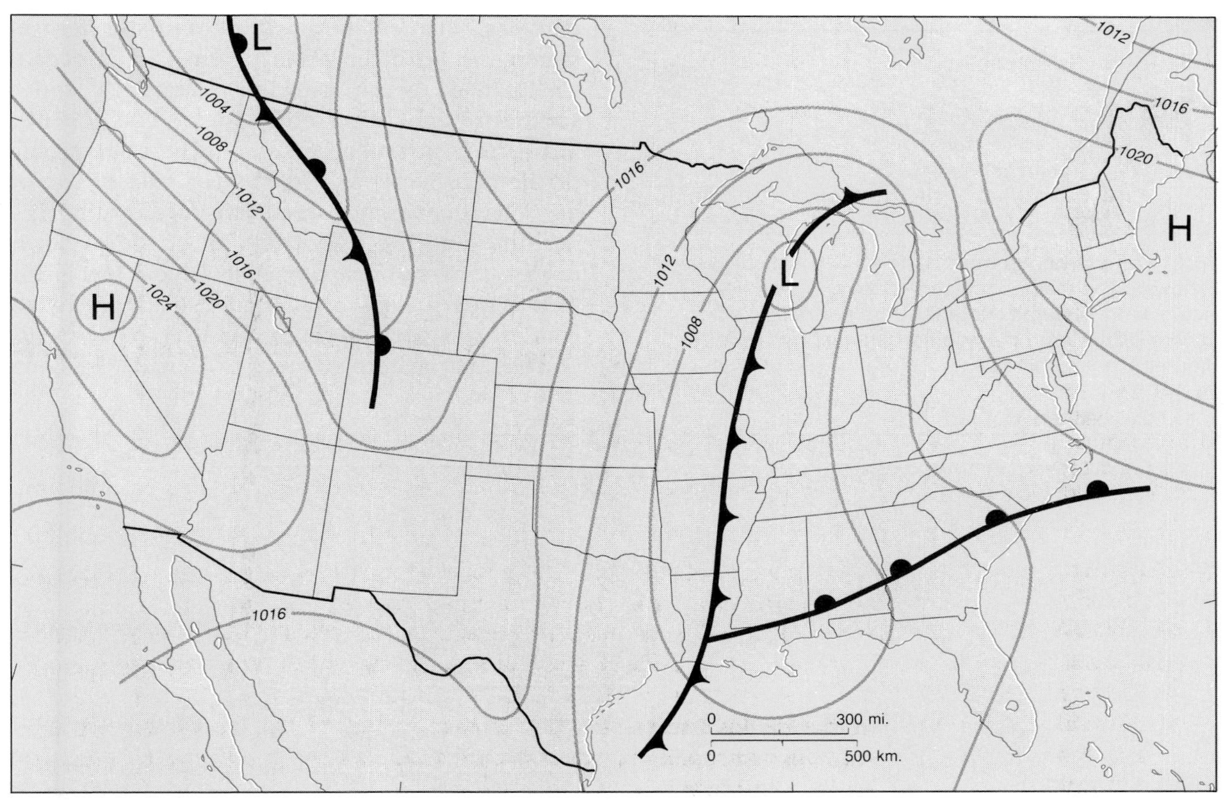

Figure B5.1 U.S. weather map showing the major pattern of barometric pressure, frontal systems, and winds. (National Oceanographic and Atmospheric Administration.)

Surface Currents

Friction caused by the circulation of the atmosphere over the water surface results in movement of the water. Therefore, the surface circulation patterns of the oceans reflect the large-scale circulation patterns of the atmosphere. The belts of prevailing westerlies and trade winds, along with polar winds, produce a pattern of large, elliptical circulation cells called **gyres.**

There is a general uniformity within the northern and southern hemispheres. Each hemisphere is dominated by a large subtropical gyre with a circular pattern that moves clockwise in the northern hemisphere and counterclockwise in the southern hemisphere. Smaller tropical gyres are present in the low latitudes of each hemisphere and move in the reverse direction of the subtropical gyres (fig. 5.9). Each hemisphere is essentially a mirror-image of the other.

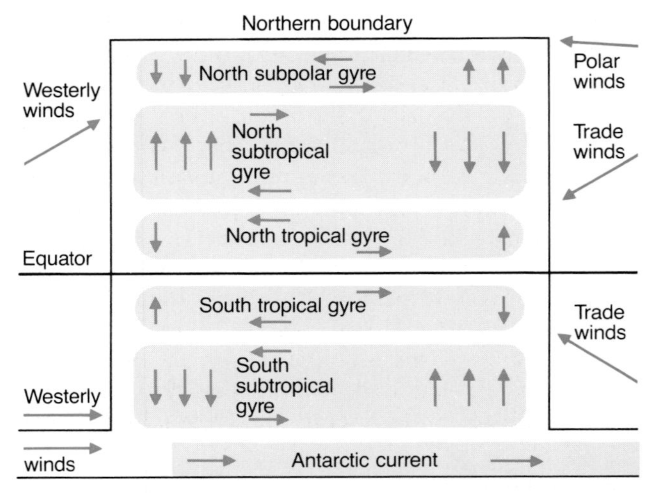

Figure 5.9 Generalized current patterns in a typical ocean basin showing the major circulation cells and the influencing wind systems. Note that the northern hemisphere differs from the southern hemisphere due to the difference in landmasses.

The location of landmasses also affects the circulation system and causes the only major difference between the northern and southern hemispheres: a circum-global current, driven by the prevailing westerlies, that moves around the Antarctic continent (fig. 5.9). It has no northern hemisphere analog.

On a global scale, there are opposite circulation patterns between the northern and southern hemispheres in both the Atlantic and Pacific oceans. In the Indian Ocean there is essentially only the southern hemisphere component. The general pattern that is produced results in the eastern margins of the continents or the western side of the ocean basin having warm waters transported to fairly high latitudes and the opposite side of the ocean basin having cold water carried to relatively low latitudes. North America serves as a good example, with the Gulf Stream carrying warm waters up to New England and the maritime provinces of Canada. By contrast,

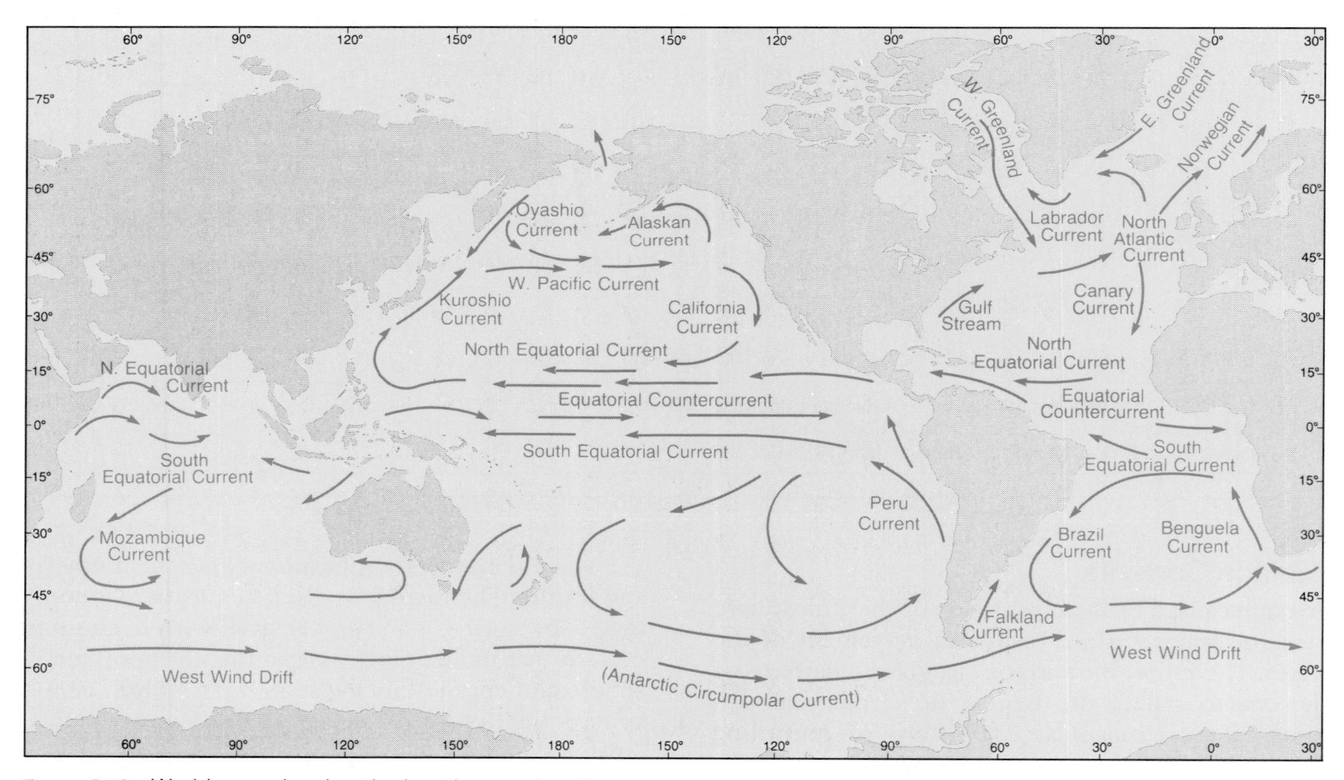

Figure 5.10 World map showing the locations and patterns of the major currents in each of the ocean basins. Compare this actual current map with the general diagram shown in figure 5.9. (Modified from Stowe, 1983.)

the California coast has cold water, in part caused by the cold waters from the northern Pacific that are carried south along the North American coast (fig. 5.10).

The west-to-east rotation of the earth in combination with the Coriolis effect influences the general circulation patterns of surface currents. The rotation causes the currents to be concentrated or squeezed together on the western sides of the ocean basins and spread out or diffused on the eastern sides because of the tendency for water to pile up against the landmasses (fig. 5.9). Note that the same situation occurs in both the northern and southern hemispheres because of the rotating earth. The latitudinal variation of the Coriolis effect comes into play where the western boundary currents, such as the Gulf Stream, are deflected toward the east. Here the Coriolis effect causes a swift and concentrated current. The opposite effect occurs in the eastern boundary currents, which are dispersed (fig. 5.9).

Ekman Spiral

Another circulation phenomenon that is related to the Coriolis effect was first recognized and modeled by V.W. Ekman. Called the **Ekman spiral,** it demonstrates that when wind blows over the water surface of the rotating earth, the Coriolis effect causes the surface current to move in a direction that is 45° to the right of the wind in the northern hemisphere.

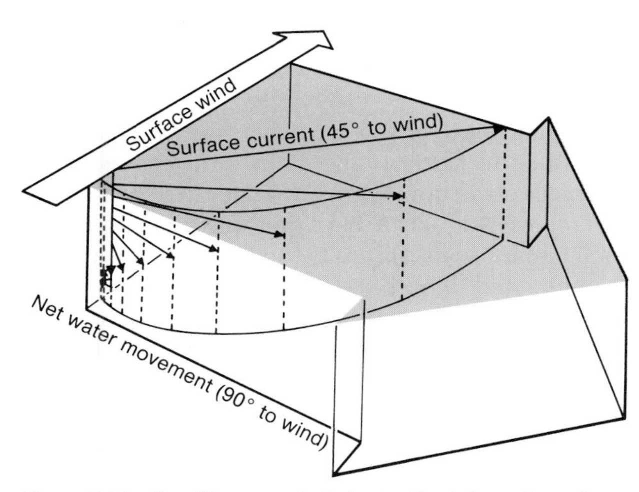

Figure 5.11 The Ekman spiral shows the interaction of wind direction and the Coriolis effect over the rotating earth and the resulting current directions. This phenomenon was first recognized by noting the relationship between wind and the direction of floating icebergs.

This surface motion influences the water immediately below in a similar fashion, so that there is motion in each water layer that is increasingly to the right of the overlying layer and has a decreasing speed. At some depth, generally near 100 m, there is essentially no wind-generated motion and the direction of motion immediately above is in about the opposite direction from that of the wind (fig. 5.11).

This relationship is a theoretical one for a homogenous water column. In nature, there is deviation from this ideal situation. In general, the net water motion or **Ekman transport** in the affected portion of the water column is about 90° to the direction of the wind blowing over the surface waters.

Geostrophic Currents

There are conditions in the circulation of fluids over the rotating earth under which pressure gradients develop. A pressure gradient occurs whenever the pressure at one location is greater than at another. Motion develops from the high pressure to the low pressure site in much the same fashion as in the atmosphere when the wind blows. When the force generated by these gradients is balanced by the Coriolis effect, which deflects the motion, it is called **geostrophic** or literally, earth-tuned. Geostrophic motion flows along the contours (fig. 5.12). Such pressure gradients can be generated by differences in water density or by variations in atmospheric pressure (wind) that deform the water surface. In the latter situation, the water surface slopes due to the wind. The result is a geostrophic current.

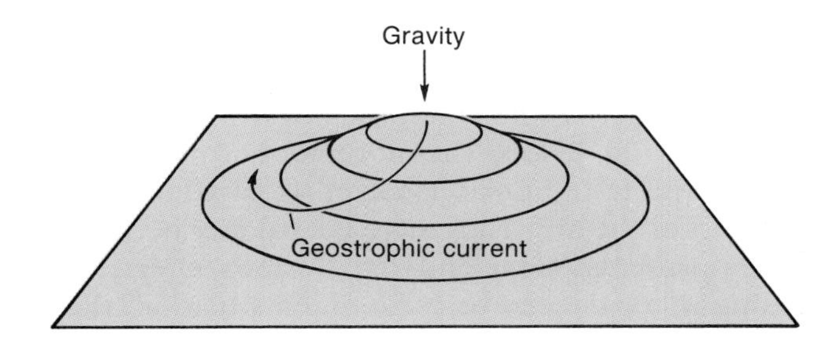

Figure 5.12 Schematic diagram of a hill of water showing the relationship between the Coriolis effect and gravity in the formation of geostrophic currents.

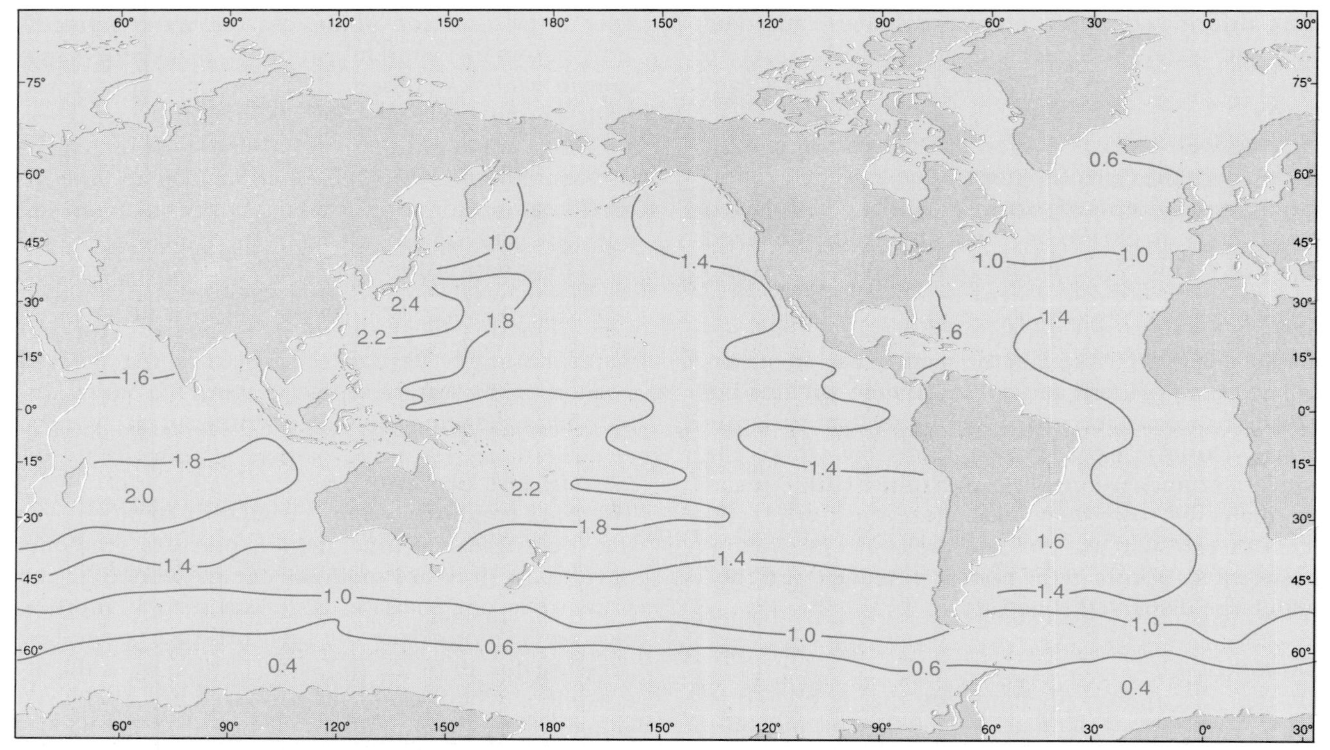

Figure 5.13 Surface elevation of the water surface for the world ocean. The values shown are given in feet and represent the uneven water surface due to the rotation of the earth. Note that highest values are on the western sides of the oceans against the continents. (From Stommel, H., 1964.)

The surface circulation of the oceans, including the Coriolis effect and Ekman transport, results in gently sloping but readily definable hills on the water surface in the central areas of the major gyres in each ocean basin (fig. 5.13). The hills provide a horizontal pressure gradient, in other words a slope, which causes the water to move away from the crest of the hill, across the contours, toward the bottom of the hill (fig. 5.12). The movement of this water, in response to the Coriolis effect, is parallel to the contours of the hill. Thus there is a balance between the pressure gradient and the Coriolis effect, resulting in a balanced or earth-tuned situation, that is, a geostrophic current.

Meanders and Rings

Until recently, our knowledge of oceanic circulation was limited to piecing together data collected from shipboard observations. Widespread use of satellites for collecting data on ocean circulation has opened new perspectives by which to observe this circulation. Primary contributions of these new tools are the ability to view large areas synoptically and also to observe short-term variations in current patterns.

One example of this new data is the recognition of small-scale patterns and short-term changes that take place. It is now known that many of the major currents of the ocean have pronounced meanders much like those displayed by rivers. These meanders are ephemeral in both space and time but they seem to be most pronounced shortly after the passage of storms. Sometimes these meanders become so pronounced that they detach and form rings. The rings have a diameter of a few hundred kilometers with maximum currents of nearly 1 m/sec. The ring itself moves a few kilometers per day.

BOX 5.2

Measuring Currents

Currents are measured using a wide range of techniques and equipment, from crude, inaccurate methods to sophisticated, accurate ones. Currents can be measured directly, where the water motion itself is monitored, or they can be measured indirectly, where some other parameter related to water motion is measured and calculations are made to determine the motion of the water.

There are two types of direct current measurement: **Lagrangian** and **Eulerian.** Lagrangian current measurement monitors the actual path of movement of the water mass, whereas Eulerian techniques measure the movement of the water mass past a fixed geographical location.

The least sophisticated Lagrangian methods use particles or objects released into the current, then they are tracked or collected after a certain length of time. These techniques use chemicals or dyes as tracers, or floats, bottles, drifters, or other objects to move with the water mass. In some situations, the tracer or float may be monitored visually and its path traced. In others, the release point and the collection point are known, as well as the time between release and collection. From this information it is possible to determine a general direction but not the actual path taken by the float. This type of current measurement has the advantage of being inexpensive and easy to carry out, but it is not accurate and provides only general information about current patterns.

Accurate Lagrangian measurements can be taken by electronically monitoring the paths of various types of floats, drifters, or drogues. These devices are tracked by either transmitting radio signals or by reflecting radar signals. Some are designed to be used below the surface, their depth determined by the density of the drifter. The path of the drifter is monitored and recorded when it reaches the desired depth. When the experiment is over, a radio signal deploys a float device that brings the instrument to the surface so it can be recovered for future use.

Eulerian current measurement generally uses a current meter to measure the rate of flow past the moored instrument. The simplest current meter has a propeller apparatus that revolves as the current flows through it. The rate at which it revolves is proportional to the current's speed. In some of these meters the revolutions are counted by the meter itself, similar to an odometer on an automobile recording the mileage. The number of revolutions over a specific period of time can then be converted into a current's speed, usually given in cm/sec.

In addition to knowing the speed of a current, it is also important to know the direction it flows. Some current meters are capable of providing this information and others are not. In some types of environments, such as tidal inlets, the current's direction is obvious and does not need to be recorded, whereas in the open ocean, direction may vary greatly with time, so both direction and speed need to be recorded.

Sophisticated current meters operate on the same principles as the simple ones, but are accurate and are equipped to be placed at a given site and left for up to months at a time. The data may be recorded internally or it may be telemetered to a base of operations. Most commonly the current data are recorded on digital tape and then input directly into the computer for processing. Some meters are placed on the seafloor whereas others are suspended from floats to record surface currents or currents at some desired depth in the water column.

Not all Eulerian current meters are based upon the principle of revolving propellers. Some have membranes or other pressure-sensitive devices that measure current speed by the amount of pressure exerted by the water mass on the membranes. These devices eliminate the potential for problems associated with friction on the propeller or problems with moving parts. Any moving parts that are reduced or eliminated in a piece of marine equipment offer that many fewer chances of corrosion.

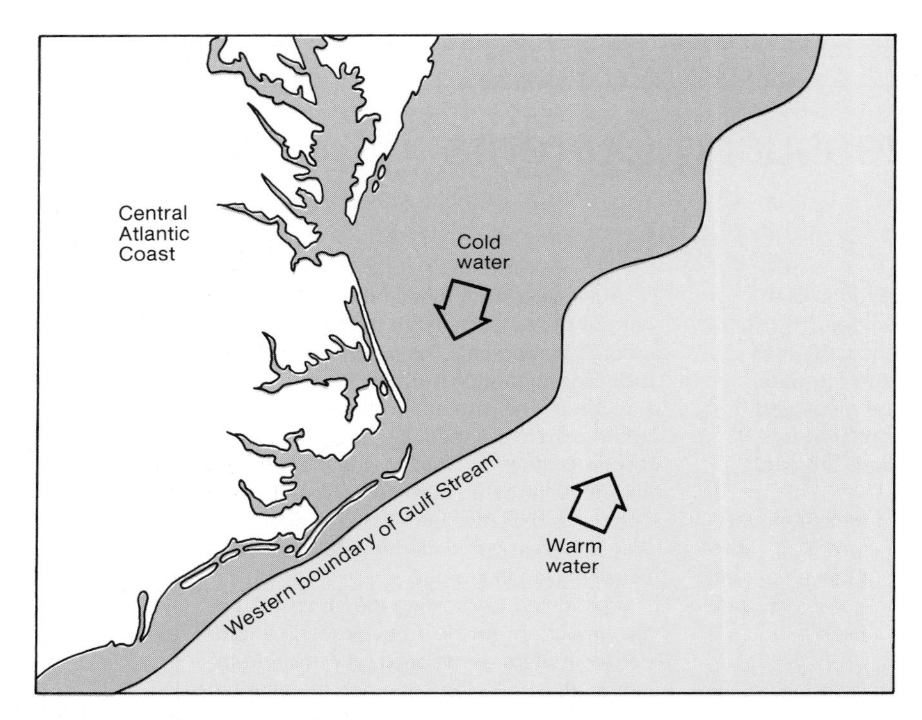

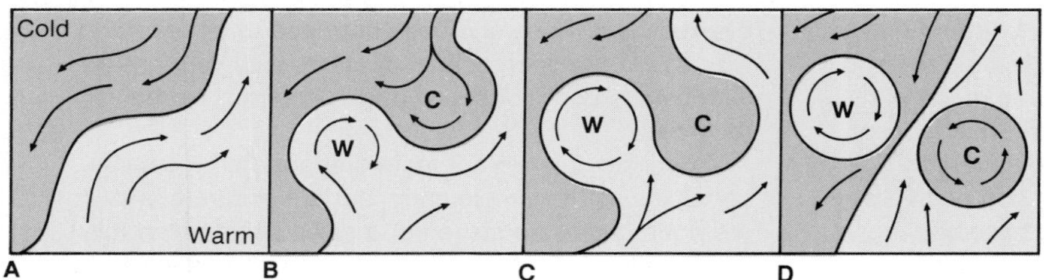

Figure 5.14 There are meanders and rings that form as the result of opposing motions along the western boundary of the Gulf Stream off the North Carolina Coast. Between the warm northerly flowing water and the cold southerly flowing water, meanders form. They eventually become rings and separate from the main water mass, thus forming isolated small water masses (a–d). (From Duxbury and Duxbury, 1984.)

Using the Gulf Stream as an example, these rings occur on both the northwest and southeast side. The former circulate clockwise and the latter counterclockwise (fig. 5.14). Those to the northwest have a warm water center surrounded by cold water and are termed warm-water rings. The rings formed on the southeast side of the Gulf Stream are the opposite with a cold water center surrounded by warm water. The rings extend to a depth of 2 km, and thus are confined to deep water away from the continental shelf. Once a ring is formed it begins to lose its energy because it is removed from the direct influence of the Gulf Stream. Some of these rings maintain a distinct circulation pattern for only days whereas others have been tracked for over a year.

Upwelling and Downwelling

Surface currents, the Coriolis effect, and Ekman transport are responsible for what seem to be anomalous temperatures along some coastal areas of the world. When a wind-driven coastal current moves along a landmass such that the combination of surface motion and net transport diverges with the coast, then the phenomenon of **upwelling** occurs. The warm surface waters are transported away from the coast causing the colder deep water to rise (fig. 5.15). This is actually a big part of the reason for the cold water along the California coast. It is not simply the cold high-latitude water being transported to the south. **Downwelling** is the reverse situation. It occurs when surface waters are blown toward the coast causing convergence with the coast due to net transport and the Coriolis effect (fig. 5.16). The result is that warm surface water is carried to much greater depths than would otherwise occur.

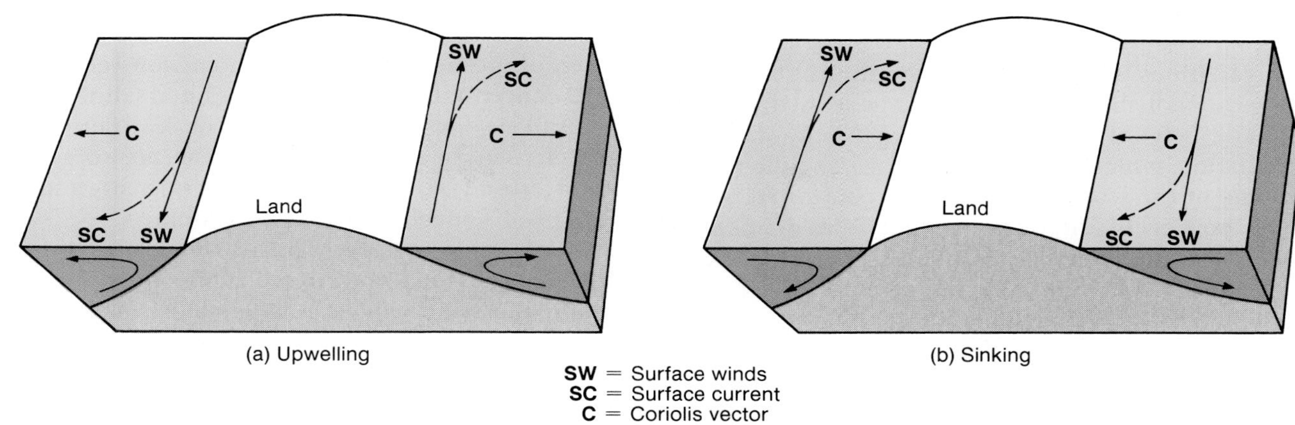

Figure 5.15 Generalized diagram showing upwelling and downwelling phenomena. When the Coriolis effect causes an offshore component for shore-parallel currents, there is upwelling of deep water to the surface (a). When there is a shoreward component due to the Coriolis effect, then downwelling of surface waters takes place (b).

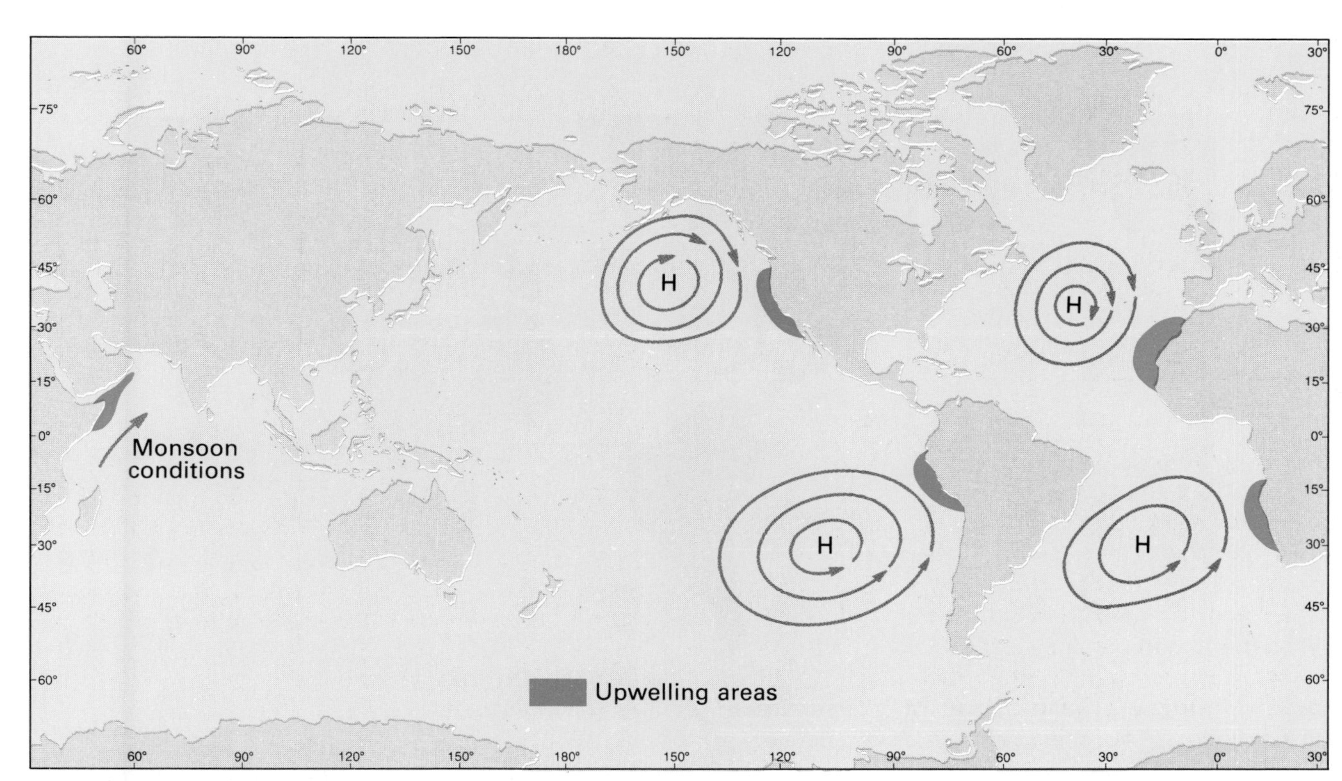

Figure 5.16 Upwelling is important in productivity and is a predominant phenomenon along the western margins of major landmasses, as a result of large weather circulation cells.

Thermohaline Circulation

The variations in temperature and salinity that produce thermohaline circulation are generated in the climatic belts of the high latitudes. Thermohaline circulation, which applies to about 90% of the total volume of the world ocean, thus is controlled by a small portion of the ocean surface. The gradients caused by differences in density of the water masses generate slow currents. These currents are due to cold, dense water sinking to the deepest parts of the ocean basin. This dense water is produced in the high latitudes as the result of direct cooling by the atmosphere or by the increased salinity that results from the freezing of surface water. Although some salts may be trapped in the ice, the ice itself is composed of fresh water. This phenomenon results in local concentration of salts and elevated salinity that in turn cause water density to increase. Thermohaline circulation responds only to the pressure gradients caused by differences in water density and to the Coriolis effect; thus it is geostrophic.

Circulation in deep waters is slow. Rates of less than 2 cm/sec or 3 cm/sec are common, with maximum values rarely above 10 cm/sec. The most rapid movement is found along the western margins of the ocean basins due to the earth's rotation and concentration of flow, similar to the situation for surface currents.

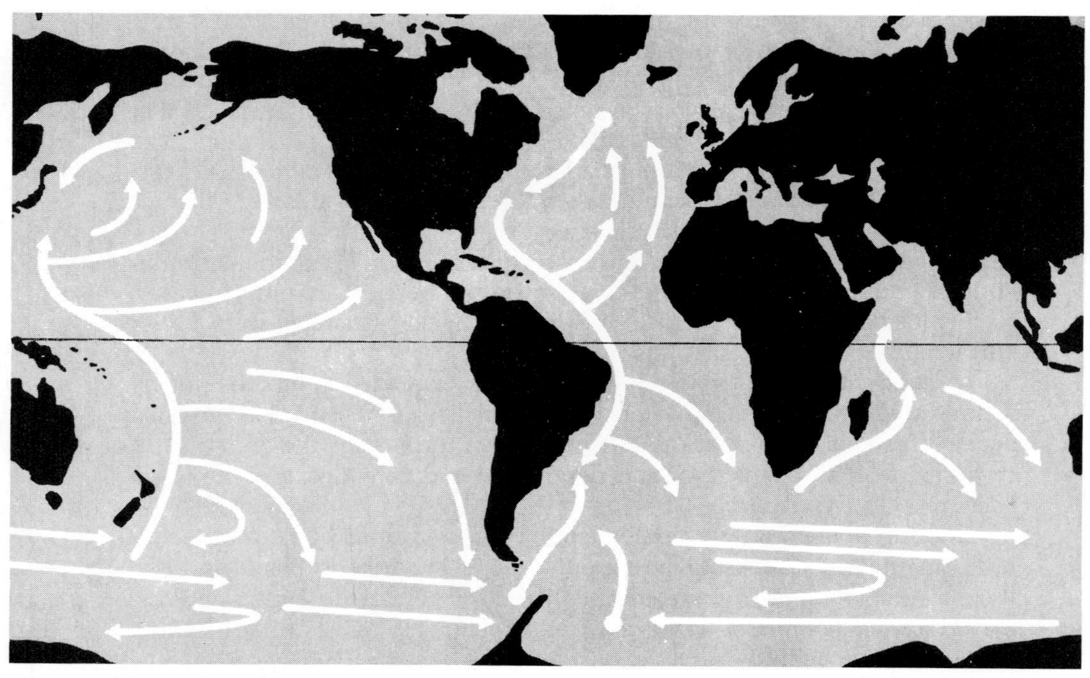

Figure 5.17 Stommel's circulation pattern of deep-ocean water. The scheme is one of water masses forming in high latitudes and moving to lower latitudes. Only near the Antarctic is there circulation between oceans.

Thermohaline circulation is generally in the form of large water masses with a characteristic density. The various water masses are well-defined and can be recognized by their temperature-salinity characteristics. The theory for general circulation pattern of deep water was developed by Henry Stommel in 1958. He showed that the high latitudes are the sources of the deep water. The model Stommel produced (fig. 5.17) has deep Atlantic water produced in the Arctic whereas the deep water of the Pacific and Indian oceans is generated in the Antarctic. This representation is only of the areal distribution of deep circulation. A cross section of the southern Atlantic Ocean shows the vertical distribution and relative motion of the various water masses (fig. 5.18). This depiction shows that there are several water masses at varying depths and that there is some interaction of the waters between these masses. This mixing between water masses is slow, however.

One of the best examples to demonstrate the nature of water masses and how they maintain their identity is the Mediterranean Sea water that enters the Atlantic Ocean through the Strait of Gibraltar. The Gibraltar Sill provides a partial barrier that restricts circulation between the Atlantic Ocean and the Mediterranean Sea. Because of the warm and evaporative climate, the water of the Mediterranean is warm and saline relative to that of the Atlantic Ocean, thus giving it a higher density than ambient water of the Atlantic (fig. 5.19a). When this water flows over the Gibraltar Sill into the Atlantic, it descends nearly a kilometer and flows as a distinct mass for a few thousand kilometers into the Atlantic Ocean (fig. 5.19b).

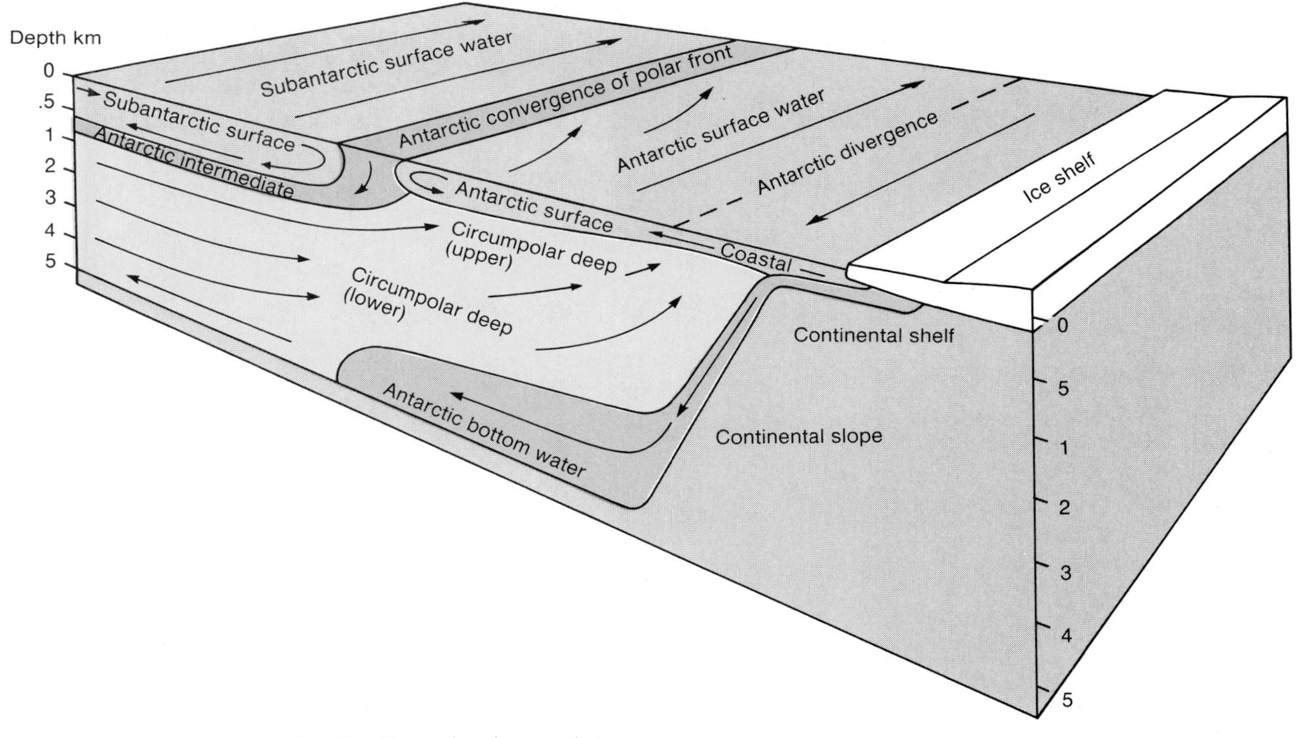

Figure 5.18 Block diagram showing the major deep-water masses for the area near Antarctica.

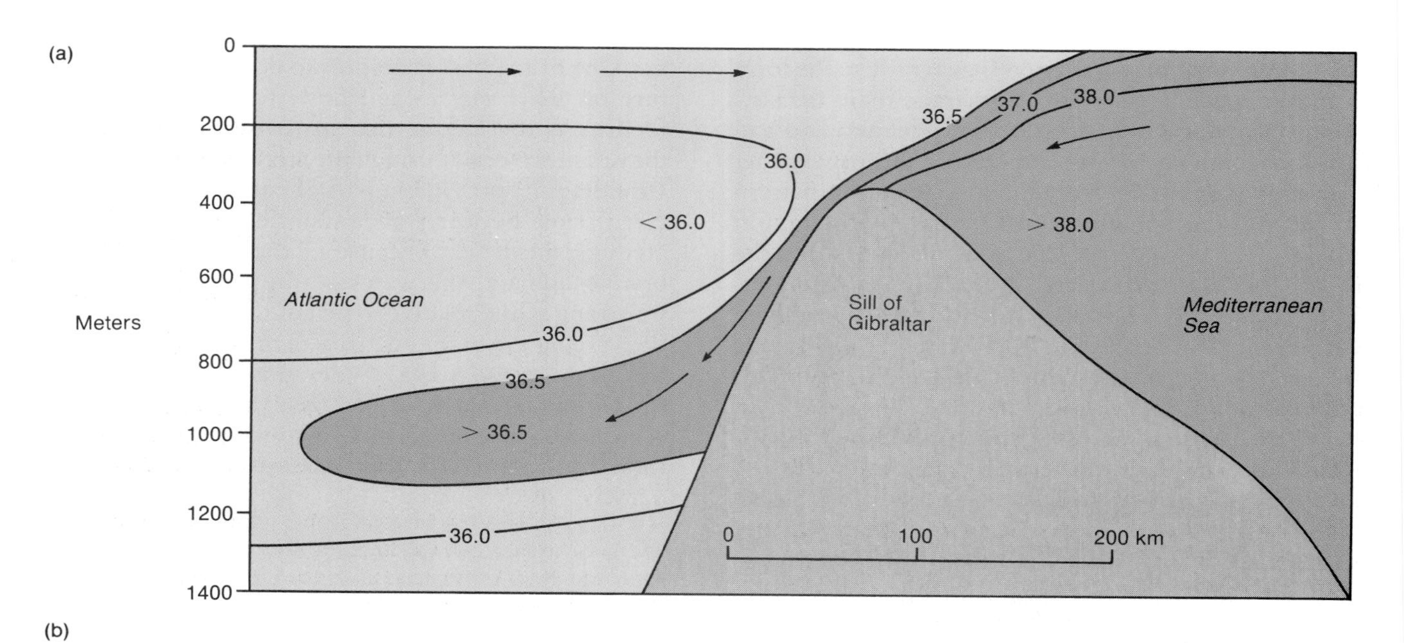

Figure 5.19 Profile showing Mediterranean seawater entering the Atlantic Ocean over the Gibraltar Sill (a) with a tongue of salty water extending thousands of kilometers into the Atlantic (b).

Summary of Main Points

The circulation of both the atmosphere and the hydrosphere is dominated by gravity or gravity-related processes. Atmospheric circulation is controlled by density gradients from high-pressure systems to low-pressure systems and is in turn strongly influenced by the earth's rotation through the Coriolis effect. The surface waters of the oceans circulate in accordance with the prevailing wind patterns of the atmosphere with influence from the Coriolis effect, the Ekman spiral, and the landmasses. Deep-ocean circulation is somewhat similar to the atmosphere in that it is dependent upon density gradients with influence of the Coriolis effect.

Each of the major ocean basins displays a similar surface circulation pattern with the northern hemisphere exhibiting a mirror image of the southern hemisphere except in the high latitudes. There the Arctic Sea and the Antarctic landmass cause major differences.

Deep-water circulation is complicated by converging water masses and the interaction of adjacent water masses resulting in water being exchanged. These water masses move very slowly and tend to move from high latitudes, where they are formed, toward the lower latitudes.

Suggestions for Further Reading

Bowditch, N. 1977. **American Practical Navigator.** vol. 1. Washington, D.C.: U.S. Defense Mapping Agency, Hydrographic Center.

Drake, C.L., J. Imbrie, J.A. Knauss, and K.K. Turekian. 1978. **Oceanography.** New York: Holt, Rinehart and Winston.

Groen, P. 1967. **The Waters of the Sea.** New York: Van Nostrand Reinhold.

Kuenen, P.H. 1963. **Realms of Water.** New York: Science Editions.

Pickard, G.L., and W.L. Emery. 1982. **Descriptive Physical Oceanography.** Oxford: Pergamon Press.

Smith, F.G.W. 1973. **The Seas in Motion.** New York: T.Y. Crowell.

Stewart, R.W. 1969. "The atmosphere and the ocean." *Scientific American,* 121:76–86.

Stommel, H. 1958. **The Gulf Stream.** London: Cambridge University Press.

6
Waves

Chapter Outline

Atmospheric circulation not only has a marked effect on the circulation of the world ocean, it also causes the perturbations of the surface that are called waves. Waves are disturbances of the surface of a fluid that typically occur on the surface between the atmosphere and a water body, but may also occur on the surface that separates two water masses. The generation of surface waves is generally the result of wind acting on the water surface but it may be due to a disturbance of the earth's crust, the container that holds the water.

There are three factors that determine the size of wind-generated waves: (1) the fetch, which is the distance of water over which the wind blows to generate the waves, (2) the duration of the wind, and (3) the speed of the wind. Increases in all of these result in increased size of waves.

Characteristics of Waves

Ideally, waves have a sinuous shape (fig. 6.1). There are a few important wave components that need to be mentioned. The crest is the high part of the wave and the trough is the low part. The **wave height** (H) is the vertical distance between the top of the crest and the bottom of the trough. The **wave amplitude** (A) is half the wave height or the vertical distance that the wave moves relative to still-water level. The **wave length** (L) is the horizontal distance between any point on a wave and the corresponding point on the adjacent wave (fig. 6.1). Waves may also be measured by their **period** (T), the time that it takes for one wave length to pass a point. This parameter is measured in seconds. Another related wave measurement is frequency (F), which is the number of waves or portions of waves per second. This is measured in hertz (cycles per second) and is applied to a variety of phenomena, not just water waves.

Wind waves are commonly grouped into one of three types, sea, swell, and surf. The term **sea** refers to those waves that are under the direct influence of the wind. These waves tend to have peaked crests and smoothly rounded troughs (fig. 6.2). Complex patterns of sea waves are common when the wind is blowing. **Swell** waves are those that have moved beyond the area where wind is actively blowing or those waves that persist after the wind ceases to blow. They tend to have low wave heights and long periods with sinusoidal shapes (fig. 6.3). **Surf** refers to breaking waves in shallow water. There are various types of surf depending upon the style of breaking. A **spilling breaker** is one in which the breaking takes place slowly over a long distance with the crest appearing to spill over the landward side of the breaking wave (fig. 6.4). **Plunging breakers** steepen gradually but break instantaneously with a crash of water (fig. 6.5). Waves that surge up on the beach under some conditions are named **surging breakers** (fig. 6.6).

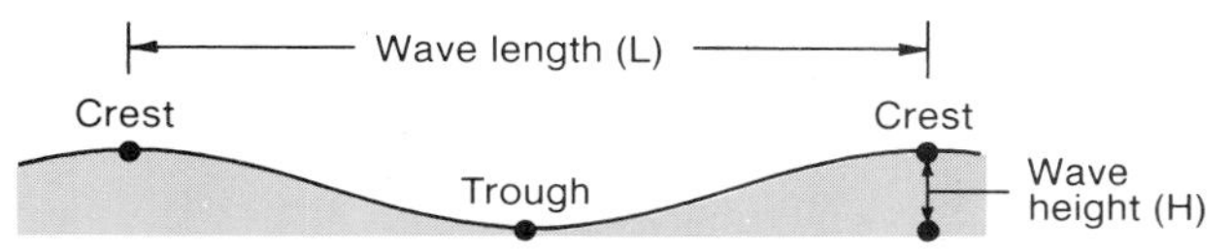

Figure 6.1 Ideal sinusoidal wave showing the basic components.

Figure 6.2 Sea waves present a complex pattern of generally peaked crests. Commonly, there is a lack of continuity of crests and multiple wave sets may be superimposed.

Figure 6.3 Swell waves appear better organized than sea waves. They tend to be long with low, undulating wave crests.

Figure 6.4 Spilling breakers are those that break over a significant distance, with the breaking lasting for several seconds. The crests appear to mimic water spilling from a container.

Figure 6.5 Plunging breakers break instantaneously, with the crest curling over and plunging. These breakers are commonly generated from swell waves.

Figure 6.6 Surging breakers are those that rush up a steep beach and break with little apparent steepening.

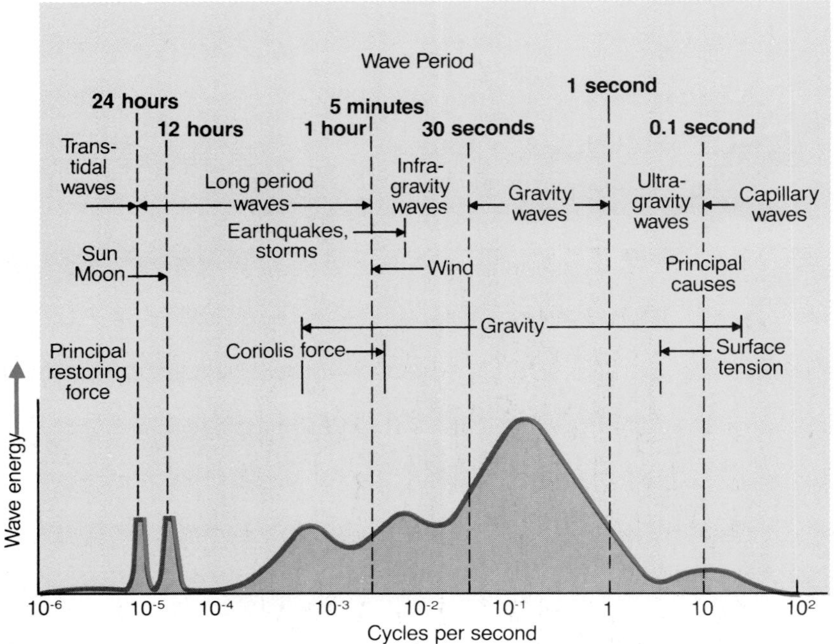

Figure 6.7 Diagram showing all types of waves ranging from tides to capillary waves. Note that most gravity waves tend to range from periods of a few seconds to about 25 seconds. (Modified from Kinsman.)

The size and shape of wind waves range greatly. The length of waves in the open ocean is commonly at least tens of meters but storms regularly generate waves over 100 m in length. Swell waves tend to be longer with lengths of 300 m being common. Some in the Pacific exceed a kilometer in wave length. In shallow water near the shore, wave lengths are typically a few tens of meters.

When someone speaks of large waves, they are generally talking about the height of the waves, not the length. Severe storms can generate high waves in the open ocean, but the typical wave height in the open ocean is only a few meters. Nearly half are less than a meter in height and more than 75% are less than 4 m. Although large waves with heights over 10 m may result from storms, these are rare. Waves in excess of 30 m have been documented. Examples are a 34 m wave measured from a ship during a storm in the Pacific in 1933 and what must have been a 40 m wave on the Oregon coast where a lighthouse had windows at that elevation broken by rocks carried by a wave. However, it is easy to over estimate the height of waves and reports of large storm waves along coasts are typically exaggerated.

The shape of waves is also a variable about which some generalizations can be made. In the discussion above, mention was made about sea waves having peaked crests and smooth troughs. They also have high **wave steepness**, the ratio of wave height to length (H/L). Short, steep storm waves make the roughest seas. Typically, swell waves have a very low steepness.

The period or frequency of waves can easily be measured and is commonly used to give some idea of wave size. Periods range from less than 0.1 sec in capillary waves to greater than twenty-four hours. Capillary waves are wind waves that are less than 1.74 cm in wave length and respond to surface tension as the primary restoring force or maintenance phenomenon. Larger wind waves are called **gravity waves** and, although they have a wide range, most have periods of 3 sec to 15 sec (fig. 6.7). In these

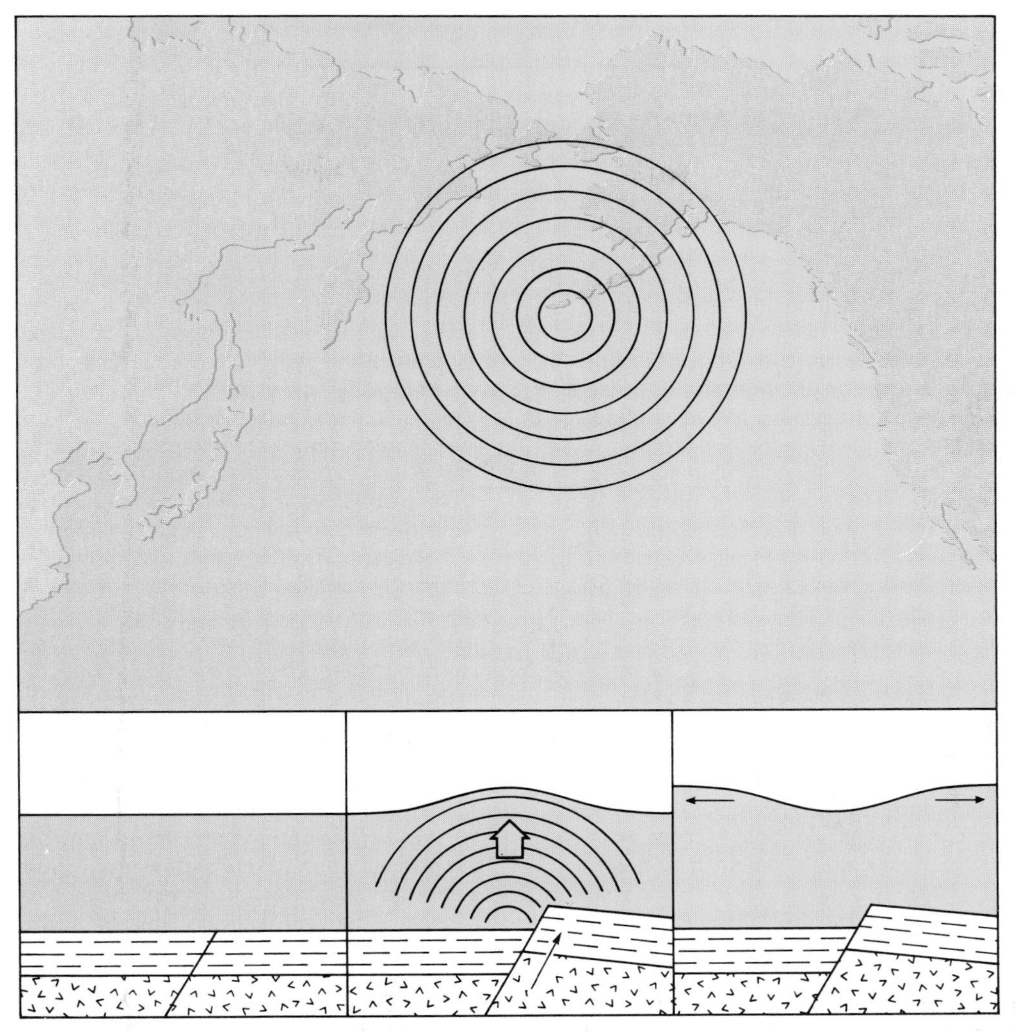

Figure 6.8 Map showing earthquake epicenter and the propagation of a tsunami across the Pacific Ocean. The tsunami can be generated by crustal movement, which causes a long, low wave in the open ocean, and may steepen to heights of tens of meters on reaching the coast. (From H. V. Thurman, 1985, *Introductory Oceanography,* 4th edition, Merrill Publishing Company, Columbus, Ohio. Reproduced by permission of the publisher.)

waves the main restoring force is the mass of water as it responds to the force of gravity. There are also long waves with periods of from 5 minutes to several hours. These may be the result of storms that have caused abrupt changes in barometric pressure. Such waves are called **seiches** and are especially common in large lakes where they may slosh back and forth from one side to the other. Earthquakes, submarine slides, and other disturbances to the seafloor may generate large waves called **tsunamis** (fig. 6.8). These waves have periods of several minutes and a length of hundreds of kilometers. In the open ocean they have wave heights of less than a meter but may steepen to heights of many meters as they strike the coast.

BOX 6.1

Measuring Waves

Of the three primary characterizing components of wind waves, two are commonly measured and the other is not. Both wave height and wave period are easily measured, whereas wave length is difficult to measure directly, except on photographs, because of the continuous motion of the wave. Wave height can be easily measured as the wave passes by any stationary pole, tower, or other calibrated standard. Wave period is measured by timing the passage of a given number of waves, then dividing the time that passed by the number of waves to determine the period. Generally, eleven wave crests are timed. This is equivalent to ten waves. If it takes fifty seconds for the eleven crests to pass, then the wave period is 5 seconds.

Almost any vertical or stationary staff can be marked to measure wave height. As the waves pass the staff, the level of the trough and the level of the crest are noted. The difference between the two is the wave height. Because waves tend to vary in size even under the same conditions, it is best to measure several and take the average in determining wave height.

Both of these approaches to measuring waves are simple but they yield fairly accurate results. They require direct observation and are therefore time-consuming and inconvenient. In the case of wave height, this method is restricted to places where a staff has been installed. In modern research, waves are generally measured and the data recorded electronically. This can be done in shallow water near the shoreline or it can be done in fairly deep water. In the surf zone, it is common to install temporary wave staffs that have electrical contacts or light sensors throughout the staff. As the wave level moves up and down the staff, these sensors or contacts either send a signal to a monitoring system on shore or to an internal recording system attached to the staff. This information can then be inserted into a computer and a complete analysis of the wave spectrum, including wave height and period, can be determined.

Various types of pressure transducers are commonly used to measure waves. These are sensors that measure the pressure change in the water column as a wave passes overhead. The crest of a wave causes greater pressure than the trough. By measuring the difference in this pressure, it is possible to calculate the height of the wave. There are also various types of similar instruments that can measure the direction of wave approach, which is an important type of information in coastal areas.

These types of pressure sensors may be placed on the seafloor or on towers or other rigid structures. The electronic data collection system that is monitoring the sensors is generally activated only for short periods on a cyclic basis, for example, several minutes at a time every several hours, so that the amount of data generated is not too voluminous. Some systems are internally recording, some are attached by cable to the shoreline where they are monitored, and in some the data are sent by radio signals to remote monitoring installations.

Wave Dynamics

Even the casual observer has noticed that objects floating on the water surface are passed by, crest after crest, as waves proceed in their direction of propagation. This is the result of the nature of the water particle motion within the wave. The floating particles appear to bob up and down, but are actually moving in a circular path, in the same way that the water particles move in a circular path. At the surface, the diameter of this path is equivalent to the height of the wave. The diameter of the orbital paths decreases with depth until there is no motion. The depth at which this takes place is equal to 1/2 of the wave length (fig. 6.9). The motion in the orbits is in the direction of wave propagation. The decrease in orbital diameter is such that for each 1/9 of a wave length increase in depth, the diameter of the orbital path is halved. For example, if a wave has a length

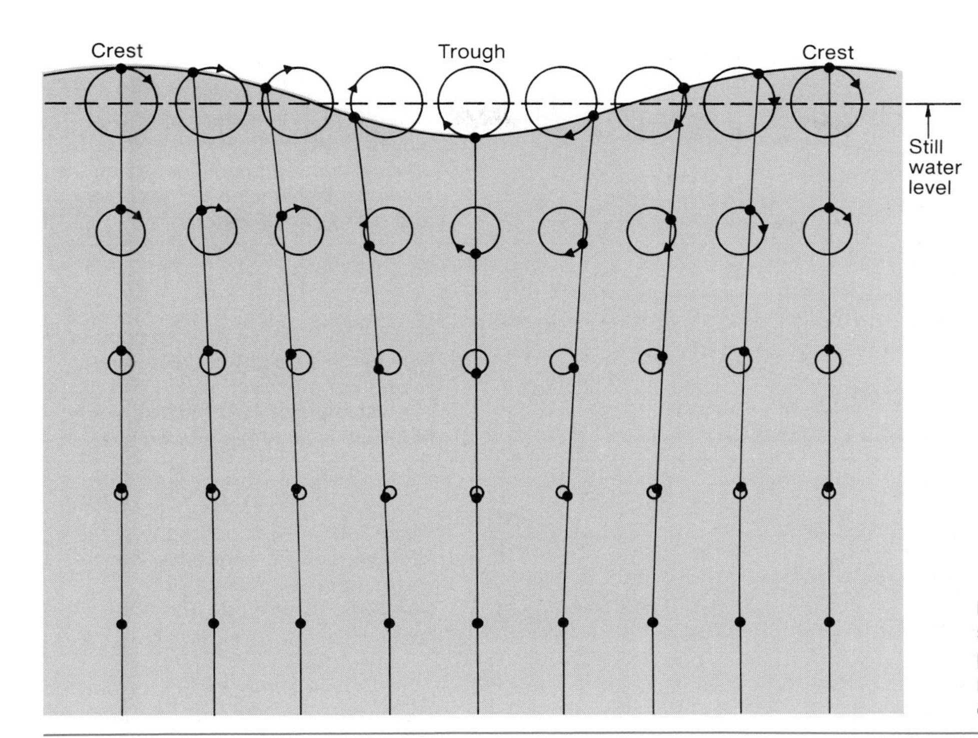

Figure 6.9 Schematic diagram showing the orbital motion of water particles in a gravity wave. This motion extends to a depth of about one-half of the wave length.

of 50 m and a wave height of 2 m, the orbital diameter would be 2 m. At a depth of 1/9 of the wave length (50/9 = 5.55 m), the orbital diameter would be 1 m; at a depth of 2/9 of the wave length or 11.1 m, the orbital diameter would be 0.5 m, and so forth.

It is possible to consider three categories of waves depending on their size relative to water depth. These are deep water waves, transitional waves, and shallow water waves. **Deep water waves** are those with water depth greater than 1/2 of the wave length. There is no interference from the sea bottom with the orbital motion within the waves. A **shallow water wave** is one in water shallower than 1/20 of the wave length. At such depths the waves are markedly deformed by interference with the bottom and may break. In between these are the **transitional waves**.

Deep Water Waves

Typically, a variety of wave sizes and shapes develops during sea conditions. This variety forms the wave spectrum from which the average wave height, wave length, and other parameters can be determined. Because of this great range in wave sizes, an average of any of these parameters would have a great standard deviation and would probably reflect the numerous small waves. For these reasons, the **significant wave height** is used as a standard measure of wave size. It is defined as the average height of the largest 1/3 of the waves.

Shallow Water Waves

When the orbital paths of water in a wave begin to interact with the bottom, there are changes that take place in the waves. This results in changes to the height and length of the waves and to the **phase velocity,** which is the speed of wave propagation. As the wave moves into shallow water, the orbits are somewhat flattened to an elliptical shape and eventually show only a back-and-forth motion. The overall

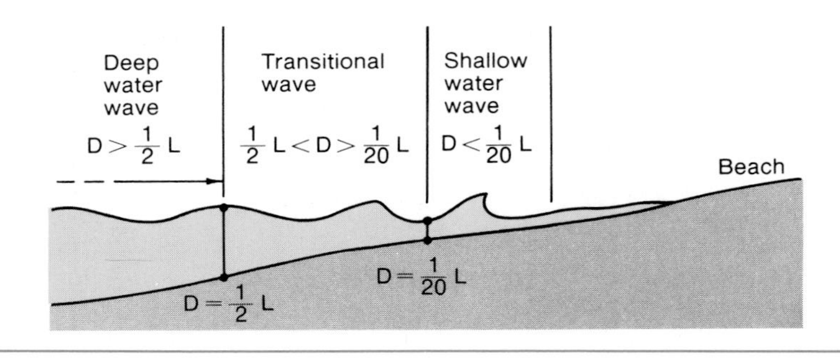

Figure 6.10 The nature of gravity waves changes from deep water to shallow water primarily based on the relationship between the wave length and the depth of water.

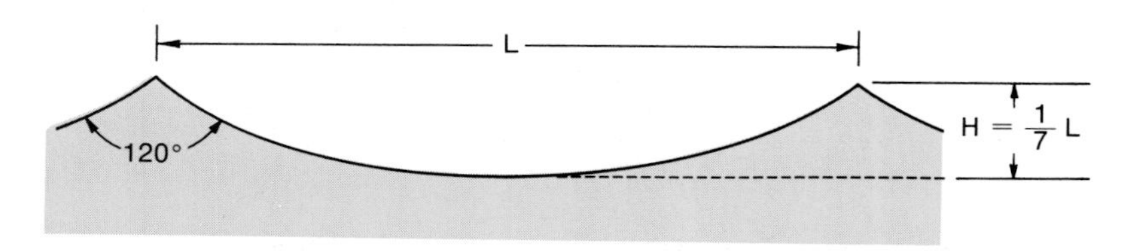

Figure 6.11 Waves may steepen until they reach a height equal to one-seventh of the wave length and an internal angle of 120°. At this point the wave is unstable and breaks.

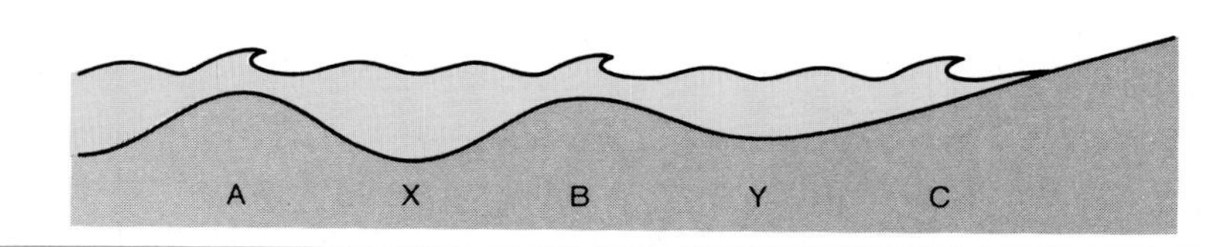

Figure 6.12 The topography of the nearshore environment may cause waves to break over shallow sandbars, then reform and break again.

result is a slowing and steepening of the wave, an increase in wave height, and a decrease in wave length, but the period remains the same (fig. 6.10).

As the wave moves into progressively shallower water, the surface orbits are moving faster than the wave propagation velocity. Eventually the wave steepens beyond the limits of stability and it breaks. The theoretical limits of wave stability are a steepness or height-length ratio of 1:7 and a 120° angle below the water surface (fig. 6.11). Even though a wave may break, it still retains some energy and may reform, then break again. This is a common circumstance along coasts with sandbars parallel to them. The bars are shallow enough to cause the wave to break, but then the trough between two bars is deep enough to permit reformation with the next landward bar also causing breaking (fig. 6.12).

Wave Diffraction, Reflection, and Refraction

Waves have many of the properties of light rays in that they can be diffracted, reflected, and refracted. **Diffraction** is a bending or spreading around objects such that energy is transmitted behind a barrier, much the way that sound travels around a wall from one room to another. As waves pass a barrier, some of their energy is transmitted along the wave, thus producing small waves behind the barrier. This occurs in places with protruding headlands along the coast, such as on the Pacific or New England coasts of the United States, or where man-made structures, such as jetties and breakwaters, protrude into the ocean. The general distribution of energy that results from wave diffraction is in the form of a parabola, with the apex at the feature that creates the diffraction (fig. 6.13). A line extended behind the barrier and perpendicular to the incoming wave will mark the location of waves with 0.5 of the magnitude of the unaffected waves. An arc away from the barrier indicates the locations of completely unaffected waves, and an arc significantly behind the barrier shows the location of waves with 0.1 of the magnitude of the unaffected waves (fig. 6.13).

Waves that meet vertical walls such as seawalls, jetties, or other similar structures are **reflected** with little energy loss (fig. 6.14). Waves that have their crests parallel to the reflecting surface may form **standing waves**, which are those that move up and down and do not progress horizontally. These are the

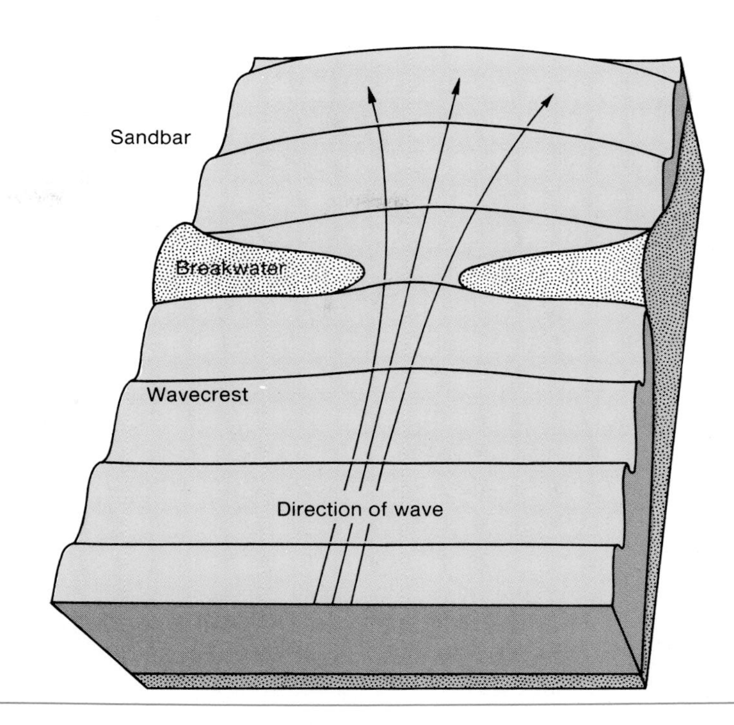

Figure 6.13 Waves diffract or spread their energy as they pass an obstruction such as a seawall, jetty, or sandbar.

Figure 6.14 Waves that strike a structure or the coast have energy reflected from the obstruction. In the case of an impermeable vertical seawall, the reflection is complete, whereas on a beach only a small portion of the energy is reflected.

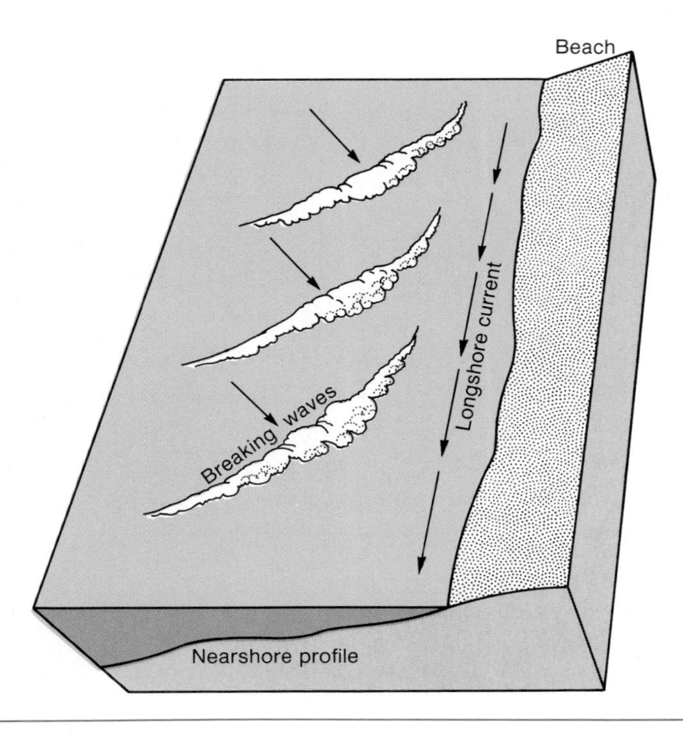

Figure 6.15 As waves enter shallow water, the shallow bottom interferes with their progress and the waves are refracted or bent. This generates longshore currents.

same waves that develop in a coffee cup, a pan of water being carried, or in the bath tub, except that in these vessels the standing wave is confined and does not progress. In a reflecting situation in an open system, the standing wave exists only for an instant as the reflected wave is reinforced by the next incoming wave. Commonly the waves approach the barrier at an angle and are reflected at the same angle according to Snell's Law, which says that the angle of reflection is equal to the angle of incidence. Beaches also present barriers to the progression of waves and reflect wave energy. Most of the wave energy is dissipated over and absorbed by the beach area so that the percentage of energy that is reflected is low. The amount of reflected energy is proportional, for those composed of sand-sized sediment, to the steepness of the beach. Gravel beaches tend to absorb rather than reflect wave energy because their coarse particles allow water to percolate into the beach so none is reflected.

Refraction of waves is the bending of wave fronts due to the effects of shallow water and its influence on the speed of wave propagation. When one part of a wave reaches shallower water before another, the part impeded by the shallow water is slowed relative to the other. When waves approach the coast at an angle, refraction occurs (fig. 6.15) and **longshore currents** are generated. Longshore currents result from the shore-parallel component of the transport of water by the refracted waves. These currents are one of the primary factors in sediment transport in the surf zone and will be considered in detail in the discussion of beaches in chapter 13.

Rarely is the bottom topography near the coast uniformly variable nor the coastline straight; they are usually complicated. As a consequence, wave refraction over such a nearshore topography may form complicated patterns. In order to assess the distribution of wave energy caused by refraction, lines or rays called **orthogonals** are constructed perpendicular to the wave crest at any point. These are lines

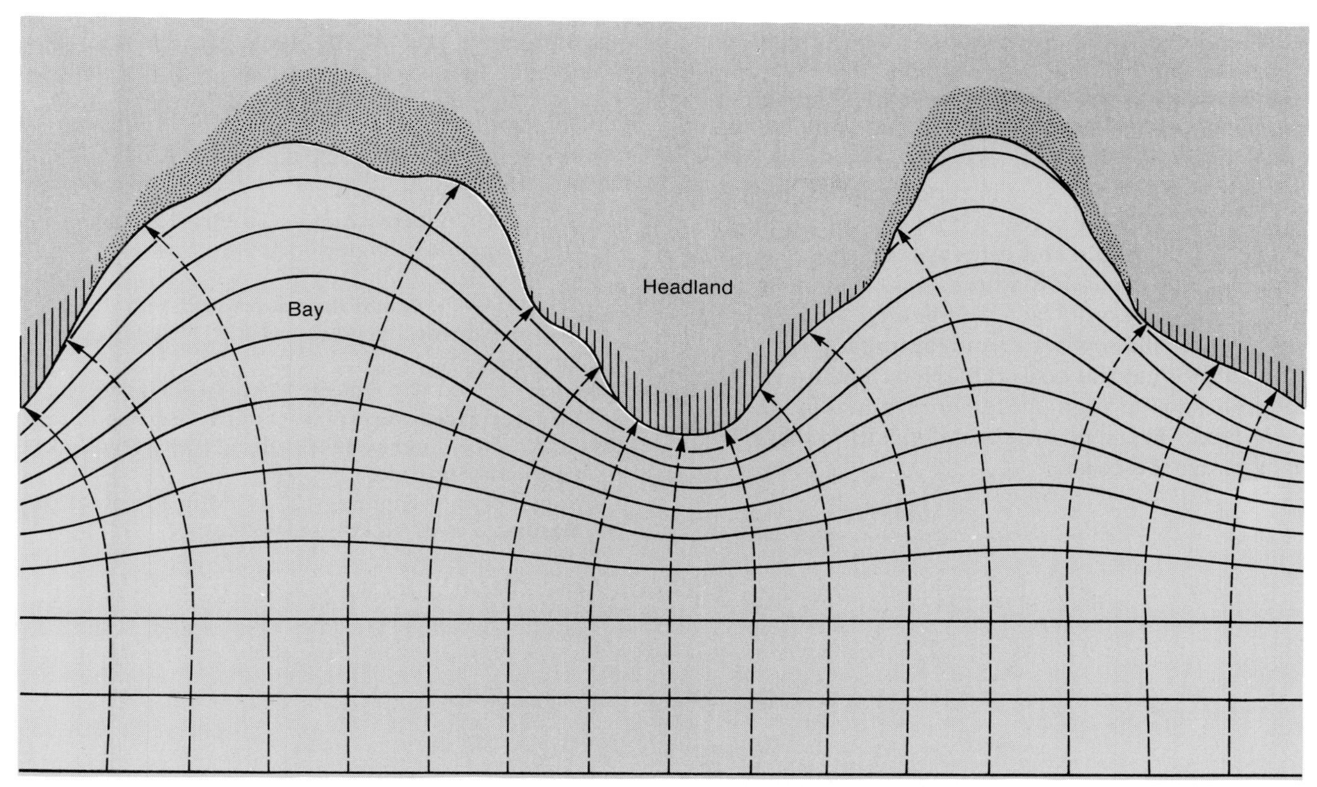

Figure 6.16 The refraction of waves causes energy to be dispersed or concentrated. This is shown by constructing orthogonals or lines perpendicular to the wave crests. Orthogonals are lines of equal energy and demonstrate where energy is concentrated or dispersed.

of equal energy and show where there are areas of energy dispersion and places where energy is focused (fig. 6.16). Such diagrams show that energy is concentrated on headlands and dispersed in the embayments.

Internal Waves

The boundary between water masses may be sharp and persistent, but usually this interface is disturbed and produces **internal waves**. These internal waves may have heights near 100 m and wave lengths of several hundred meters, with periods of at least a few minutes. Internal waves can be generated by the same phenomena that cause tsunamis: earthquakes, submarine volcanic eruptions, and submarine slides.

Because the concept of internal waves is difficult to grasp, a simple example may be helpful. Water and oil are liquids of contrasting densities that do not mix well. If we take a vessel and fill it half with water and the other half with oil, the result will be two distinct layers of liquid with a sharp boundary. A slight disturbance to the container or to the interface between the liquids will cause irregularities to the surface that are analogous to waves. This is a simple version of internal waves.

Summary of Main Points

Waves are disturbances to the interface between two fluids. On the sea surface these are usually generated by wind and are called gravity waves. They are typically from a few to several seconds in period, up to several meters in height, and up to hundreds of meters in length.

The actual motion of water particles in surface waves is orbital with a decrease in orbital diameter to a depth of about one-half of the wave length where orbital motion is absent. Upon entering shallow water, the orbital motion of waves is influenced by the bottom, causing a slowing and steepening of the wave. Eventually the wave steepens to the point that it is no longer stable and it breaks. This occurs near the shoreline in shallow water.

Waves may be diffracted, reflected, or refracted as they enter shallow water and/or impact on coastal structures or natural coastal barriers. The transfer of wave energy through these three phenomena is a major factor in the erosion and deposition that takes place along the coast.

Suggestions for Further Reading

Bascom, W. 1980. **Waves and Beaches,** 2d ed. Garden City, New York: Anchor/Doubleday.

Fox, W.T. 1983. **The Edge of the Sea.** Englewood Cliffs, New Jersey: Prentice-Hall.

Knauss, J.A. 1978. **Introduction to Physical Oceanography.** Englewood Cliffs, New Jersey: Prentice-Hall.

Komar, P.D. 1975. **Beach Processes and Sedimentation.** Englewood Cliffs, New Jersey: Prentice-Hall.

Pickard, G.L., and W.L. Emery. 1982. **Descriptive Physical Oceanography.** Oxford: Pergamon Press.

Ross, D.A. 1979. **Energy from the Waves.** New York: Pergamon-Elmsford.

U.S. Army Corps of Engineers. 1984. **Shore Protection Manual,** 2d ed. Vicksburg, Mississippi.

7
Tides

Chapter Outline

The regular and predictable rise and fall of sea level is called the **tide.** This change in water level is the result of a forced wave caused by the mass attraction of the earth, moon, and sun. Newton's Law of Universal Gravitation states that any two objects are attracted to each other by a force that is related to their masses and the distance between them such that

$$F = G\frac{M_1M_2}{d^2} \qquad (7.1)$$

where F is the gravitational force, G is the gravitational constant, M_1 and M_2 are the respective masses of the objects, and d is the distance between the centers of the objects.

This relationship applies to all objects. In the case of celestial bodies affecting the earth, only the moon, because of its close proximity, and the sun, because of its great mass, have an appreciable effect.

Centripetal force is also an important factor in tides. It results from the acceleration of a mass along a curved path and is directed at 90° inward along this path, in this case the surface of the earth. Because of the difference in the sizes of the moon and the earth, the center of revolution of this paired system is nearer to the earth than it is to the moon. Because of the difference in distance between the moon and one side of the earth as compared to the other, there are unbalanced forces acting on this system. The gravitational attraction exceeds the centripetal force on the side closest to the moon but the centrifugal force exceeds the gravitational on the far side of the earth. These forces produce the bulges on each side of the earth that are called tides (fig. 7.1). Because water is readily deformed, the amount of distortion is significant and readily visible; however, there are also tides in the solid earth. They are small and undetectable without sophisticated instrumentation.

Lunar Cycles and Tidal Species

The moon revolves around the earth in what is essentially a 28-day cycle during which there are various spatial relationships between the sun, moon, and earth that cause variations in the tides at any given location. During new moon and full moon conditions, the sun, earth, and moon are aligned, thus providing maximum distortion of the water envelope and therefore maximum tides, which are called **spring tides** (fig. 7.2). First quarter and third quarter conditions have the sun and moon at right angles to the earth, thus working in opposition to each other

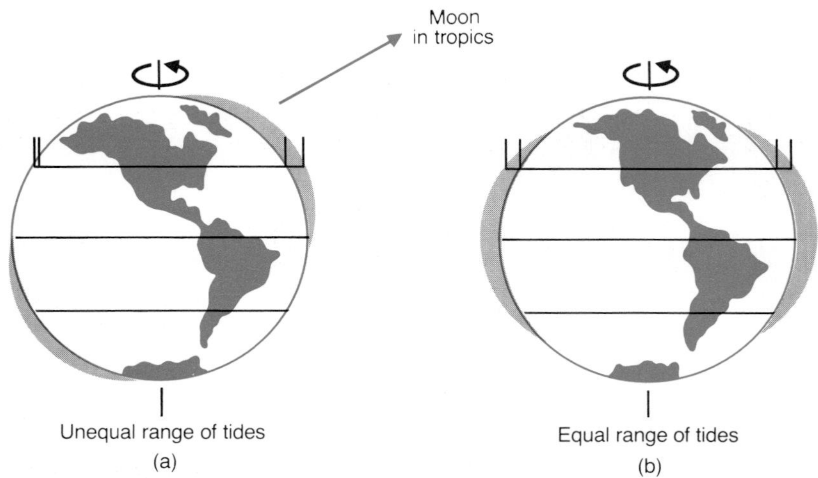

Figure 7.1 Bulges of the water envelope over the earth due to centrifugal force.

and producing minimal tidal ranges called **neap tides** (fig. 7.2). Each of these circumstances occurs twice per lunar cycle or once every two weeks.

A lunar day is 24 hours and 50 minutes in duration, which means that the moon passes a given point on the earth once in that period of time. Because of the bulges in the water surface, there would be expected to be two highs and two lows at any given place on the earth for each lunar day. In effect, there are two wave lengths for the circumference of the earth.

Some other phenomena must also be considered. The earth is tilted 23° on its axis and the moon's position relative to the equator ranges through 28.5° to the north and to the south. Because the tidal waves respond to the movement of the moon, there is great variation in tidal range at a given location as the moon changes position relative to the earth. It might be expected that the greatest tidal range would occur when the moon is directly in line with a given location, either above or below.

We must remember that the earth is rotating at a great speed, which causes the bulges to be carried forward. If there was no friction between the envelope of water and the earth, then the bulges would be directly under the moon. In fact, however, the

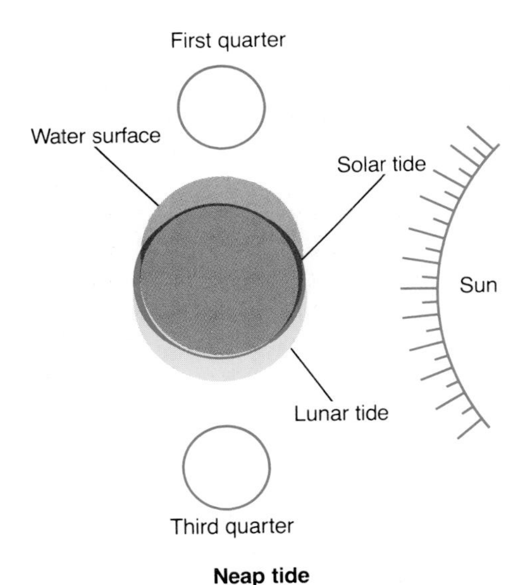

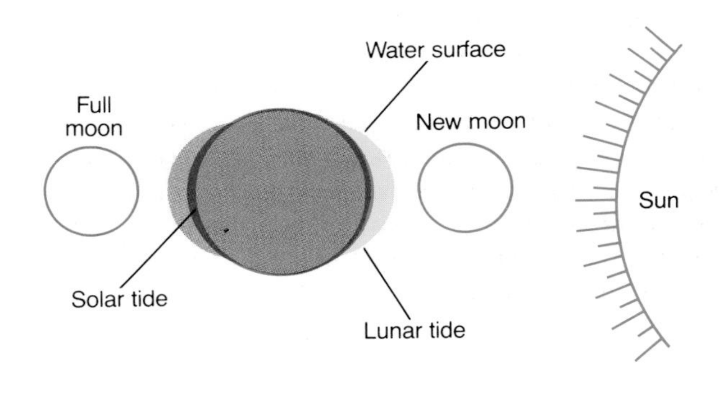

Figure 7.2 Tidal bulges that develop on the water envelope of the earth during different lunar stages. Note that spring tide conditions develop when the sun, moon, and earth are aligned, whereas neap tide conditions develop when the sun, moon, and earth are at right angles to one another.

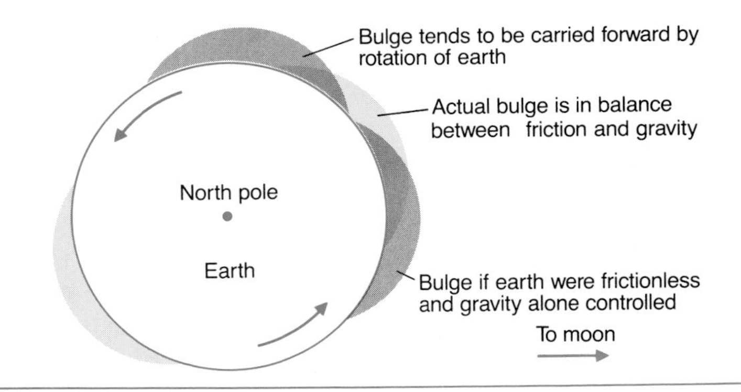

Figure 7.3 Bulges of the water envelope on the earth showing the effect of moon and friction.

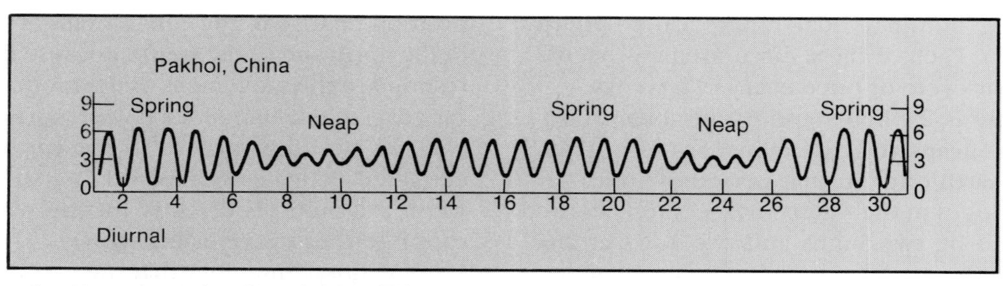

Figure 7.4 General pattern shown by diurnal tides. This example from China has a large variation between spring and neap conditions.

TABLE 7.1
Selected Principal Tidal Harmonic Components

Name of Component	Symbol	Period (hrs)	Relative Influence M_2 = 100
Semidiurnal			
Principal lunar	M_2	12.42	100
Principal solar	S_2	12.00	46.6
Larger lunar elliptic	N_2	12.66	19.2
Lunisolar semidiurnal	K_2	11.97	12.7
Diurnal			
Lunisolar diurnal	K_1	23.93	58.4
Principal lunar diurnal	O_1	25.82	41.5
Principal solar diurnal	P_1	24.07	19.4
Long-period components			
Lunar fortnightly (spring and neap tides)	M_f	327.86	17.2

tidal bulges are located between these extremes due to the balance between the attraction by the moon and the friction causing the water to rotate with the earth (fig. 7.3).

Tidal waves can be broken down into their individual **tidal constituents** or species. These represent the various influences the sun and moon have on the earth at different positions within the various astronomical cycles. Although there are more than fifty of these constituents, only four account for just over two-thirds of the tidal variation. There are three primary families of periods for these constituents: (1) those that are semidiurnal, or occur twice daily, (2) those that are diurnal, or occur daily, and (3) those of periods that are in excess of one day (table 7.1). The primary lunar component, M_2, has a period of 12 hrs 25 min (12.42 hrs). It is just over twice the amplitude of the main solar constituent, S_2, which has a period of exactly twelve hours. These are the two most important components affecting the spring and neap tidal conditions. When they are in phase, it is a time of spring tides and when they are out of phase, neap tides result. The N component is from the elliptical path of the moon and the K_2 component is the result of the changes in declination of the sun and the moon in their orbital cycles. Both of these have periods close to the M_2 and S_2 components respectively (table 7.1).

The diurnal tidal components all have periods very close to twenty-four hours (table 7.1). Of the three components, the K_1, which is due to the earth's declination, is typically the largest. The lunar component, O, reflects the greater tidal range on the side of the earth nearest to the moon. The solar component, P, produces the same effect for the sun.

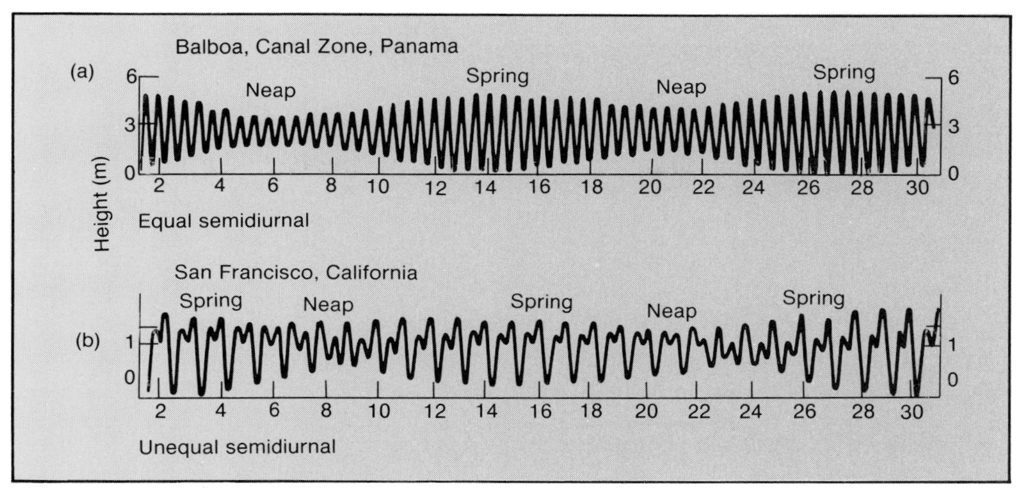

Figure 7.5 Pattern of semidiurnal tides showing an example from Panama, where there is equality, and the more common inequality pattern, with an example from San Francisco, California.

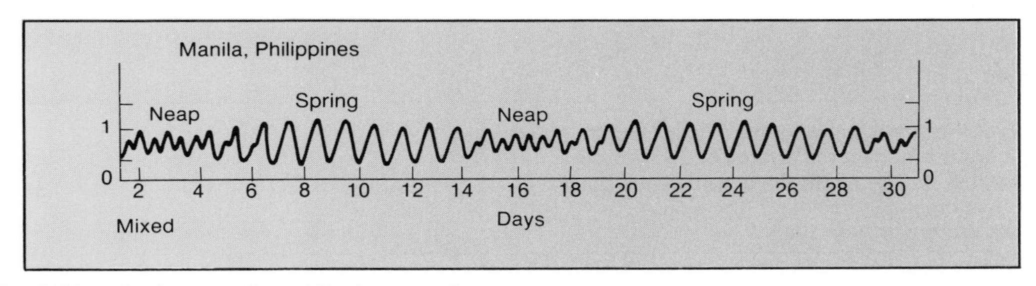

Figure 7.6 Mixed tides display a portion of the lunar cycle when diurnal tides occur and a portion when semidiurnal tides occur.

At any given geographic location, tides may be dominantly diurnal, semidiurnal, or a mixture of the two. The pattern exhibited at any location is the result of the combination of all of the tidal constituents. A diurnal tidal pattern is produced when the diurnal components are amplified and the semidiurnal components show little amplitude. The example shown is from Pakhoi, China, where not only are the tides diurnal but there is great difference between neap and spring ranges (fig. 7.4). By contrast, semidiurnal tidal patterns from Balboa in the Panama Canal Zone show great influence of semidiurnal components with little amplitude from the diurnal components (fig. 7.5a). In this example, the difference between spring and neap tide is less than the previous example. Unequal semidiurnal tidal cycles are also present at some locations as shown in the San Francisco example (fig. 7.5b). The semidiurnal and diurnal components interact to produce mixed tides that have periods of diurnal and semidiurnal tides during each lunar cycle as in the Manila example (fig. 7.6). Actually, the diurnal tides are simply the result of minimal influence of the semidiurnal components. The tidal record shows great inequality in the semidiurnal phase. By increasing this inequality, diurnal tides are produced.

Figure 7.7 Map of the world showing the cotidal lines for each of the numerous amphidromic systems. Each of the Roman numerals that radiates from the amphidromic point represents the tidal crest at that hour in the tidal cycle, from I through XII. (After von Arx, 1962.)

Tidal Circulation and Amphidromic Systems

Landmasses complicate the progression of tidal waves as the earth rotates. Ocean bottom topography can also have an impact. The consequence is that the world ocean is subdivided into basins displaying a tidal wave that progresses in a circular pattern and that is impacted by the Coriolis effect. By synoptically mapping the high tide at any given place in a basin at regular intervals throughout the tidal cycle, it is possible to show the complete pattern of the crest of the tidal wave. The most striking aspect of the pattern is that the crests radiate from a central place like spokes on a wheel (fig. 7.7). Each line represents a tidal crest at a given time and is called a **cotidal line**. There is a point from which the cotidal lines emanate called the **amphidromic point**. This system is the result of the combined effects of the progressing tidal wave, the earth's rotation, and the horizontal currents in the ocean. An amphidromic circulation pattern occurs not only in the major ocean basins but also in the North Sea, the Mediterranean, and the Great Lakes (fig. 7.8). Coastal embayments or estuaries have amphidromic motion that diagrammatically shows the water level tilting about the amphidromic point as the tide floods and ebbs (fig. 7.9).

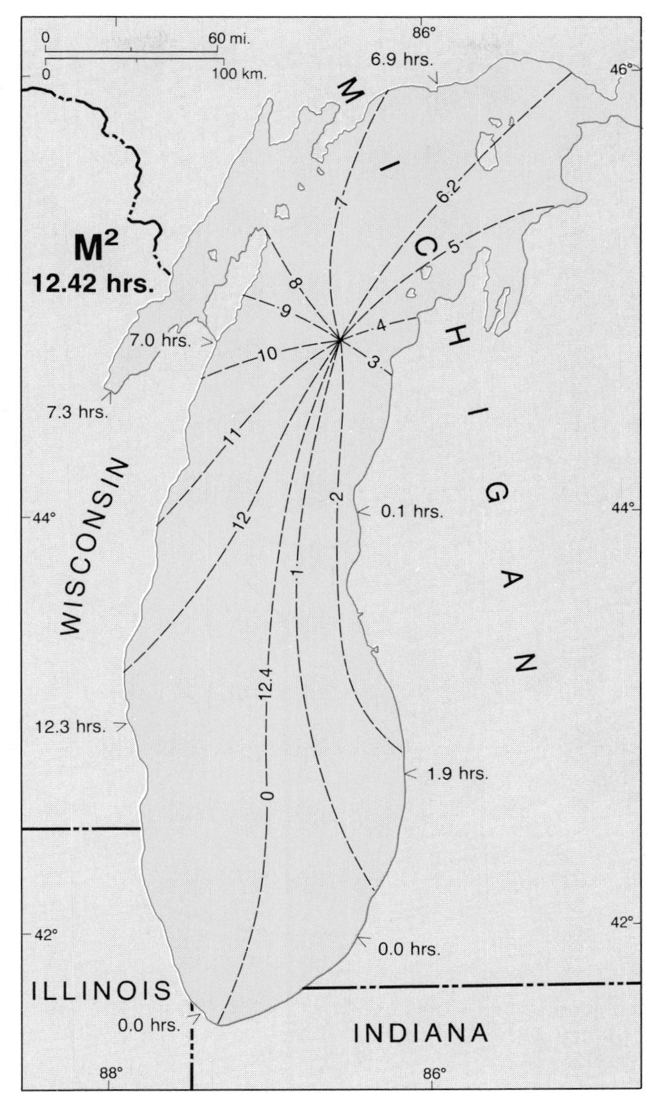

Figure 7.8 Amphidromic system in Lake Michigan.

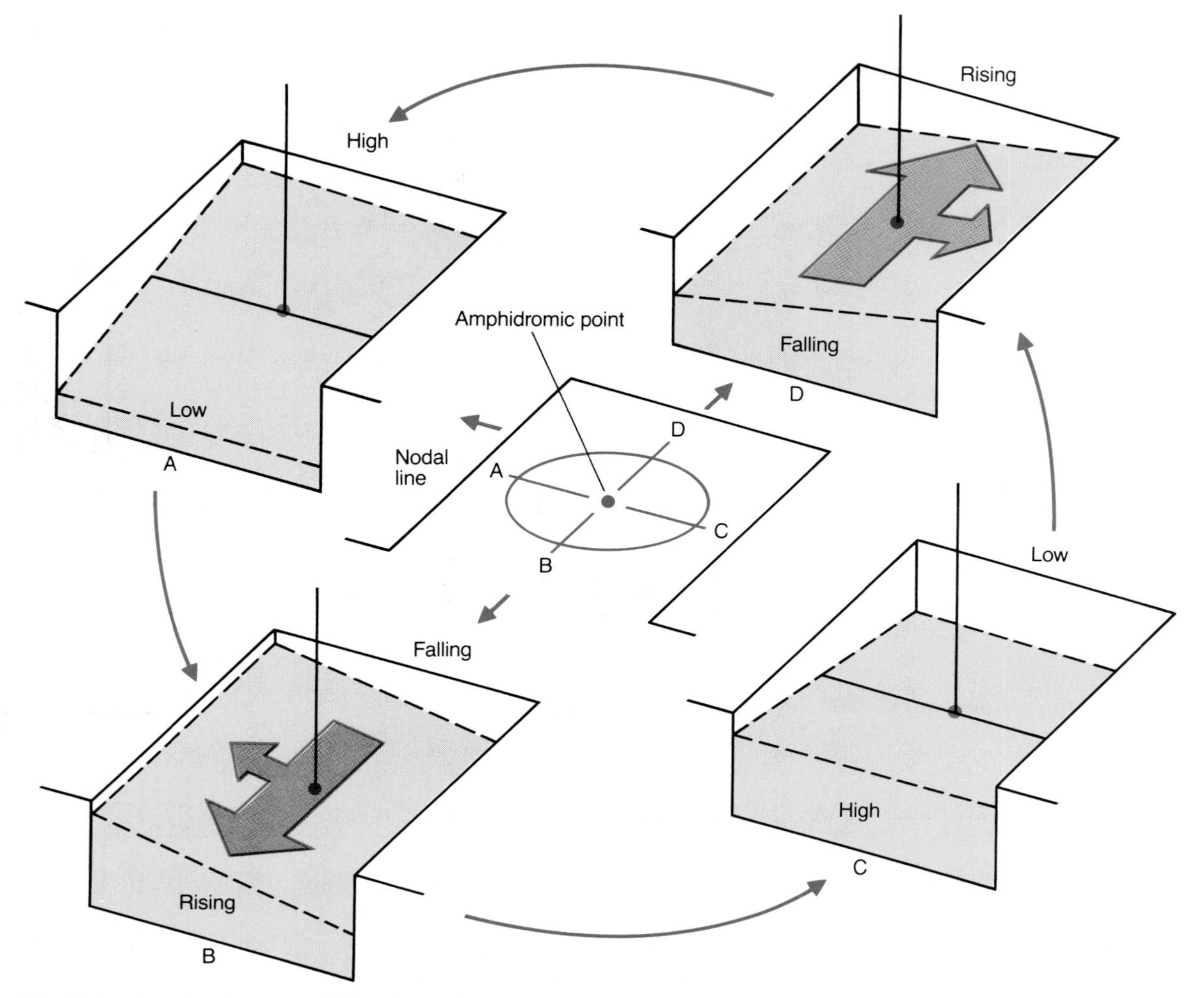

Figure 7.9 Estuaries also have amphidromic motion as shown in this diagram. The water surface tilts around the amphidromic point as the tidal cycle progresses.

Tidal Range and Prism

The tidal record for any given location is much like the fingerprints of a person—no two are the same. Each coastal location has its own unique coastal configuration, bottom topography, and geographic position. Obviously, adjacent and geographically juxtaposed sites show similar tidal records in terms of range, patterns, and time for highs and lows. It does, however, take time for a tidal crest to progress so the time of high or low tide differs as the tidal wave moves along the coast. Likewise, those areas behind a barrier or in the upper reaches of an estuary experience a given tidal stage much later than the adjacent open water.

Because tidal range shows great variation, coasts of the world are classified according to this phenomenon. Microtidal coasts have ranges of <2 m, mesotidal coasts 2–4 m, and macrotidal coasts are those

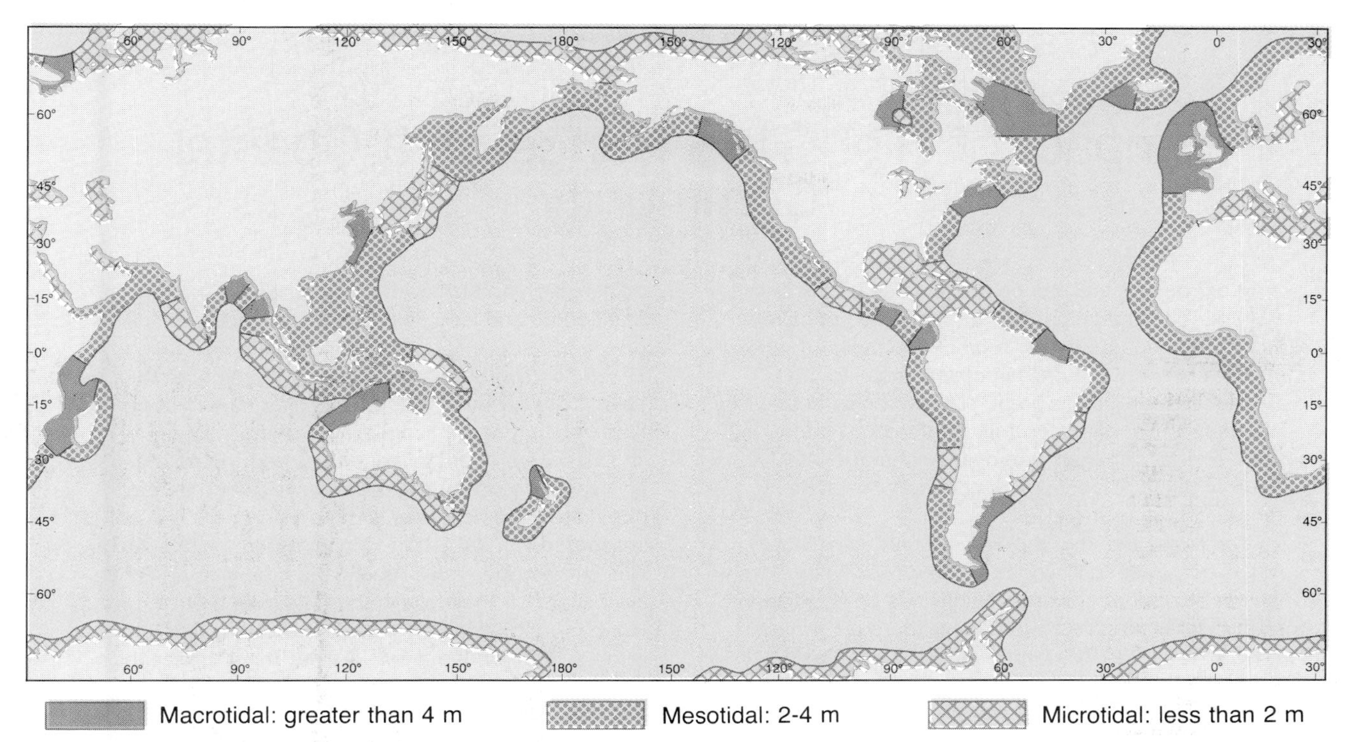

Figure 7.10 The distribution of tidal range around the world. There is great variety; however, mesotidal coasts are the most common situation. (After Davies.)

with ranges of >4 m (fig. 7.10). Coasts that are somewhat funnel-shaped or those where water tends to pile up due to the forced tidal wave entering the embayment have high tidal ranges. Two of the highest are the Bay of Fundy, where the Minas Basin (fig. B7.1a) experiences spring tides of over 16 m, and the Bay of St. Malo on the Brittany coast of France (fig. B7.1b), where the range is slightly less. In the United States, much of New England, the Georgia Bight, and the coasts of Oregon and Washington as well as much of the southern coast of Alaska have at least mesotidal ranges.

When the tide rises and falls, it does so because there is water being transported to and from an area. This is not as apparent on the open beach as it is in an inlet or the mouth of an estuary. Much water passes back and forth with each flood and ebb of the tide. The water flux or budget is called the **tidal prism,** which is the volume of water that moves in and out of an embayment or any defined area during each tidal cycle. There is a natural tendency to relate the tidal prism to the tidal range, but this is not necessarily a direct relationship, because the area of the embayment, lagoon, or other defined area is equally important. Tidal prism is the area of the embayment multiplied by the tidal range. If a large embayment has a tidal range of only a meter, the prism is greater than an embayment one-fourth the size but with twice the tidal range. This is an important aspect of tides, especially in barrier islands where inlets connect lagoons and estuaries with the open ocean.

BOX 7.1

Impact of High Tidal Range on a Coastal Community

Most people that live on a marine coast are aware of tidal fluctuations but the tides do not have a significant impact on their recreation, transportation, or work. This is not the case in coastal communities where tidal ranges are great, such as those in the Bay of Fundy in Canada, along the Brittany coast of France, or at Turnigan Arm near Anchorage, Alaska. Each of these places experience tides that have a range of 10 m or more.

The Minas Basin of the Bay of Fundy (fig. B7.1a) has the highest tidal ranges in the world: 16 m. Here it is imperative that any and all activities on or adjacent to the water be scheduled around the tides. At low tide there is little water in the entire Minas Basin with only a few natural tidal channels still submerged. As a result, any type of boating activity must take place during or near high tide conditions. This includes recreation, commercial fishing, and commercial transportation. A good example is at the boat dock near the Avon River where gypsum from nearby mines is loaded on freighters for transport. At low tide the dock is high and dry with no water at all. The ship must come and load its cargo quickly and depart before the tide stage is too low for safe navigation. These conditions dictate that a small freighter must be used, both because of the water depths and in order that loading can be completed in a short time.

Because of the tidal conditions and the restrictions on boating, much of the commercial fishing in tide-dominated coastal areas is done using nets that are set to trap fish carried by tidal currents. This technique was initiated by native Americans centuries ago. Fish are caught by their gills in the nets during ebbing tides, then the fishermen go onto the tidal flats to retrieve them during low tide. They must collect their fish and make any adjustments to the nets before the next flood tide covers the area.

These are a few of the types of activity that must be coordinated around the rise and fall of the tide. In such areas, the tides—not daylight—dictate the work schedule.

Meteorological Tides

Barometric pressure variations and winds can also cause changes in water level. These changes are neither cyclic nor predictable but they are important. The most pronounced elevated water levels are those associated with hurricanes or other severe storms. The strong winds blowing for several days cause water to pile up on the coast and may result in storm tides of a few meters height. This situation is one of the biggest causes of damage and destruction resulting from coastal storms. There can also be negative tides from strong, persistent offshore winds, but these rarely cause problems except for inhibiting navigation due to the unusually shallow water.

Tidal Currents

The movement of masses of water from one location to another and back again, such as occurs during the flooding and ebbing of the tides, causes currents. These currents may be strong, with the power to erode and transport sediment long distances, or they may be slow and have no noticeable effect on the bottom. As a general rule, tidal currents are negligible out beyond the continental shelf and reach up to 15 cm/sec or so across the shelf. In inlets and some estuaries with high tidal ranges, tidal currents commonly exceed 1 m/sec. Along the open coast, tidal currents are slow and are masked by waves so that they are unnoticed. The currents are rapid any place

BOX 7.1

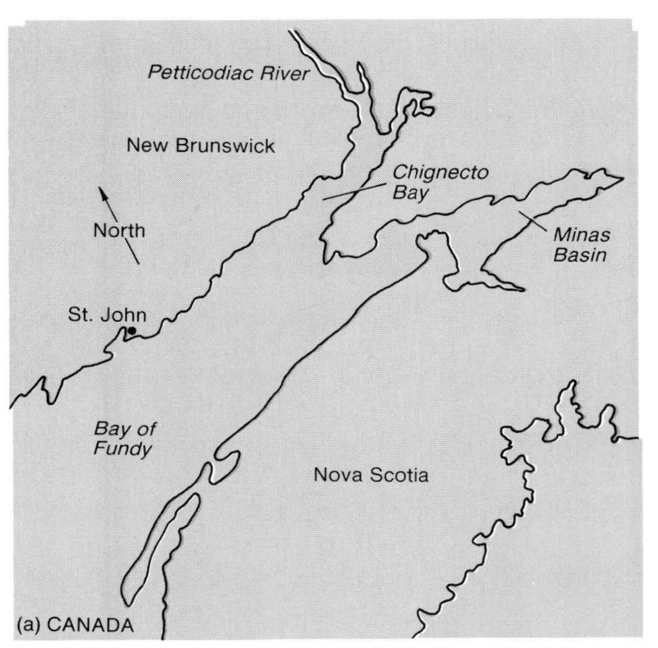

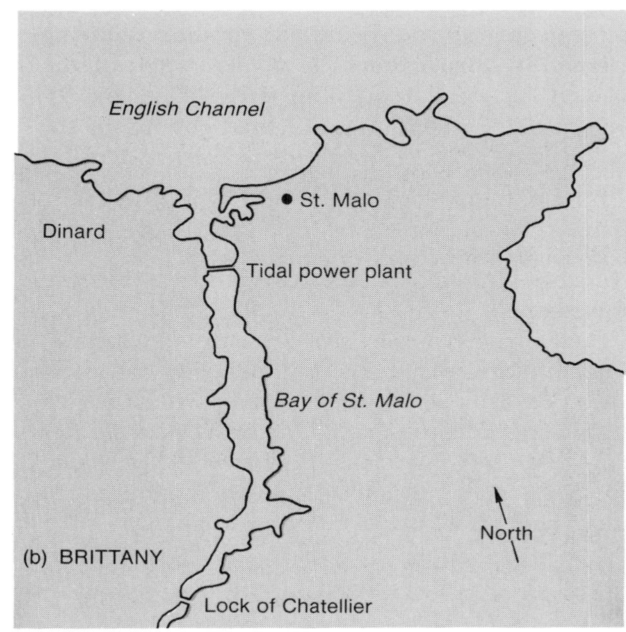

Figure B7.1 Schematic maps of (a) the Bay of Fundy in Canada and (b) the Bay of St. Malo on the north coast of France. These areas are among the highest tidal ranges in the world.

where constrictions are present, even if the tidal range is low, because the entire tidal prism must pass through the constriction in a six-, or at most, a twelve-hour period. The great power of these tidal currents is a source of energy that is being utilized in a few areas. It offers much potential for energy production in the future.

Many places along the coast where tidal currents are prominent, such as in estuaries or river deltas, are also sites of freshwater discharge. The currents produced by the discharge affect the tides by retarding flood tidal currents and by reinforcing ebbing currents. In some streams tidal influence may extend hundreds of kilometers upstream from the mouth. Albany, New York, for example, experiences tides on the Hudson River, even though it is about 150 km from the open ocean. Probably the most extreme example of this phenomenon is the Amazon River, where tides are present nearly 500 km upstream from the mouth.

In some locations, the flooding tide is accompanied by a wave called a **tidal bore**. The specific cause of this progressive wall of water is not known but it is related to the interaction of the discharge of the stream combined with the flooding tide. The shallowing and funnelling of the water as the wave proceeds are also contributing factors. These bores are typically a few tens of centimeters in height but some are much larger. The Fuchun Kiang River in China has a bore of about 7 m.

Summary of Main Points

All objects are attracted to one another by the relationship expressed in Newton's Law of Universal Gravitation. The water envelope around the earth reacts to such attraction with the sun and moon to produce tides. The astronomic cycles of the moon, sun, and earth produce cyclic variations in the tides that result in neap and spring tide conditions during the monthly lunar cycle. The configuration of the coast and the ocean floor, along with the general location, provide a unique set of tidal conditions for each geographical location on the earth. Some locations experience diurnal tides, some, semi-diurnal tides, and others experience mixed tides. Generally the range of tides is low on island coasts and high on broadly embayed coasts, where tidal waves are funnel-shaped.

Tidal currents may be strong locally, such as embayments where tidal range is high or in tidal inlets where the tidal prism is large. Tidal currents are essentially imperceptible along the beaches or similar open coast environments. A few rivers experience tidal effects great distances from their mouths. In some of these tidal bores may develop.

Suggestions for Further Reading

Bascom, W. 1980. **Waves and Beaches**, 2d ed. Garden City, New York: Anchor-Doubleday.

Defant, A. 1958. **Ebb and Flow: The Tides of the Earth.** Ann Arbor: University of Michigan Press.

Fox, W.T. 1983. **The Edge of the Sea.** Englewood Cliffs, New Jersey: Prentice-Hall.

Knauss, J.A. 1978. **Introduction to Physical Oceanography.** Englewood Cliffs, New Jersey: Prentice-Hall.

Picard, G.L., and W.L. Emery, 1982. **Descriptive Physical Oceanography.** Oxford: Pergamon Press.

Redfield, A.C. 1980. **Introduction to Tides.** Woods Hole, Massachusetts: Marine Science International.

8
The Marine Biological Environment

Chapter Outline

The marine environment supports an abundant and diverse suite of organisms. They occupy nearly every nook and cranny of the world ocean. For such extensive and varied life to exist, it is necessary for the proper conditions to exist. This requires an appropriate chemical environment, proper nutrients and food supply, and the right physical conditions.

The following discussion will consider the conditions that are necessary to support life in the marine environment and also the broad marine environments that are present in the world ocean. This sets the stage for a consideration of the organisms themselves, presented in the following chapter.

Biogeochemical Cycles in the Ocean

Although phosphorus and nitrogen are not among the most abundant elements in the oceans, they are certainly among the most important because of their role in the organic cycle. They are essential nutrient elements for living organisms. These elements are present in various states and concentrations. This variety is due to the uptake by living organisms and release by decaying tissue, and also to circulation within the oceans. Even though there is not a constant relationship between the nutrient elements and the major elements dissolved in seawater, there is a constant relationship between nitrogen and phosphorus. Nitrogen is seven times more abundant than phosphorus by weight. Along with carbon, these elements are critical for production of organic material; however, carbon is abundant relative to phosphorus and nitrogen. The other two elements may be limiting factors in the organic cycle of the oceans.

Silicon is also an important trace element in the oceans. It is necessary for the skeletal material (SiO_2) of important organisms such as the diatoms and radiolarians.

Organic Cycle

The oceans have an organic cycle that is similar to that of any environment, aquatic or terrestrial. Carbon, the basic constituent of all living compounds, and nutrient elements, such as nitrogen and phosphorus, are combined during photosynthesis to produce plant tissue, which serves as food for animals. Both the plants and animals produce organic matter when they expire. This material decays to produce the raw materials necessary to start the cycle all over again (fig. 8.1). A more detailed discussion of the elements of the organic cycle is presented later in this chapter under *Productivity*.

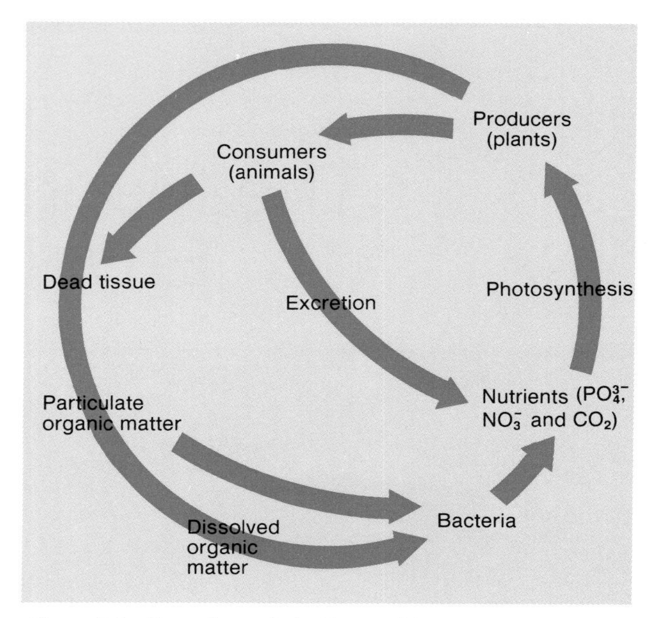

Figure 8.1 Organic cycle in the world ocean.

Phosphorus Cycle

Phosphorus occurs in a variety of states in the marine environment. Some is in organic compounds, such as proteins and sugars, some is in inorganic calcium and iron phosphate particles, and some is dissolved as inorganic phosphate. The last form is the most abundant and may reach 90% of the total phosphorus during times when organic production and, therefore, uptake are low. This typically occurs in the winter. In summer, when productivity is high, the opposite situation exists and inorganic phosphate is less than 50% of the total.

Orthophosphates are the inorganic phosphate compounds that are most abundant in the phosphorus cycle. They are produced by bacterial breakdown of organic phosphates from decaying tissues (fig. 8.2). Because this is a relatively simple and rapid process, it takes place relatively high in the water column and thereby provides phosphorus for plant uptake. Even though phosphorus occurs in concentrations much below nitrogen, the fact that it is readily made available within the zone of light penetration generally prevents phosphorus from being a limiting factor in marine productivity.

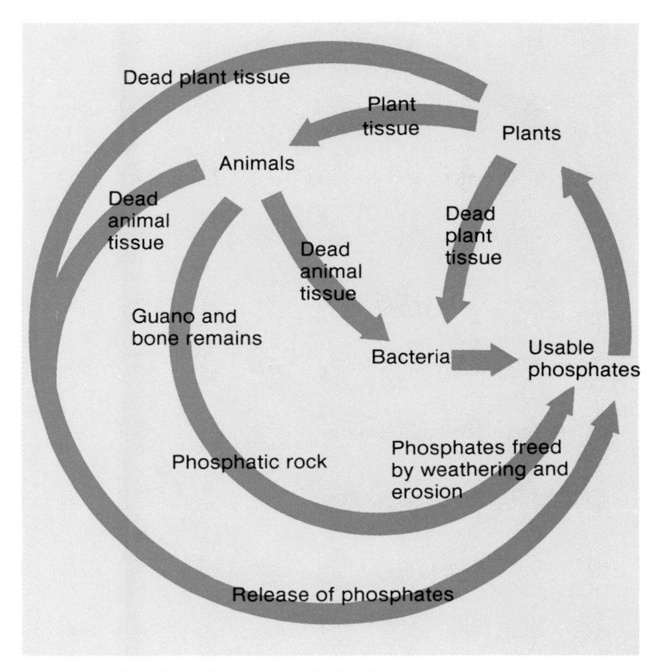

Figure 8.2 Phosphorus cycle in the oceans.

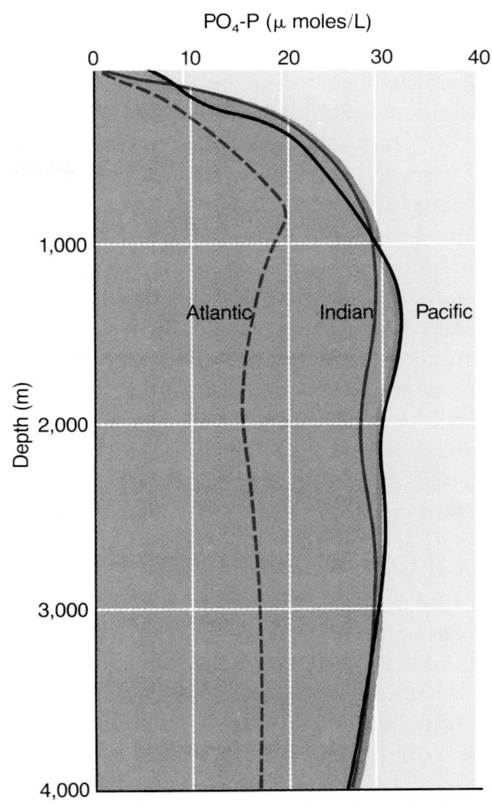

Figure 8.3 Vertical distribution of phosphate in the water column. Concentrations in the Indian and Pacific Oceans are similar, but are much higher than in the Atlantic. (From Sverdrup/Johnson/Fleming, *The Oceans,* © 1942. © renewed 1970, pp. 18, 19, 241, 242, 245. Reprinted by permission of Prentice-Hall, Englewood Cliffs, New Jersey.)

Within the phosphorus cycle there are numerous interactions between plants and animals, inorganic and organic compounds, and between the water column and the surface and substrate. For example, some animals release considerable amounts of soluble phosphorus in their feces. The phosphorus is then dissolved in the water column and is thereby made available for uptake by plants. Some inorganic phosphate compounds are precipitated as minerals on the sea floor (fig. 8.2).

The mean concentration of phosphorus in the oceans is 70 micrograms per liter (0.07 ppm). There is a wide variation from this, however, especially near the surface where values are low (fig. 8.3). The distribution of phosphorus concentrations throughout the water column is predictable. It is similar for all of the oceans except that concentrations in the Atlantic Ocean are much lower than in the Indian and Pacific Oceans, which are very similar.

It is possible to break down the vertical distribution of phosphorus into distinct layers. The surface layer has the minimum concentration due to high uptake as the result of great organic productivity. The second layer extends to depths of a few hundred meters and shows a great increase in concentration, which results from a decrease in uptake due to organic activity and because this is the zone where release by organic decay begins. Below that is the layer of highest concentration which results from decay of dead organisms as they descend through the water column, generally about 500-1,000 m. There is some decrease in abundance to the thick bottom layer.

There are also temporal changes in the distribution of phosphorus. The shortest of these are **diurnal** in nature and are most noticeable in shallow coastal waters. They are linked directly to sunlight and plant photosynthesis. During mid-day, when plant activity is greatest, phosphorus concentrations are minimal because the phosphorus is being taken up by plants. It reaches the maximum diurnal concentration just before sunrise, after a long period of no photosynthesis.

Seasonal variation in phosphorus abundance can be pronounced. It is tied to latitudinal belts because of climatic changes. In the mid-latitudes, where there is maximum seasonal change, there is also maximum change in the concentration of phosphorus. In winter, when day length is short and plant productivity is low, the concentration is greatest, whereas in the summer much of the phosphorus is consumed in the production of organic tissues.

Upwelling (see fig. 5.15) is also a source of both spatial and temporal variation in phosphorus. The phenomenon is best developed along the edge of the continental margin in the mid-latitude of the eastern side of the ocean basins. Deep, cold, nutrient-laden water rises and provides large, localized increases in phosphorus that commonly lead to blooms of plant productivity.

Nitrogen Cycle

Many of the generalities for phosphorus are also true for nitrogen. There are, however, also many differences between these critical nutrient elements. Nitrogen occurs in a variety of states in the ocean, the most abundant being molecular nitrogen (N_2) which is many times more abundant than either nitrite (NO_2) or nitrate (NO_3), but is not in a form that is useful to organisms. Ammonia (NH_4) is also present in the oceans.

Nitrogen serves a critical role in the organic cycle in production of the amino acids from which proteins are made. In the nitrogen cycle plants take up inorganic nitrogen in one of the combined forms or as molecular nitrogen (fig. 8.4). These plants make protein that is consumed by animals and converted to animal protein. Dead organic tissue is broken down by various bacteria. Included are nitrogen-fixing bacteria that fix molecular nitrogen into various combined forms (NO_2, NO_3, NH_4) and denitrifying bacteria that do the opposite (fig. 8.4). Nitrogen escapes to the atmosphere and also is taken from the atmosphere during this cycle. There is not a significant amount of nitrogen that is incorporated into minerals through precipitation on the ocean floor.

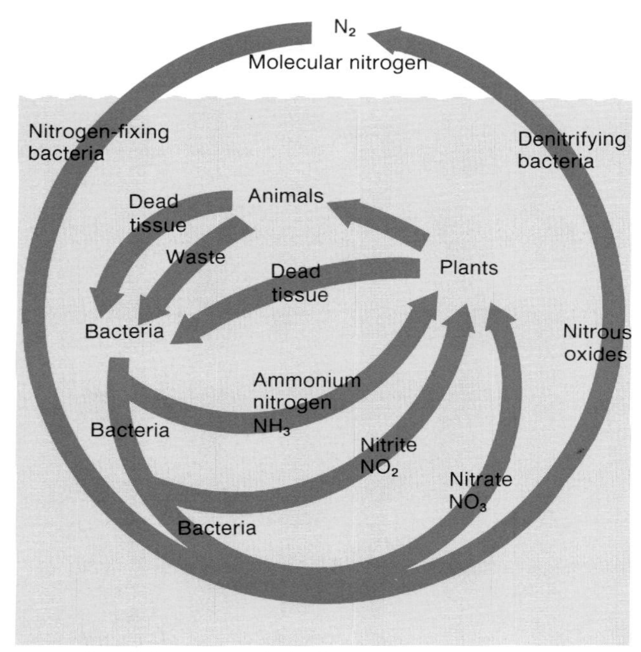

Figure 8.4 Nitrogen cycle in the oceans.

The distribution of nitrogen through the water column in the ocean is much like that of phosphorus. The overall concentration is different, but the shape of the curve and the distribution among the three oceans is similar (fig. 8.5).

Variations occur in the vertical distribution of the different species of nitrogen in the oceans. Nitrate (NO_3) is most abundant in the upper waters, ammonia (NH_4) is nearly uniformly distributed, and nitrite (NO_2) is concentrated near the thermocline. The interactions between various states of organic nitrogen and bacteria are such that by the time the nitrogen has been converted to inorganic varieties the materials have descended below the thermocline. This causes some problems in the availability of nitrogen because the thermocline is a barrier to nitrogen migration and in fact, nitrogen availability may be a limiting factor in productivity in the oceans.

During spring, when the seasonal thermocline disappears temporarily, there is upwelling and overturn, and nitrogen is made available for use in the organic cycle in the photic zone. The spatial and temporal variations in nitrogen distribution generally parallel that of phosphorus.

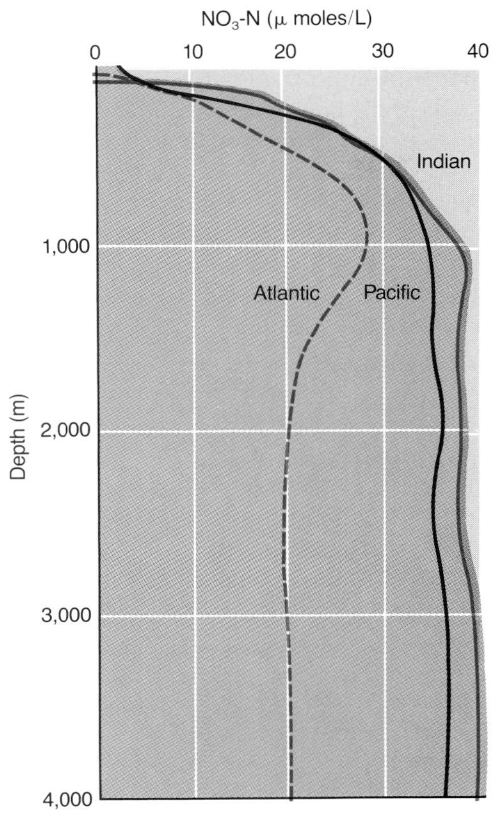

Figure 8.5 Vertical distribution of nitrogen in the water column. The concentrations closely follow those of phosphate, with the Atlantic Ocean low, relative to the Indian and Pacific Oceans. (From Sverdrup/Johnson/Fleming, *The Oceans,* © 1942. © renewed 1970, pp. 18, 19, 241, 242, 245. Reprinted by permission of Prentice-Hall, Englewood Cliffs, New Jersey.)

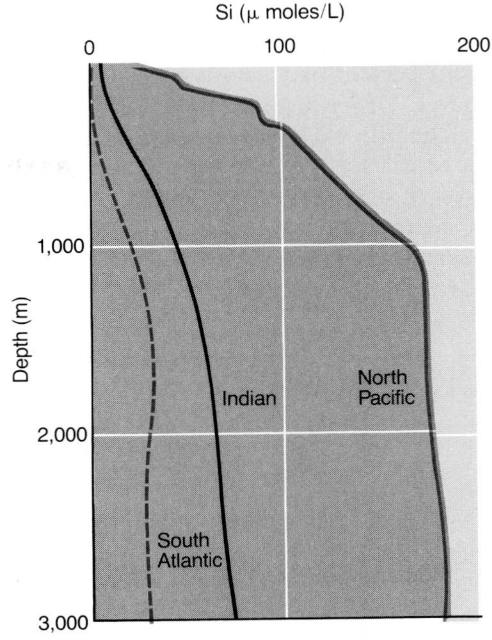

Figure 8.6 Vertical distribution of silicon in the water column. Although silicon is not a nutrient element, it is important due to its incorporation in diatom tests; therefore it tends to parallel the trends for nitrogen and phosphorus, except in the Indian Ocean. (From Sverdrup/Johnson/Fleming, *The Oceans,* © 1942, © renewed 1970, pp. 18, 19, 241, 242, 245. Reprinted by permission of Prentice-Hall, Englewood Cliffs, New Jersey.)

Silicon Cycle

Although silicon is not a nutrient element, it is necessary for the skeletal material in **diatoms,** one of the most important producer groups in the oceans. It is also necessary for **radiolaria,** which are small, floating, one-celled animals. Because of its role in diatom production and the importance of the diatoms, silicon follows a cycle much like that of the nutrient elements. It is derived from the weathering of rocks of the earth's crust and occurs in various states in the ocean, including SiO_2 and in clay minerals.

The concentration of silicon ranges widely, from near zero at the surface to as much as 8.4 ppm. The Pacific Ocean has the highest concentration and the Atlantic the lowest (fig. 8.6). This is partly due to produced by it. Overall, the concentration of silicon the volcanic activity in the Pacific and the silicon in the oceans is well below saturation levels. This low abundance is probably related to the incorporation of silicon in various minerals that are precipitated on the seafloor and its incorporation into diatom and radiolarian tests.

Productivity

The previous section of this chapter considered the nutrient elements, their distribution, and cycles in the marine environment. The next logical step is to incorporate these nutrient elements into living organic tissue through **photosynthesis**. This process occurs both in the atmosphere and the aquatic environment; in both fresh water and marine water. In this process, sunlight combines with essential components from seawater to produce living plant tissue:

$$CO_2 + H_2O + \text{minerals} + \text{sunlight} \rightarrow \text{organic matter} + O_2 + \text{heat} \quad (8.1)$$

This reaction takes place in all photosynthetic organisms and forms the basis for all life in the oceans except for certain bacteria. It provides the primary food source, either directly or indirectly, for all consumers. The process is called **primary production**. The above reaction is reversed during **respiration** as oxygen combines with organic matter to produce the raw materials for photosynthesis. These two processes serve as the fundamental components of the organic cycle (fig. 8.1).

The photosynthetic process in the oceans leads to production of plant tissues that serve as the fundamental building blocks of life in the sea. These plant tissues are consumed by herbivorous animals that are in turn consumed by various carnivorous animals. At each level there is excretion and death with accompanying decay. These processes release nutrient materials and carbon dioxide, which are then incorporated, along with sunlight, into the photosynthetic process.

BOX 8.1

Measuring Productivity

Many types of biological research try to determine the rate at which organic tissue is being produced, that is, productivity. This can be done in a variety of ways with different levels of sophistication and accuracy. The aquatic scientist has an advantage in productivity determinations because there are many techniques that can be used in water that cannot be used in terrestrial environments.

Productivity can be determined directly, by measuring the amount of organic tissue formed, or indirectly, by measuring one of the components in the photosynthetic process and using it as an index of productivity. We will look at examples of each method.

The harvest method is probably the most common direct technique for measuring productivity. A small area, such as a square meter, is designated for study and all plants are collected, weighed, and replaced. The growth or increase in mass of a population of bottom-dwelling organisms is determined. This procedure is repeated at predetermined intervals to measure the rate at which the plants are growing. This method has serious disadvantages, including the likely disturbance of growth patterns of the individual plants and the fact that it is restricted to shallow bottom-dwelling plants. The harvest method does provide a good estimate of growth rates, however.

More commonly, productivity is determined indirectly by measuring changes in one of the major components in the photosynthetic reaction, such as oxygen production, carbon dioxide assimilation, or nutrient uptake. The most widely used of these is oxygen production, which is directly related to the production of organic tissue (equation 8.1). Because photosynthesis, and therefore oxygen production, varies diurnally, it is necessary to measure oxygen in the water over a twenty-four-hour period in order to make an accurate determination of productivity. The lowest oxygen content is just before sunrise, that is, after a long period of darkness during which no oxygen is produced but a large amount is utilized by organisms. The highest oxygen content is generally about mid-afternoon, when sunlight is most efficient and the rate of photosynthesis is greatest.

Measurement of the uptake of carbon dioxide or nutrient elements will also provide similar indirect information on productivity. Both are consumed by the photosynthetic reaction. The change in carbon dioxide content of water over a day shows the opposite trends to those shown by oxygen content with highest values before daylight and lowest during maximum photosynthesis in the mid-afternoon.

Primary productivity is the rate at which photosynthesis or carbon fixation takes place. The total amount of organic material is the **standing crop** or **biomass.** In an analysis of a given environment it is important to consider both the gross and the net productivity. There are situations where productivity is high but because of consumption by herbivorous organisms, the biomass is low.

The primary producers are efficient because they provide a great deal of energy to the ecosystem relative to the amount used in the production process. Each successive **trophic level** or position in the food chain, such as herbivores or carnivores, becomes less efficient. An ecological pyramid illustrating relative biomass for each of these trophic levels shows this relationship well (fig. 8.7). The top carnivore, for example, a shark, requires a biomass that is orders of magnitude greater than its own to support it.

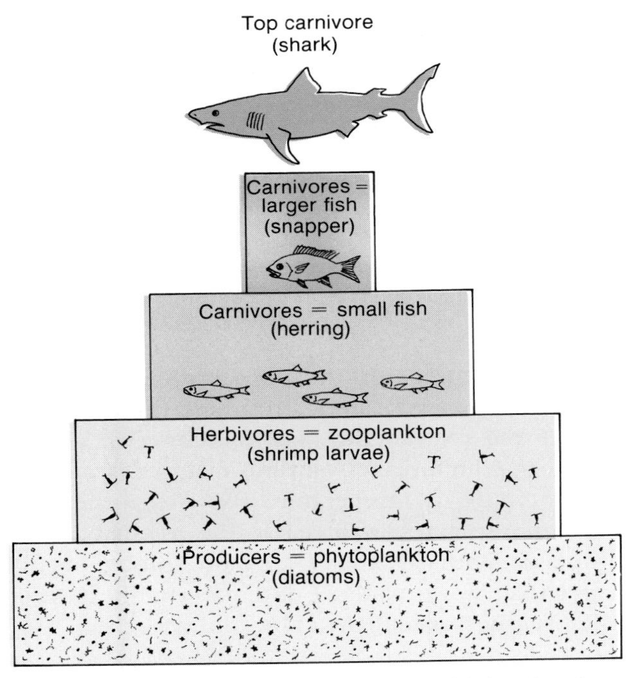

Figure 8.7 A generalized ecological pyramid showing the trophic levels in the marine food chain. The size of the boxes gives the general relative mass of one level as compared to others. In other words, it takes an enormous mass of diatoms to support one shark.

Factors in Productivity

Because sunlight is a necessary ingredient in the photosynthetic process, anything that affects sunlight will affect productivity. Some factors are temporal and periodic, some are aperiodic, some are related to weather, and others result from activity of organisms. Seasonal cycles of day length for example, are important. In the high latitudes there is great variation in day length from winter to summer. Photosynthesis may take place almost continuously above the Arctic Circle during the summer but not at all in the winter. A similar, but less dramatic effect occurs in the mid-latitudes due to seasonal changes in day length. Availability of nutrient materials for photosynthesis may also be seasonal because upwelling is most common in spring and this phenomenon aids greatly in the high productivity at that time.

The most important cycle in productivity is the diurnal one. Photosynthesis takes place during daylight, with the peak rate in the early afternoon. During most of the daylight hours there is an excess of photosynthesis over respiration so that a net production of organic tissue occurs. At night, the opposite situation occurs as respiration takes place without any productivity (fig. 8.8).

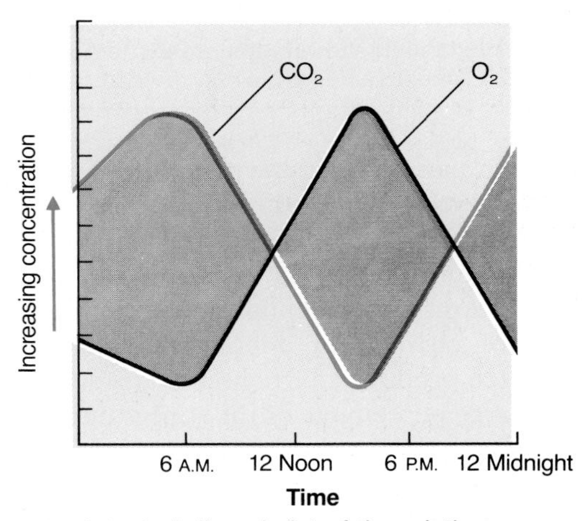

Figure 8.8 A typical diurnal plot of the relative concentrations of oxygen and carbon dioxide in the photosynthetic zone. Oxygen increases during the daylight hours and carbon dioxide increases during the night, when respiration is greatest.

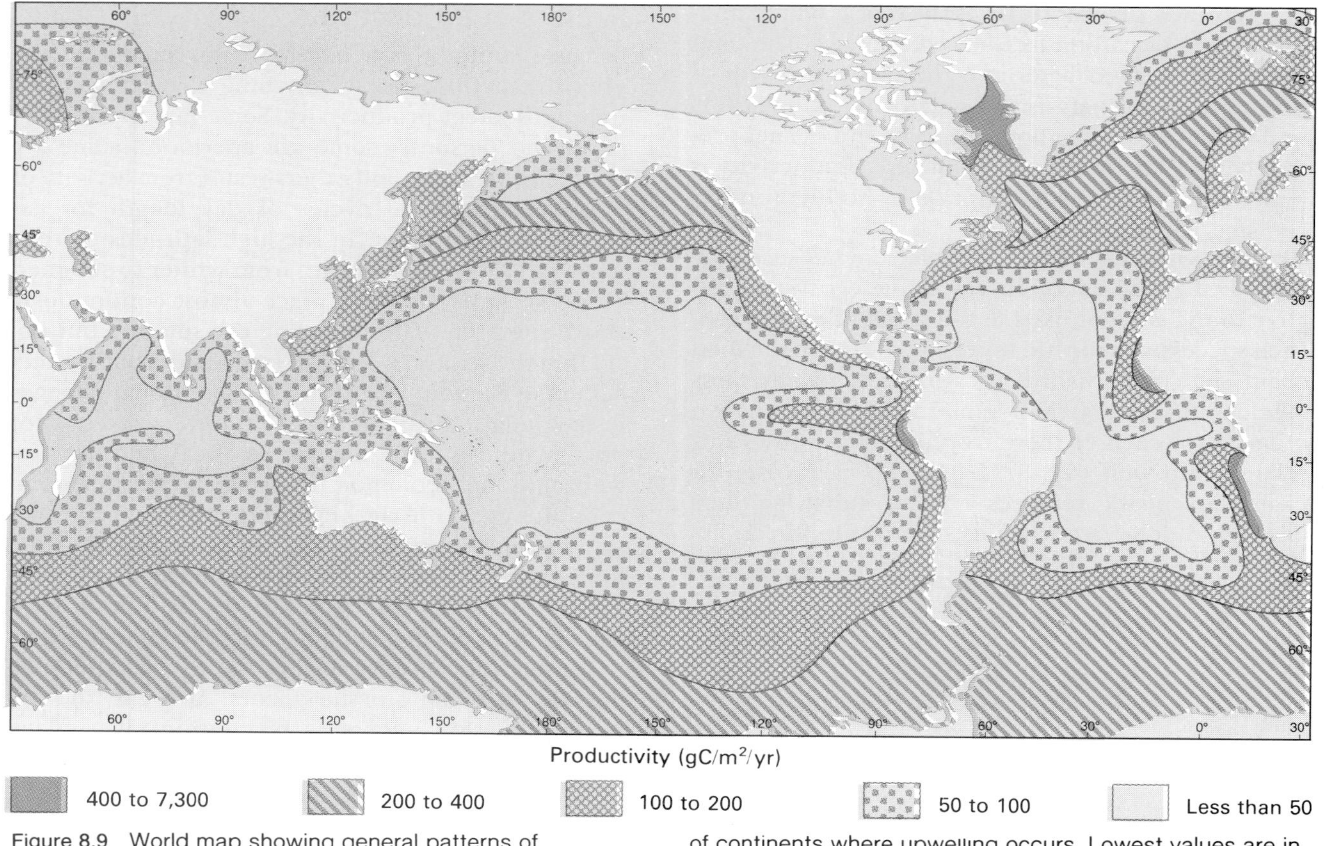

Figure 8.9 World map showing general patterns of productivity. Highest values are along the eastern margins of continents where upwelling occurs. Lowest values are in low-latitude areas where little mixing occurs.

Weather may affect primary productivity through cloud cover, wind, and indirectly through the temperature. Cloud cover inhibits sunlight, and therefore slows photosynthesis; it is, however, typically a local, short-term phenomenon. Wind can bring about a variety of changes in photosynthesis. It causes waves, which produce an irregular water surface that reflects a large percentage of the sunlight it receives, relative to a smooth water surface. Waves may also interact with the bottom in shallow water and produce turbidity, due to suspended sediment, which in turn reduces the rate of photosynthesis. In the other direction, the wind can increase the circulation of water, which induces upwelling and thereby provides abundant nutrients for photosynthesis.

Temperature aids the photosynthetic process because seasonal variations result in the deterioration of the thermocline and thereby give rise to overturn, which makes nutrient materials available for productivity. Temperature also affects the solubility of the gases involved in the photosynthetic processes. Both CO_2 and O_2 are more soluble in cold water than warm water. Consequently, the rate of photosynthesis is enhanced by cold water.

Distribution of Primary Productivity

Photosynthesis is not directly proportional to the intensity of light. The upper 10–15 m of the water column experience a lower rate of productivity than the zone from about 15–30 m because light at the surface is too intense for many organisms, which may be harmed by ultraviolet rays. Photosynthesis extends to depths near 100 m, where light intensity is about 1% of what it is at the surface.

In general, the open marine environment has low productivity relative to many other aquatic environments (fig. 8.9). This is largely because the open ocean is far removed from land, where much nutrient material originates, and also because of the great volume of the oceans, which dilutes the concentration of nutrient materials. **Eutrophic** environments are those with a great deal of nutrient material. These environments include shallow lakes, ponds, and marshes in the freshwater environment and coastal bays and estuaries, which are among the most productive of all marine environments. The combination of high nutrient levels from runoff with shallow, well-circulated waters makes the ideal situation for high levels of production. The high content of suspended sediment in shallow areas may, however, inhibit light penetration, which can be a limiting factor in deep coastal bays.

Oligotrophic environments are those with little productivity, such as the open ocean, large deep lakes like the Great Lakes of North America, and some coastal lagoons where circulation is restricted.

Biozones

Although numerous people recognized that organisms were restricted in their distribution in the ocean, it was not until the efforts of Edward Forbes, in 1834, that an attempt was made to zone the distribution of organisms. His work was based on a systematic dredging of the Irish and Aegean Seas. Even though his work was restricted to shallow ocean margins, Forbes is given credit for the discovery of the zonation of life in the ocean. He did believe, however, that life was not present at great depths due to intense pressure and the absence of light.

Numerous varieties of marine zonation have been proposed since Forbes. Eventually, primarily through the efforts of the National Research Council, the terminology associated with biological zonation of the oceans was standardized in the 1950s. There are two broad oceanic environments: the **pelagic** or water environment and the **benthic** or bottom environment. Each of these broad categories contains a variety of environments within it.

Pelagic Environments

All organisms that live in the marine environment but do not live on the seafloor are termed pelagic. Their environment extends from the shoreline to the greatest depths of the oceans. The basic parameter for subdividing the pelagic environment is depth; however, there is also a separation that is essentially physiographic in nature. The water environment over the continental shelf is termed the **neritic province** and the remainder of the pelagic environment is the **oceanic province** (fig. 8.10). There is no single depth boundary associated with this separation because of the geographic variation in the depth of the shelf/slope break, although 150–200 m is an approximation of the boundary.

There are important reasons for separating the waters over the shelf from those over the rest of the ocean basins. A fairly distinct community of organisms occupies the neritic province largely because of the abundant nutrient levels. The chemistry of the water over the shelf is also different than it is over the oceanic province, primarily due to the great variety of dissolved materials that are carried into the marine environment from runoff of the land. Neritic waters experience considerable variation in both space and time as compared to oceanic waters. This is especially true fairly close to the landmasses where salinity may vary greatly and where there is an influx of various dissolved constituents derived from land. Light distribution, suspended sediment, and physical energy within the water column are also different over the continental shelf as compared to oceanic waters.

The water column in the oceanic province is usually subdivided into four layers. The **epipelagic zone** is the upper 200 m and is nearly equivalent to the **photic zone,** which is the region of effective light penetration for photosynthesis. In some areas, especially over the continental shelf, light penetration may be markedly less than in the open ocean due to suspended sediment. The epipelagic zone encompasses both the neritic and oceanic provinces (fig. 8.10). Below it, in descending order, are the **mesopelagic,** the **bathypelagic,** and the **abyssopelagic** zones. The mesopelagic zone extends to a depth of 1,000 m. This range (200–1,000 m) corresponds well with the zone of great temperature

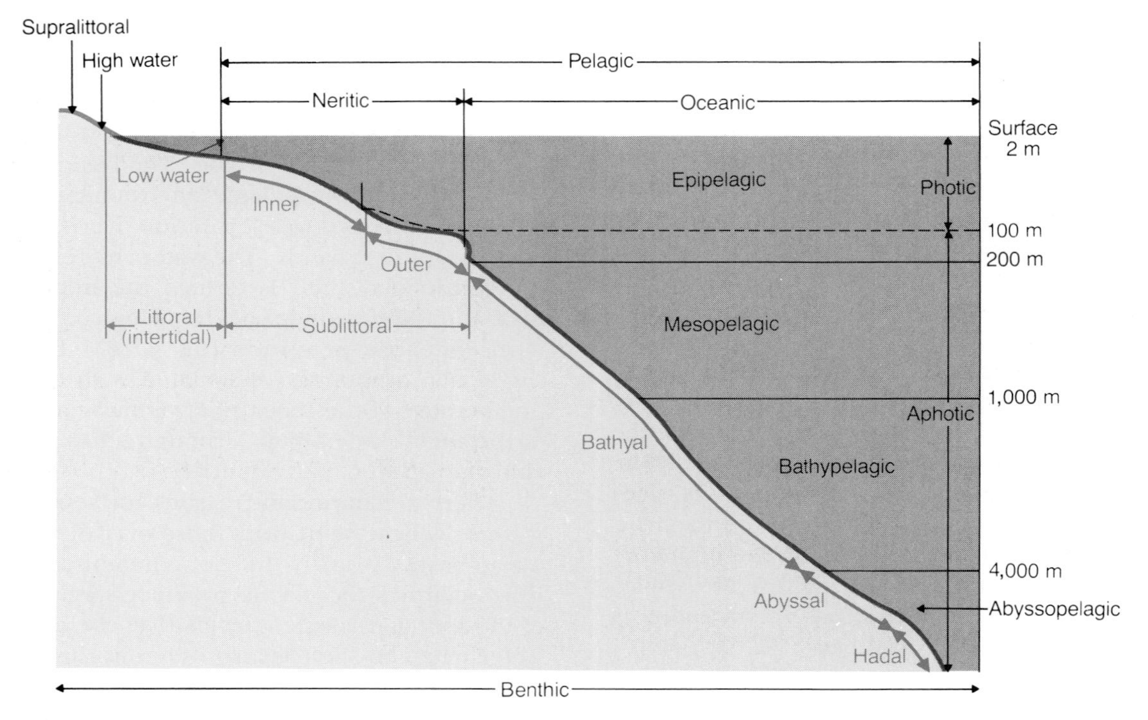

Figure 8.10 Diagram showing the major environments of the oceans.

change and contains the thermocline. Because it is below the photic zone, no primary productivity takes place here. This zone is occupied mostly by primary consumers that take in detritus descending from shallower waters. The bathypelagic zone extends to 4,000 m, or the depth of the deep ocean floor. It is uniform in terms of physical properties. The abyssopelagic zone extends to the deepest parts of the oceans. It is, for practical purposes, the trench environment. Organisms occupying both of these deep pelagic environments are adapted to not only the absence of light and cold temperatures, but also to great pressures.

Benthic Environments

In addition to the variables that affect organisms within the pelagic environments, organisms in the benthic environments also must contend with variables presented by the bottom itself. These include the firmness of the substrate, particle size of the sediment, composition of the sediment, small-scale relief of the bottom, and the interaction of the ocean floor with various physical processes, such as waves and currents. Those that live on the bottom within the intertidal zone have additional problems because some of this area is exposed to the atmosphere for certain periods during the lunar tidal cycles.

The intertidal zone is commonly referred to as the **littoral zone**. It extends from high water to low water, the level of which varies during the lunar cycle (see chapter 7). Above the high water or high tide line is the **supratidal** zone, also called the **supralittoral** zone (fig. 8.10). Organisms living in the supratidal zone must have special adaptations to withstand prolonged periods of exposure. This area is only inundated during storms.

The **sublittoral** zone covers the continental shelf. It is an environment of abundant and diverse organisms because of light penetration and the availability of nutrients. Some of the sublittoral zone may be below the photic zone due to suspended sediment or water depth, either of which can limit light penetration. There is much variation, both in space and time, of the sublittoral substrate. Its mobility also varies due to storms, which can cause massive sediment movement, even at the edge of the shelf. Additionally, there are currents across the shelf that can move sediment. The **bathyal** zone extends from the edge of the shelf to 4,000 m, essentially equivalent to the mesopelagic and bathypelagic zones (fig. 8.10). Forbes originally believed that this was the azoic zone because he considered it too deep for life. This zone includes the slope and rise, which may experience short periods of substrate mobility such as submarine slides and flows but where the bottom is relatively static and uniform as compared to shallower benthic zones.

The **abyssal** and **hadal** zones are essentially equivalent to the abyssal plain and oceanic trench environments, respectively. The substrate is uniform relative to other zones and consists of either unconsolidated sediment or lithified rock. As time passes and more data are collected, it becomes apparent that there is a great diversity of life on the bottom of these remote and rigorous environments.

Summary of Main Points

Although not very abundant, the nutrient elements play a major role in the world ocean's organic cycle. The true nutrient elements phosphorus and nitrogen occur in much different absolute concentrations; however, the relative distribution of each is much the same, both globally and throughout the water column. Both also exhibit some cyclic variation seasonally and diurnally. While not considered a true nutrient element, silicon is important because of its utilization in many primary producers such as diatoms. The distribution of silicon tends to parallel that of nitrogen and phosphorus.

Photosynthesis is the basic process that serves as the base for all productivity. Rates of productivity vary greatly in time, both diurnally and seasonally. Weather, turbidity, and temperature all affect the rate of photosynthesis and therefore, productivity.

Zonation of the various environments that comprise the world ocean is based primarily upon depth of water. There are two basic systems: the pelagic or water environment and the benthic or bottom environment. They experience essentially the same variables except that the benthic environment also shows variation in the makeup of the bottom. Most organisms in the ocean are restricted to one or, at most, a few of the depth zones.

Suggestions for Further Reading

Cushing, D.H., and J.J. Walsh. 1976. **The Ecology of the Seas**. Philadelphia: Saunders.

Harvey, H.W. 1957. **The Chemistry and Fertility of Seawater,** 2d ed. Cambridge: Cambridge University Press.

Hedgepeth, J.W. (ed). 1957. **Treatise on Marine Ecology and Paleoecology**, v. 1, Ecology. Geological Society of America, Memoir 67.

Levinton, J.S. 1982. **Marine Ecology.** Englewood Cliffs, New Jersey: Prentice-Hall.

Raymount, J.E.G. 1980. **Plankton and Productivity in the Oceans,** 2d ed., v. 1, Phytoplankton. Elmsford, New York: Pergamon.

Tait, R.V. 1968. **Elements of Marine Ecology.** London: Butterworths.

9
Life in the Sea

Chapter Outline

The volume and diversity of life in the world ocean is amazing. Scientists are still discovering some of the inhabitants of the planet earth, especially those in the remote and inaccessible environments of the ocean. The range in conditions within the ocean is broad and its inhabitants are diverse, but there is an order to the distribution of these creatures.

This chapter concentrates on the diverse adaptations of organisms to certain environmental characteristics, to factors that control distribution of organisms, and to some of the general interactions of marine organisms with their environment. Characteristic organisms of specific environments and communities will be deferred to the chapters that treat the various marine environments (chapters 11–19).

Life in the sea ranges from submicroscopic, single-celled organisms to whales over 10 m in length. Hundreds of thousands of species are known, and all niches of the marine environment are occupied by some living thing. In most places there is a community of many different types of organisms interacting with one another, but in some parts of the ocean there may be very few kinds or numbers of organisms due to food or environmental constraints.

Types of Organisms

Although there is a great diversity of marine life, it is commonly separated into only three major categories: **plankton, nekton,** and **benthos.** These subdivisions are based solely on the general habit of the organisms, and have nothing to do with their scientific classification, their size or complexity, or whether they are plant or animal.

The plankton are organisms that live within the pelagic zone and float, drift, or swim feebly, that is they cannot control their position against currents. The plankton include plants, which are called **phytoplankton,** and animals, which are called **zooplankton.** Nekton are those organisms that swim. Only animals are included in this group. The benthos are those organisms that live on or within the bottom, the benthic environment.

The plankton are the most diverse and numerous, with the benthos not too far behind. Many groups of organisms spend a portion of their life cycle in more than one of these modes of life. It is common for a particular group to be planktonic in the larval or juvenile stage and then nektonic or benthonic as adults.

TABLE 9.1
Classification of Organisms Using the Common Oyster as an Example

Kingdom	Metazoa (animals)
Phylum	Mollusca (shell animals)
Class	Pelecypoda (bivalves)
Order	Fillibranchia (filter-feeding, bottom dwellers)
Family	Ostriedae (oysters)
Genus	*Crassostrea* (euryhaline oysters)
Species	*Crassostrea virginica* (common edible oyster)

Scientific Classification of Organisms

Before proceeding with a discussion of various groups of organisms that live in the sea and their general ecology, it is appropriate that some consideration be given to the way in which these organisms are classified and named. Our system of biological nomenclature dates back to the eighteenth century when Carolus Linnaeus, a Swedish naturalist, devised a means of classifying all living organisms. This has become known as the Linnaean system of nomenclature.

This system is based on morphological similarities with a hierarchy of categories. There are seven primary categories. In descending order, they are kingdom, phylum, class, order, family, genus, and species. Originally, Linnaeus named only the plant (now called Metaphyta) and animal (now called Metazoa) kingdoms; however, five are now designated with the addition of the Monera (single-celled without membrane), Protista (single-celled with membrane), and Fungi. The system is best understood by considering an example (table 9.1) such as the common oyster (fig. 9.1). This organism belongs to the kingdom of animals, the Metazoa, which are all multi-celled animals. They are in the phylum Mollusca, which includes the common shell animals. The class is Pelecypoda or bivalve mollusks.

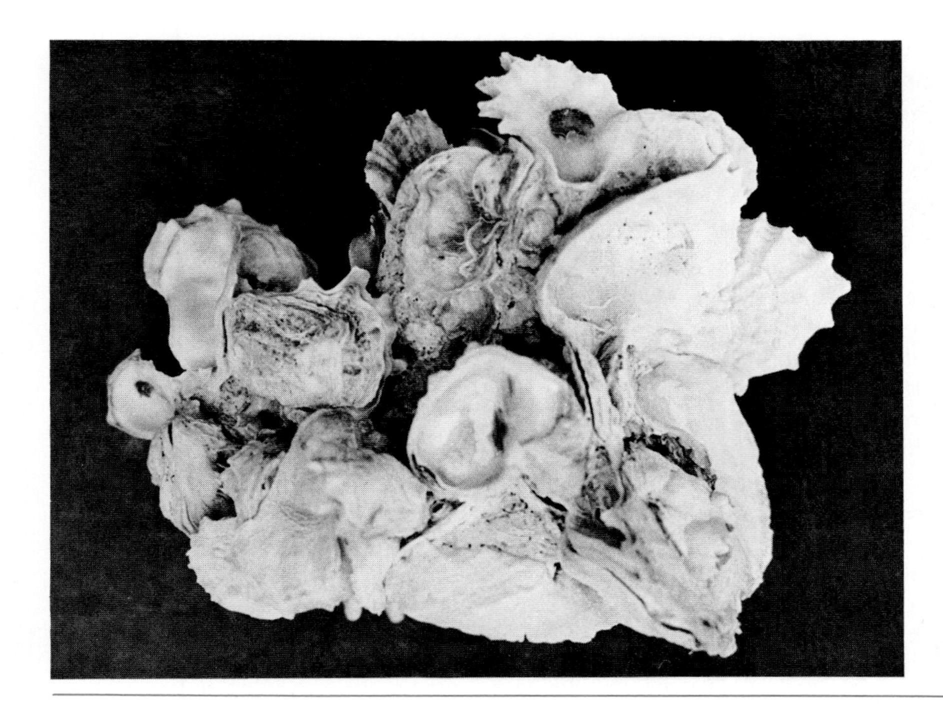

Figure 9.1 The common edible oyster, *Crassostrea virginica.*

Oysters, along with mussels, belong to the order Fillibranchia, which are filter-feeding bottom dwellers. The family is the Ostriedae, which separates the oysters from the mussels. The genus is *Crassostrea*, which are oysters that can tolerate low salinities, and the species is *virginica*, the common commercial oyster that tastes good whether raw, steamed, or fried. When giving the formal, scientific name for a species, both the genus and species names are included, the genus is always capitalized, and both the genus and species are italicized, for example, *Crassostrea virginica.* A classification list for all common marine organisms is given in Appendix B.

Plankton

Floating organisms comprise the bulk of marine life, both from the standpoint of biomass and the number of species. The combination of the primary producers, the phytoplankton, and the herbivores or first-level consumers plus the larval and juvenile stages of other types of animals creates a tremendous volume and diversity of life. For the most part the plankton are restricted to the upper few hundred meters of the ocean.

There is a wide range in the size of various types of planktonic organisms. The smallest are the **ultraplankton,** which are less than 0.005 mm (5 microns) in diameter. Included are some bacteria and the very small diatoms. The **nannoplankton** are less than 60–70 microns in diameter and are too small to be caught in standard plankton nets (see box 9.1). They are caught by allowing large quantities of water to stand for long periods of time, then recovering them from the bottom of the vessel or by centrifuging them from the water. They are also called centrifuge plankton. **Net plankton** or **microplankton** range up to a few millimeters in size and are those organisms typically caught in various types of plankton nets. There are also large plankton, both in the phytoplankton and zooplankton.

Some animals occupy the planktonic mode of life for only a portion of their life cycle, typically during the larval or juvenile stages. These are called **meroplankton**. Those organisms that are planktonic throughout their life span are **holoplankton**.

Phytoplankton

Although the phytoplankton account for a great deal of the biomass in the world ocean, they are represented by only a few phyla. Most of this group is unicellular and microscopic. They are in the phylum

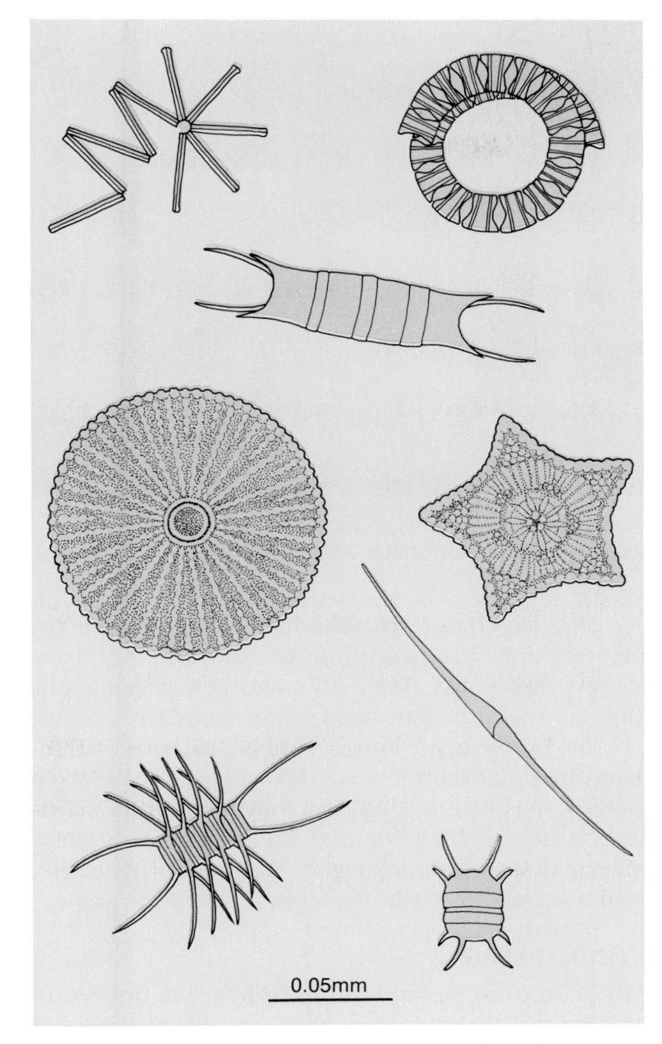

Figure 9.2 Selected examples of diatoms, the primary marine phytoplankton.

(a)

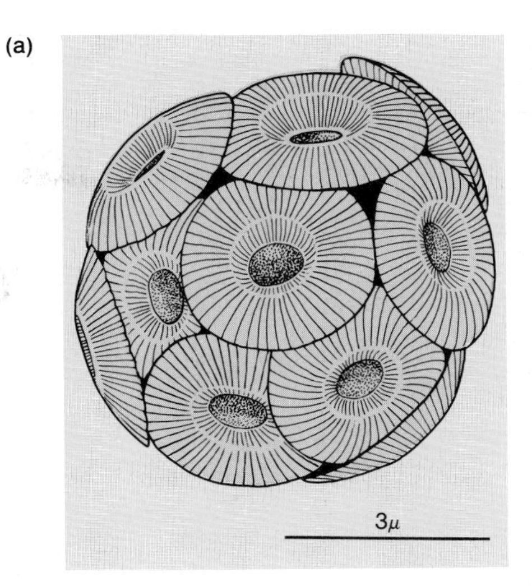

(b)

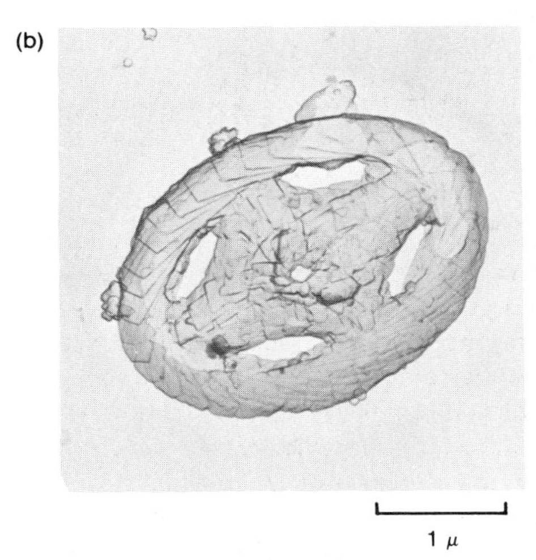

1 μ

Figure 9.3 Examples of coccolithophores (coccoliths), which are calcareous nannoplankton. The living individual has many articulated plates (a) but these separate (b) shortly after the organism expires.

Chrysophyta, the yellow-green algae, which includes the diatoms and coccolithophores. Some species in the green algae (*Chlorophyta*), the blue-green algae (*Cyanophyta*), the brown algae (*Phaeophyta*), and the large group of dinoflagellates (*Pyrrophyta*) also are phytoplankton.

Diatoms are the most abundant primary producers in the marine environment. They occur throughout the ocean but are most abundant in areas of upwelling and in high latitudes where there is cold, nutrient-laden water. These single-celled organisms are commonly called the golden brown algae because of the color they exhibit. Diatoms are all small but have a wide range in size, from only a few microns to about a millimeter. Their siliceous **tests,** or external skeletons, typically display complicated, delicate shapes and patterns (fig. 9.2).

The coccolithophores are nannoplankton, only a few microns in diameter. The living individual includes numerous calcareous plates held together by organic tissues (fig. 9.3). Upon expiration of the individual, the plates disaggregate and may fall to the ocean floor to become part of the sediment. Coccolithophores occur in shallow warm water.

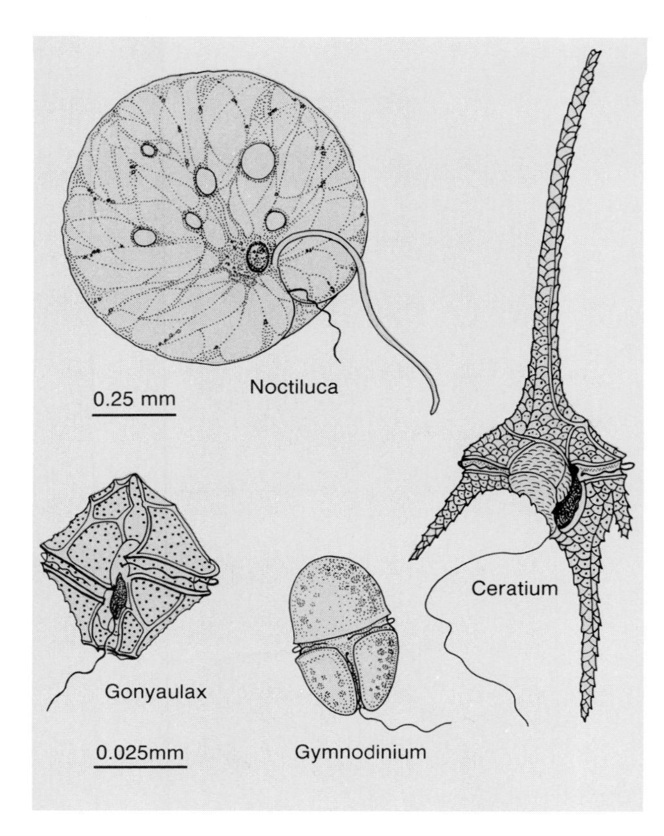

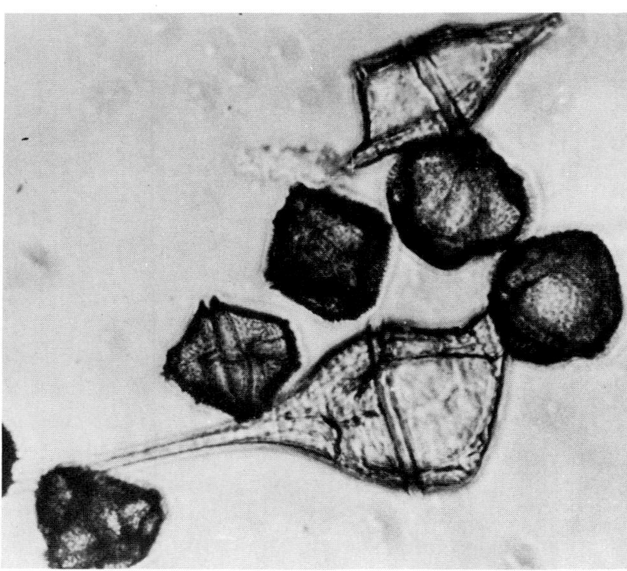

Figure 9.4 Sketches and photo of representative dinoflagellate forms, each of which contains a whiplike flagellum.

Figure 9.5 *Sargassum* is a brown algae and is one of the macroscopic phytoplankton.

Dinoflagellates are also diverse, but unlike the diatoms and coccolithophores, they do not have mineral skeletons. They do have whip-like flagella (fig. 9.4), which give them slight mobility.

The brown algae *Sargassum* is a relatively abundant phytoplankton that is macroscopic. These large floating masses (fig. 9.5) include branching stems and clusters of bladders that act as floats. It is these spherical structures that give *Sargassum* its name, which comes from the Portuguese word for grape.

Zooplankton

Although zooplankton include fewer numbers of species and less biomass than the phytoplankton, they are a more diverse group. At least nine phyla have representatives in this group and they range in size from microscopic to more than a meter in diameter. Some are meroplankton and others are holoplankton. Nearly all marine animals are plankton during some stage in their life cycle.

One of the most diverse of the zooplankton phyla is the Protozoa (fig. 9.6), which includes single-celled microscopic animals. The most important groups within the phylum are the Foraminifera, which have calcareous tests, and the Radiolaria, which have siliceous tests. Both are important contributors to deep-sea sediments.

The Coelenterata not only include some of the largest of the zooplankton but also some that are planktonic only during their larval stage. This

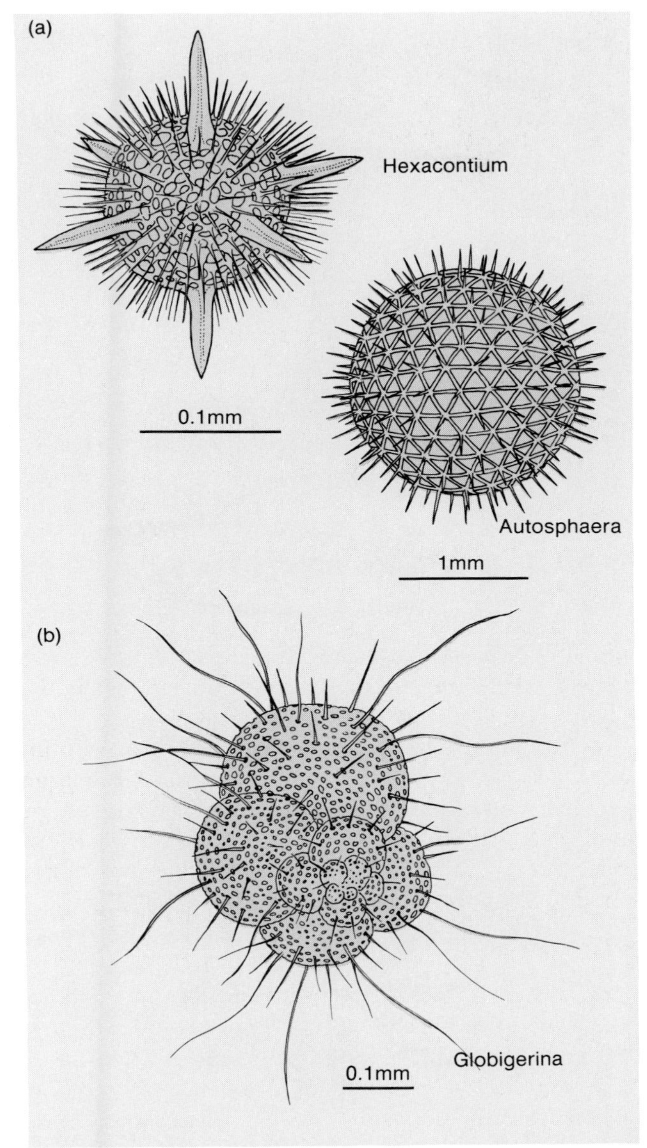

Figure 9.6 Planktonic protozoans include (a) Radiolarians, which have siliceous tests, and (b) Foraminifera, which have calcareous tests.

phylum includes the jellyfish (fig. 9.7), corals, and anemones. Corals and anemones are benthic organisms during much of their life cycle, but during their medusa or free-moving stage they are planktonic. The jellyfish of various types float at or near the surface and may have tentacles that are meters in length. Some possess weak locomotive powers in that they can contract and relax their tissues to move vertically in the water column.

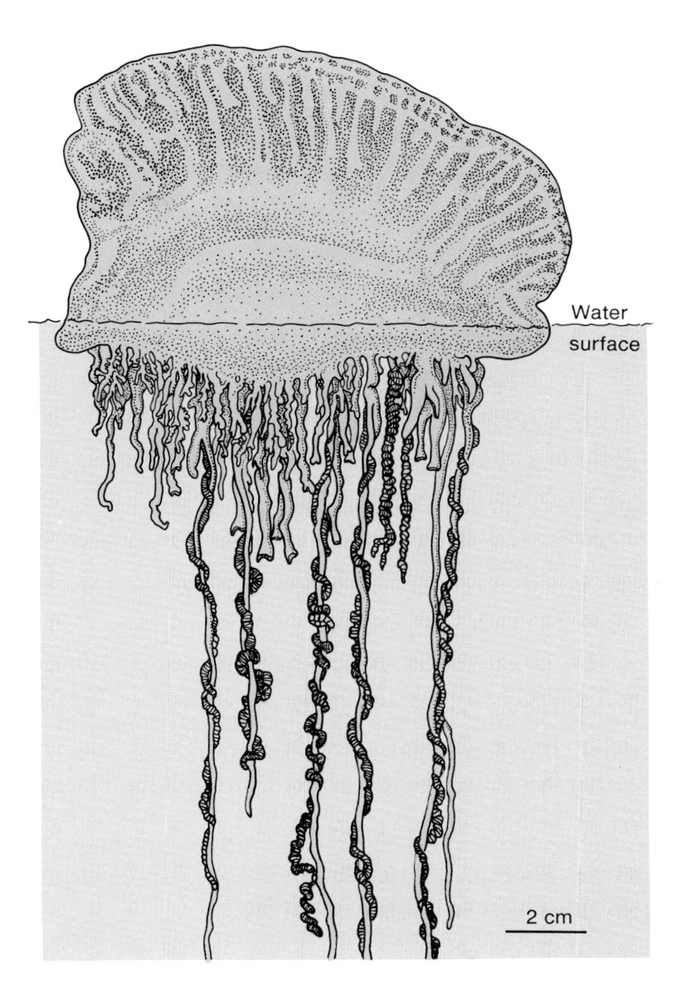

Figure 9.7 The Portuguese man-o-war, the genus *Physalia* is a common and hazardous zooplankton. The float not only keeps the animal on the surface, but also acts as a sail for transportation.

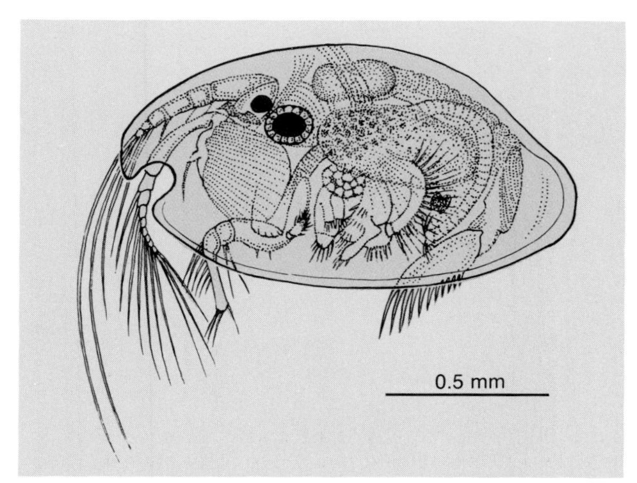

Figure 9.8 Ostracods are microscopic representatives of the planktonic arthropods and are among the few in that group that have calcareous hard parts.

The larval forms of many roundworms (Annelida), shelled animals (Mollusca), and spiny-skinned animals (Echinodermata) are prominent contributors to the zooplankton. These numerous, small, floating larvae can be distributed throughout a large area in a short time by wind-driven currents. As the larvae develop and settle to the bottom they may or may not encounter suitable conditions for development into mature benthic organisms.

The animals with jointed appendages (Arthropoda) are the most abundant and diverse phylum in the zooplankton. This phylum contains many large groups of terrestrial, freshwater, and marine organisms, including insects. Nearly all of the marine varieties are in the class Crustacea, which includes ostracods, copepods, and malacostracans. Ostracods are small (<5 mm) bivalves that live in the neritic zone (fig. 9.8). They may contribute their calcium carbonate valves to the sediment.

Copepods comprise the bulk of the zooplankton and are the keystone group in the marine food chain. The more than 4,000 species exceed all of the rest of the zooplankton combined in both diversity and abundance. These holoplankters are from about 0.2 mm to 20 mm in length. Copepods have numerous long, delicate appendages (fig. 9.9), some of which are used for feeble locomotion, some for feeding, and others for both functions. Rapid, jerky motion of appendages permits some movement, primarily in a vertical direction for feeding purposes. Feeding is by filtering diatoms and particulate organic matter from the water. Copepods serve as a primary food source for larger carnivorous zooplankton and for some fish.

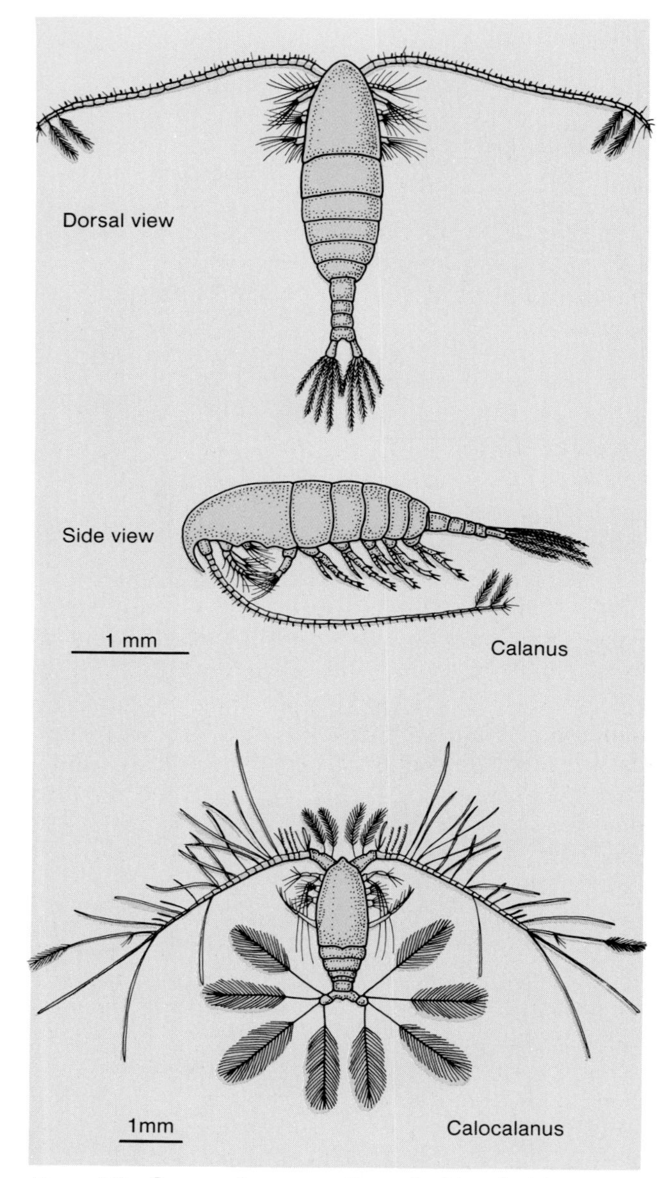

Figure 9.9 Copepods are small zooplankton that have numerous delicate appendages. These small herbivores are a major food supply for shrimp.

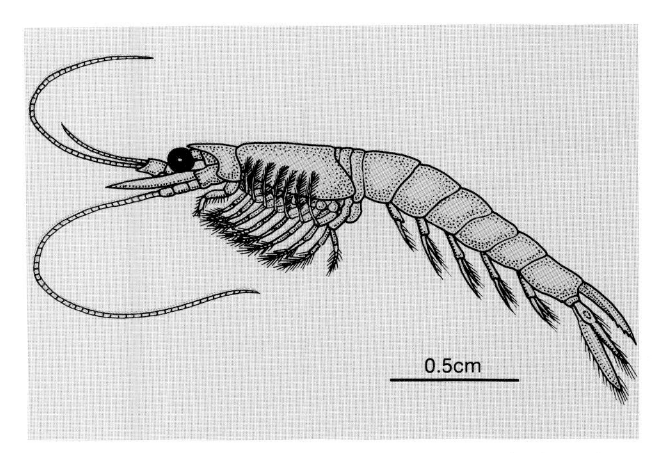

Figure 9.10 Mysids are small shrimp-like zooplankton that are a major food supply for small fish.

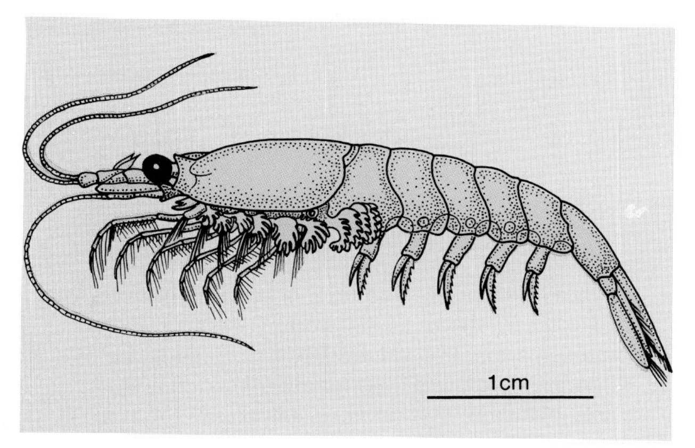

Figure 9.11 The euphausiids are crustaceans that form one of the primary constituents of the diet of baleen whales.

The malacostracans are the largest and the most economically important zooplankton group. The isopods and amphipods are primarily benthic with only a few planktonic forms. Some are holoplanktic and others spend the day as benthos and become planktonic at night. Mysids are typically larger, generally a few centimeters in length, and look somewhat like shrimp (fig. 9.10). Many are benthic but some are planktonic, living primarily in the neritic zone. They serve as food for many of the shallow-water fishes.

Euphausiids are some of the most important zooplankton because they serve as a primary food source for many fish and also for some varieties of whales. They are all marine, about 2–5 cm in length, and also look like small shrimp (fig. 9.11). The term **krill** is applied to these organisms, which may comprise more than half of the total zooplankton by weight in the world ocean They are particularly abundant in the waters surrounding Antarctica. These great numbers provide a potential for a tremendous fishery. At the present time several countries are harvesting the krill but the market is limited because the size of the organism limits its use to a protein food additive or as a shrimp paste.

The decapod malacostracans are currently the most important economic zooplankton group. Included are the shrimp (fig. 9.12), crabs, and lobsters. The latter two groups are only planktonic as larvae but the shrimp are essentially holoplankters. Some might argue with this statement because shrimp have swimming capabilities, but they are generally at the mercy of currents. There are also benthic shrimp, but the few species utilized by the commercial shrimping industry are typically planktonic. Individuals may reach 20 cm long.

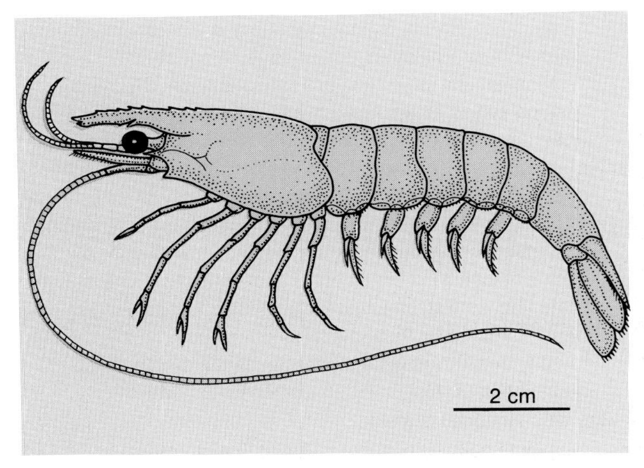

Figure 9.12 The common shrimp, *Pinnaeus,* is one of the largest crustacean zooplankton and is of great economic value.

The only significant representives of the vertebrates that occur as plankton are hatchlings of fish. They remain in this mode of life until they can swim well enough to control their location.

BOX 9.1

Collecting Plankton

To a large extent, the type of equipment used to collect plankton is dictated by the size of the organisms that are to be collected. Three types of apparatus are used: that for ultraplankton and nannoplankton, that for net plankton, and that for the large floating organisms. The smallest of the plankton are not retained by even fine mesh nets and therefore are typically obtained by first collecting a large volume of seawater, which is pumped through a filtering mechanism or is centrifuged to concentrate the small organisms. This is a time-consuming operation, taking several hours or even a few days to process a sample.

Most of the plankton are net plankton. There are various types of apparatus and techniques for collecting these organisms. The most common piece of equipment for plankton sampling is a plankton net (fig. B9.1). Although there is some variety in the details of the plankton net, all consist of some type of conical, fine mesh net mounted on a circular frame with a bottle or can for collection of the sample at the small end. The net can be towed behind a vessel or can be hauled vertically through the water column. It is usually towed, commonly at a predetermined depth using some type of weighting in combination with a particular boat speed.

It is possible to obtain at least semi-quantitative information on the abundance of the plankton. This is done by determining the volume of water that has passed through the plankton net and comparing that to the mass of sample collected. The easiest way to obtain this kind of information is by towing the net a known distance and then multiplying that number by the area of the net opening. For example, a common size is a 0.5 m net, which if towed for a kilometer, will pass 500 m^3 of water through it (1,000 m x 0.5 m). More sophisticated plankton nets have flow meters with propellors attached to the frame to record the flow of water, thereby giving a better method of determining the volume of water that passes through the net.

The plankton net is a somewhat fragile piece of equipment that must be towed slowly. This is time-consuming and therefore expensive. Some sturdier types of plankton samplers have been devised that have rigid metal housings that allow them to be towed at the cruising speed of the ship. In this way almost no time is expended solely for plankton sampling. These types of apparatus typically have a streamlined shape with only a small opening for plankton to enter. They also have devices for determining water volume to enable quantitative sampling.

A problem with all types of microplankton sampling is the damage that occurs to the individual specimens. These organisms are delicate and typically have many appendages. It is not uncommon for some of the specimens to be damaged to the extent that they cannot be identified.

The large plankton can be easily collected with the same type of heavy nets that are used in trawling for fish or shrimp. They are discussed in box 9.2.

Plankton Ecology

Plankton are distributed throughout the upper waters of the world ocean. They are restricted by food supply, temperature, and light (in the case of phytoplankton). Phytoplankton are most abundant just below the surface in shallow water and zooplankton are most abundant just below the phytoplankton. Some zooplankton move vertically on a diurnal basis for feeding. They move toward the surface during the night and to greater depths in the daytime. Euphausiids are among these organisms.

In the zooplankton, there is a significant difference between the oceanic community and that of the shallow neritic environment. In shallow water, the larval forms of the adult benthos are a much greater proportion of the total than in the open ocean. Also, the ostracods tend to be most prominent in the

BOX 9.1

Figure B9.1 Conical plankton net for the collection of small organisms.

shallow shelf and coastal waters. Both phytoplankton and zooplankton tend to have peak concentrations with seasonal upwelling along the outer limits of the continental margin. The phytoplankton bloom as the result of a large supply of nutrients and the zooplankton respond to this large food source.

Although these organisms are present throughout the world, only a very few species are themselves worldwide. As is typically the case, the low latitudes support a wide variety but modest populations, whereas the high latitudes have few taxa but tremendous numbers of individuals.

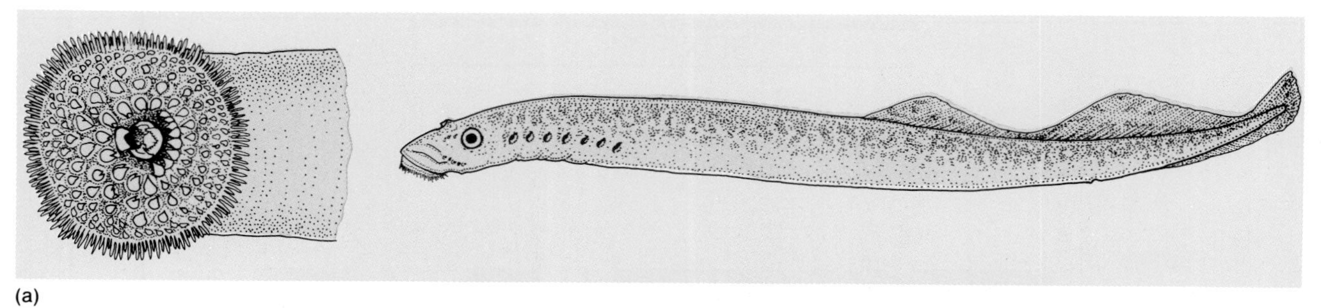

(a)

(a)

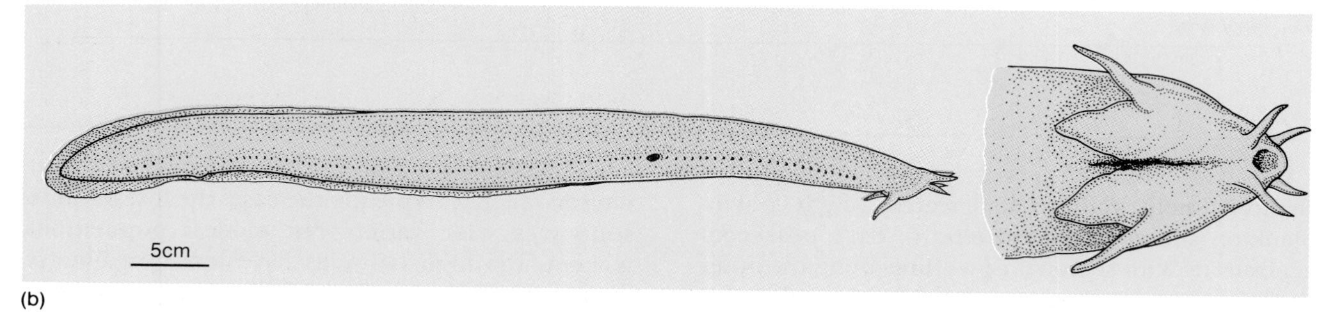

(b)

Figure 9.13 (a) The sea lamprey and (b) the hagfish are examples of jawless fish. They live attached to rocks or other organisms from which they suck blood for nutrients. Both have rasping or sucker-like mouths.

Nekton

The swimming animals of the oceans have long been of interest because of their tremendous economic importance as a food source for humans. The nekton are not diverse relative to the other two modes of marine life. The only groups included are cartilaginous fishes, bony fishes, sea turtles, and marine mammals among the vertebrates, along with the squid, which are mollusks. There are no plants with true swimming capabilities.

Fish

The vast majority of all fish are nekton but there are some variations in their living habits. Several varieties of fish live on the bottom although they have good swimming abilities. These are the **demersal** fish. Those that do not spend any time on the bottom are called pelagic. Demersal fish generally feed and lay eggs on the bottom. In some cases, the adults can settle just under the sandy substrate for camouflage.

Fish may also be discussed taxonomically. Most present-day fish are bony fish (Osteichthyes), but there are also various primitive types that still exist, the so-called living fossils. These include the jawless fish (Cyclostomata) and those that have cartilaginous skeletons (Chondrichthyes). Included in the latter group are sharks and rays.

Non-bony Fish

The jawless fish are lampreys and hagfishes. They tend to be long and slender with sucker-like mouths that are used to feed and, in some species, also for attachment. Some feed by rasping, such as the hagfish and the lampreys (fig. 9.13). Others move across the substrate ingesting debris like a vacuum cleaner. Hagfish are about 20–25 cm long, whereas the lamprey may reach 1 m in length.

The cartilaginous fish lack an air bladder and do not have paired gills but do have gill slits. The absence of an air bladder means that they must stay in motion or they sink to the bottom. Many of the skates and rays have some demersal habits; their bodies are flat (fig. 9.14) for this type of activity. Sharks, on the other hand, are streamlined for rapid movement (fig. 9.15) and they have a tailfin morphology that not only helps to propel them forward, but also upward. These fishes must move through the water to keep a continual supply of oxygen flowing through their gill slits.

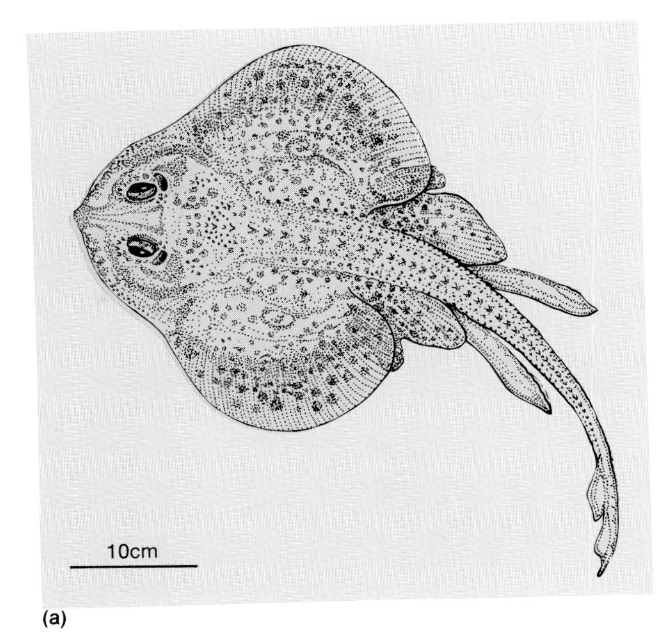

(a)

(b)

10 cm

Figure 9.14 (a) Sketch and (b) photo of a common ray, an example of a bottom-dwelling cartilaginous fish.

The demersal varieties of cartilaginous fish may exhibit coloration and markings that obscure their presence. Both sharks and rays can be huge, with the "wing span" of the giant manta ray being 2–3 m. The whale shark can reach 20 m in length, which makes it the largest of all fish.

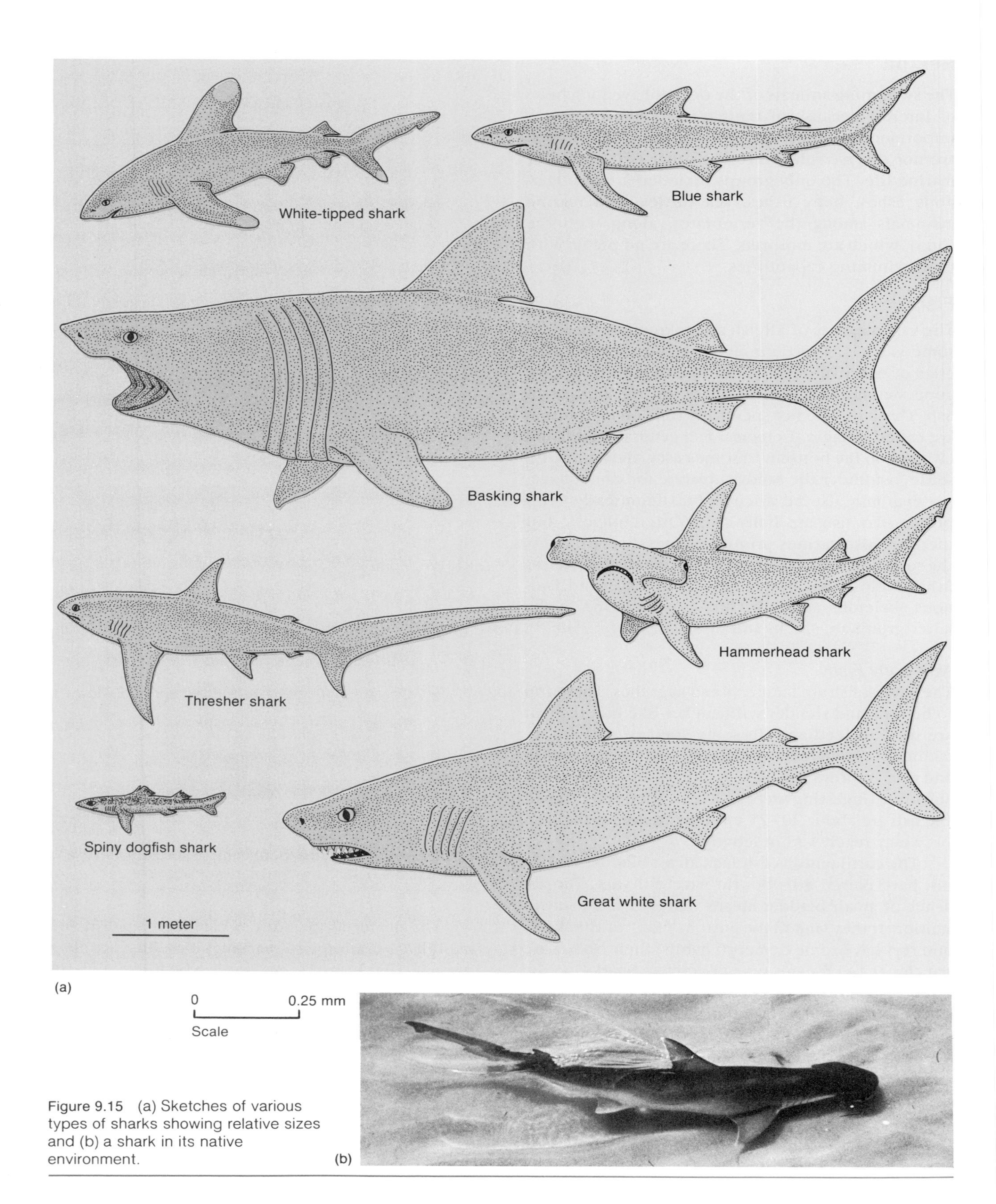

Figure 9.15 (a) Sketches of various types of sharks showing relative sizes and (b) a shark in its native environment.

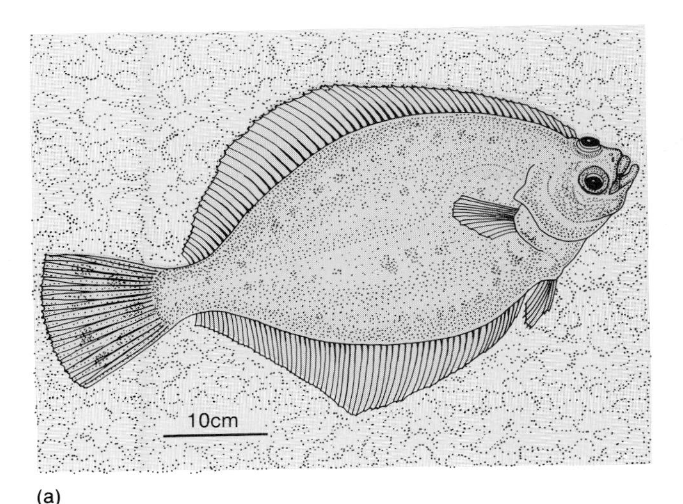

(a)

(b)

Figure 9.16 (a) Sketch and (b) photo of typical bottom-dwelling flatfish. These fish have a shape and coloration that is well suited for demersal life.

Figure 9.17 Example of a parrotfish, an abundant reef-dwelling fish that feeds on corals of the reef.

Bony Fish

The modern fishes are diverse, including about 100 families and over 30,000 species. They range widely in size and in their ecology, so that at least one species occupies nearly every niche in the pelagic realm. For purposes of discussion here, the bony fish will be divided into those that are of economic value and those that are not. Most economically important varieties are in shallow neritic waters.

Bony fish have air bladders, paired fins, gills with a gill cover, and scales. These all provide advantages over the more primitive types discussed earlier. The paired fins provide ease of locomotion and stability, the gills provide a circulation of water across the tissues so that oxygen can be extracted, and the scales give protection.

Many bony fish live on the bottom for a significant part of their lives even though they possess good swimming capabilities. The **fry** or hatchlings of these types are pelagic but they adapt to demersal life as they mature. The bodies of demersal fish are most commonly flat. These flatfish have become so adapted to bottom dwelling that they experience a marked transformation in morphology as they mature from fry to adults. The young flatfish looks like most fish. As it increases in size, the body changes to a flat, somewhat elliptical shape (fig. 9.16). Coincident with this is the shifting of one eye so that both eyes occupy one side of the fish. Pigment is lost from one side and the fish lays on the substrate on its side with the non-pigmented side on the bottom and the pigmented side facing up.

Numerous other fish feed and lay their eggs on the bottom, such as those in the cod group. Also, many reef fish, such as the parrotfish, feed on the substrate of the reef structure itself (fig. 9.17).

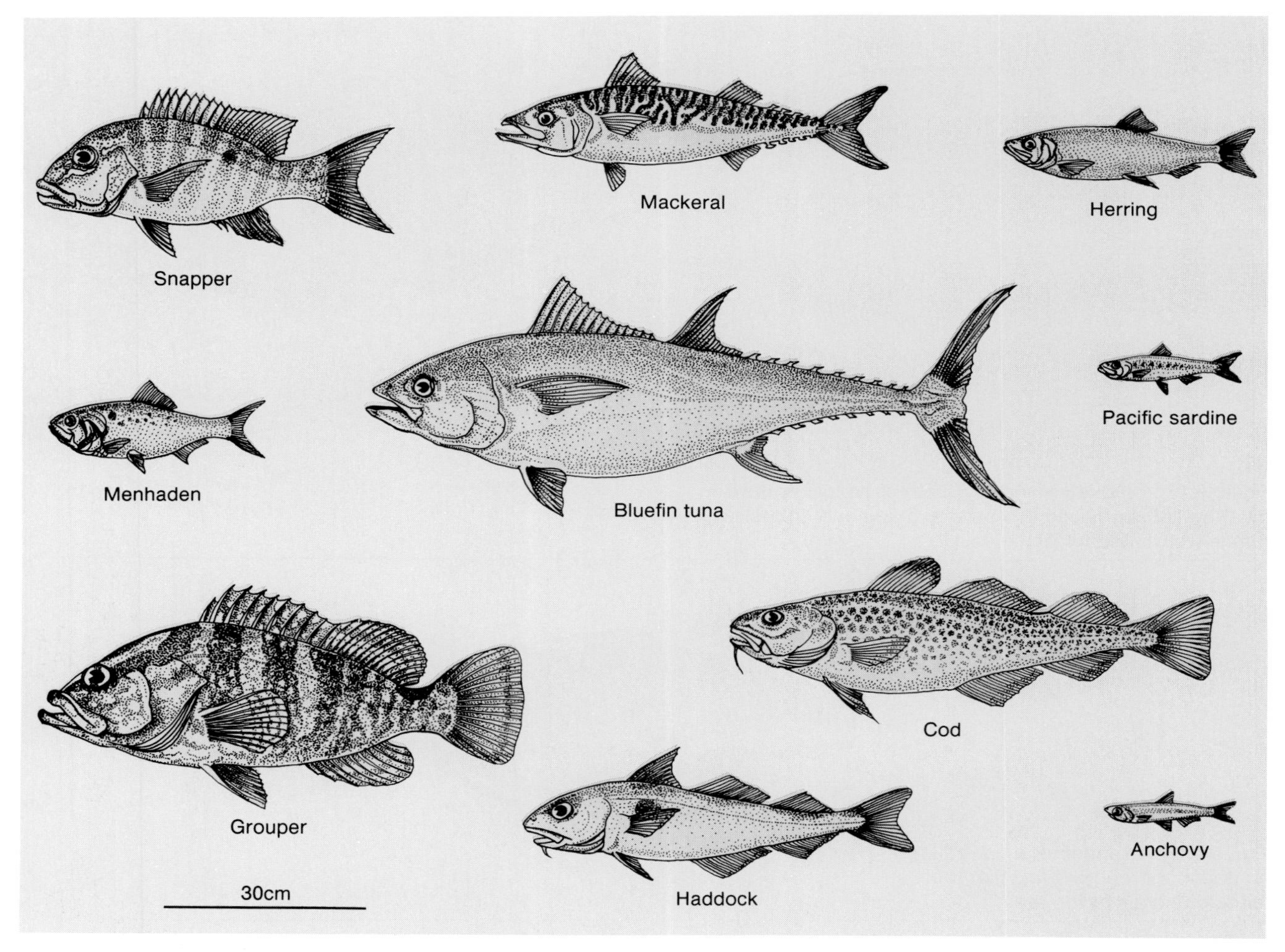

Figure 9.18 Selected types of commercial fish. Note the size differences.

Economically important fish include both demersal and pelagic varieties. Most of the world's fisheries are on the continental shelves, where there is an abundant food supply. Valuable demersal fish include the flounder, sole, and halibut, all of which are excellent table food. Pelagic fish (fig. 9.18) include more varieties, such as cod, haddock, snapper, grouper, anchovies, sardines, mackerel, tuna, and others, all of which are also excellent for human consumption. There are many valuable pelagic fish that are useful for other purposes. Menhaden, for example, are extremely numerous and are used for pet food, for oil, and for fertilizer.

There are non-economic fish that occupy the same shelf environment, but are too small, too large, too bony, or taste too strong for consumption by people. Most of the fish that are not in demand by the fisheries industry are oceanic and many live at great depths. These varieties have special adaptations for the harsh conditions that exist there, including special sensory organs to replace functional visual organs, large mouths to aid in food intake, special organs to enable them to withstand the extreme pressures, and small size (fig. 9.19). Most of the deep-sea fish do not have functional eyes, but they may have various types of luminous organs or bacteria that emit light. A few have touch-sensitive appendages that replace visual organs.

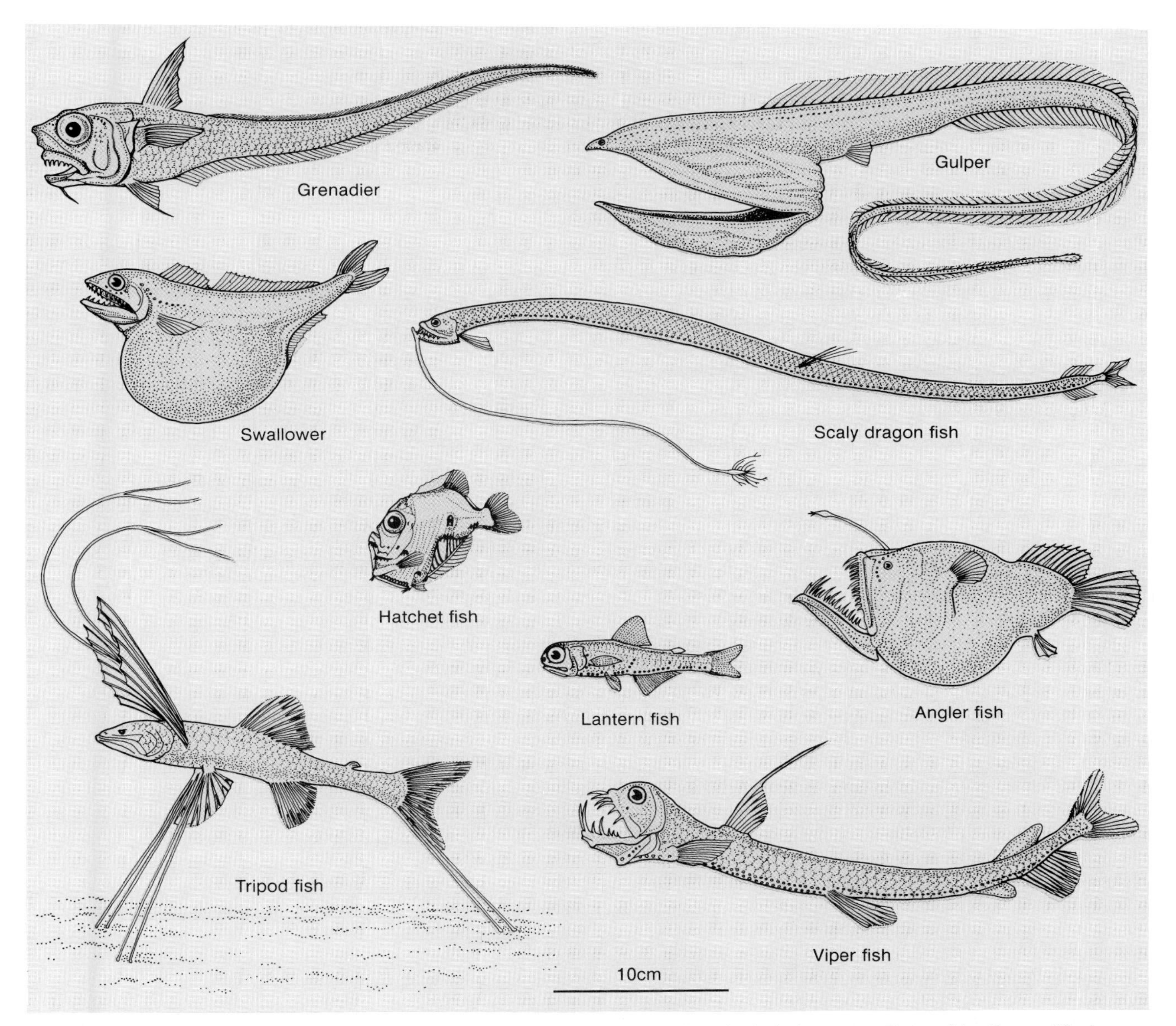

Figure 9.19 Various types of deep-sea fish showing anatomical adaptations to this rigorous environment. These adaptations include large mouths and teeth, modified appendages to serve as lures, and large stomachs.

Their small size is the result of the scarcity of food. Most species are less than 10 cm in length. The large mouths contain inordinately large teeth for efficient hunting. Deep-sea varieties are typically carnivorous. The disproportionately large mouth is to enable the individual to ingest large prey, sometimes bigger than the fish itself. Such activity is necessitated by the sparse populations in the deep sea; it may be several days between meals! Some species have unusual structures, such as luminescent appendages, to attract prey. These are some of the special adaptations that deep-sea fish have evolved that permit them to survive and, in some cases, be abundant. In fact, certain species of them are among the most abundant species in the ocean.

BOX 9.2

Collecting the Nekton

Nektonic organisms are collected primarily for commercial fishing and for scientific research. This discounts sportfishing, but that is strictly a hook-and-line type of activity. Most methods of collecting nekton involve large numbers of individuals; in terms of the fishing industry, the larger the better. There are only a few major types of collecting apparatus that are used, but each has subtle variations that have been developed over years by various users to improve efficiency.

There are essentially two approaches to collecting fish and other nekton: one takes the net or other apparatus to the areas where the fish are, and the other places the net where the fish are expected to be, for them to move through. The former is the most widely used.

Bottom or near-bottom devices include the **beam trawl** and the **otter trawl**. The beam trawl consists of a coarse mesh net held open by a rigid frame. This apparatus was widely used in earlier years but is no longer common in the fishing industry because the rigid frame restricts its size and makes it difficult to handle on the ship (fig. B9.2a). Additionally, this type of trawl is dragged along the bottom and can be a destructive factor in benthic communities. The otter trawl is composed of a similar mesh net but with otter boards replacing the rigid frame (fig. B9.2b). These boards are rigged to keep the net open as it is towed through the water. This allows for much larger nets than the beam trawl and has much less effect on the bottom communities.

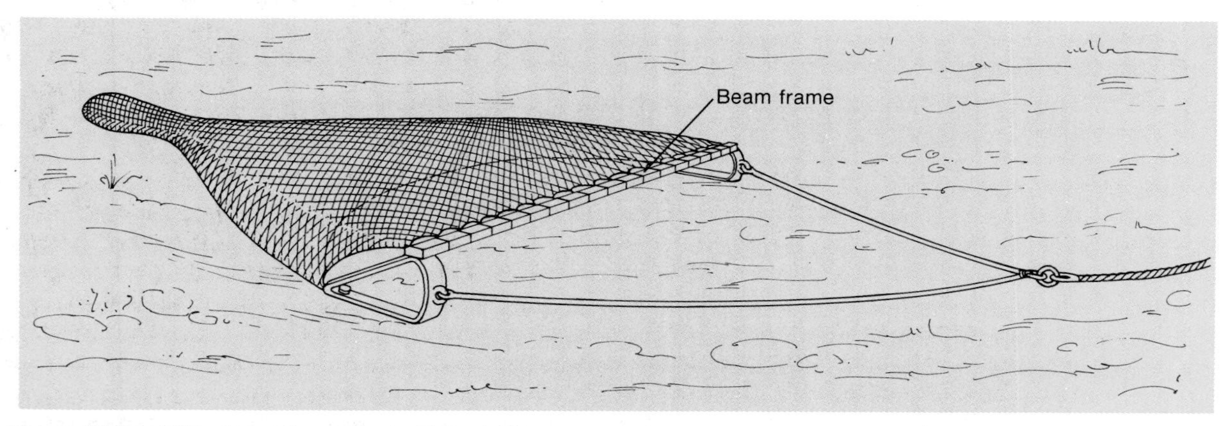

Figure B9.2a Old-fashioned beam trawl with its rigid frame. This apparatus was towed along the floor of the ocean to collect fish.

BOX 9.2

Nektonic organisms that are near the surface are commonly collected by very large nets called **purse seines**. This type of net has floats on the top to keep it at the surface and weights at the bottom. The net is deployed in a large circle, hopefully trapping fish, then a cable that acts like the drawstring on a purse is pulled in to close off the net. The entire net (with the catch) is then hauled onto the ship. This type of net may be deployed directly from the ship or it may be handled from a number of small skiffs working with a large ship.

Some collecting operations use either **gill nets** or **long lines**. The gill net is a linear net that hangs several meters below the surface from floats and has large mesh to catch the gills of fish when they swim into it. The gill net is positioned so a school of fish will swim into it, then the net and its catch are retrieved. A long line apparatus is a long wire or nylon line with hundreds of hooks, each attached at a particular interval. Each is baited as the line is payed out from the ship. After a predetermined period of time, the line is reeled in on a big spool and the fish taken from the hooks.

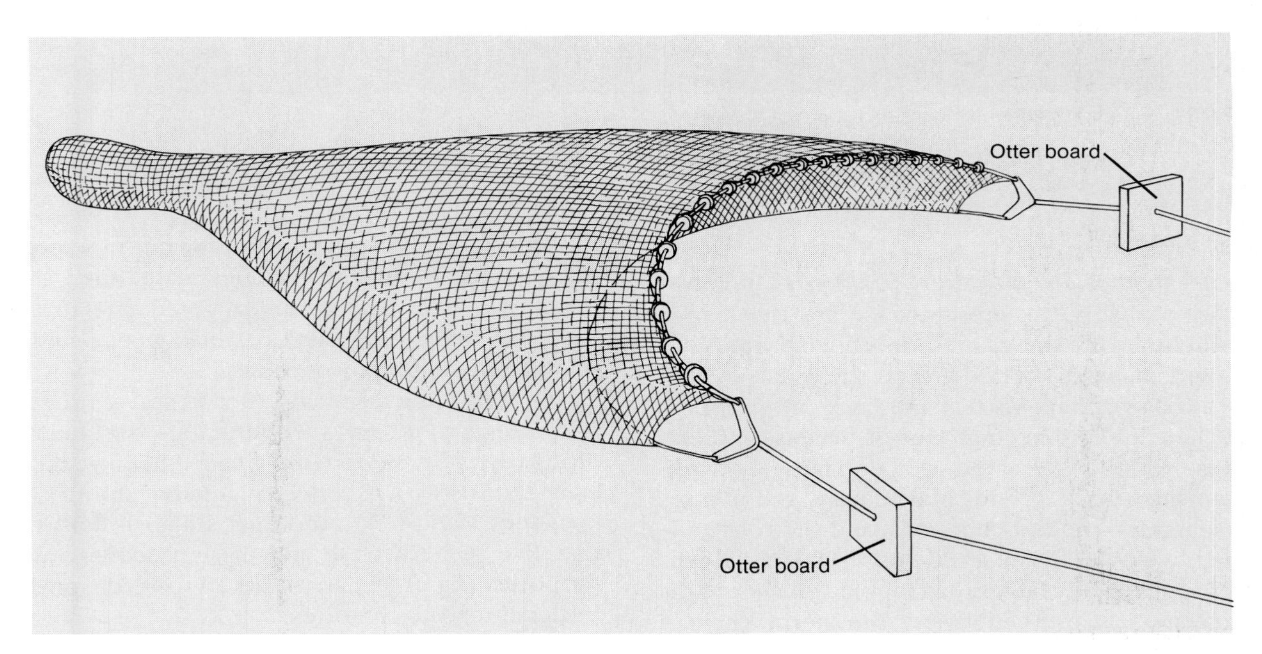

Figure B9.2b The otter trawl enables a much larger net to be used. The otter boards keep the net open during the trawling procedure.

Figure 9.20 Filtering mechanisms in the mouth of a baleen whale.

Other Nekton

Although there is a wide variety of other swimming animals, probably the most diverse are the mammals. Included are the whales, dolphins, porpoises, seals, and manatees. They are all fairly large and some, such as certain whales, are huge, up to 30 m long. There are two main groups of whales: **baleen whales,** which have large, yet delicate, filtering mechanisms (fig. 9.20) in their mouths and whose diet is largely krill and copepods; and the **toothed whales,** which feed on squid and fish. The baleen whales are generally larger than toothed whales, with the largest of the toothed variety, the sperm whales, about the same size as the smallest of the baleen type, the gray whales (fig. 9.21).

Various reptiles also inhabit the ocean. Probably the best known are the varieties of sea turtles (fig. 9.22) that have become endangered due to overharvesting. There are also marine iguanas and sea snakes, which are poisonous.

Other marine mammals, such as seals (fig. 9.23), porpoises, and walruses are voracious fish eaters. Additionally, they are, like the whales, open water, pelagic animals with excellent swimming abilities. The manatees (fig. 9.24) are large, sluggish herbivores that live in both fresh and marine coastal waters. Porpoises (fig. 9.25) are believed to be the most intelligent marine creatures.

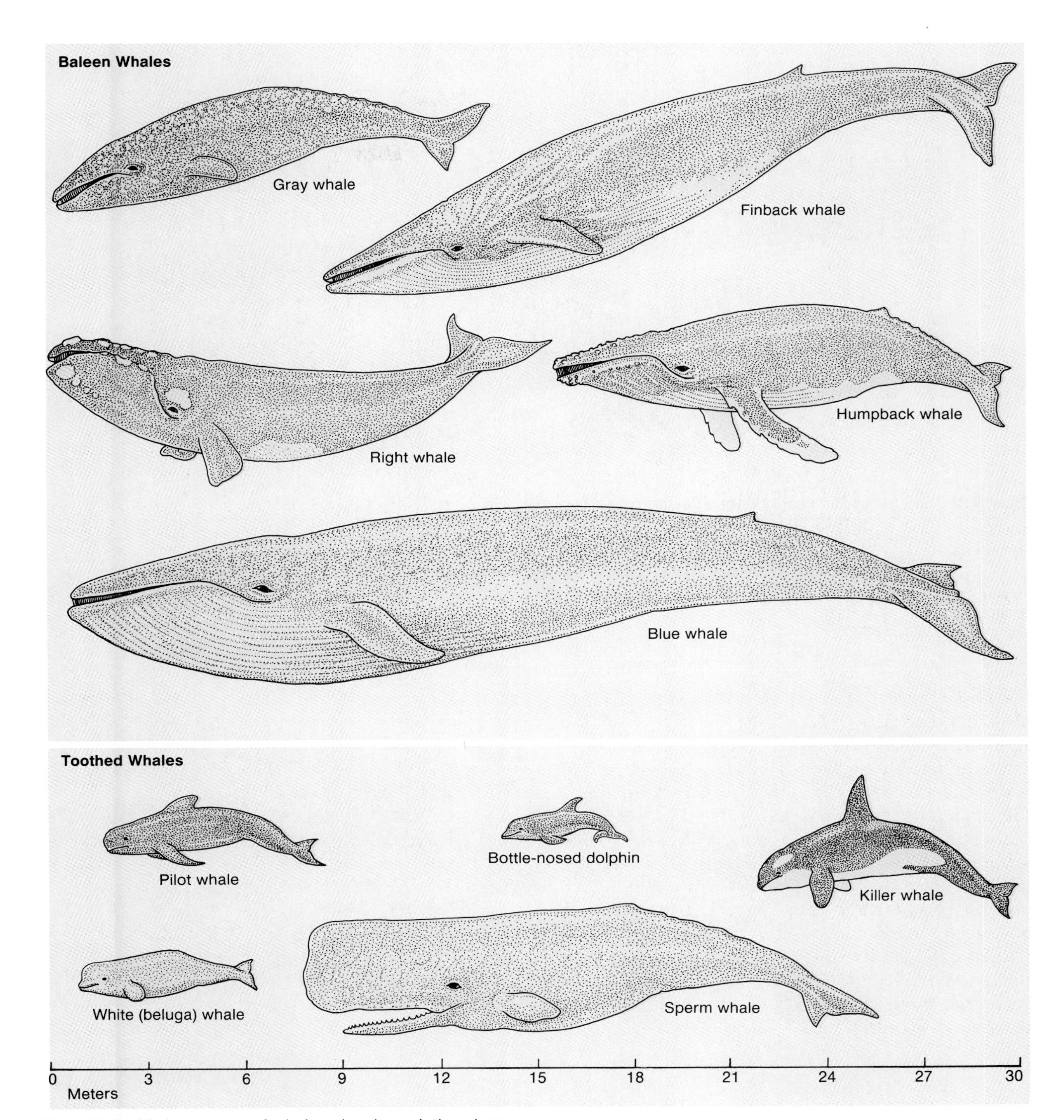

Figure 9.21 Various types of whales showing relative sizes. Note the general size difference between baleen and toothed whales.

Figure 9.22 Sea turtles on a sandbar of a Caribbean reef.

Figure 9.23 Seal colony along the rocky coast of California.

Figure 9.24 Manatee cow and calf in a Florida estuary.

Figure 9.25 Porpoises swimming in tandem. These are among the most intelligent of all animals.

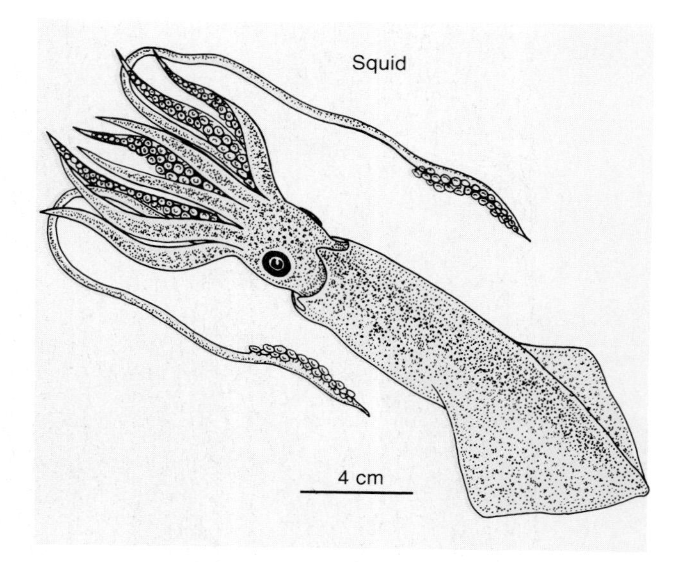

Figure 9.26 The squid are among the few truly nektonic invertebrates.

The only important invertebrate nekton are the squid, which are excellent swimmers, including the giant squid. It is the largest known invertebrate and may exceed 30 m in length. Commercial squid (fig. 9.26) are typically 10–30 cm in length and are harvested from neritic waters.

Benthos

Bottom-dwelling organisms occupy a variety of niches and are nearly as diverse as the plankton. There is a variety of both plant and animal types but no vertebrates are true benthos. These bottom-dwelling organisms are in direct contact with the substrate, which is an additional variable to the environmental conditions that may control the distribution of certain species. Such factors as the composition of the bottom, the particle size, its firmness or resistance to penetration, its mobility, and the food that it may contain, all may limit the distribution of organisms.

There are various modes of life that bottom-dwelling organisms can occupy. Some are **sessile** or attached. They have no mobility and rely on currents or other mechanisms to bring food to them. Plants, such as benthic algae and sea grasses, and animals, such as corals, may be sessile.

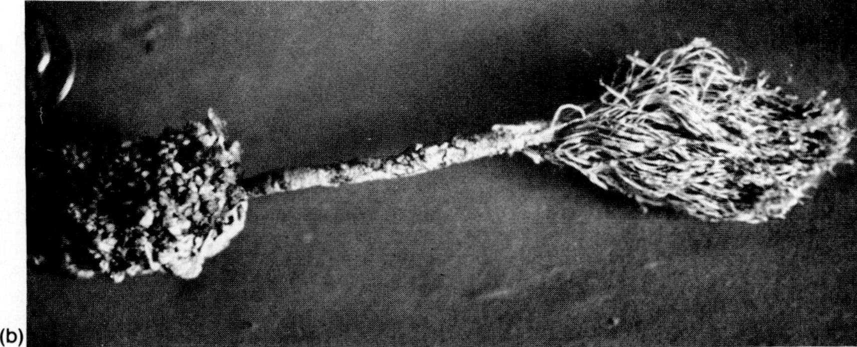

Figure 9.27 Examples of calcareous benthic algae: (a) *Halimeda* and (b) *Penecillus.*

The **vagrant** benthos have locomotive powers, although some can move rapidly and others move slowly. Only animals are included here. Other modes of life include the **infaunal,** in which the organism lives within the substrate, and **epifaunal,** in which the organism lives on the substrate. These terms are applied regardless of whether the substrate is unconsolidated sediment or hard rock. Organisms that penetrate the unconsolidated substrate are said to be **burrowers,** whereas those that penetrate the rock substrate are **borers.**Various modes of feeding exist among the vagrant types. Some scavenge across the substrate for organic debris, some are aggressive carnivores, others plow through the sediment ingesting unselectively, still others burrow into the sediment and take in food through a filtering system of inhalant and exhalant siphons.

Plants

The variety of photosynthetic organisms on the seafloor is restricted to many types of algae and some angiosperms or sea grasses. The algae include the Cyanophyta (blue-green), Chlorophyta (green), Phaeophyta (brown), and Rhodophyta (red), in increasing order of diversity. All range from microscopic to large except for the blue-greens, which are all microscopic. They commonly have a filamentous character and occur in clusters or mats. Green algae include several calcareous species (fig. 9.27) that are important contributors to the sediment substrate, especially in the shallow waters of the low latitudes. The brown algae include the largest varieties, the kelp, which are very abundant in shallow, cold

Figure 9.28 Marsh grasses: (a) *Spartina alterniflora,* the low marsh grass and (b) *Juncus roemerianus,* the common high marsh grass. Both examples shown are from the Georgia coast.

waters. Most of the benthic seaweeds belong to this phylum. Many are of commercial value because they contain **algin**, a gelatinous material that is used as an emulsifier in ice cream, paint, drugs, and cosmetics. The red algae include more species than all other phyla of algae. They may extend to relatively deep water. Some are calcareous and are important constituents of low-latitude reefs.

There are only a few common types of marine grasses. Some of these occupy the intertidal marsh environment. *Juncus* is the high marsh variety and *Spartina* is the low marsh type (fig. 9.28). Subtidal grasses, including *Thalassia*, commonly called turtle grass (fig. 9.29), and *Halodule,* live in shallow waters of the low latitudes and occupy similar environments. In the higher latitudes *Zostera,* or eel grass, replaces these as the dominant sea grass type.

Animals

Benthic animals comprise the most diverse major group of marine organisms, with about 150,000 species known. The vast majority of these are epifaunal. Because of the numerous modes of life and diversity of the organisms, it seems most appropriate to discuss the benthos through their life-styles.

Figure 9.29 (a) Sketch and (b) photo of *Thalassia,* commonly called turtle grass.

Vagrant Epifauna

This group probably is the largest in terms of both the numbers of individuals and the diversity of types. They have two primary features in common: all live on the surface of the ocean floor and all have at least some ability to move. They may live on a hard substrate, firm sand, or soft mud. There is a wide range in size from microscopic to over a meter in length. Some move extremely slowly whereas others move very quickly.

The smallest and probably the slowest of the benthic animals are the single-celled types, which include the Foraminifera. These tiny creatures are typically less than a millimeter in diameter with variously shaped tests or shells, most of which are multi-chambered (fig. 9.30). Most foram tests are composed of calcium carbonate, although some are comprised of sediment particles held together by organic material. They are a significant contributor to marine sediment. Mobility is nearly nonexistent in forams; however, the individual foram can extend its protoplasm through pores in the test and then contract it to pull the test, thus moving slowly. They feed by engulfing particulate organic matter.

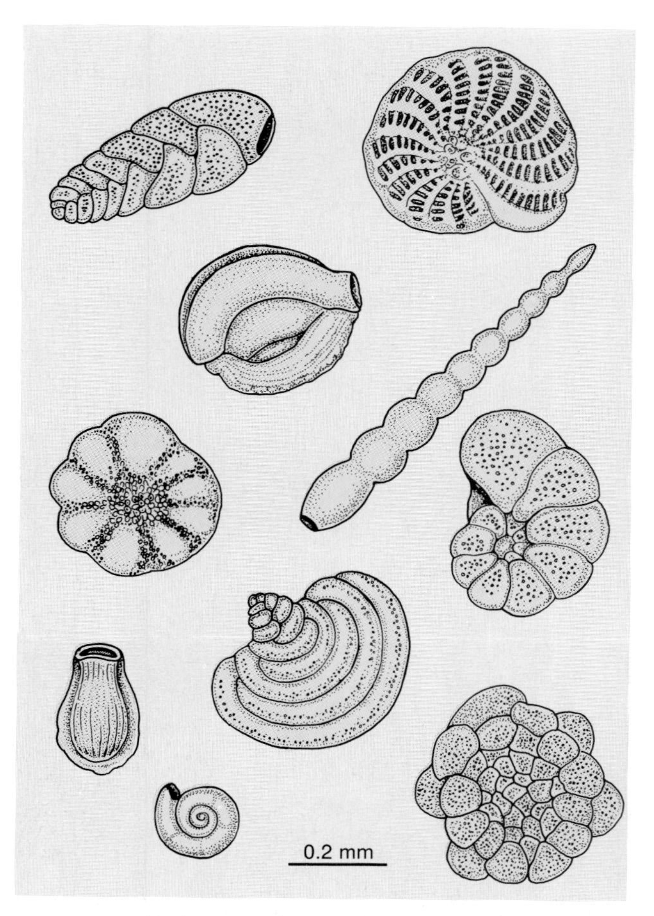

Figure 9.30 Various species of benthic foraminifera.

Most vagrant benthos are shelled macroinvertebrates, although there are some worms that crawl over the substrate. The shelled invertebrates are dominated by three groups: (1) the phylum Mollusca, which includes the clams, snails, chitons, and octopuses; (2) the phylum Echinodermata, which includes the sea urchins, sand dollars, starfish, and sea cucumbers; and (3) the crabs and lobsters of the phylum Arthropoda.

Nearly all vagrant benthic mollusks have external shells and move slowly, on the order of millimeters or centimeters per minute. Some, such as the chitons (fig. 9.31), are completely protected by their shell and will move across a hard substrate rasping and scraping food from the surface. A few snails occupy a similar niche. Most vagrant clams and snails move slowly through the sediment, ingesting whatever material they encounter, with little or no selection. They digest the nutrient material and excrete the mineral sediment in the form of pellets which become an important contributor to the volume of sediment. Generally these animals have thick, heavy shells to protect them from predators and from the elements. Most have somewhat bulbous shapes (fig. 9.32); those that live on soft sediment may have special adaptations in shell morphology to prevent sinking into the mud (fig. 9.33).

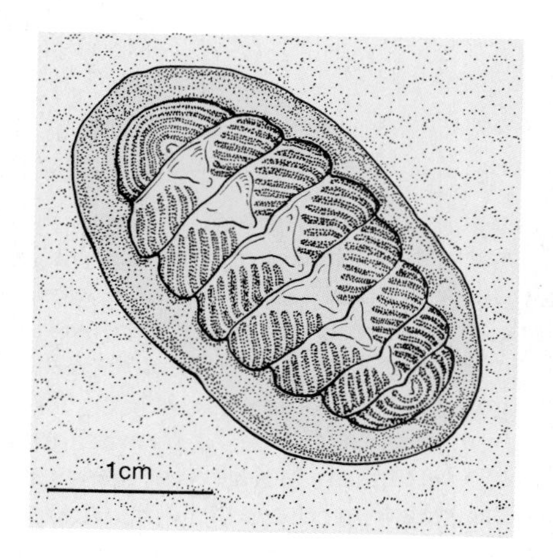

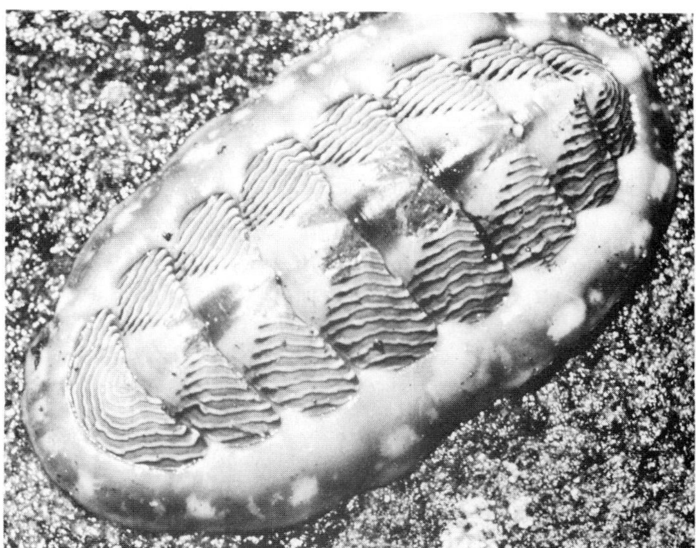

Figure 9.31 Chitons, slow-moving mollusks that feed by scraping food from rocks and other hard surfaces.

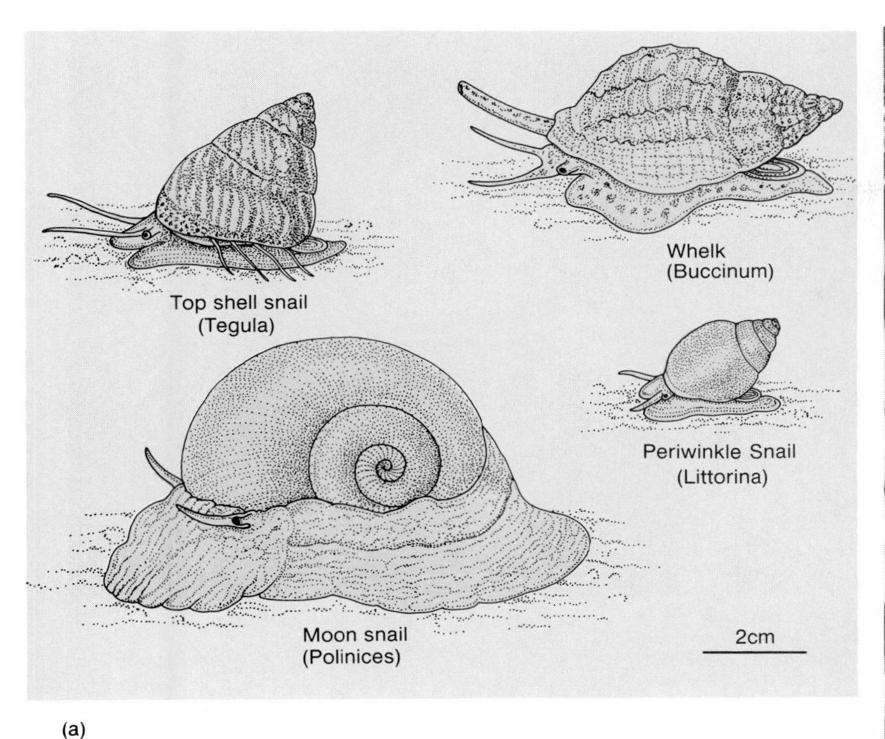

(a)

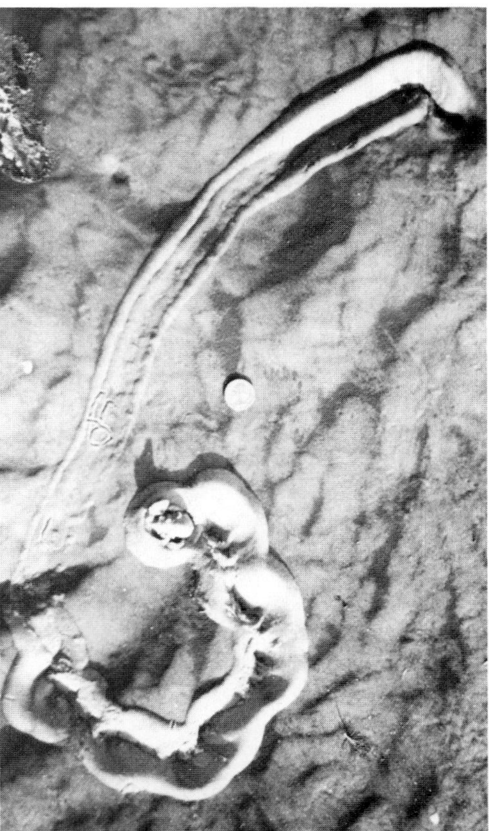
(b)

Figure 9.32 (a) Various examples of bulbous gastropods that have an epifaunal mode of life and (b) a *Polinices duplicatus* on a sandy tidal flat.

0 2 cm

Figure 9.33 Gastropod with shell modified to cover a large area and thereby adapt to epifaunal life on a soft substrate.

(a)

(b)

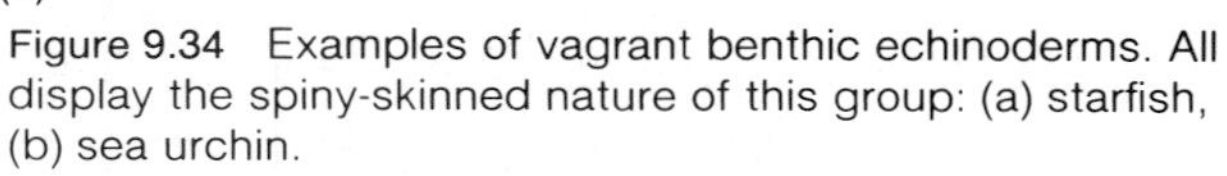

Figure 9.34 Examples of vagrant benthic echinoderms. All display the spiny-skinned nature of this group: (a) starfish, (b) sea urchin.

The echinoderms, or spiny-skinned animals, have bulky shapes, such as the various sea urchins, sea cucumbers, and many of the starfish (fig. 9.34). They have numerous appendages in the form of sucker feet or spines that are used for locomotion. The sea urchins live on hard substrates where they feed on debris attached to the rocks or they may live on the unconsolidated sediment. Sand dollars have the poorest mobility but they can move slowly by the whisker-like feet that surround their body. These animals may be partially infaunal in that they can bury themselves just under the sediment surface for protection.

Except for the octopus, the crabs and lobsters (fig. 9.35) are the largest of the vagrant benthos and the fastest. These jointed-leg animals can move rapidly; they use this ability along with their hard exoskeleton for protection from enemies. In addition, they have some swimming ability, using either their tails and/or specially adapted legs. Many species in this group live in the shelter of rocks, ledges, or other cover. They are scavengers and will eat almost anything that is available.

Sessile Epifauna

Many organisms are attached to the substrate throughout their maturity and have no mobility at all. Included are both solitary organisms and colonial ones, with as many as hundreds of individuals merged into large condominium-style skeletal complexes. Some can be torn up and moved, then reattached and still carry on, whereas others expire when uprooted. Virtually all sessile epifauna are filter feeders, relying on currents to carry their nourishment to them.

There is a variety of sessile solitary invertebrates that attaches to hard substrates, typically bedrock. These include barnacles, some brachiopods and mussels, sponges, sea anemones, and sea lilies. Although all are macroscopic, there is some range in size, from sponges only a few millimeters across to large sea lilies with arms that may be meters long.

Brachiopods and mussels (Pelecypoda) are both bivalves and filter feeders. Brachiopods attach with a stemlike foot that extends from near the hinge line

(a) (b) (c) (d)

Figure 9.35 (a) A New England lobster and (b) spiny lobster and (c) a blue crab, which is the common commercial variety, and (d) a horseshoe crab, which is a living fossil.

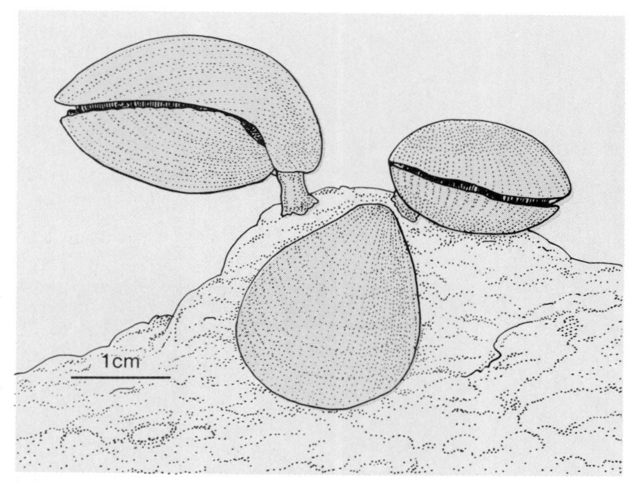

Figure 9.36 A sessile brachiopod or lampshell with its foot attached to a hard substrate.

(a)

(b)

Figure 9.37 (a) Sketch and (b) photo of *Mytilus,* the common edible mussel that attaches to firm substrates.

that holds the shells together (fig. 9.36). Mussels are about the same size and they attach themselves to a hard surface with strong threadlike structures that develop at the hinge line (fig. 9.37). The mussels are especially well attached and can withstand vigorous wave and current action.

Anemones (Coelenterata) and sea lilies (Echinodermata) belong to different phyla and have markedly contrasting anatomies, but there are some similarities in their feeding activities. Both have multiple appendages that serve as the primary food-gathering mechanisms. Anemones (fig. 9.38) are carnivores: they grasp their prey, then envelop and digest it. Some have sticky substances or toxins in their appendages to aid in capture and submission of their prey. By contrast, the sea lilies (fig. 9.39) have a sophisticated system of circulating plankton and organic debris to their centrally located mouth.

(a)

(b)

Figure 9.38 (a) Sketch and (b) photo of a sea anemone, which is a sessile, carnivorous coelenterate.

Figure 9.39 A crinoid or sea lily, which is one of the stalked varieties of echinoderms.

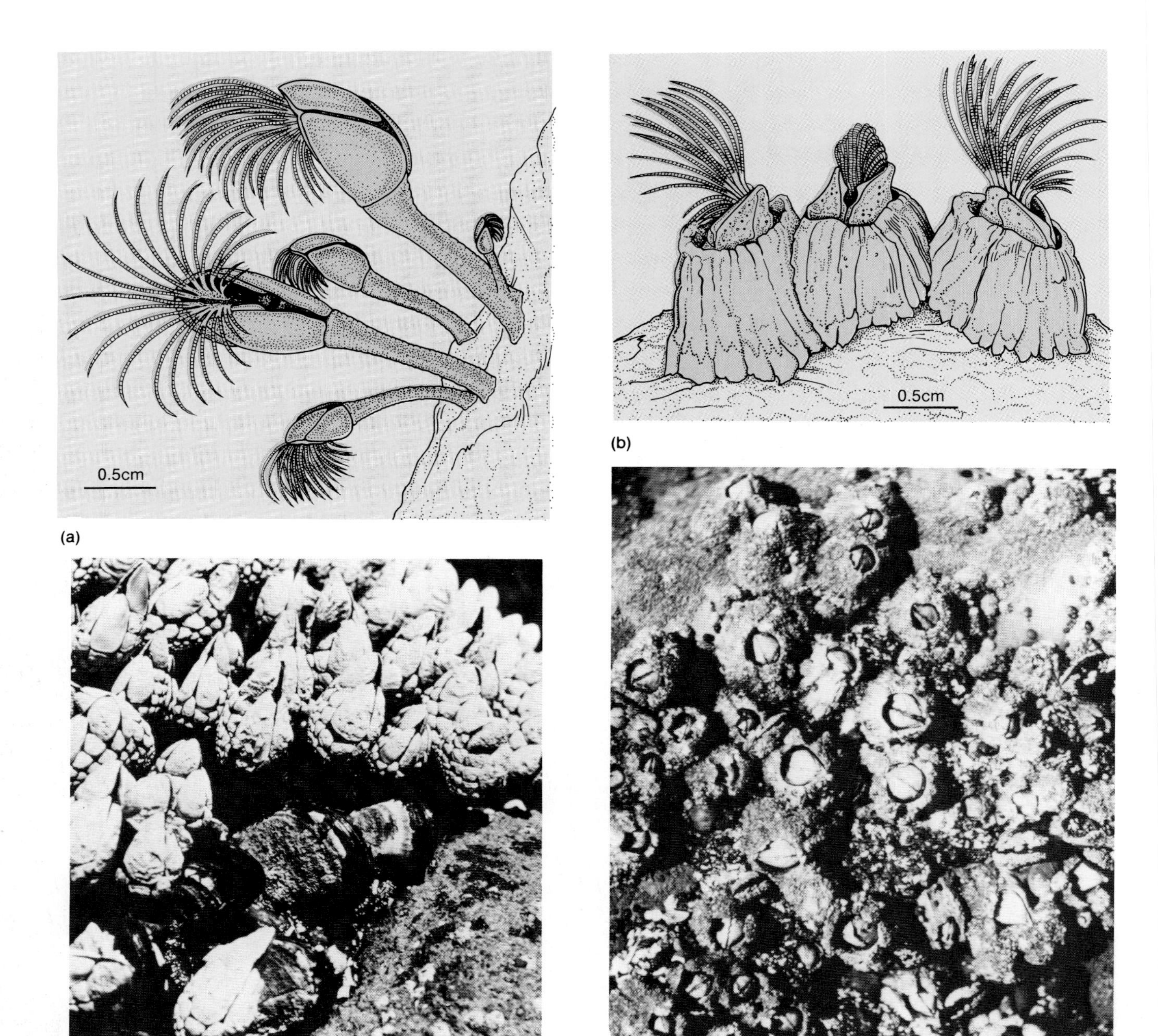

Figure 9.40 Examples of (a) encrusting barnacles and (b) goose-necked barnacles. Both are types of sessile benthos, and are filter feeders.

Barnacles are crustaceans (Arthropoda) that exist in two different forms, both sessile. Encrusting barnacles (fig. 9.40a) have their calcareous shells attached directly to the hard substrate, their soft part extended during feeding and retracted for protection. These are the barnacles that encrust boats, bridges, and other marine structures. The goose-necked barnacles (fig. 9.40b) have fleshy, stalklike structures that attach to the substrate and emanate

Figure 9.41 Sea whip with numerous flexible arms and a holdfast. It looks like a plant, but it's an animal.

from the shell that contains the soft parts of the organism. Both feed by removing small particles of organic debris from the water.

There are also colonial varieties of sessile benthonic animals. Primary members of this group are the corals, sea whips, sea fans, and bryozoans. Bryozoans are small and may be encrusting or delicately branching. They have calcareous external skeletons forming lacy structures that have given this phylum the nickname "moss animals." The corals, sea whips, and sea fans are all coelenterates, the same phylum as the sea anemones. Sea whips (fig. 9.41) and fans (fig. 9.42) do not have a hard, articulated skeleton and so they disintegrate upon expiration, whereas the corals have massive calcareous skeletons that house numerous individuals. All feed by filtering plankton and organic debris from the water.

Figure 9.42 Sea fan, a fairly stiff relative of the sea whip.

Infaunal Organisms

Many of the groups that are epifaunal also have representatives that burrow or bore into the substrate. Included are various snails, clams, worms, sea urchins, and crustaceans. Some groups are entirely infaunal, such as the tusk shells (scaphopods). Infaunal organisms occupy two different modes of life. Some graze or plow through the sediment and others construct extensive burrow complexes that they occupy and in which they move about. There are also those that burrow or bore near the surface and simply occupy that place; they do not move from place to place unless uprooted by waves, currents, or other organisms.

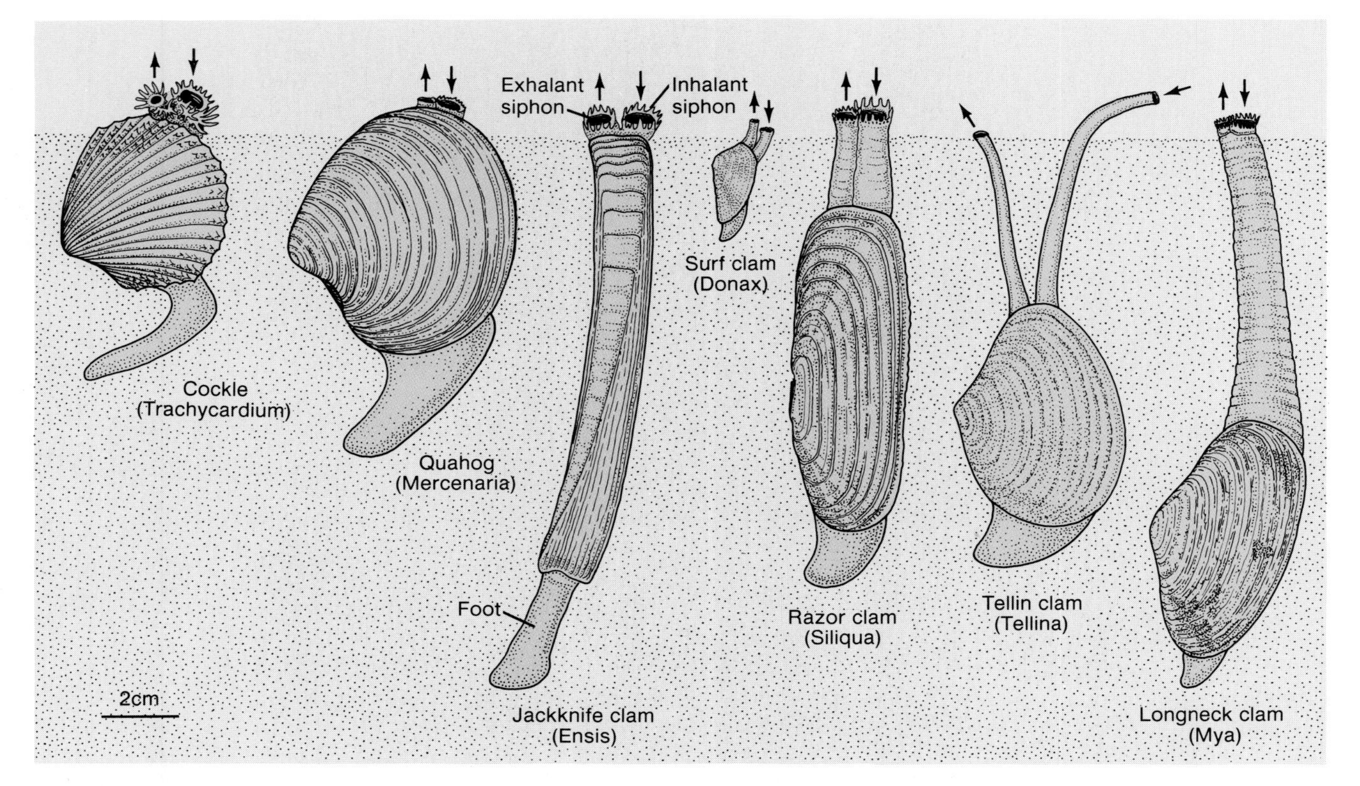

Figure 9.43 Sketches of various burrowing clams showing the muscular foot used for burrowing.

Grazing or plowing organisms include some sea urchins, snails, and clams. These organisms have shells that are streamlined for this type of activity. In the case of the sea urchins, they have short, stubby spines. Such animals ingest large quantities of sediment, extract the organic debris, and then excrete the sediment in pellet form.

A few types, like the ghost shrimp, have great burrowing abilities and may be found over 2 m beneath the substrate. They take in suspended particles and digest the organics, then excrete the mineral sediment. This feeding style is also used by the numerous clams that burrow near the surface. They have an inhalant siphon and an exhalant siphon that are used for circulating the water through their digestive systems (fig. 9.43). Numerous varieties of worms also occupy this mode of life. These types of infaunal organisms typically move only when they are exhumed from their burrow.

There are some organisms, such as certain clams and sponges, that can bore into solid rock or shells (fig. 9.44). This is done through a combination of physical rasping and chemical reaction between substances secreted by the organism and the substrate.

Summary of Main Points

Marine organisms are tremendously abundant and diverse. They can, however, be divided into three major groups based on their life-styles: the floating plankton, the swimming nekton, and the bottom-dwelling benthos.

Plankton include both plants (phytoplankton) and animals (zooplankton). Although both groups include a spectrum of sizes, most are small. The phytoplankton serve as the foundation of all life in the

Figure 9.44 Boring clams that use both physical and chemical means to penetrate hard substrates, such as rock or concrete.

world ocean because they are the producers. The zooplankton are generally the primary consumers and they, along with the phytoplankton, live in shallow water.

The true swimmers are the least diverse group of the three major oceanic life-styles. Fish are the dominant types, but mammals, reptiles, and squid (the only true swimming invertebrates) are also in the nekton. Most of the large, shallow-water forms live in the neritic waters over the continental margin, whereas smaller, deep-water forms are in the oceanic environment.

Benthic organisms are diverse and occupy a variety of modes of life. The sessile types include all of the plants plus such groups as the corals, oysters, barnacles, and sponges. Many of the benthic organisms, such as crabs, have good powers of mobility; however, most move slowly. The burrowing organisms may do so for shelter, for food, or for both.

Suggestions for Further Reading

Barnes, R.D. 1980. **Invertebrate Zoology**, 4th ed. Philadelphia: W.B. Saunders.

Dawes, C.J. 1982. **Marine Botany.** New York: John Wiley & Sons.

Fell, H.B. 1975. **Introduction to Marine Biology.** New York: Harper & Row.

Life in the Sea. 1982. Readings from *Scientific American*. San Francisco: W.H. Freeman.

Sumich, J.L. 1984. **An Introduction to the Biology of Marine Life**, 3d ed. Dubuque, Iowa: Wm. C. Brown.

Wheeler, A. 1975. **Fishes of the World: An Illustrated Dictionary.** New York: Macmillan.

10
Marine Geology and Geophysics

Chapter Outline

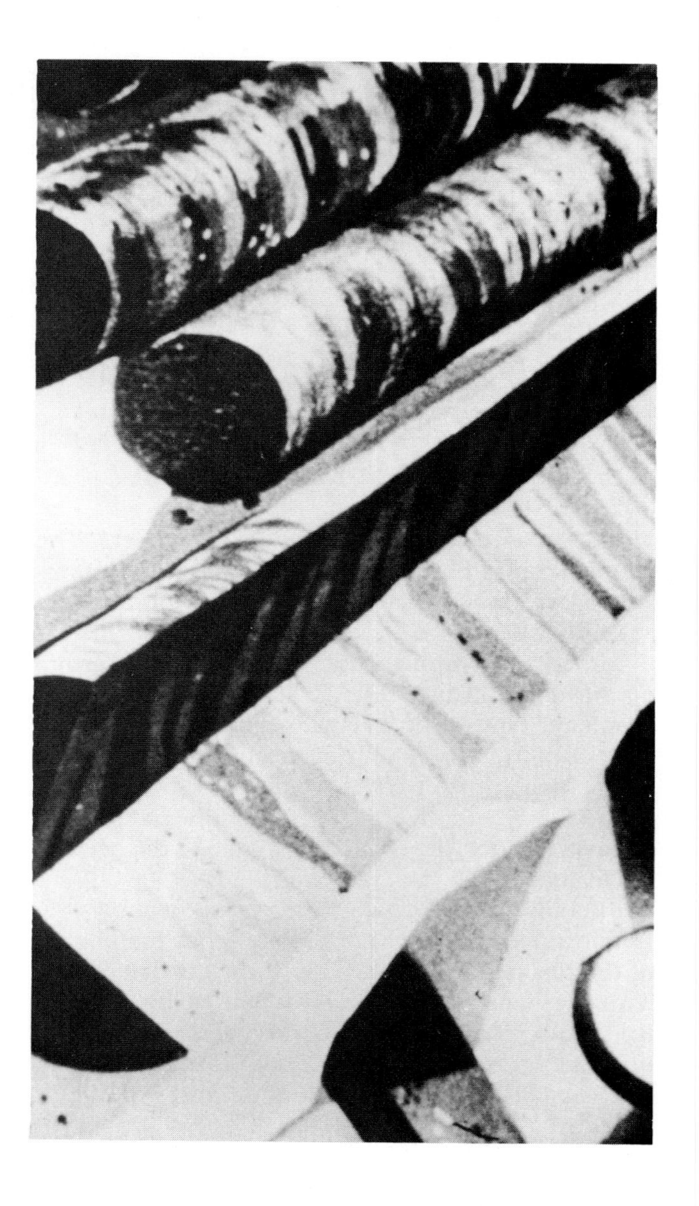

In order to fully understand the world ocean, it is necessary to consider the container that holds the water and its biota. The large-scale or global considerations were covered in chapter 3 in the discussion of plate tectonics. Here it is necessary to include not only the large-scale features of the ocean basins but also the sediments that come to rest on the ocean floor, where they come from, how they arrive at their destination, and how they interact with the ocean system. Additionally it is important to consider the history of sediment accumulation in the oceans. Although only about 180 million years of the earth's history is recorded in sediments of the ocean basins, many clues as to ancient conditions of the marine environment can be gleaned from studying the layers of sediments. Changes in life, salinity, temperature, circulation patterns, and tectonic activity are detectable.

A knowledge of the distribution and nature of strata beneath the ocean floor will aid in the understanding of the earth as a whole. Because much of this material is both deep below the sea surface and also at a significant depth below the seafloor, it can best be studied by remote and indirect means. The marine geophysicist who conducts such studies makes interpretations based on various physical properties of the rocks, such as density, magnetics, heat flow, and the transmission of seismic waves.

Marine Sediments

There is great variety in the sediment that comprises the floor of the ocean. This variation includes the composition, source, size and shape of the particles, mode of transport to the basin of deposition, and also the movement of sediment within the basin. Although there are exceptions, the rate of accumulation generally is greatest in the shallow coastal waters and decreases into the deep ocean basins with the abyssal plain having the slowest rate of sedimentation.

Origin and Composition of Marine Sediments

There are essentially four major sources of sediment that eventually find their way to the floor of the marine environment. These are (1) from previously existing rocks, primarily from land, as products of erosion and runoff, (2) from skeletal remains of marine organisms, (3) from direct precipitation from seawater in the marine environment, and (4) from extraterrestrial sources beyond the earth's atmosphere. The first two of these comprise the vast majority of all marine sediment.

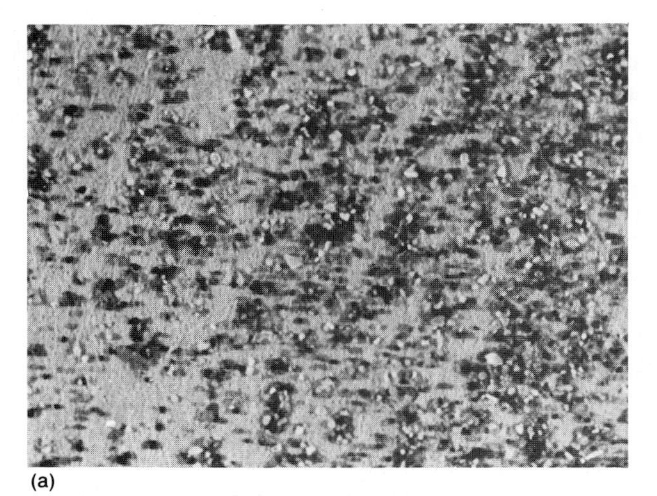

(a)

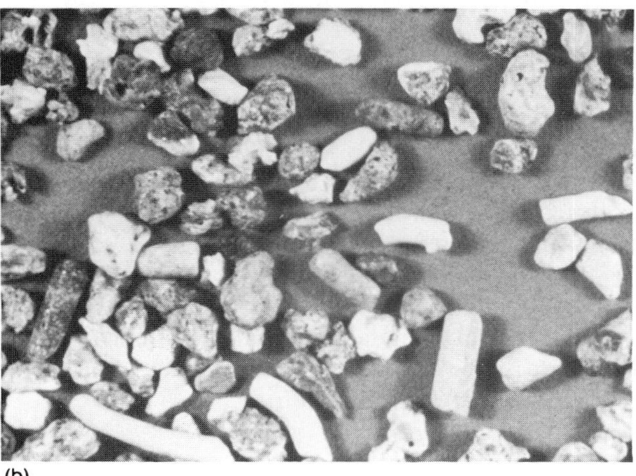

(b)

Figure 10.1 Close-up photograph of (a) terrigenous sand and (b) biogenic sand.

Land-derived sediment is primarily provided by stream runoff but also may result from windblown sources or from direct erosion of the coast by wave and tidal actions. Sediment particles from the land are almost infinitely varied, but a large percentage of them consists of only a few types (fig. 10.1a): clay minerals, which are largely weathering products of other minerals; quartz, which is a stable and resistant mineral; feldspar, which is also fairly resistant and the most abundant mineral in the earth's crust; and rock fragments, which are particles comprised of aggregates of small mineral grains. These four particle types constitute at least 90% of the material eroded from the landmasses and carried to the sea. All are silicate minerals; that is, they contain silicon and oxygen along with various cations. Volcanic islands and eruptions are also an important source of sediment.

Marine-produced sediment is largely **biogenic,** consisting of the skeletal material of marine organisms (fig. 10.1b). Most of this is either calcium carbonate ($CaCO_3$) or silicon dioxide (SiO_2), which is quartz. Most skeletal debris originates as the external shell of planktonic or benthonic organisms. Nekton provide only a small amount of biogenic sediment.

Marine sediment is also derived from direct precipitation of mineral material in the marine environment, including calcium carbonate, manganese nodules, some clay minerals, and evaporite minerals such as halite (salt).

Extraterrestrial sediment comprises only a trace amount of the total received by the oceans. It is in the form of small metallic or silicate spherical particles, or pieces of chondrite meteorites, which are of Fe-Mg silicate composition.

Sediment Accumulation Styles

There are two primary mechanisms whereby sediment makes its way to the seafloor. Much of the sediment that is derived from land, especially the relatively coarse material, is carried onto or near the seafloor by currents. These sediments accumulate by filling in low areas and may not be at a common depth over a wide area (fig. 10.2a). By contrast, **pelagic sediments** are the fine-grained particles that fall through the water column like snow through air and come to rest in relatively uniform layers on the seafloor (fig. 10.2b). Both fine land-derived particles and fine skeletal debris from planktonic organisms comprise pelagic sediment.

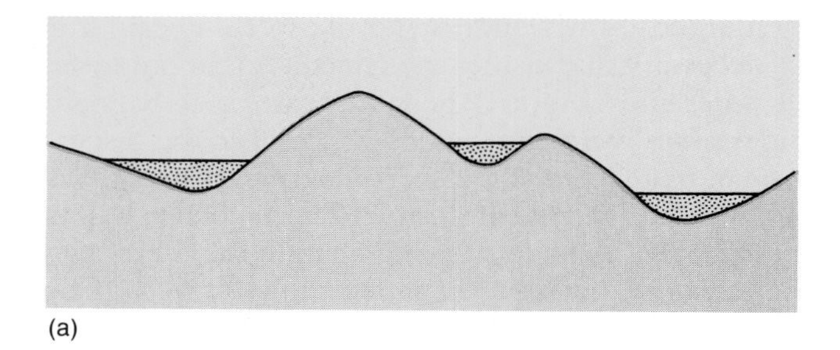

(a)

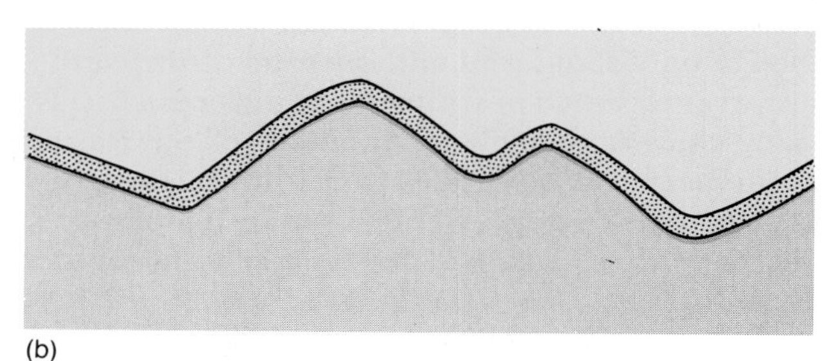

(b)

Figure 10.2 Sediment accumulation styles in the deep ocean: (a) land-derived sediment, which is transported and deposited along the ocean floor and (b) pelagic sediment, which settles from the water column in uniform layers.

Sediment Texture

The texture of sediments can provide clues to its direction and mode of transport, the nature of the environment of deposition, and the changes that have taken place since it accumulated. Most commonly, the size and shape of grains is considered, but their fabric or arrangement relative to one another is also important.

Because of the great range in particle size, from microns to meters, it is necessary to classify the grain size of sediments by a logarithmic scale. Scientists use what is known as the Wentworth scale as the standard. The scale is metric with definite size ranges for such commonly used categories as clay, sand, pebbles, and boulders (table 10.1). All sediment particles, regardless of size and composition, can be fitted into an appropriate category. The term **mud** is a grain-size term that includes subequal mixtures of clay and silt.

Depending on the grain size, a small volume of marine sediment may contain thousands of sediment particles. Typically, the grain size for such a sample

TABLE 10.1
Simplified Wentworth Grain-Size Scale with the Major Textural Categories

Particle Diameter μ	Particle Diameter mm	Size Class	
	mm	Boulders	
	256		
		Cobbles	Gravel
	64		
		Pebbles	
	4		
		Granules	
	2		
μ		Very coarse	Sand
	1		
		Coarse	
500	1/2		
		Medium	
250	1/4		
		Fine	
125	1/8		
		Very fine	
62	1/16		
		Silt	Mud
2	1/512		
		Clay	

is based on the average or **mean** particle size. In order to obtain this information, the sediment sample is sieved through a set of ordered mesh sizes or settled through a column of water where the largest grains settle first and so on. Both techniques allow for fairly rapid measurement of many grains. Most commonly a particular sediment is described in terms of its mean or average grain size and its **sorting value,** which is simply an expression of the spread of particle sizes within a given sample (see fig. 10.3). In a well-sorted sediment most of the particles are about the same size and in a poorly sorted sediment there is a wide range of particle sizes. Both of these parameters may have implications for the depositional environment of a sediment. For example, coarse sediments like gravels are typically restricted to high-energy environments, such as a beach, whereas very fine-grained muds suggest low-energy environments. Sediments that are well sorted imply some mechanism or mechanisms to achieve that sorting, such as waves or currents.

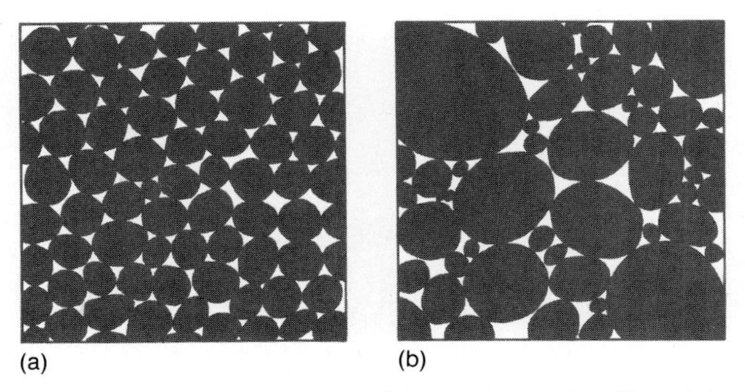

Figure 10.3 Diagram showing (a) a well-sorted sediment, in which most particles are the same size and (b) a poorly sorted sediment, with a range of particle sizes.

Sediment Transport

Great quantities of sediment are transported from land to the marine environment. In addition, there is considerable sediment produced within the marine environment. It is likely the vast majority of the sediment that makes its way to the ocean will be transported many times during its lifetime. This transportation is accomplished primarily by waves and currents, although wind and organisms may also move particles. These processes, their duration, intensity, and location, determine the nature of the sediment that accumulates in any given place within the marine environment. Grain size, sorting, and fabric are all the result of the processes to which sediments have been subjected.

Probably the most fundamental concept in sediment transport is that demonstrated by Hjulström's curve, which shows the relationship between grain size and current velocity in transporting sediment (fig. 10.4). As expected, the larger the particle, the greater the velocity required to transport it. This is only partially correct, however, because some very fine sediment particles require substantial currents to move them. The curve shows that for fine sediments with grains less than about 0.1 mm, an increasingly strong current is required to erode them. This is due to the cohesion of the sediment particles. Mud is more cohesive than sand, which is the reason for the nonlinear relationship. Once a particle is picked up and is under transport, the relationship between velocity and particle size is linear (fig. 10.4).

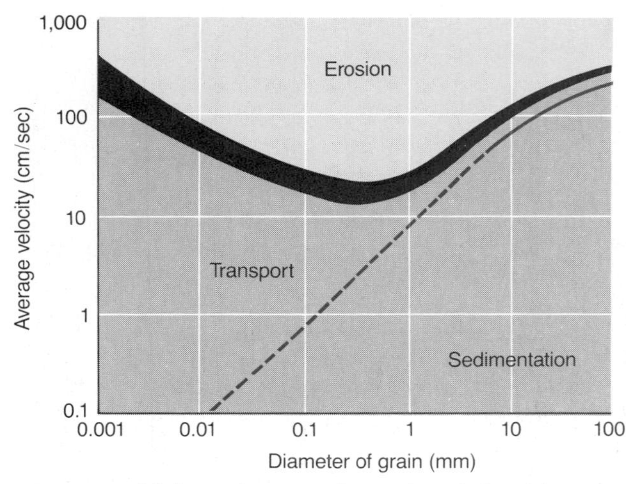

Figure 10.4 Hjulstrom's curve shows the relationship between current velocity and grain size with respect to erosion, transportation, and deposition of sediment.

Classification of Marine Sediments

The four main sources of sediment mentioned above serve as the basis for a simple but useful classification of marine sediments. These can be considered (1) lithogenous, those derived from rocks; (2) biogenous, those derived from organisms; (3) hydrogenous, those derived or precipitated from seawater; and (4) cosmogenous, those sediments derived from outside the earth's atmosphere.

Lithogenous Sediments

Any sediment that is eroded from a previously existing rock or sediment and carried into the marine environment is considered lithogenous. The vast majority of lithogenous sediments are carried by rivers to the marine environment where they may accumulate as a delta, or be carried along the coast, or be transported into deeper water. Other lithogenous sediments are derived from erosion of rocks exposed along the coast, or from islands in the marine environment, or are carried to the marine environment by glaciers and then deposited when the ice melts. Wind may also account for some lithogenous sediment, although most people believe that this is a volumetrically insignificant contribution.

Relatively coarse lithogenous sediment is usually deposited along or near the coast. Sand-sized sediment forms beaches, barrier islands, and other coastal sediment bodies. The lithogenous sediment that reaches the deep part of the ocean is typically fine sediment, such as silt and clay, that is carried in suspension and then settles slowly to the ocean floor under conditions of low physical energy. Glaciers and ice-rafting are mechanisms for transporting coarse sediment particles into deep water. Currents may carry sediment-laden icebergs for hundreds of miles before they melt and free the trapped sediment. Relatively coarse sediment also accumulates at the base of steep slopes in ocean basins, where density currents and gravity (see chapter 5) transport sediment.

Biogenous Sediments

Both plant and animal skeletal remains are important sediment constituents in the marine environment. Virtually any place in the ocean is likely to receive some type of biogenous sediment. With rare exception, this material is composed of calcium carbonate or silica. Phosphate material from bones of fish and other vertebrates is present but not in significant quantities.

Near the coast and on the shelf most of the skeletal material is calcareous and is derived from benthic invertebrates and algae. Primary contributors are mollusks, echinoderms, coelenterates, and green and red algae. In oceanic waters small planktonic organisms are the primary contributors to the sediment and, in fact, they are the dominant sediment constituent throughout much of the deep-sea environment. Diatoms are the primary contributor of siliceous sediment (fig. 10.5) and radiolarians (fig. 10.6) are a distant second in abundance. Most calcareous deep-sea sediment is provided by **tests** or skeletons of foraminifers (fig. 10.7), with coccoliths (fig. 10.8) important contributors in shallow, warm ocean basins.

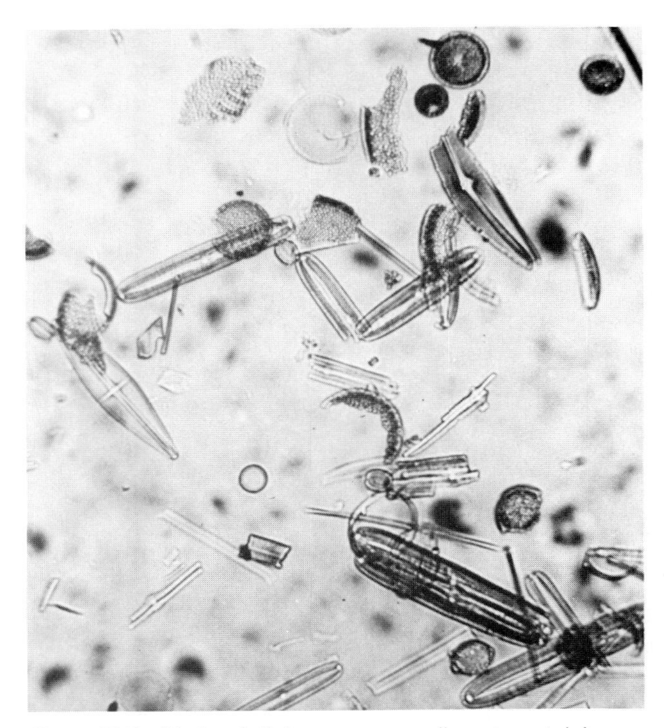

Figure 10.5 Photo of diatomaceous sediment containing various species.

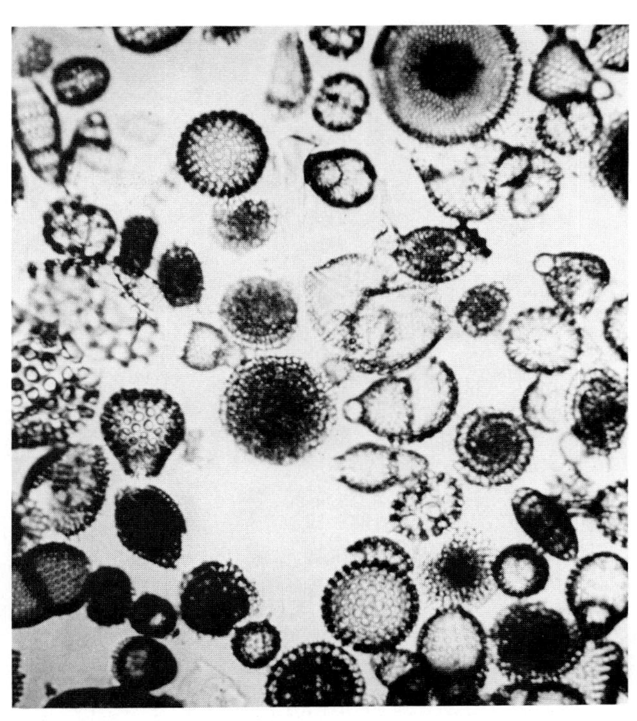

Figure 10.6 Photo of radiolarians, a fairly common siliceous deep-sea sediment.

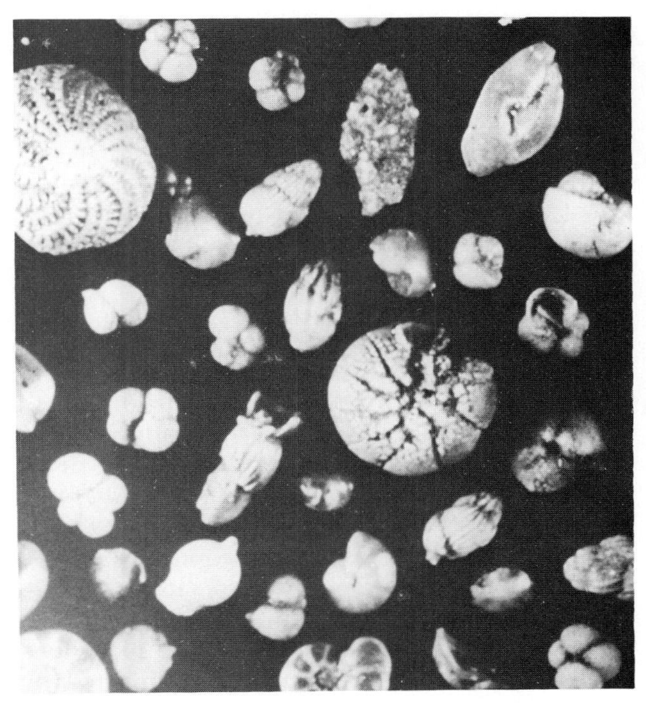

Figure 10.7 Photo of deep-sea ooze with abundant foraminifera.

Figure 10.8 Photo of coccolith sediment showing numerous plates, each of which is only a few microns in diameter.

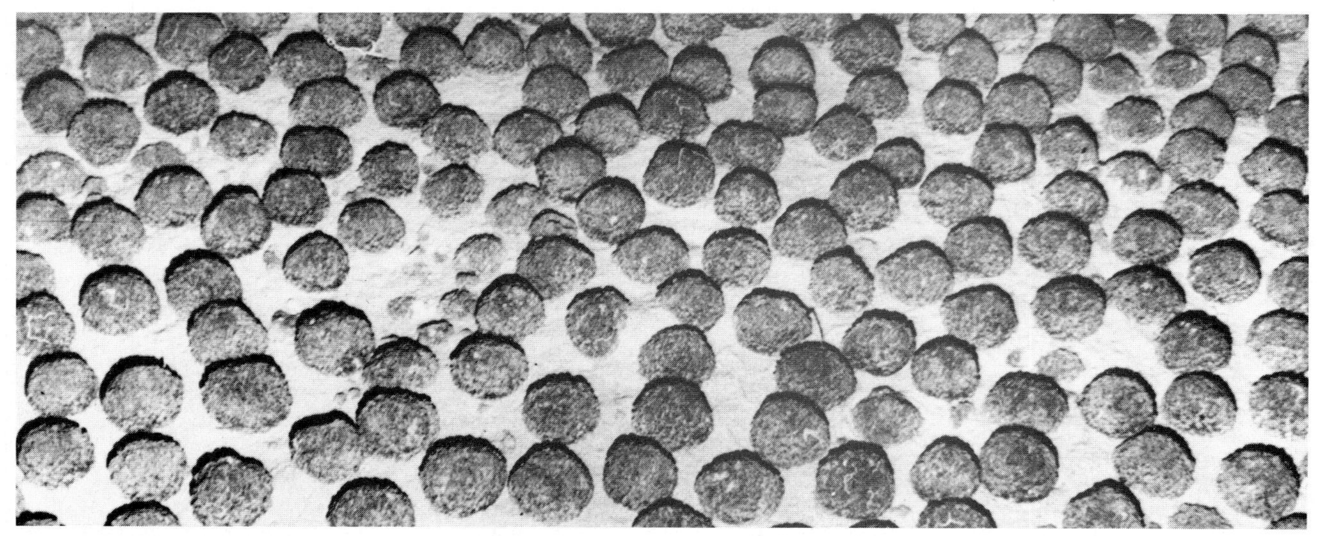

Figure 10.9 Manganese nodules from the ocean floor. These may range from only a few millimeters in diameter to over a meter.

Hydrogenous Sediments

Direct precipitation of minerals from seawater provides for locally abundant accumulations on the seafloor. **Manganese nodules** (fig. 10.9) are amorphous masses of manganese oxide and iron oxide that are widely distributed on the deep ocean floor. They range from a few millimeters to several centimeters in diameter and are most abundant where the rate of sedimentation is low. These nodules are of considerable interest because of their potential economic importance.

Phosphorite nodules are also of potential economic importance and are typically present along the outer portion of the continental margin. Phosphorite nodules and encrustations are concentrated in areas where upwelling brings large quantities of phosphorus up along the continental slope (see chapter 5). The nodules may be up to 30% phosphate by weight.

Carbonate minerals ($CaCO_3$) may precipitate from seawater. This is generally confined to warm shallow water where the pH is at least 8.1. The precipitation of carbonate is facilitated by photosynthesis because that process removes carbon dioxide from the water and thus raises the pH. Increased water temperature brings about the same set of circumstances.

Cosmogenous Sediments

Outer space provides a small amount of unusual and interesting sediment to the earth. Tiny, magnetic nickel-iron spherules comprise one of the major cosmogenic ocean sediments. These particles are typically less than 30 microns in diameter. Thousands of tons of these fall to the earth each year but they are so dispersed that they are undetectable in most sediments.

The other primary type of cosmogenous marine sediment is small fragments of stony meteorites called **chondrites**. These particles range up to nearly a millimeter in diameter and are iron and magnesium silicate in composition.

Distribution of Marine Sediment

Broad categories of marine sediment distribution are easily related to the two major zones of the ocean basins (chapter 8): the neritic, or shallow, continental shelf environment and the oceanic, or deep, margin and ocean basin floor sediment. Although both may contain all of the four sediment types listed above, there are certain generalizations that can be made about the two categories. For example, neritic sediment is typically coarser than oceanic sediment and it accumulates more rapidly. The biogenous component of neritic sediment is dominated by skeletal remains of such benthic macro-invertebrates as mollusks, corals, and echinoderms, whereas

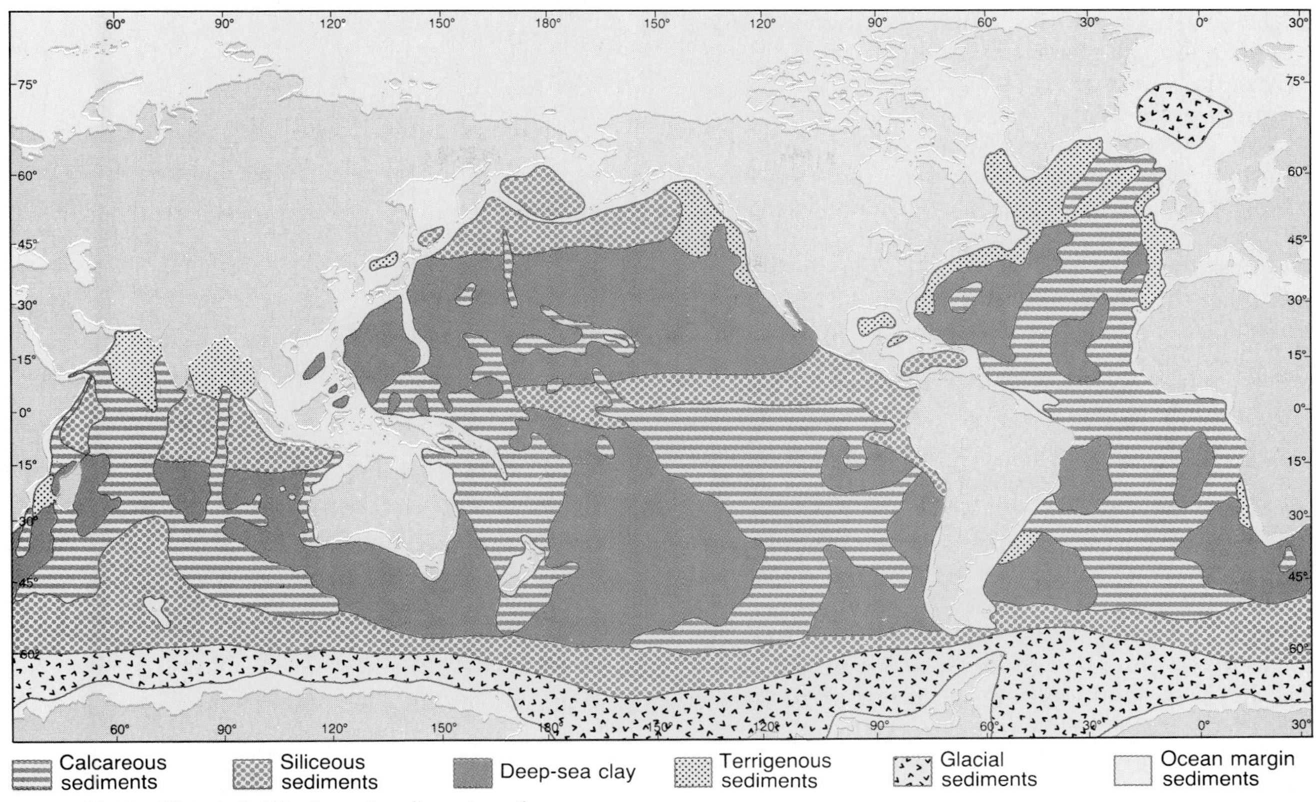

Figure 10.10 Global distribution of sediment on the ocean floor.

in oceanic sediments planktonic microorganisms such as foraminifera and diatoms are dominant.

The neritic sediment is largely sand and mud. Distribution depends on the nature and intensity of physical processes along the landmasses and on the mechanism of transport to the marine environment. A great deal of sediment is trapped in coastal bays, estuaries, and lagoons before reaching the open marine environment. Some reaches the ocean and is carried both parallel and perpendicular to the coast by waves and currents. Large concentrations of sediment called deltas form at the mouths of some rivers. Therefore, the coastal zone displays a complex pattern of sediment distribution.

The sediments of the continental shelf display a patchy distribution of sand, mud, and mixtures of both. In high latitudes, glacial sediments may be present across the shelf. These include gravel up to boulder size as well as finer sediment. Some is deposited directly from the glacier and some is rafted out from the shore by way of icebergs.

Sediment at the interface between the neritic and oceanic environments includes components of both provinces. The continental slope and rise contain thick accumulations of sediments that are primarily lithogenous and mostly move by downslope transport due to gravity. Some sediment is deposited as pelagic sediment, both lithogenous and biogenous types.

Sediments of the deep sea are relatively homogenous in comparison to other marine environments. The vast majority are pelagic in nature but some are of glacial origin or are from gravity-generated density currents. Pelagic **oozes**, sediments that have at least 30% by weight planktonic skeletal remains, are the most widely distributed. The low- and mid-latitudes of all ocean basins except the North Pacific are dominated by calcareous ooze (fig. 10.10). The North Pacific basin is covered primarily by pelagic brown clay. In high latitudes of the southern hemisphere, siliceous ooze predominates, and glacial sediments surround the Antarctic landmass (fig. 10.10). Abundant manganese nodules cover the seafloor in the south-central Pacific and Indian Oceans and just west of South America.

This general picture of deep-sea sediment distribution is oversimplified due to the enormous extent of the ocean basins. Nevertheless, deep-sea sediments are laterally homogenous in comparison to their continental counterparts. This is simply a reflection of the environments in which the respective sediments are deposited.

Deep-sea Stratigraphy

The study of the layers of rock and sediment of the earth's crust is called **stratigraphy**. On the continents this type of investigation is accomplished largely through the exposure of the strata at the surface with some additional information supplied through subsurface data provided by wells for petroleum and water. In the ocean there is little opportunity to observe exposures of strata and most data are provided by coring the layers of rock and sediment. Until recently this was restricted to various types of piston corers that operated by gravity (fig. 10.11). This type of coring apparatus is capable of sampling only about 20 m into the strata of the ocean floor.

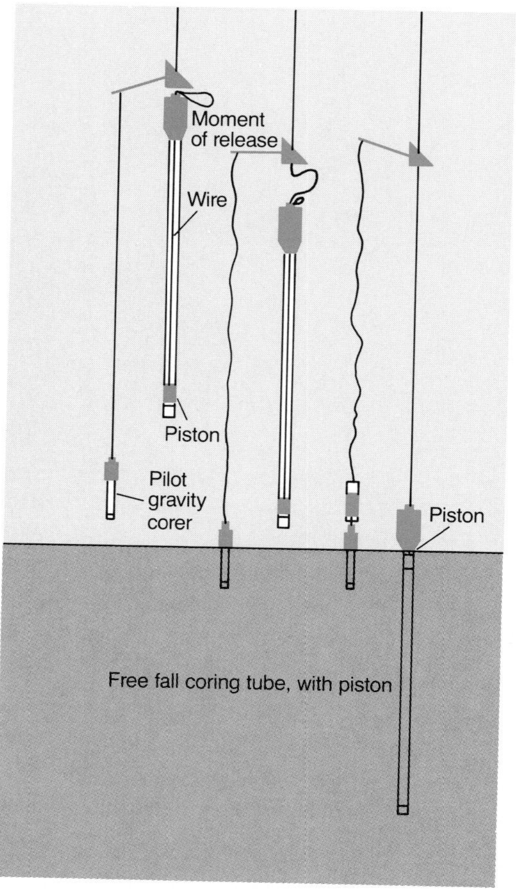

Figure 10.11 Diagram showing various stages in the operation of a piston coring device. The pilot corer triggers the main corer as it hits the seafloor. As the main core falls, the piston inside assists the core in penetrating the sediment.

Drilling on the Ocean Floor

The initiation of the Deep Sea Drilling Project (DSDP) in 1968 opened the door to deep penetration and recovery of the sediment layers beneath the deep ocean floor. Major accomplishments of the project include drilling successfully in water 7,000 m deep and drilling one hole to a depth of over 1,700 m (more than a mile) beneath the ocean floor. More than 900 holes have been drilled at over 600 sites with a cumulative depth of over 200 km. These cores and holes have provided a great deal of information about the history of the ocean basins but they really only scratch the surface. The density of holes on a global basis is less than one per 500,000 km^2 and the deepest hole penetrates less than 25% of the thickness of the oceanic crust.

Perhaps the most important contributions of the DSDP are the data that confirm the concept of plate tectonics and continental drift (chapter 3) and contribute extensive knowledge of the geologic history of ocean basins, including the evolution of oceanic species. The project demonstrated that the oldest

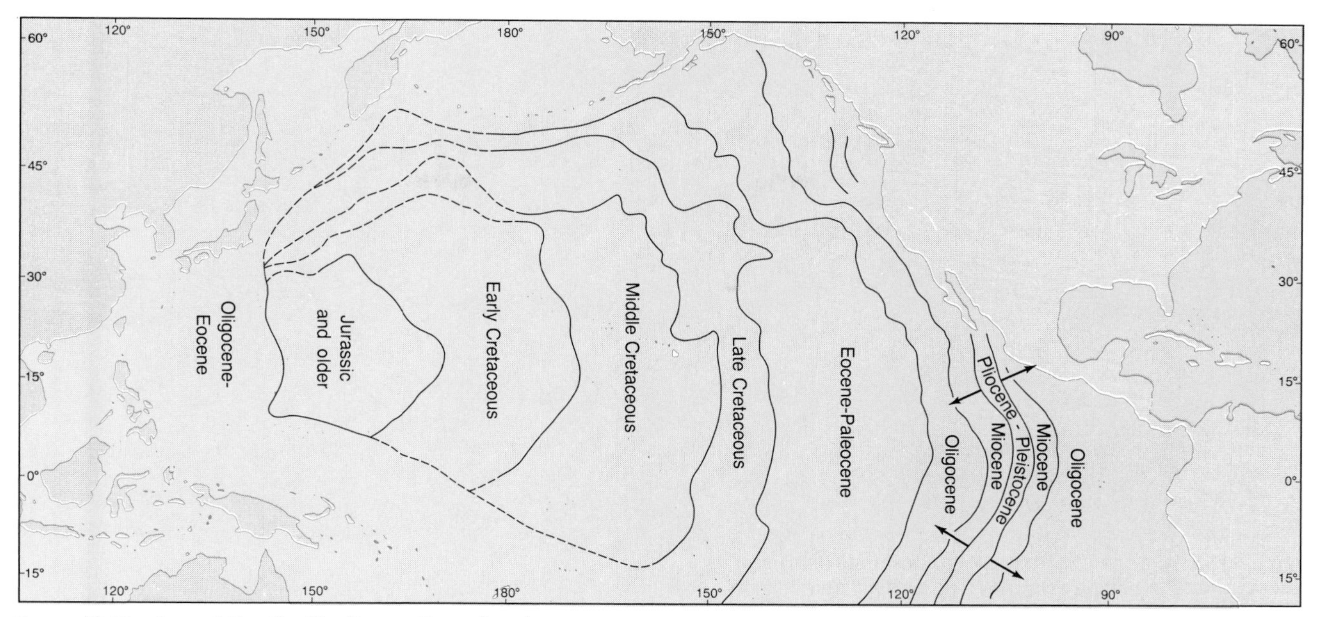

Figure 10.12 Age of the Pacific Ocean floor. Age increases away from the oceanic ridge.

sediments in the ocean basins are less than 200 million years old, whereas prior to this discovery many scientists thought that the ocean basins had accumulated sediments for billions of years.

The DSDP has now been replaced by the Ocean Drilling Program and the *Challenger* has been replaced by the *JOIDES Resolution*. This new project will largely be a continuation of the deep drilling initiated under DSDP but with much greater technology. Deeper holes in deeper water are possible with the new ship.

General Stratigraphy

The plate tectonics model for the earth and the information gained from Deep Sea Drilling Project cores provide a good general stratigraphic model for ocean basins. There is a general thickening of oceanic sediment layers away from the spreading centers or oceanic ridges. Additionally, the age of the basalt layers increases away from these spreading centers. Both the Atlantic and Pacific Oceans conform to this pattern. Strata only a few million years old are at or near the oceanic ridge and age increases away from the ridge, culminating with strata over 150 million years old at the margin of the ocean basin (fig. 10.12).

In cross section, the oceanic crust acts like a conveyor belt moving slowly from the spreading center, sinking as it goes. The pelagic sediments of the deep sea accumulate slowly on this surface thus producing a very thin wedge of strata (fig. 10.13). The thickness of these strata tends to show an inverse relationship to the rate of spreading from the center.

Marine Geophysics

Techniques for studying the earth's crust beneath the ocean floor are similar to those employed on land. In order to study the deep part of the earth's crust it is necessary to apply the principles of seismology, gravimetrics, and magnetics to the study of the oceanic crust. **Seismology** is the study of earthquakes or earth vibrations, including those that are

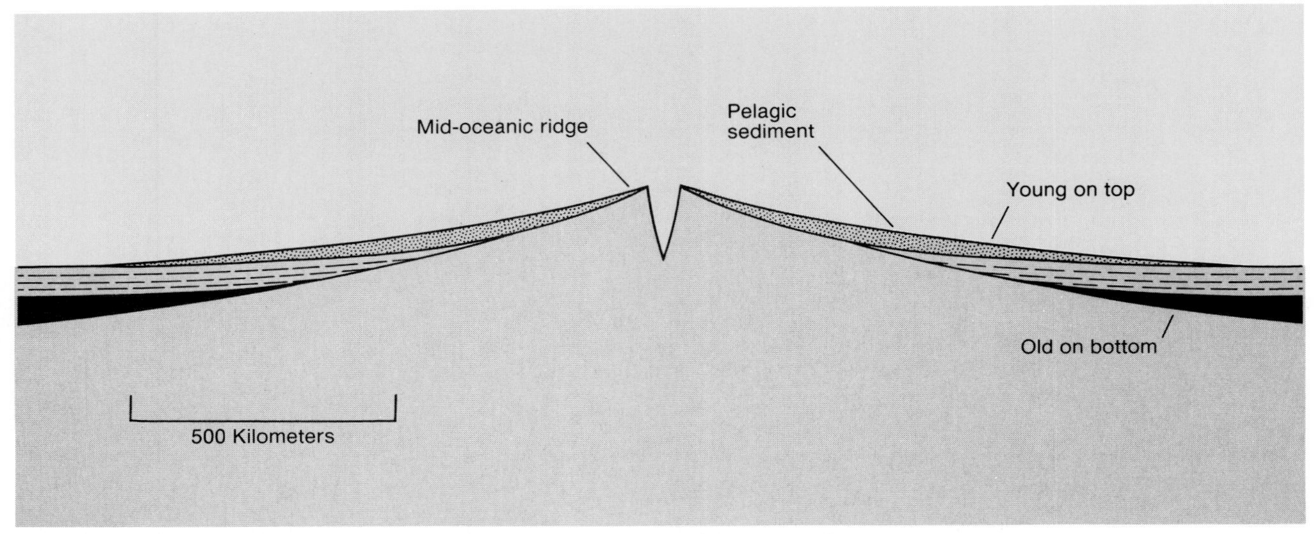

Figure 10.13 Simple stratigraphic diagram for deep-sea sediment accumulation relative to spreading center. Note the thickening away from the ridge.

artificially generated. By studying these waves and their transmission through the crust, it is possible to interpret the type, thickness, and arrangement of sediments and rocks in the crust. **Gravimetrics** is the measurement of the earth's gravity field. Subtle variations in the pull of gravity over the globe reflect differences in the type and the thickness of the rocks. **Magnetics** is the study of the earth's magnetic field, which also varies in intensity and can be related to rock type.

Seismic Reflection and Refraction

Sound impulses were first used in precision depth recorders in charting the bathymetry of the ocean floor. The PDR, as it is called, became a prominent tool during World War II both for surveying the ocean and in antisubmarine warfare. Its operation is based on the measurement of the time that it takes for sound impulses to travel from the ship to the seafloor and return. The frequency of vibration of these sound impulses is generally about 12 kHz (12 kilohertz). A **hertz** is one vibration per second, so 12 kHz is 12,000 vibrations per second.

Reflection profiling of the sediment layers beneath the ocean floor involves a variety of seismic frequencies, depending upon the nature of the data desired. The basic technique is similar to that for depth recorders except that the sound waves penetrate the seafloor and are reflected off the various layers of sediment. Reflection is caused by differences in sediment and rock properties such as density, porosity, and compaction. The vibrations may be generated by explosion, mechanical impact, or electronic impulse. The third method is the most common one at the present time.

In its simplest form, seismic profiling is accomplished by emitting a sound source from the surface. The waves travel to and through the sediment layers until they strike a surface that reflects them, generally a hard surface. They are reflected back to the surface where they are accepted by a receiver (fig. 10.14). The data are interpreted and printed on a recorder, which displays a profile or cross section (fig. 10.15). These data are collected continuously over the area of study.

The penetration and the detail provided by the seismic profiles depend largely upon the frequency of the signal that is sent out from the ship. In general, the higher frequency signals provide excellent resolution but have little penetration. Low-frequency signals yield deep penetration but with

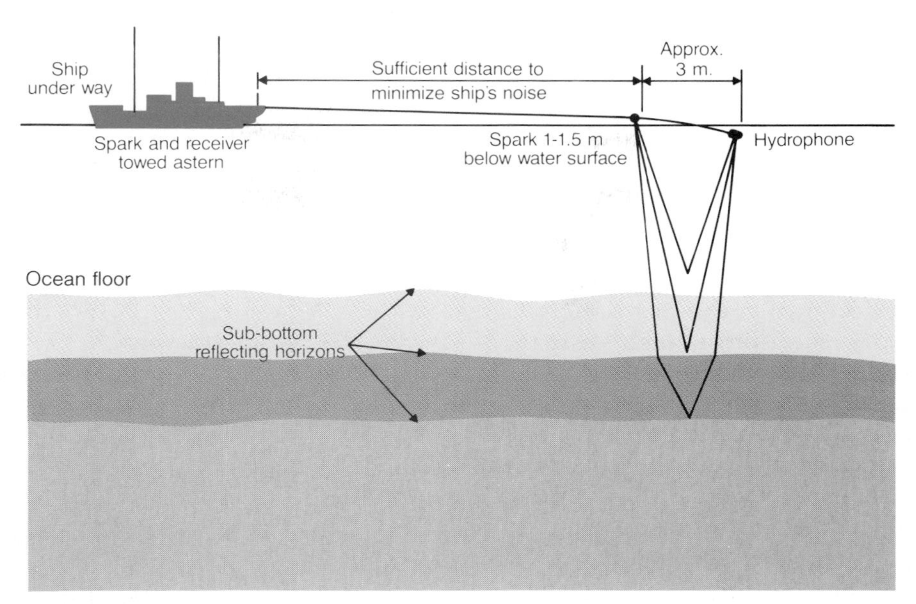

Figure 10.14 Diagram of apparatus used in continuous seismic reflection profiling.

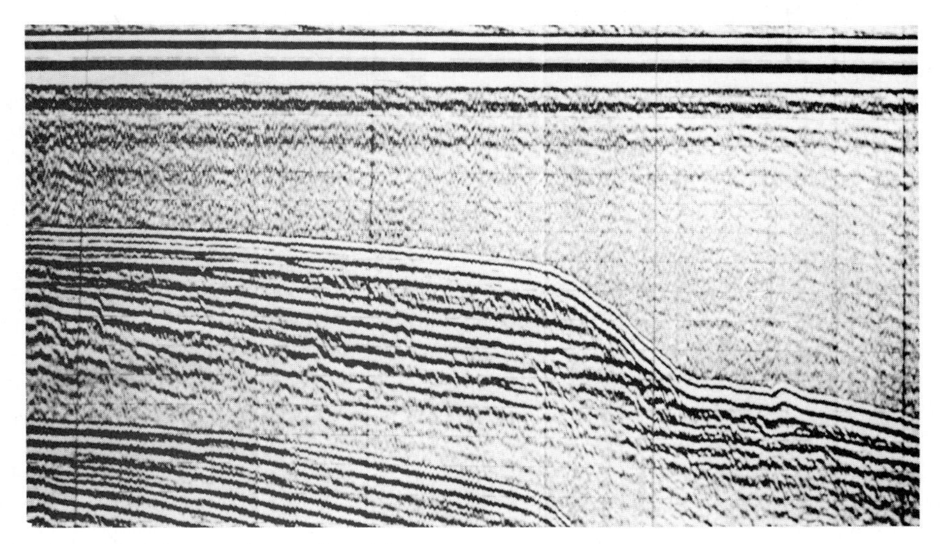

Figure 10.15 A seismic profile taken on the continental shelf off the north coast of Puerto Rico.

only broad detail. For example, the 3.5 kHz unit is desirable for penetrations of a few tens of meters because it provides excellent detail, but frequencies of a few hundred hertz are required for deep penetration of the sediment layer. Generally the low-frequency systems utilize some type of compressed air impulse (air gun) or an electric spark. In either case, an impulse is released at regular intervals and the data are picked up and recorded just as in the higher frequency systems.

In recent years the technology of **seismic reflection** has advanced greatly, largely due to the activities of industries involved in petroleum exploration. A major innovation has been the development of multichannel seismic systems. This type of system employs long streamers, some over a kilometer in length, containing numerous hydrophones or receivers. Such a system enables many signals to be received from each reflection point beneath the ocean floor. Among other things, this technique provides for a clearer and more detailed profile record (fig. 10.16).

Seismic refraction involves similar principles and equipment as does reflection but in this case the refraction or bending of waves that pass through sediment and rock layers is recorded. Seismic reflection only penetrates into the upper sedimentary layer of the oceanic crust. In order to penetrate the entire crust, it is necessary to use refraction techniques. Because they produce the desired low-frequency waves, explosives are commonly used to generate the seismic waves. A string of hydrophones is used to pick up the refracted signals at the surface. This can be done with two ships, using one for detonation of the explosives and the other to receive the signals, or with one ship, utilizing a long string for hydrophones (fig. 10.17). This type of seismic study can penetrate several kilometers.

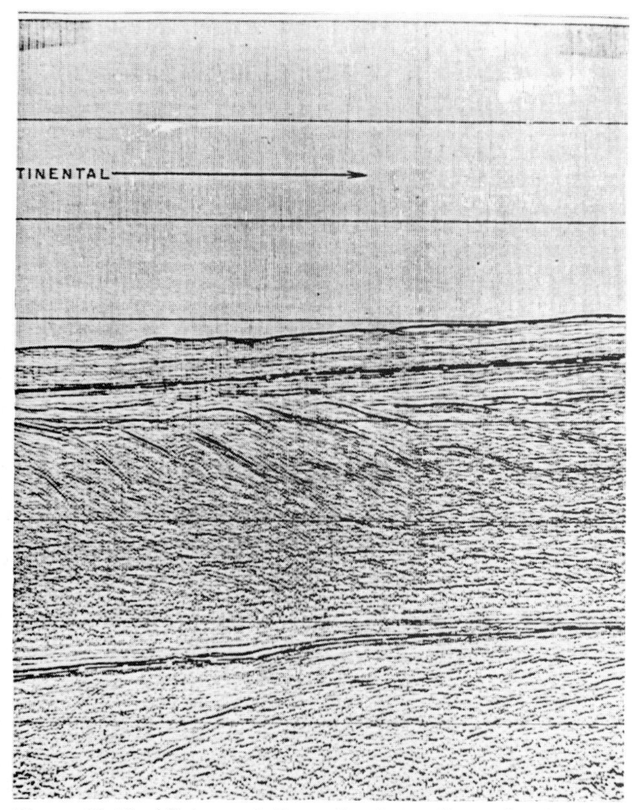

Figure 10.16 High-resolution seismic profile under the continental shelf showing details of the stratigraphy, including dipping sequences of sediments.

Gravimetrics

We generally think of the pull of gravity as uniform over the earth's surface. In order for that to be true, the earth would have to be a smooth sphere with either uniform layers or without layers. This, of course, is not the case and as a result there are subtle but important differences in the earth's gravity field over various parts of the earth's surface. A simple example is that caused by the bulge of the earth at the equator. It causes a decrease in the pull of gravity so that a person weighs less at the equator than in the United States. Extreme differences in elevation cause the same effect. Something on a mountain top weighs less than it does at sea level.

Extremely accurate measurements of gravity are provided by a **gravimeter,** which is a delicate instrument that measures the pull of gravity on a calibrated spring. These instruments also must be calibrated for elevation relative to sea level, temperature, and latitude, all of which can affect the readings on the gravimeter. Shipboard gravimeters are specially adapted to eliminate the movement of the ship due to waves and thus are more complicated than those used on land. It is also possible to use gravimeters in airplanes, which is an efficient method of collecting data.

Places where the pull of gravity is markedly different than the normal or expected values are called **gravity anomalies**. In the ocean basins the most pronounced gravity anomalies are over the oceanic trenches and associated with the oceanic ridges. Some other features that show anomalies are fractures, seamounts, and volcanic islands. The anomalies associated with the trenches are gravity lows or

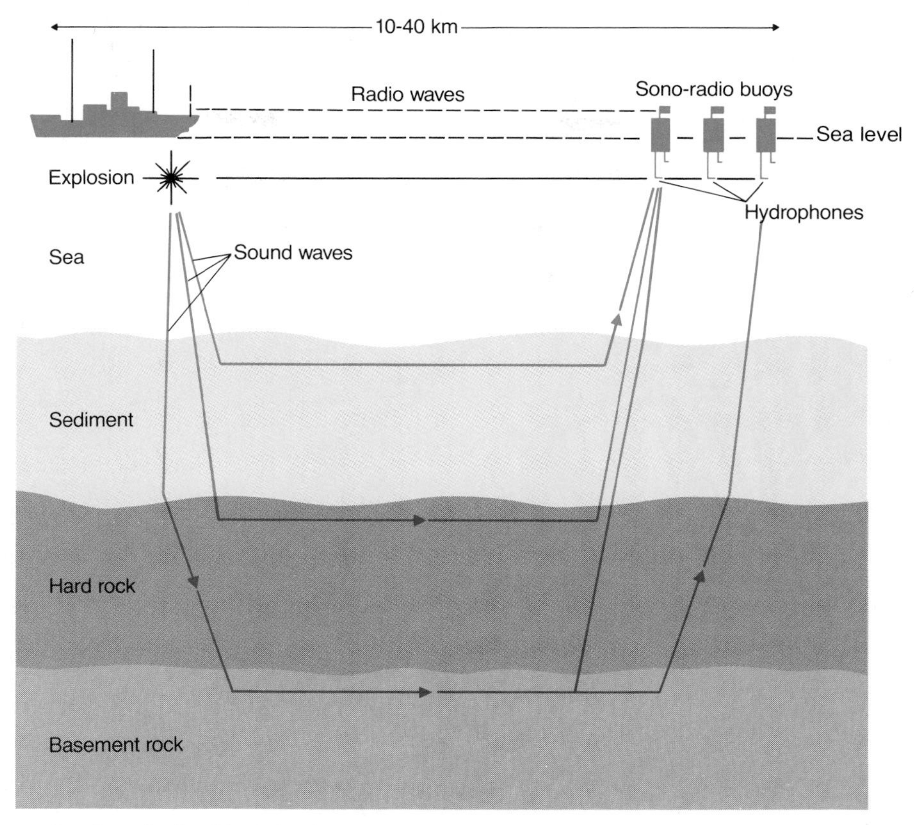

Figure 10.17 Diagram showing single-ship seismic refraction.

negative anomalies, which means that the pull of gravity is lower than normal in these regions. Gravity highs are associated with unusually thick and/or dense rock bodies, such as volcanic islands or seamounts.

Magnetics of the Ocean Basins

The earth is a giant magnet and as such, it influences the crystallization or accumulation of all magnetic particles. This occurs in basalt, where magnetite crystallizes directly from a melt and also when magnetic sedimentary particles settle. Knowledge of the magnetism of the rocks of the earth's crust is valuable to supplement gravimetric and seismic data.

Measurement of the magnetic field is accomplished with a proton precession **magnetometer**. This instrument consists of a wire coiled around a vessel of water. A current passed through this coil causes the hydrogen ions or protons to align themselves along the axis of the coil. Variations in the earth's magnetic field cause the protons to generate a signal that is monitored on the ship towing the instrument.

The most important data produced by the earth's magnetic patterns are the reversals in magnetism (see chapter 3) that create strip-like patterns and enable the definition of large-scale structural features, such as fractures in the oceanic crust (fig. 10.18).

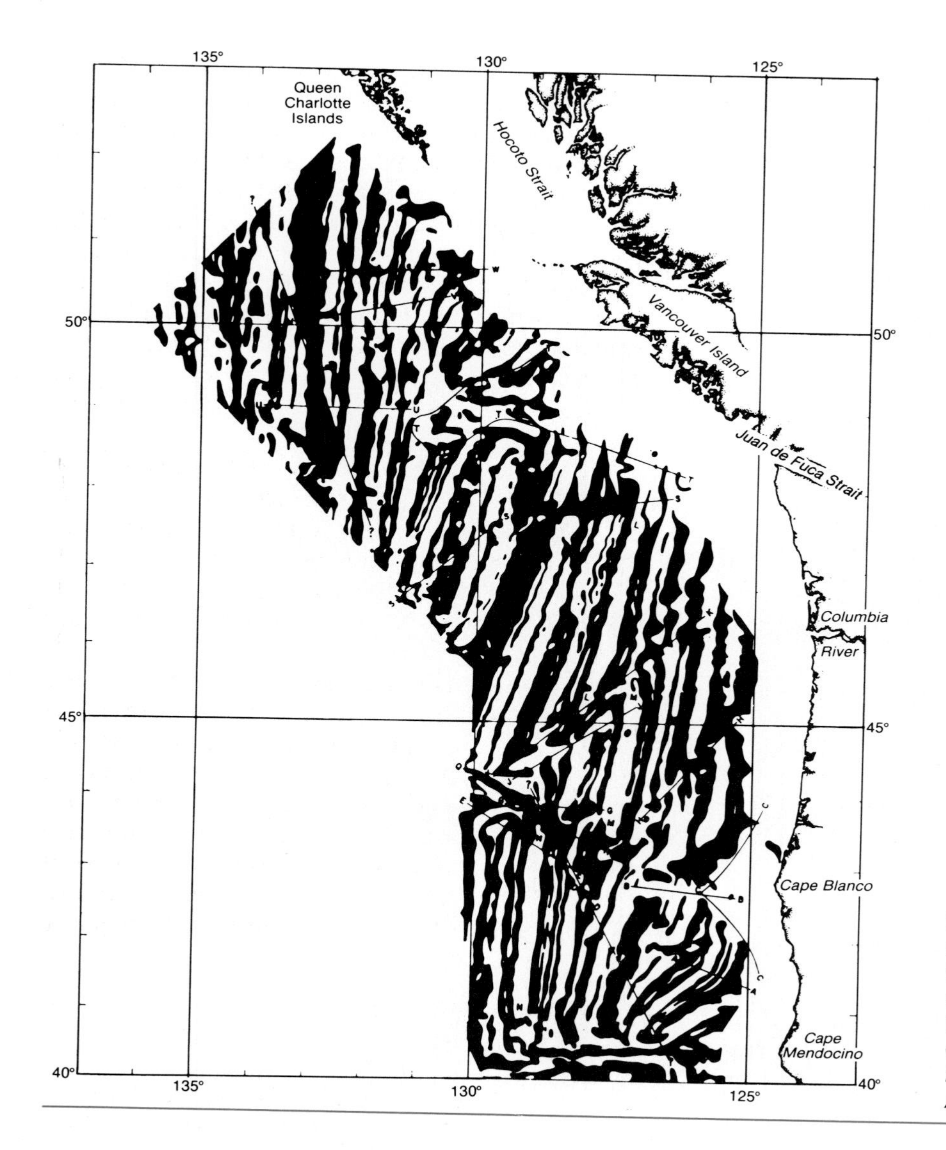

Figure 10.18 Complex patterns of magnetic reversals off the west coast of North America. (After Raff, A. D., and R. G. Mason, © 1961, "A Magnetic Survey of the West Coast of North America, 40° N to 52½° N," *Bulletin of the Geological Society of America,* 72, 1260.)

Layers in the Oceanic Crust

Based largely on seismic reflection and refraction it is possible to generalize about the layers of the oceanic crust. Beneath the sea's surface are the water column, and three layers of sediment and rock in the crust. Seawater obviously has the lowest density and the slowest velocity of transmission of seismic waves (fig. 10.19). Beneath the floor of the ocean is a layer of unconsolidated sediments about 500 m thick, much less than was expected by most marine geologists prior to the Deep Sea Drilling Project. This thin, slow layer rests upon Layer 2, which is a combination of consolidated sediments and basalt. It is 1.75 km thick, has a high seismic velocity relative to Layer 1 and has a density near that of the continental blocks (fig. 10.19).

Layer 3 comprises the bulk of the oceanic crust and consists of basaltic and gabbroic rocks. These iron- and magnesium-rich rocks are about 5 km thick and have a density of about 3.0 gm/cc. Thickness varies locally with the thinnest areas near the trenches and the thickest places at the oceanic ridges. A marked discontinuity in seismic properties occurs

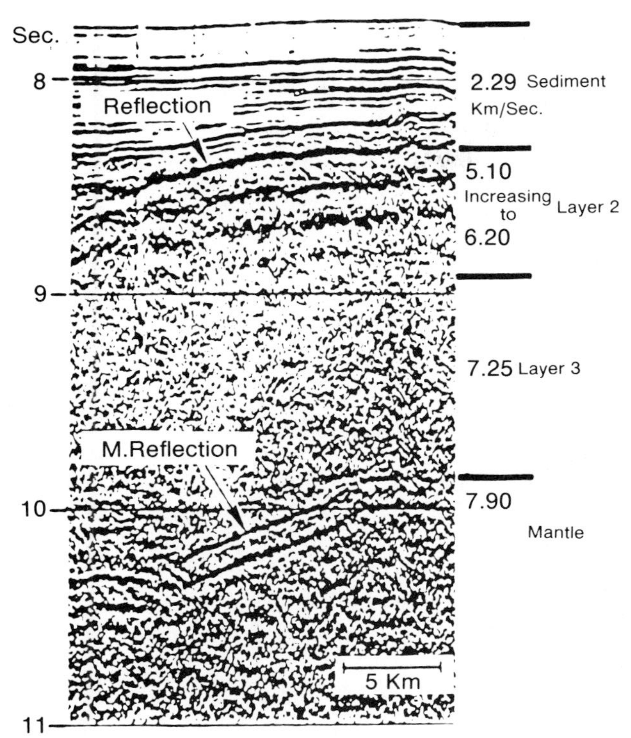

Figure 10.19 Seismic reflection profile of the oceanic crust showing the major layers.

between Layer 3 and Layer 4. The velocity abruptly changes to just above 8 km/sec and the density at the top of Layer 4 is interpreted to be about 3.3 gm/cc. This discontinuity is called the Mohorovicic Discontinuity or Moho after its discoverer. It represents the boundary between oceanic crust and the upper mantle. Upper mantle rocks are thought by some geologists to represent a phase change from the lower crust rather than a change in composition. There is still debate on this question.

Summary of Main Points

Marine sediments vary in origin and composition. They may originate as weathering products of previously existing rocks or sediments on land masses, they may originate from the skeletal material of organisms, they may precipitate directly in the marine environment, or they may come from outer space. The first two types of sediment comprise the vast bulk of marine sediments.

There are general global patterns to the distribution of sediments in the ocean basins. The continental margins are dominated by sediment derived from the adjacent landmasses. The high latitude areas of the deep ocean are covered primarily by sediments that originated from glaciers. Adjacent to this in the slightly lower latitudes are siliceous biogenic sediments. Calcareous biogenic sediments are the dominant sediment on the deep seafloor and occur throughout most of the low and mid-latitude areas except in the deep parts of the ocean where they are dissolved by intense pressures and low temperatures.

The layers of sediments and sedimentary rocks in the ocean basins show a distinct pattern that is related to plate tectonics and seafloor spreading. The strata are thinnest and youngest at the oceanic ridges; they thicken and increase in age away from the ridges. The oldest strata in the ocean basins are about 180 million years.

Geophysics enables us to obtain data on the deep parts of the crust under the oceans. Seismic reflection and refraction provide continuous cross sections of the thickness and geometry of deep layers of the crust. Gravity studies enable generalizations about the overall nature of the crust but do not give good data on individual layers. Magnetic properties of the earth can tell us about the general composition of the rocks and also about the earth's magnetic field during the past.

Suggestions for Further Reading

Blatt, H., G.V. Middleton, and R.C. Murray. 1980. **Origin of Sedimentary Rocks**, 2d ed. Englewood Cliffs, New Jersey: Prentice-Hall.

Garrels, R.M., and F.T. MacKenzie. 1971. **Evolution of Sedimentary Rocks.** New York: W. W. Norton.

Kennett, J.P. 1982. **Marine Geology.** Englewood Cliffs, New Jersey: Prentice-Hall.

Pettijohn, F.J. 1975. **Sedimentary Rocks,** 3d ed. New York: Harper & Row.

Shepard, F.A. 1973. **Submarine Geology,** 3d ed. New York: Harper & Row.

Shepard, F.A. 1977. **Geological Oceanography: Evolution of Coasts, Continental Margins and the Deep Sea Floor.** New York: Crane, Russak.

PART THREE

Coastal Environments

11
Estuaries and Related Environments

Chapter Outline

Estuaries are included in the broad spectrum of coastal embayments present along all types of coasts in the world. These embayments are adjacent to low-lying coastal plains as well as to coasts with high relief and rugged mountains. They occur in essentially all latitudes and climates, and they may be controlled by tectonic setting or by erosional topography. An **estuary** is any coastal embayment that is influenced by freshwater runoff, such as a stream or river, and that typically displays tidal circulation with the open ocean.

By their nature, estuaries have salinities below that of the open ocean and display a salinity gradient from that of fresh water at the location of the fresh water input ($<0.5‰$) to nearly normal marine levels ($35‰$) at the seaward limit of the influence of fresh water. Estuaries may be small or they may be large, such as Chesapeake Bay, Delaware Bay, or Pamlico Sound. Many are shallow, especially those adjacent to low-relief coastal plain areas, but some are very deep such as the **fjords** of the northern latitudes (fig. 11.1), which are simply a special type of estuary that has been carved out of the coast by glaciers.

The location and nature of the estuarine environment make this an extremely productive environment for marine life. The runoff from land provides a continuous supply of nutrients that combine with tidal circulation to produce a somewhat eutrophic but well-oxygenated system. The generally shallow conditions are beneficial for photosynthesis because light can penetrate through most or even all of the water column in an estuary. This combination results in a highly productive system, typically the highest of all coastal environments. Estuaries serve as a nursery for many marine organisms including various types of fish and shellfish of economic value. Some of the latter, such as oysters and crabs, make the estuary their home throughout their life span.

The runoff from land and the tidal interaction of estuarine systems also make this environment a place where sediment tends to accumulate, a so-called **sediment sink**. These embayments act as efficient traps for sediment entering from both the landward and seaward directions. As a result, estuaries tend to fill in with sediment and are short-lived in terms of

Figure 11.1 Photograph of a fjord, a special type of estuary formed by glaciers, along the coast of Alaska.

geologic time. Their life expectancy normally is on the order of hundreds to thousands of years rather than millions.

Estuaries may form in a variety of ways. Presently this type of environment enjoys widespread distribution on the earth's surface due to the rise in sea level over the past several thousands of years as the glaciers have melted. This rise in sea level has literally drowned the coastlines of the world and any irregularity caused by such phenomena as erosion, tectonics, or glaciation, will produce a coastal embayment. Many of these embayments, in fact most of them, are estuaries.

Estuarine Processes

Both physical and biological processes affect the estuarine environment. Tides are the dominant physical factor; waves are locally important, especially in large estuaries, where there is a significant fetch. Benthic organisms, both infaunal and epifaunal types, interact with the substrate by burrowing or crawling over the surface to produce bioturbation effects.

Tides

Tidal currents dominate the circulation in estuaries. Some influence is also exerted by streams entering the system and by wind-generated currents and waves. Coasts are classified according to their tidal range: **microtidal** coasts have a range of <2 m, **mesotidal** coasts have a range of 2–4 m, and **macrotidal** coasts have ranges >4 m. The mouth of the estuary or its opening with the ocean is typically affected by the relative impact of tides and waves on the coast. Generally, microtidal coasts tend to have waves as the dominant coastal process. This commonly results in the mouth of the estuary being nearly closed by sand spits, which are generated by sediment transported along the coast by wave-generated longshore currents (fig. 11.2a). Along macrotidal coasts, tides dominate and the mouth of the estuary tends to be wide and the estuary itself may be funnel-shaped (fig. 11.2b). Mesotidal coasts may be sub-equally influenced by tides and waves.

An important factor in the circulation of an estuary is the **tidal prism,** which is somewhat like a tidal budget. It is the total amount of water that

Microtidal estuary

Longshore current

Waves

(a)

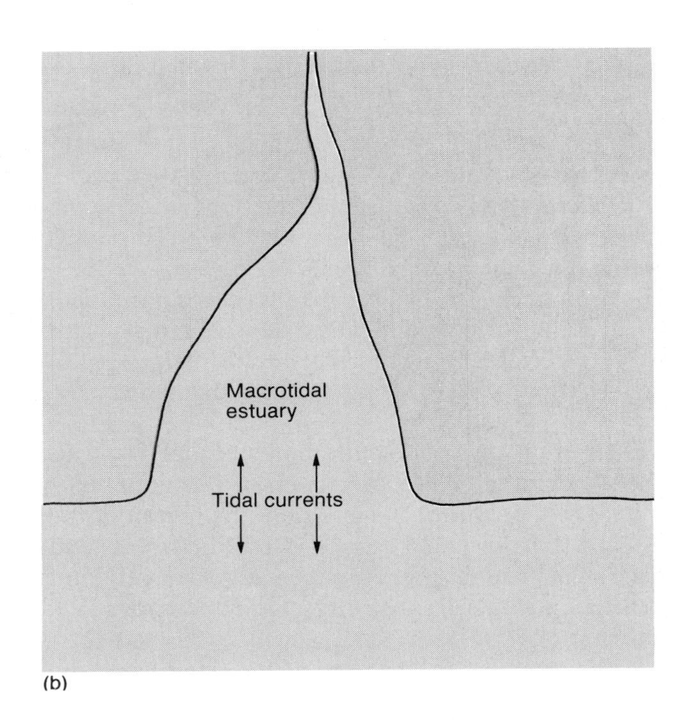

(b)

Figure 11.2 Generalized diagram of (a) a microtidal estuary with spit across the mouth and (b) a macrotidal estuary kept open by strong tidal currents.

(a)

(b)

Figure 11.3 Photographs of (a) a tide-dominated microtidal estuary: the mouth of the Crystal River, Florida, where tidal range is nearly a meter and (b) a wave-dominated estuary with tidal range near 4 m: Willapa Bay, Washington.

passes into and out of the estuary during a tidal cycle. It is determined by multiplying the tidal range by the area of the estuary. The prism may be more important in determining the dominant physical process operating within an estuary than is tidal range alone. For example, a geographically extensive estuary with only a microtidal range may generate a larger prism than a small estuary with a mesotidal range (fig. 11.3).

The tidal prism is important because the currents generated by the movement of this volume of water tend to control the nature of the mouth of the estuary. A large prism results in rapid currents and deep tidal channels, whereas a small prism will tend to produce sluggish tidal currents with perhaps a lack of channels.

The semi-enclosed nature of the estuary and the rotation of the earth produce a tidal circulation system that is similar to the amphidromic system in the open ocean. The Coriolis effect causes flooding tides in the northern hemisphere to display a water surface that is tilted up to the right (fig. 11.4a). The ebbing tide in the northern hemisphere exhibits the same effect for the same reason; the water surface slopes up to the right as the water flows back to the ocean (fig. 11.4b). In diagrammatic form, the circulation shows cotidal lines rotating about an amphidromic point (fig. 11.4c). Extreme freshwater discharge into the estuary may mask the tilting of the water surface created by the Coriolis effect simply by providing an overwhelming mass of runoff. Another aspect of this circulation system is the lower salinity on the left side because of the Coriolis effect.

The incursion and excursion of a saltwater wedge into and out of the estuary is associated with the flooding and ebbing of the tides. The denser ocean water moves into the estuary beneath the less dense fresh water from the river discharge (fig. 11.5). The extent to which this landward movement of the salt

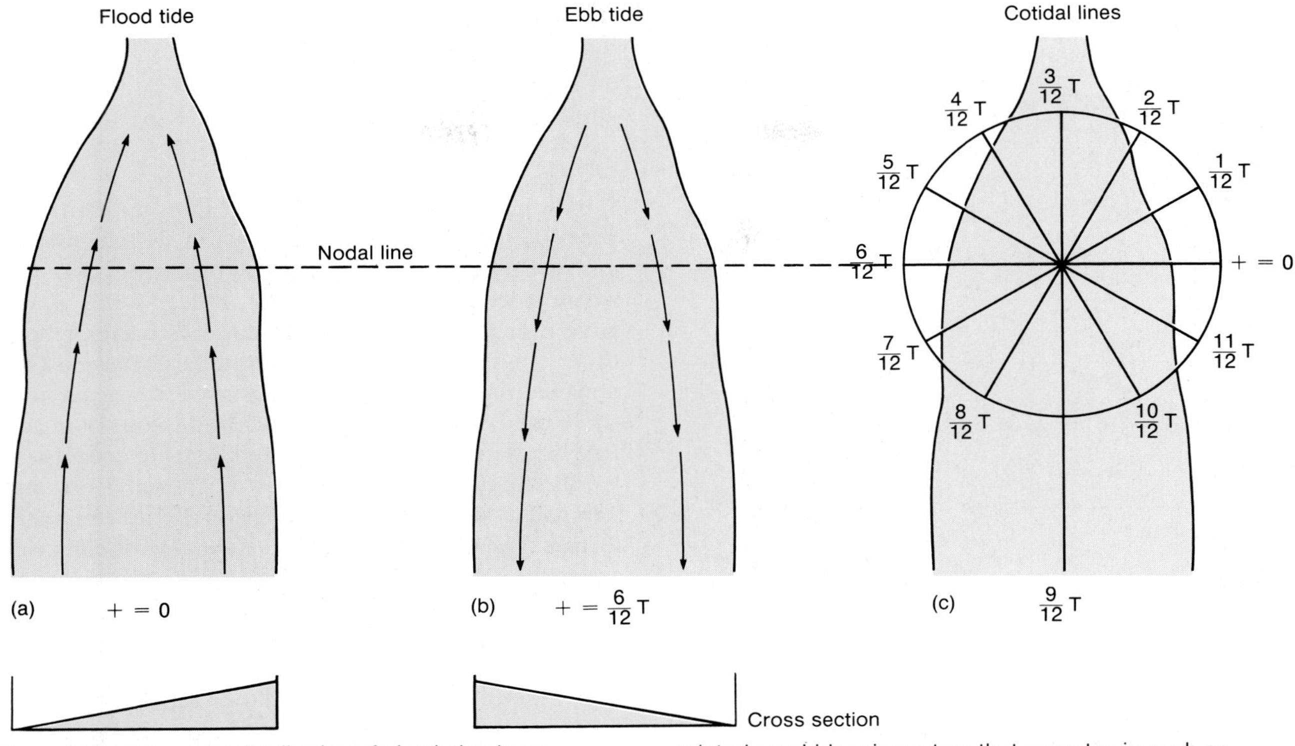

Figure 11.4 The aerial distribution of circulation in an estuary (a) during flood tide, (b) during ebb tide, and (c) the related amphidromic system that operates in such an estuary.

wedge occurs is related to the rate of freshwater runoff, the tidal range, the lunar tidal stage, and the configuration of the estuary. In some estuaries, such as the mouth of the Mississippi River, the salt wedge moves over 100 km inland during times of low discharge, whereas during river flooding the great discharge may reduce the penetration of the salt wedge to only 1–2 km.

Circulation Classification

The interaction of freshwater runoff and tidal conditions may also serve as the basis for a classification of estuaries. Low-energy estuaries, which are characterized by low tidal ranges, sluggish tidal currents, and flow conditions dominated by freshwater input, produce **stratified estuaries**. Under such conditions the freshwater layer retains its character and moves seaward over the saltwater wedge, as described above. The low-energy conditions result in little mixing and the **isohalines** display a nearly horizontal pattern (fig. 11.6a). These have also been called Type A estuaries.

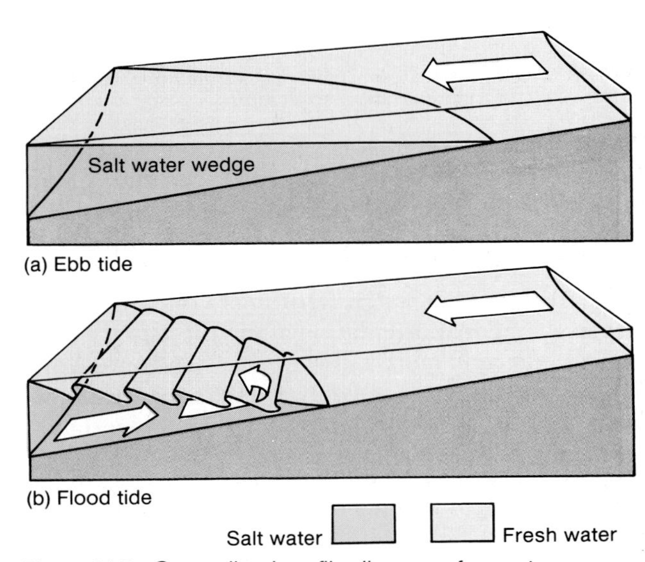

Figure 11.5 Generalized profile diagram of an estuary, showing circulation (a) during flood tide and (b) during ebb tide.

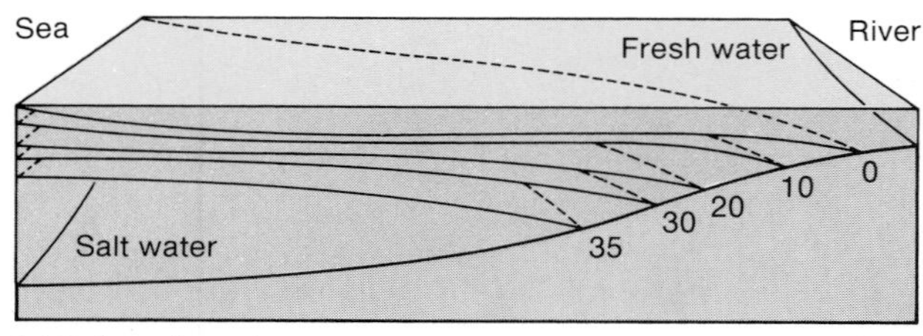

(a) Salt wedge

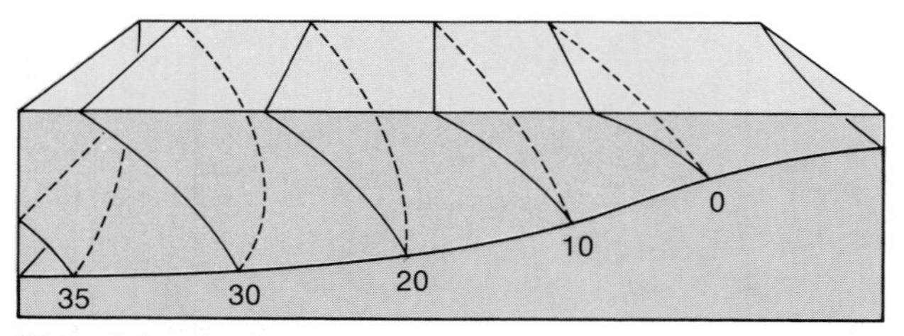

(b) Partially mixed

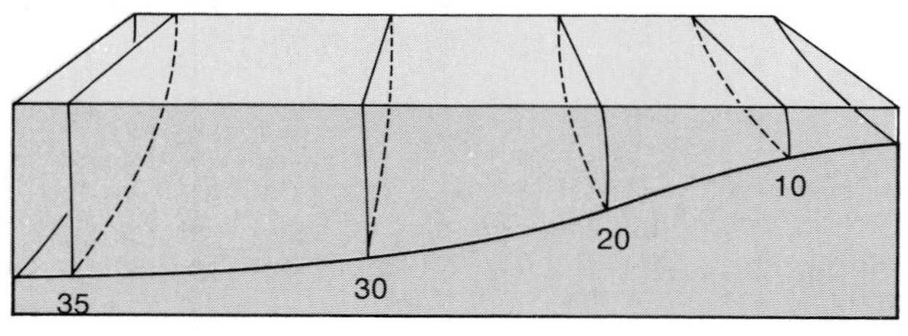

(c) Fully mixed

Figure 11.6 Profile diagrams of typical estuarine water masses showing (a) a stratified estuary, (b) a partially stratified estuary, and (c) a well-mixed estuary. The concept of this classification is from A. H. Postma.

Conditions under which tidal currents exert subequal influence with riverine discharge on estuarine circulation produce **partially mixed** or Type B estuaries. There is a modest amount of mixing of the fresh and salt water, resulting in isohalines that display a diagonal orientation along the estuary (fig. 11.6b). There is more seaward than landward flow on the surface and more landward than seaward flow in the bottom waters. These isohalines move landward and seaward with the flood and ebb of the tides. The Chesapeake Bay is a partially mixed estuarine system.

Estuaries in which tides are the dominant process are **totally mixed** or Type C estuaries. This may be caused by minimal runoff and/or strong tidal currents. There is a prominent salinity transition from the landward part of the estuary to the ocean with the water column isohaline at any location (fig. 11.6c). Delaware Bay and the Bay of Fundy are examples of totally mixed estuaries.

Other Processes

Wind can also be an important factor in estuarine processes. The two primary ways in which this occurs are by generating waves and by causing wind or storm tides (also called storm surge). Wind waves have a major impact only in large estuaries where there is sufficient fetch to generate large waves. This is particularly important in shallow estuaries where these relatively large waves will feel the bottom and interact with both sediments and benthic organisms. Pamlico Sound on the North Carolina coast and Tampa Bay on the west coast of Florida are examples of this type of large, shallow estuary. Most estuaries are small enough so that waves do not become large, so they affect only the shoreline. These waves may, however, cause shoreline erosion. This is a serious problem along much of Chesapeake Bay in Virginia.

Estuaries are excellent places for the development of storm tides. High-velocity winds that persist for up to several days may change the water level due to friction between the wind and the water surface (see chapter 5). Estuaries with funnel shapes or that are located adjacent to broad, shallow shelves are particularly susceptible to these storm tides. Low atmospheric pressure and the resulting strong onshore winds tend to elevate the water level, which is one of the potentially most destructive aspects of a hurricane. Low-lying areas adjacent to large coastal bays may experience increases in water level of up to 5 m, enough to flood large areas including the surrounding cities. Offshore winds may cause negative storm tides, but they are not of serious concern because water levels are lowered. They may have an impact on the fauna and flora of the estuary in that a normally subtidal environment may be exposed for an inordinately long period of time.

Biological processes can also have an impact on estuarine environments. Such processes are dominated by the activities of benthic invertebrate animals. There are two that are important: (1) **bioturbation** or physical rearranging of sediment and (2) production of pellets from suspended particulate material. Bioturbation by benthic animals results from grazing or burrowing in search of food or burrowing for shelter. The unselective ingesting of sediment by organisms results in the production of abundant pellets as the organisms rid their digestive tracts of unwanted mineral mud. Benthic organisms, both the infaunal sessile types and vagrant epifaunal grazers, are common in estuaries. As a result bioturbation is extremely common and widespread on the floor of the estuary.

The soft sediment that characterizes estuarine substrates provides an excellent habitat for burrowers. These animals disrupt the sediment to depths of as much as a meter below the surface, whereas animals that move over the surface are only modest contributors to this bioturbation. The production of pellets by organisms contributes significantly to sediment accumulation on the floor of the estuary. In fact, some estuaries have most of their sediment pelletized.

The level to which biological processes affect an estuary is inversely related to the level of physical energy in the system. That is, biological processes are most influential in quiet, low-energy estuaries where organisms are not moved or disturbed by waves or tidal currents. In high-energy estuaries the rapid currents and waves cause the substrate to be very mobile and uninhabitable for most benthic organisms.

Estuarine Sediments

Although nearly any type or size of sediment particle can be found in the estuarine environment, there are certain generalities that can be made about estuarine sediments. Grain size is typically that of mud and/or sand. The gravel-sized material is from skeletal material produced by organisms that live in the estuary.

Origin of Estuarine Sediments

Quartz and clay are the dominant minerals present in estuaries, but that can vary greatly because the composition of sediments is affected by four primary factors: (1) the drainage basin or basins that feed it, (2) the composition of the surrounding shore, (3) the adjacent marine environment, and (4) the sediment produced in the estuary by organisms living there.

Coastal plain estuaries are usually fed by large rivers with extensive drainage basins that provide a wide variety of minerals to the estuary. Estuaries along a rugged coastline, such as along the west coast of the United States, tend to be fed by streams from small, high-relief drainage basins. These streams deposit their coarse sediment load at or near their mouths with the fine sediment carried beyond. Rates of sediment influx from land runoff typically show seasonal variations due to seasonal patterns in rainfall, which affect discharge. The wet season, commonly the spring in the mid-latitudes, produces high discharge and therefore much sediment for the estuary. Extreme events, such as tropical storms, also have an impact on sediment influx from land in that they may produce great amounts of rainfall. It is not uncommon for a hurricane to be accompanied by 30 cm or more of rainfall. This in turn produces a great deal of sediment as the large volume of runoff carries the sediment to the estuaries after the passage of the storm.

Although there was some disagreement several years ago about how much sediment was contributed to estuaries from the open marine environment, it is now agreed that much sediment comes from that source. The most common mechanisms of deposition are tidal currents and waves. Additionally, sediment from barrier spits and bars may wash over or blow over into the estuary. The net result is permanent accumulation of sediment from the adjacent shelf into the estuary. Although tidal currents move sediment in both directions, there tends to be a net transport by flood tides over ebb tides. Extreme conditions may alter these general trends. For example, a low-energy estuary with a small tidal prism and a high rate of runoff will result in much sediment being carried past the estuary and onto the adjacent continental shelf. Conversely, a hurricane may sweep a large amount of sediment into an estuary and speed up the infilling process.

Many varieties of estuarine organisms produce skeletal material that contributes to the sediment of the estuary. Most common are the shells of benthic mollusks, such as clams, snails, and especially oysters. Ostracods, foraminifera, and others are also contributors. For the most part, these skeletal contributions to the sediment are not transported but instead accumulate where the organisms lived.

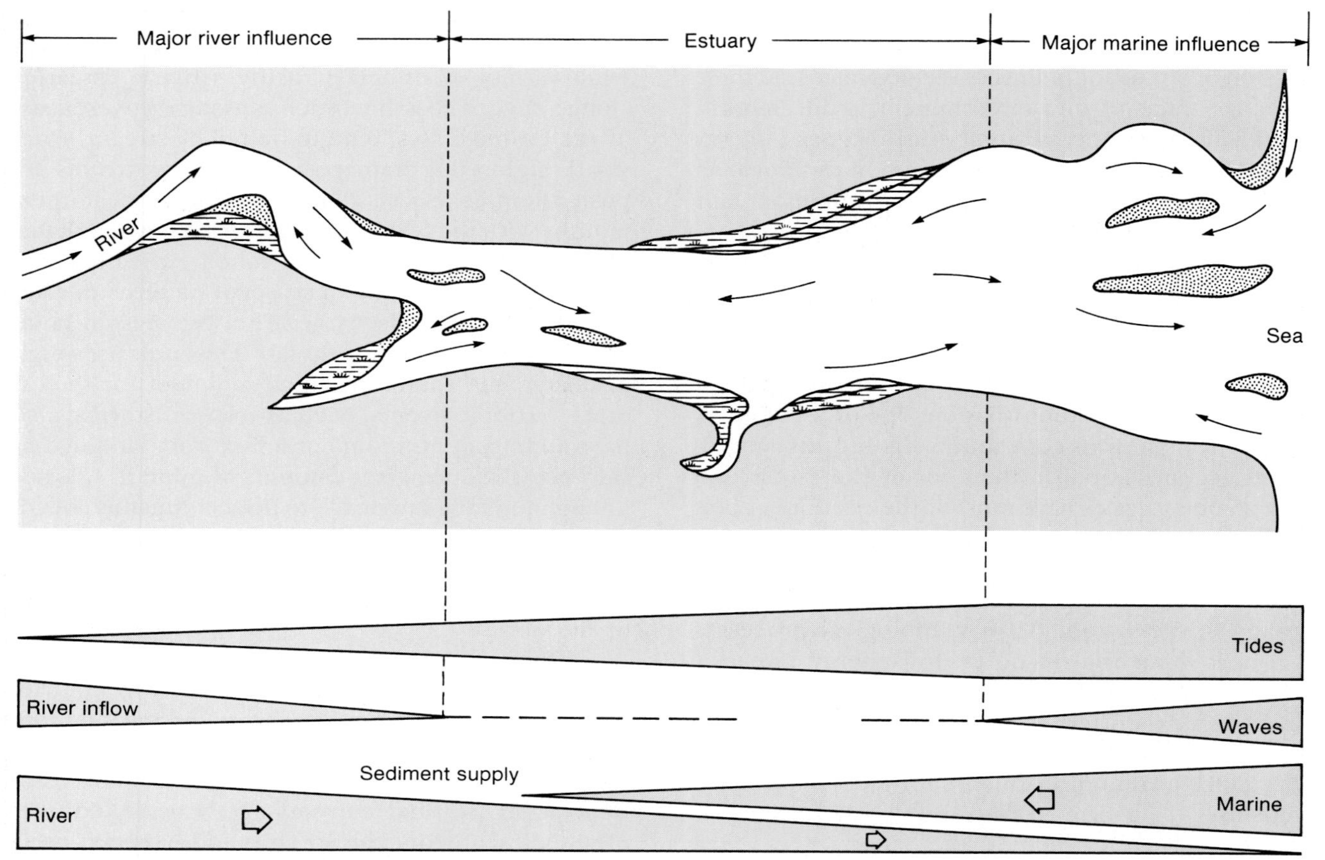

Figure 11.7 Schematic diagram showing the various interrelationships of processes within the estuarine system with both riverine and marine influences.

Estuarine Sediment Distribution

The distribution patterns of estuarine sediment vary greatly, depending upon the relative contributions of the three sources, the dominating physical process, and the configuration and setting of the estuary. A general model can, however, be used to show how sediments and processes are related. This model utilizes the common funnel-shaped estuary with a single river contributing the land-derived sediment (fig. 11.7). Tidal currents are shown as the dominant process, having a diminishing effect upon the estuary. The influence of waves and rivers is usually restricted to the areas immediately adjacent to the open marine and river environments respectively. The sediment is derived from both major sources, more or less equally, with overlap in the central part of the estuary (fig. 11.7). Shell material produced within the estuary may display a variety of patterns, but, generally, estuarine organisms are most abundant in the central region that is both away from the fresh water at the river mouth and away from the normal marine salinities at the estuary mouth.

Grain-size distribution of the sediment tends to reflect the two major sediment sources. There is a general decrease in grain size of the nonskeletal portion toward the central part of the estuary (fig. 11.7). Sand or muddy sand typifies the river mouth and estuary mouth whereas mud is most common in the

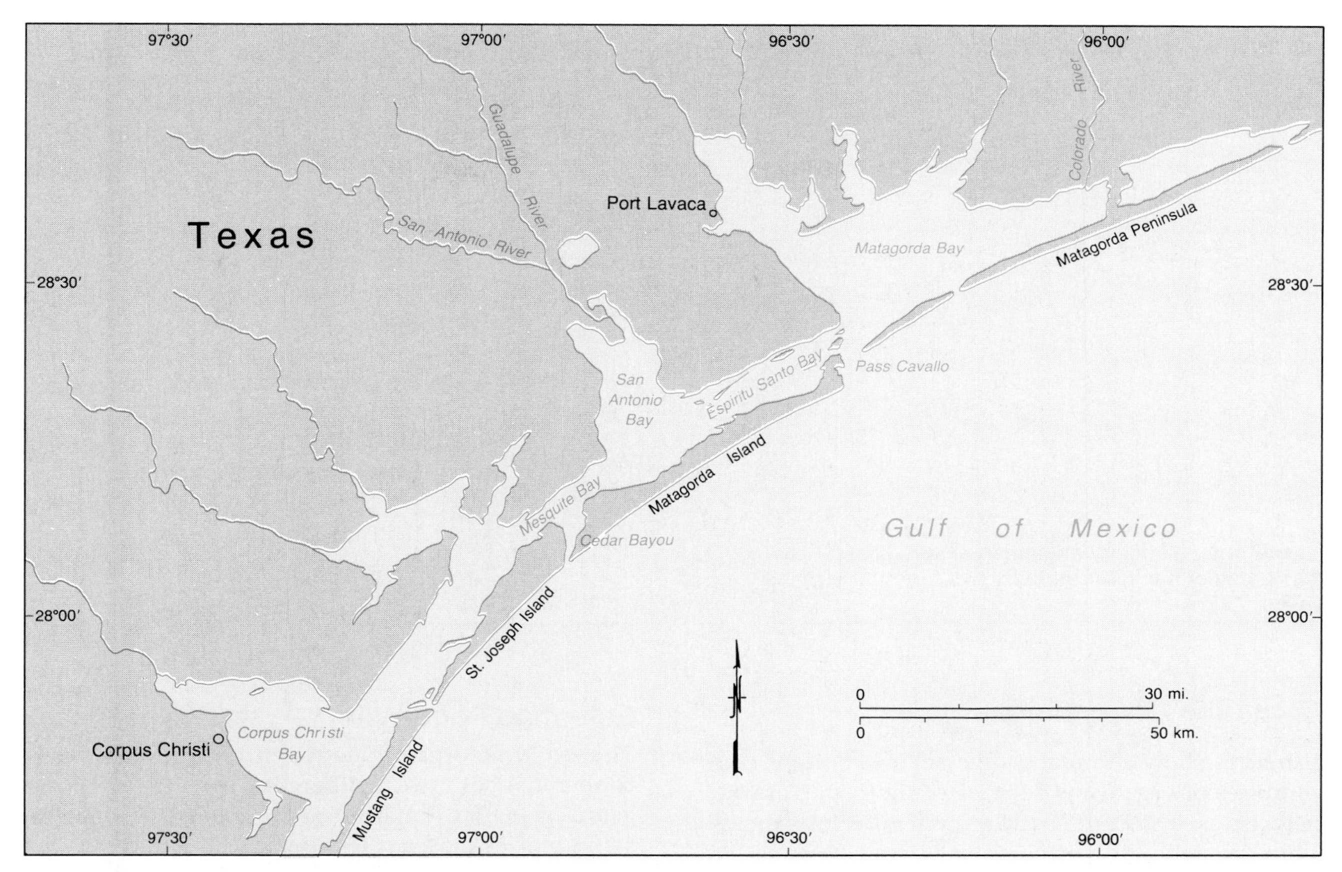

Figure 11.8 Outline map of estuaries along the central Texas coast with multiple streams emptying into the coastal system.

central estuary. Coarse shell material may occur throughout, but is most concentrated in the central area, thus producing a bimodal sediment, with a fine mud portion and a coarse shell portion.

Some of the possible variations on this simplified theme include multiple sources of runoff, differing tidal ranges and prisms, circular or digitate estuaries rather than funnel-shaped, and the amount of closure or circulation restriction at the mouth of the estuary. Numerous rivers entering an estuary will not only tend to fill it with sediment relatively quickly, but also will give a different pattern to the sediment distribution than the one considered in figure 11.7. There will be a tendency for sediment to be coarse along the shore and fine toward the middle, especially in a more circular estuary. This is the case in some of the Gulf Coast estuaries, especially along the Texas coast (fig. 11.8).

High tidal range tends to be associated with few or no spits or barriers across the mouth of the estuary. Consequently, there tends to be more relatively coarse sediment contributed by the open shelf. This is in contrast to a low-tidal range, wave-dominated system where there is a barrier across the estuary mouth and only one or two inlets (fig. 11.9). Such an estuary receives most of its shelf-derived sediment from waves and wind carrying sand over the barrier rather than from tidal circulation.

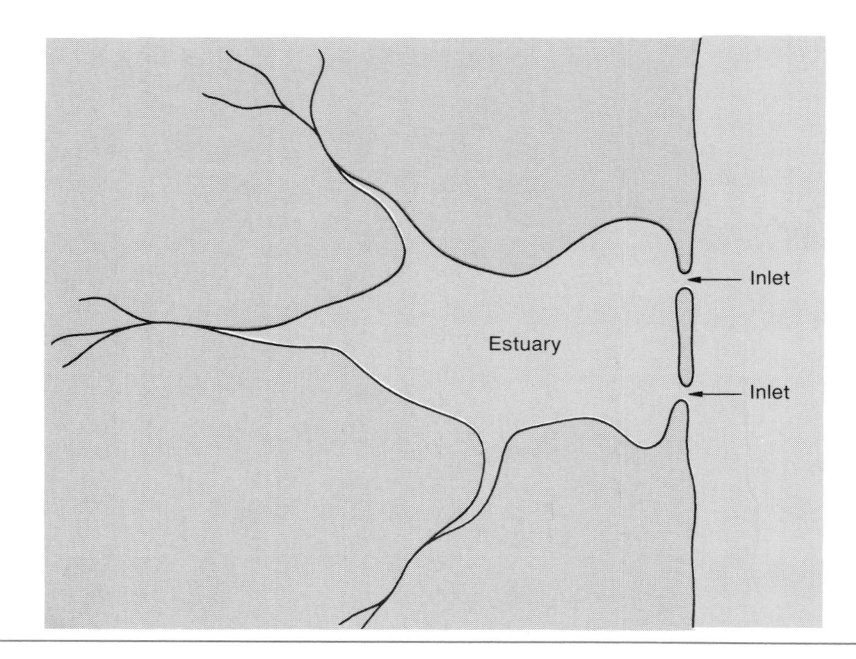

Figure 11.9 Sketch of an estuary that has a barrier bar interrupted by two inlets.

Estuarine Communities

Estuaries show abundant, diverse, and unique communities of organisms because of their special characteristics. They are shallow, generally low-energy, nutrient-rich, and have intermediate salinities. Because of their transitional nature, estuarine communities have some characteristics of both freshwater and marine environments. There is overlap among the types of organisms as a result. Some have affinities to fresh water, some to the marine environment, and many are specially adapted to the brackish estuarine environment.

The organisms that live in estuaries must tolerate the fluctuations in conditions that occur there. These fluctuations include the regular changes in salinity caused by tidal cycles and by seasonal changes in runoff. Additionally, there may be major storm events that cause changes in salinity, turbidity, and rates of sediment accumulation or erosion. In short, estuaries are environments with varied conditions and all organisms that occupy these environments must be able to adjust to such changes.

Estuarine Food Web

The cycle of organic matter in the estuarine environment is not greatly different from the general organic cycle for the marine environment. Although the salinity of estuaries is less than that of the open marine environment, estuarine water should not be considered simply diluted seawater. For example, various nutrient elements, such as phosphorus, nitrogen, and silicon, are significantly more concentrated in estuaries than in the open ocean. This is also true for some trace elements and for the concentration of dissolved organic carbon, that is, carbon that originated in organic tissues. In the case of the latter, the combination of relatively high concentrations in streams plus the high amount produced by plants within the estuarine environment results in high levels of both dissolved and particulate organic matter in estuaries.

A schematic representation of the organic cycle (fig. 11.10) illustrates the primary relationships and pathways for cycling organic material in the estuarine environment. Photosynthesis, as usual, is the basis for productivity. In the estuary, the primary limiting factor in this process is light penetration.

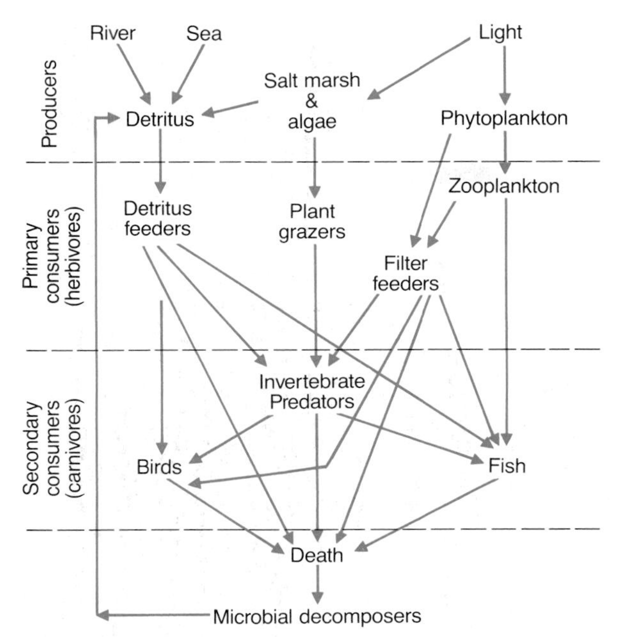

Figure 11.10 The organic cycle in an estuary showing the interrelationships among the various trophic levels and the environment.

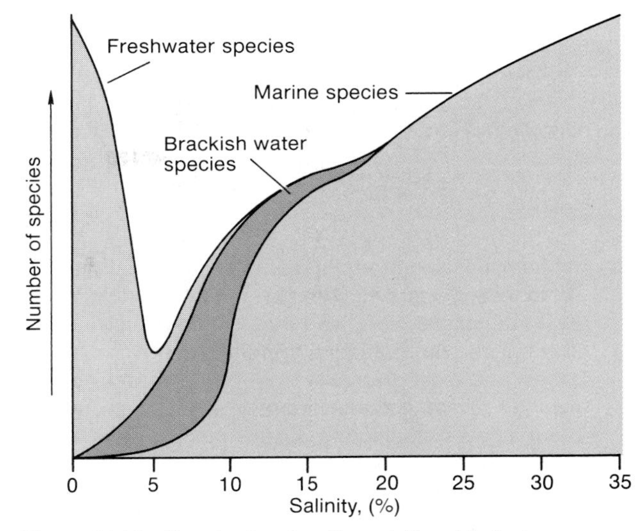

Figure 11.11 Graph showing the relationship between salinity gradient and species diversity throughout the estuarine environment.

Many estuaries have so much sediment in suspension that the resulting turbidity greatly limits light penetration, sometimes to less than a meter. Both organic detritus and dissolved organic carbon are products of expired primary producers. This material is utilized as food by bacteria, protozoans, and herbivores, which also consume the plants directly (fig. 11.10). Waste products of these organisms plus those of the carnivores can also be recycled back to first-level consumers.

Organisms

The diversity and abundance of organisms in the estuarine environment is great and there is considerable variation from the freshwater end of the estuary to the marine portion. This is well illustrated by a relationship comparing the number of species with the salinity (fig. 11.11). At the freshwater side of the diagram the number of species is great, but there is a marked decrease at a salinity of about 5‰. As salinity increases, the number of species increases until there are at least as many in the marine environment as in the freshwater environment. In this same relationship it is convenient to consider three types of organisms: freshwater, intermediate or brackish, and marine. The brackish species are only numerous in low salinities (5–10‰), whereas the marine species dominate at salinity levels of 12–15‰ and above (fig. 11.11). This indicates that freshwater species have much less salinity tolerance than the marine species. Organisms that have high salinity tolerance are called **euryhaline** and those with little or no salinity tolerance are **stenohaline**.

Phytoplankton in the estuary are the important food source for primary consumers. Their abundance tends to reflect runoff and is therefore seasonal. Great quantities of nutrient material during periods of discharge commonly produce blooms of diatoms and dinoflagellates. It is interesting to note that the abundance of zooplankton in the estuary does not follow this pattern. The reason for this is the generally inefficient utilization of the phytoplankton as a food source by the zooplankton. Another type of periodicity to phytoplankton abundance is due to density stratification in some estuaries. This causes nutrients to be isolated in the deeper and colder layers until they are mixed, in which case a great bloom of phytoplankton production occurs. A fjord is an example of where this might occur.

BOX 11.1

Chesapeake Bay, a Large and Diverse Estuary

The rise in sea level during the past several thousand years has resulted in the drowning of low-lying coastal plains and their drainage basins. The Chesapeake Bay estuarine system is one of the largest and most diverse in the world. This system includes several estuaries from large rivers emptying into the bay, including the Susquehanna, Potomac, Rappahannock, York, and James Rivers. These rivers and their numerous tributaries have a shoreline that is several thousand kilometers long.

Chesapeake Bay is a microtidal estuarine system with tidal ranges of 0.2–0.9 m. Most of the bay is a low-energy estuary that accumulates mud, but at the mouth are swift tidal currents and high energy, due to the huge tidal prism that passes through this relatively narrow opening. The mean depth is 8.4 m, deep enough for navigation, but shoals near the mouth are irregular and shallow, making navigation dangerous. Tides extend over 300 km up the rivers from the ocean and noticeable salinities penetrate nearly that distance.

Sediment distribution in Chesapeake Bay conforms well to the standard estuary model. The mouths of the rivers contain a mixture of sand and mud, the central part is dominated by mud with shells from benthic organisms, and there is mostly sand near the mouth of the estuary. Bioturbation is widespread except near the mouth, where strong currents inhibit benthic organisms.

Pollution has become a major problem to the Chesapeake Bay environment. This estuary has been one of the largest commercial fishing grounds in the United States for centuries. The impact of large cities, such as Baltimore, Maryland, Washington, D.C., and Norfolk, Virginia, has had a detrimental effect on the quality of the estuary. Chemical toxicants and organic pollution have caused serious problems in the food chain, especially for the shellfish. Clams and oysters are among the most important products of the bay, but many areas have been closed to commercial harvesting. Extensive agricultural activity and population growth have resulted in organic pollution and increased sedimentation locally in the estuary.

There is a great variety of benthic photosynthesizers, both plant and bacteria. The latter are particularly important in estuaries because they serve as an important food source for detritus feeders. Macroscopic algae and sea grasses are abundant bottom dwellers in estuaries where turbidity does not prohibit light penetration and thus photosynthesis. *Zostera* or eelgrass (fig. 11.12) is among the most common, in part because of its wide salinity tolerance.

Benthic animals comprise the bulk of the consumers in the estuary. The soft, organic-rich floor of the estuary provides an excellent substrate for infaunal and grazing animals alike. Especially abundant are some of the sessile epifauna, such as oysters and mussels. Both of these organisms thrive in estuaries where tidal currents carry nutrient-laden waters to their filtering mechanisms. The burrowing polychaete worms feed in a similar fashion. These organisms may reach concentrations of more than 10,000 individuals per square meter.

Crawling grazers, such as some of the gastropods, are also common in estuaries but are not as abundant as the previously mentioned types. The blue crab (*Callinectes*) is of great economic importance as a food source and thrives in shallow estuaries where it ingests detrital organic debris.

Some estuaries are subjected to strong tidal currents that keep the substrate in a nearly constant state of flux. The result is that benthic organisms, especially infaunal varieties, cannot live in such an environment. A good example is the Minas Basin in the Bay of Fundy, Canada, where sediments on the estuary floor are disrupted to depths of a meter or more during each tidal cycle. Virtually no benthic organism can survive under these conditions.

Figure 11.12 Photograph of eel grass (*Zostera*) on the bottom of an estuary.

The higher level of consumers in the estuarine environment includes large fish, sharks, predatory birds, and, of course, humans. Some fish spend their entire life cycle in the estuary, some live in the estuary only as juveniles, and others migrate into and out of the estuary seasonally. Because of their mobility, fish can move to avoid undesirable salinity changes. There are some, for example the killifish (***Funulus***), that can also adapt to a wide range of salinities by **osmoregulating** to bring their body fluids to the salinity of their environment.

Tidal Flats

A special environment that deserves a separate discussion surrounds many of the estuaries of the world. This is the typically gently sloping, typically unvegetated, unconsolidated, intertidal part of the estuary called the **tidal flat** or the intertidal flat (fig. 11.13). This environment, by definition, extends from low tide to high tide and includes the **supratidal** flat that is flooded aperiodically during storms. These tidal flats may extend for only a few tens of meters along some coasts but they may exceed 10 km in width along others. Examples of extensive tidal flats are in the Bay of Fundy and along the North Sea coast of Germany.

Tidal flats have their own unique set of physical processes and sediment distribution and they are home to a variety of organisms especially adapted for life in such a rigorous environment. At any given locality, the width of exposure of the tidal flat environment changes with the lunar tidal cycle. During spring tide it is at its maximum and during neap conditions it is of minimal width, just as the tidal range changes. The width of the tidal flat environment is also controlled by the gradient of the estuary margin. In a fjord, for example, there is essentially no intertidal zone. The same may be true along high-relief, unglaciated coasts. Broad, low-relief coasts with at least mesotidal ranges provide the best setting for development of tidal flats.

Processes

As might be expected, currents associated with the rise and fall of the tides are the dominant process on tidal flats, but other processes also have significant impact on this environment. Tides produce currents that flow over the tidal flat and they also allow waves to influence the tidal flats by submerging them on a regular and predictable basis. The bioturbation caused by numerous benthic organisms is also an important process. It will be considered in a later section.

(a)

(b)

Figure 11.13 Photographs taken at low tide showing extensive tidal flat environments (a) north of Adelaide, Australia, and (b) in the Bay of Fundy near Windsor, Nova Scotia, Canada. Both areas experience macrotidal conditions.

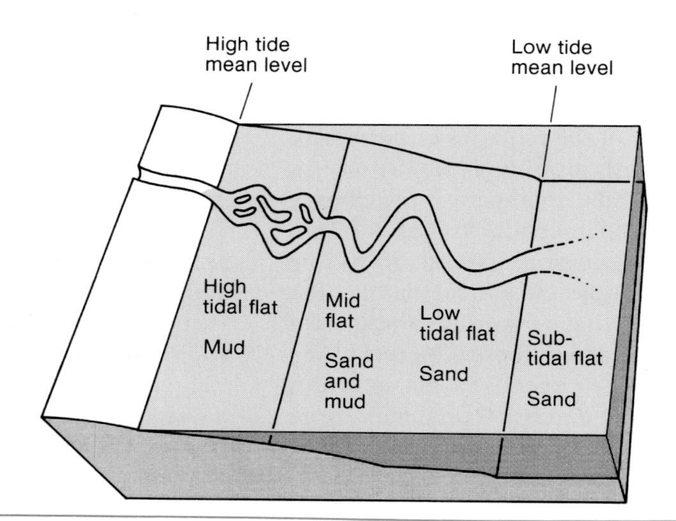

Figure 11.14 Generalized diagram of a tidal flat showing the trend in sediment grain size from low tide, where sediment is coarse, to high tide, where it is relatively fine.

The base of the intertidal zone is inundated by water almost all of the time, whereas the high-tide area is exposed most of the time. That being the case, the low-tide region is subjected to tidal currents and wave activity nearly all of the time, whereas only during high tide is the uppermost part of the tidal flat subjected to the same processes. As broad tidal flats are inundated by the forced tidal wave, flooding takes place very fast and a person walking on the tidal flat may find it necessary to run to escape the incoming water. After high tide, the water level subsides due to gravity with the flow following topography. Such ebbing tidal conditions may form small channels in the topographically low areas, where runoff is concentrated.

Wave activity is typically not intense in estuaries that include the intertidal flat environment. Only during storm conditions are waves an important process here, especially when the storm coincides with high tide or, more importantly, during spring high tide. Some of the most severe property loss and damage around estuarine shores has been the result of storms, especially hurricanes, having their landfall coincident with spring high tide. Conversely, a storm that peaks at spring low tide will have minimal effect because the intertidal zone serves as a buffer for the flooding and devastation of the storm.

Sediments

The sediment deposited upon the intertidal flat tends to have a predictable distribution and a somewhat restricted grain-size variation. There is a common association between tidal flats and mud; the term mudflat is even applied in many cases. Although this is a widespread occurrence, there are also many tidal flats, perhaps even most, that are dominantly sand-sized material. Gravel is not a typical tidal flat sediment except in situations where erosion of bluffs adjacent to the tidal flat produces very coarse particles or in situations in which the tidal flat is composed of shell material from organisms that live on or in the tidal flat.

The usual situation is that the sediments on a tidal flat are carried there and deposited by currents associated with the flooding and ebbing of the tides. That is, sediment comes to the tidal flat from the estuarine basin, not from the land surrounding the estuary. These sediments typically display a grain-size trend from coarser at the low-tide line to finer at the high-tide line (fig. 11.14). This gradation trend may be from medium sand to very fine sand at high tide in some locations or it may range from coarse sand to mud as it does around the Bay of Fundy (fig. 11.15).

Figure 11.15 A tidal flat in the Bay of Fundy, Nova Scotia, Canada, with coarse sand at the low-tide zone, grading up to mud at the high-tide level.

(a)

(b)

Figure 11.16 Sandy tidal flats tend to be characterized by various types of ripples. They may be small (a) or they may be large with smaller ones superimposed (b).

Figure 11.17 The upper part of the tidal flat commonly has mud cracks due to the alternate wetting and drying that takes place there.

Figure 11.18 Rill marks are common features of tidal flats. They are small drainage features formed by water flowing down a gentle slope.

Numerous sedimentary structures may be formed on tidal flat sediments. These include such products of physical processes as ripples, mud cracks, rill marks, and raindrop imprints, and such biogenic structures as tracks, trails, and burrows. Ripples are regular undulations of the sediment surface formed by currents (fig. 11.16). Mud cracks and raindrop imprints are related to climate and form near high tide, especially between neap high tide and spring high tide. Mud cracks are the result of shrinkage when fine sediment is dried after being saturated (fig. 11.17). Raindrop imprints are, as the name implies, the result of the impact of raindrops on the mud. Drying the mud preserves them. Rill marks are small branching furrows that form as water runs over sand or mud (fig. 11.18).

Various organisms that inhabit the tidal flat environment also leave their marks. Animals such as birds, reptiles, or mammals that feed on the tidal flat may leave tracks (fig. 11.19) and grazing invertebrates, such as snails, leave trails (fig. 11.20). A wide variety of burrows, some distinct and others appearing as mottled or blotchy sediment, is produced by infaunal organisms (fig. 11.21).

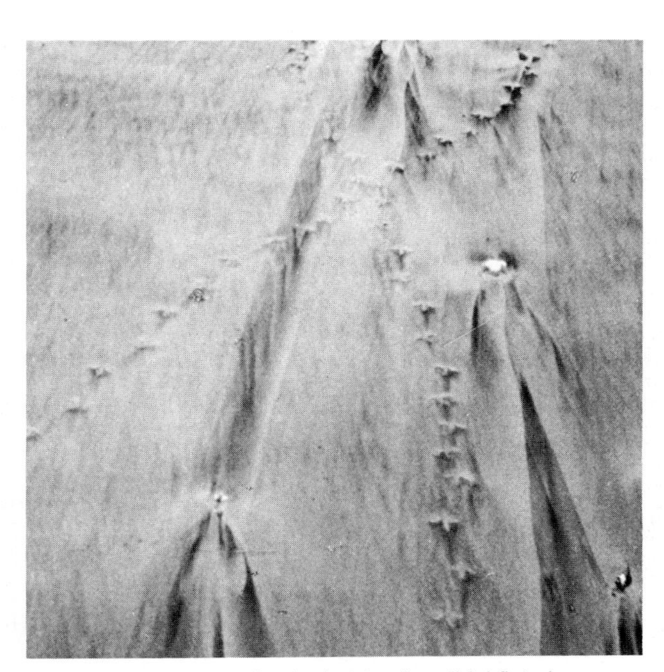

Figure 11.19 Many animals that feed on tidal flats leave tracks, such as these of birds.

Figure 11.20 Numerous grazing snails leave their trails in the mud on the upper part of the tidal flat environment.

Figure 11.21 Infaunal organisms that inhabit the tidal flat develop numerous, varied burrows in the sediments.

Figure 11.22 Oysters (*Crassostrea virginica*) are common in the intertidal flat where clumps (a) of individuals may occupy the intertidal area near the base of the salt marsh or (b) the mangrove community.

Organisms

Tidal flats are teeming with organisms, both vagrant and sessile. Economically important shellfish are abundant on these intertidal environments, including oysters, mussels, and soft-shell clams, or steamers as they are known in New England. These are perhaps the most obvious because of their large numbers and size (fig. 11.22), but they may not be the most numerous. Burrowing ghost shrimp and smaller arthropods are also common in this environment. The great influx of feeding birds at low tide testifies to the abundance of life on the intertidal flats.

Because of the special circumstances that cause the alternation of flooding and exposure of tidal flats, organisms that live here must have adaptations to these conditions. The most important of these adaptations is the ability to withstand exposure of up to several hours without the organism expiring. Virtually all tidal flat animals have some type of gill mechanism that permits the organism to extract its needed oxygen directly from seawater. Tidal flat organisms have the ability to seal out the atmosphere to prevent drying of tissues and also to prevent suffocation due to absence of water. This is usually accomplished either by closing their shells, the way that clams, oysters, and snails do, or by burrowing into the saturated tidal flat sediments and thus keeping within the salt-water environment. Some snails crawl across the wet tidal flat surface (fig. 11.23) so that the feeding and gill apparatus is in water-saturated sediment even though the shell is exposed to the atmosphere. In the case of filamentous algae that may be found on the upper part of the tidal flat, the tissues have the ability to store water and provide protection from desiccation during the period of exposure.

Salt Marsh

Many estuaries are in part bordered by vegetated intertidal flats called salt marshes. Certain types of plants are adapted to life in the intertidal environment and form the landwardmost edge of the estuarine environment where present. Marshes are the last stage in the infilling of the estuary and as such, they occupy the upper part of the intertidal zone. They are not restricted to estuaries, but may border lagoons, marine bays, or the landward side of barrier islands. Marshes may develop in a wide range of salinities and tidal ranges.

Marshes are distributed throughout the mid- and high-latitudes. In subtropical and tropical coasts this niche is occupied by **mangrove swamps,** with various species of mangrove trees (fig. 11.24) replacing the marsh grasses. In some areas, such as peninsular Florida, these vegetation types are mixed within the intertidal zone.

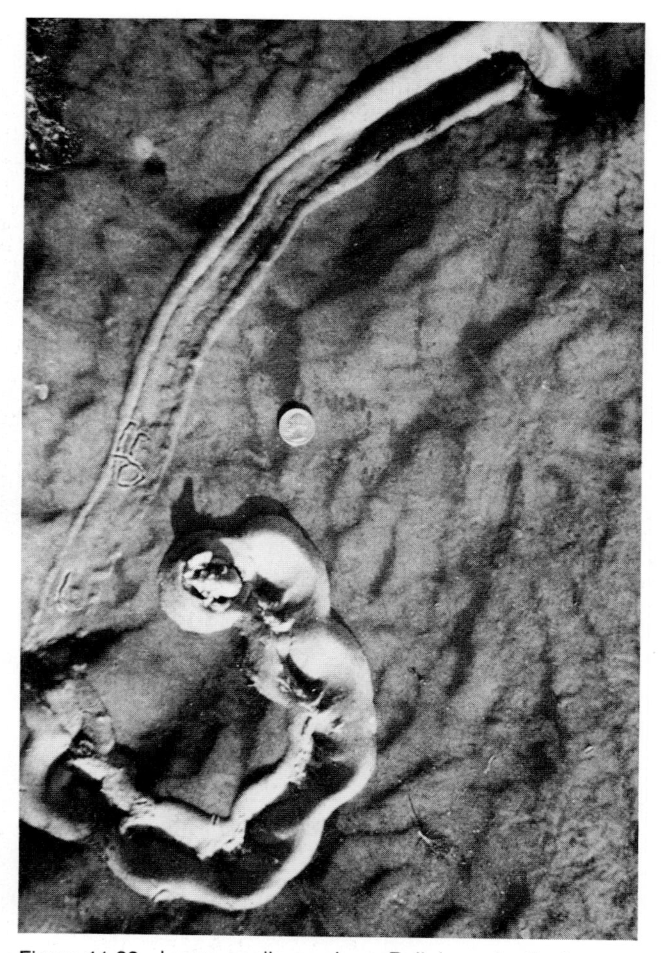

Figure 11.23 Large snails, such as *Polinices duplicatus,* travel over the wet sand of the intertidal flat to keep moist during low tide.

Marsh Zonation

Except for tidal creeks, the overall relief on the marsh environment is very low, commonly only a few tens of centimeters across the entire marsh. The plants that inhabit this environment respond to this extremely subtle relief both in the species that persist and also in the growth forms of a given species. It is common for the differences in the height of marsh grasses on slightly different elevations to be much greater than the relief itself (fig. 11.25). Tallest grass tends to be associated with the natural levees along tidal creeks because that is a place of relatively high elevation and it is also a place where nutrients are abundant.

Figure 11.24 Certain mangrove trees tend to develop near neap high tide, such as those shown here from Western Port Bay, Victoria, Australia.

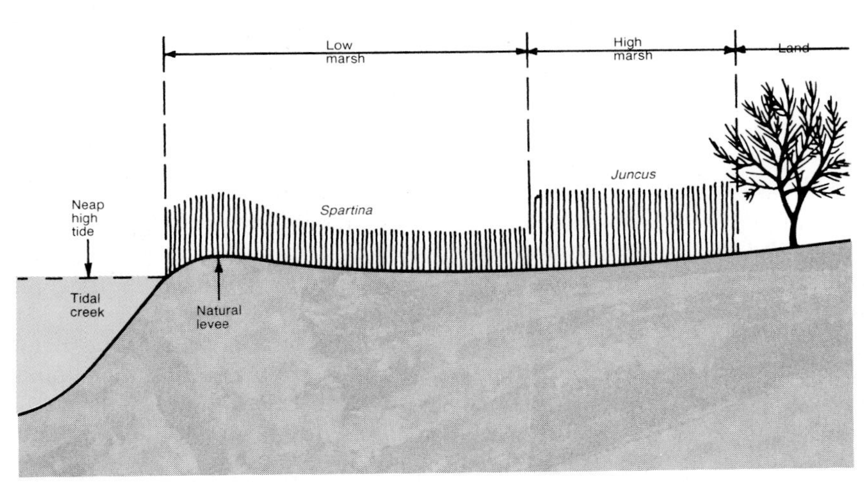

Figure 11.25 The marsh is zoned according to elevation and vegetation, with the low marsh being characterized by *Spartina* and the high marsh by *Juncus*.

Figure 11.26 An example of a low marsh from the coast of North Carolina.

The marsh starts at about the neap high-tide level and is conveniently subdivided into low marsh and high marsh. The low marsh (fig. 11.26) extends up to almost spring high tide and is characterized by *Spartina* (cordgrass). This part of the marsh is flooded fairly regularly. The high marsh is inundated only during spring tides or during storm conditions. It is characterized by the succulent *Salicornia* (glasswort) and *Juncus* (needle rush) (fig. 11.27). The relative amounts of low marsh and high marsh are a general indication of the maturity of the marsh. An immature marsh is dominated by low marsh and a mature marsh is dominated by high marsh.

Marsh Sedimentation

Sediment accumulation in the marsh may take place in two distinctly different styles in two separate environments. One is in the tidal channels and creeks that commonly network the marsh. Here tidal currents carry sediment and deposit it much like that in rivers. These tidal creeks meander through the marsh with sediment accumulations concentrated at the outside of the meander loops. This sediment is typically sand and occurs in crescent-shaped accumulations that conform to the meanders of the tidal creeks.

The vast majority of the marsh environment consists of the nearly flat, vegetated zone above neap high tide. Sediment reaches this environment only as the result of being carried in suspension by water that floods the marsh. Consequently the sediment is generally fine-grained clay and fine silt, and accumulates very slowly. Marsh vegetation assists in the sedimentation process by (1) physically trapping some of the suspended sediment and (2) slowing the tidal currents that pass over the marsh and thus facilitating settlement of the sediment particles from suspension. During storm conditions, extreme tides elevate water much above normal and waves may cause more than the normal amount of sediment to be carried in suspension. Consequently, there may be rapid sediment accumulation associated with such a storm event, perhaps equivalent to several years of sediment accumulation without benefit of storms.

Figure 11.27 An example of a high marsh from the Gulf Coast of Florida.

Abundant, dense vegetation on the marsh is also an important contributor to marsh sediment. Plant debris from expired marsh grass is typically trapped in place and accumulates on the floor of the marsh. In fact, in many marshes this is the major contributor to marsh sediment with the result being **peat** or at least organic-rich mud. The rate of plant accumulation relative to inorganic sediment, the rate of decay of the plant material, and consumption of the plant debris by other organisms are factors in determining the nature of the marsh sediment that accumulates. There is a wide range of sediment types in marshes. In Georgia, an area of extensive, dense marshes, essentially no peat or peaty sediment accumulates, whereas in some of the coastal marshes of New England, peat dominates. The difference results from the relative abundance of plant debris as compared to mineral sediment and also from the oxidation of this plant debris. In New England there is little sediment entering the estuaries and lush marsh growth prevails. On the Georgia coast the sediment influx is high and oxidation of plant debris is rapid.

The Marsh Community

Coastal marshes are the most productive portion of the estuarine system and are among the most productive of all marine environments. Abundant nutrient material is dissolved in estuarine waters, providing food for marsh plants. Upon expiration, the decay of the great quantities of plant material from the marsh adds to this already high level of nutrients. Studies have found that plant debris contributes from 100–200 gm/m^2 per month. Primary production in the estuary proper is high. In addition to the dense grasses, algae of various types are also abundant on the marsh substrate. Primary consumers are varied in size and type, including small mollusks and arthropods, as well as herbivorous birds and mammals that feed on the marsh grasses. Common carnivores in the marsh are some small invertebrates, such as oyster drills, birds, and insects. Birds of prey such as osprey or hawks are the top carnivores.

BOX 11.2

A Georgia Marsh

The coast of Georgia and southern South Carolina contains some of the most extensive marsh environments in the world. These marshes extend from landward of the barrier islands across to the mainland coastal plain (fig. B11.2), in some areas covering distances of 20–30 km. There is little open water between the mainland and the barrier islands except for the tidal channels and rivers.

The coast is in the mesotidal range with spring tides of about 3 m. The combination of a high tidal range with a gently sloping coastal area produces a wide intertidal environment. Considerable fine-grained sediment is available from rivers emptying onto the coast and the tidal currents have redistributed this sediment onto the tidal flats. At the point that the tidal flats accumulate enough sediment to reach the neap high-tide level, then marsh vegetation becomes established. The Georgia coast has reached this stage in the development of the intertidal environment almost throughout.

There is extensive low marsh grass (*Spartina*) and high marsh grass (*Juncus*) development in this region. The low marsh is restricted to the areas immediately landward of the barrier islands and the high marsh borders the land areas both on the barriers and the mainland. High marsh extends many kilometers up the tidal creeks and coastal plain rivers as well. This vegetation type is the most widespread of the two primary varieties of marsh grass.

The rate of vertical accumulation of sediment on these marshes is low. Only during spring tide does the water cover all of the *Spartina* marsh.The *Juncus* marsh is covered by water only during storms when the surge extends to the supratidal environment. During such situations the suspended sediment carried in the estuarine waters can settle onto the marsh floor or become trapped by the blades of marsh grass.

These extensive marsh areas on the Georgia coast support a tremendous fauna. There are abundant invertebrates, such as fiddler crabs, snails, worms, and crustaceans, that live within the marsh. These in turn support abundant carnivores, especially birds. The marsh also provides shelter for birds and small mammals.

(a)

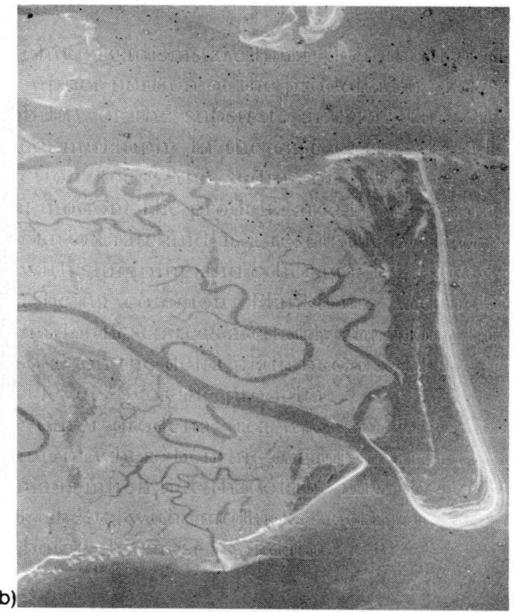

(b)

Figure B11.1 Photo of (a) coastal marsh and (b) aerial photo of Georgia marsh showing the extensive distribution of this coastal environment.

Figure 11.28 Fiddler crabs on an intertidal flat.

Figure 11.29 Photo of a tidal creek bank in the marsh showing numerous burrows formed by fiddler crabs.

The most prominent biological process on the marsh is bioturbation, including that caused by the root systems of the various grasses and that caused by abundant burrowing invertebrates, especially fiddler crabs (fig. 11.28). These organisms are so numerous and are such active burrowers that their activities can be an important factor in the erosion of tidal creek banks (fig. 11.29). A significant consequence of bioturbation in marshes is the general absence of any bedding in many marsh deposits. Although the sediment accumulates in thin, widespread layers (fig. 11.30), these layers are destroyed by roots and burrowers.

Figure 11.30 Photo of marsh sediment layering caused by alternation of organic-rich sediments with those having little organic matter.

Summary of Main Points

Estuaries are coastal embayments where seawater is diluted by freshwater runoff from the land. The recent rise in sea level caused by melting glaciers has drowned much of the coast of the world's landmasses, producing extensive estuaries. The interaction between fresh water and seawater provides for a means of classifying estuaries based upon the degree to which mixing takes place.

The location of estuaries provides for abundant nutrients and thus this environment supports a large and diverse population. Estuaries also serve as a nursery for many organisms that occupy the open ocean during their adult stages.

Tides tend to be the dominant physical process in most estuaries; in fact, the highest tidal ranges in the world occur in estuarine environments. In extreme situations, tidal currents may be so strong that they inhibit the development of benthic communities because the substrate is so unstable.

The intertidal margins of estuaries are typically rimmed by tidal flats and/or marshes. Both are intertidal environments that contain communities that are specially adapted to alternate exposure and inundation. As rivers that empty into the estuary fill it with sediment, the marsh and tidal flat environments migrate toward the center of the estuary. It is possible for the estuary to completely fill with sediment.

Suggestions for Further Reading

Barnes, R.S.R., and J. Green, eds. 1972. **The Estuarine Environment**. London: Applied Science Publ.

Clark, J.R. 1977. **Coastal Ecosystem Management**. New York: John Wiley & Sons.

Lauff, G.H., ed. 1967. **Estuaries**. Washington, D.C.: Amer. Assoc. Adv. Sci., Publ. No. 83.

Redfield, A.C. 1980. **Introduction to Tides**. Woods Hole, Massachusetts: Marine Science International.

Schubel, J.R., ed. 1971. **The Estuarine Environment: Estuaries and Estuarine Sedimentation**. Amer. Geol. Inst. Short Course Lecture Notes. Washington, D.C.: Amer. Geol. Inst.

12
Deltas

Chapter Outline

The **delta** is a large accumulation of sediment at the mouth of a river, which is the place where the river discharges into a standing body of water. The term delta comes from the Greek letter Δ and was first applied to the sediment accumulation at the mouth of the Nile River, which has a triangular or delta shape. In fact, however, most river deltas do not have this shape.

Deltas are in many ways much like estuaries in that they have freshwater influx, are affected by tides, and typically contain tidal flats and marshes. Also, both deltas and estuaries are sediment sinks. Whereas an estuary is a brackish water coastal embayment, a delta is a brackish water coastal protrusion. Deltas form wherever there is more sediment dumped at the coast by a stream than can be dispersed by coastal marine processes. It is also necessary that there be a shelf environment on which the deltaic sediment can accumulate. A look at the world distribution of deltas (fig. 12.1) shows a general absence along leading edge active margin coasts (see chapter 2). This is due to small drainage basins and the absence of a shelf on which a delta can develop.

Deltas are of great economic importance because they are productive environments for brackish and marine organisms including various types of commercial fish, shrimp, and oysters. Deltaic deposits in the ancient geologic record are among the greatest producers of petroleum and coal. The delta environment also provides land where there was none.

The delta is one of the most complicated and diverse of all marine systems in that it contains many different environments, ranging from freshwater to normal marine conditions. Some environments are well protected, whereas others are exposed to high-energy wave and tidal conditions. Deltas may also rapidly change in size and shape as the result of the interactions between the river feeding the delta and the marine processes on the ocean side, thus reshaping or eroding the delta. Typically deltas tend to increase in size overall but some areas are being eroded while others are expanding as sediment accumulates. Some deltas have recently been experiencing an overall decrease in size due to the activities of man. This phenomenon is primarily caused by the

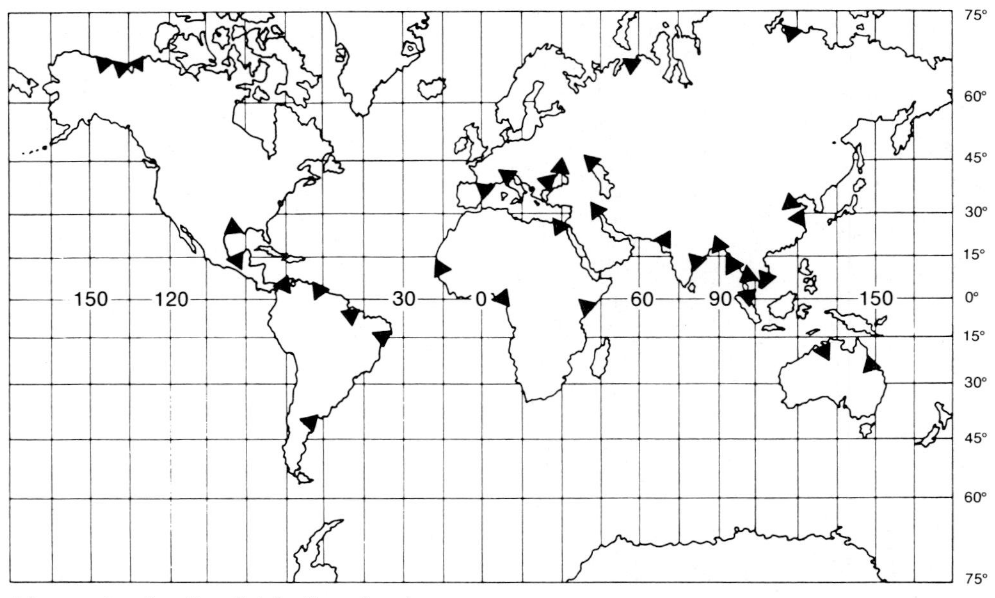

Figure 12.1 World map showing the distribution of major river deltas. Note that most are confined to areas of passive continental margins.

construction of dams that stop the flow of some of the sediment and reduce the discharge of sediment at the river mouth. One of the most notable of these circumstances has occurred on the Nile Delta where the huge Aswan Dam has caused a great reduction in sediment supplied to the delta. As a result, the delta has been eroding because waves and currents in the Mediterranean Sea are carrying away more sediment than is being contributed to the delta system.

Major Elements of a Delta

A delta is comprised of three major segments: the delta plain, the delta front, and the prodelta (fig. 12.2). The **delta plain** is the landwardmost portion of the delta and is partly fresh water, partly subaerial (exposed), and partly marine (submerged). It is by far the most complex deltaic segment and includes all of the environments that occupy the low-relief upper surface of the delta. The **delta front** is the

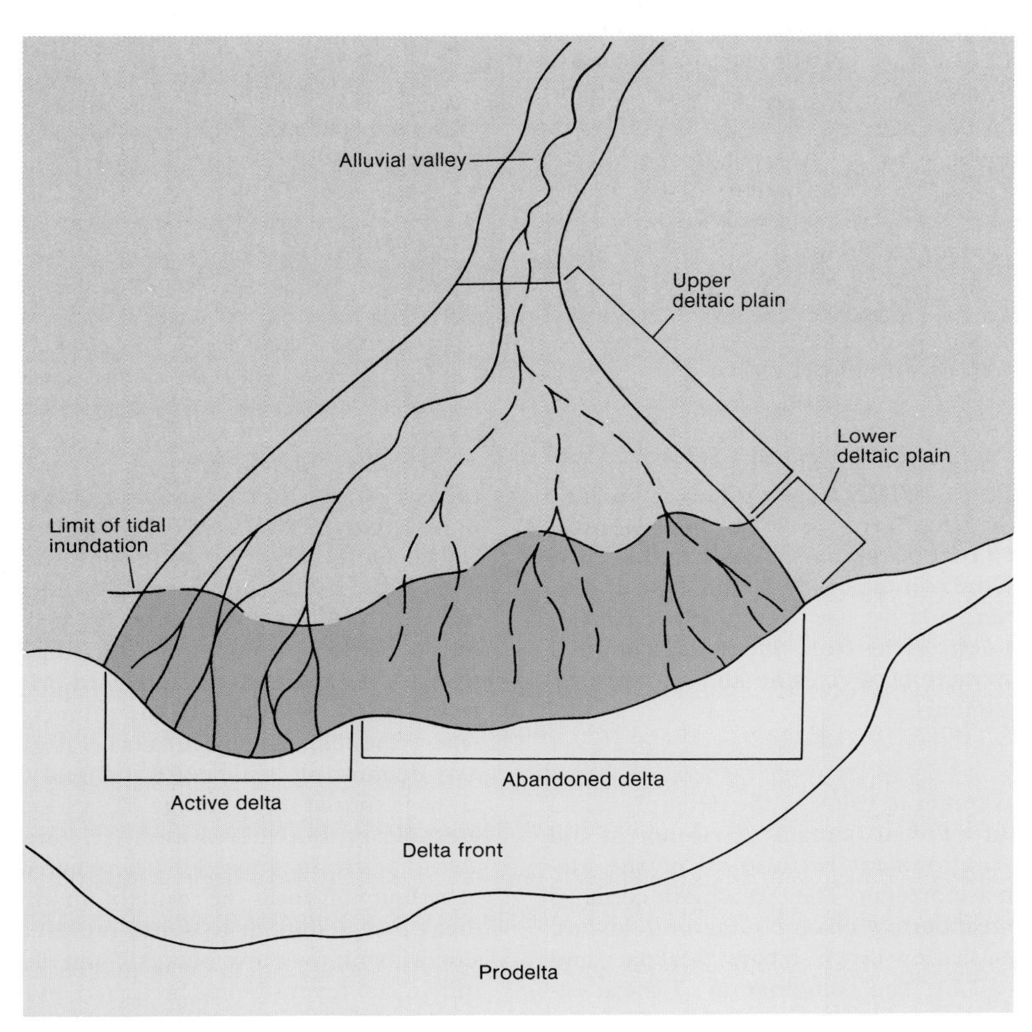

Figure 12.2 Generalized diagram showing the three main elements of a delta system: delta plain, delta front, and prodelta.

Figure 12.3 Oblique photo of a Mississippi Delta plain area showing interdistributary bays and marshes.

segment that is adjacent to and oceanward of the delta plain. It is essentially marine and is dominated by marine processes. The **prodelta** is the sloping, relatively deep part of the undersea delta that comprises most of the volume of the deltaic system. The overall shape and relative size of each of these major segments of a delta varies from one delta to another due to the interaction of riverine and marine processes and sediments.

Delta Plain

It may be convenient to think of the delta plain as a combination of a fluvial (stream) environment and an estuarine environment because all of the elements of both are present. The delta plain contains floodplains, meandering channels, natural levees, crevasse splays, shallow brackish bays, tidal flats, and marshes (fig. 12.3). The combination of these environments serves as the complex transition from the freshwater river environment to the marine environment.

The delta plain represents the coastal area where the river loses its competence and capacity to transport sediments very quickly. When the river empties into a standing body of water it is suddenly unconfined and the flow slows rapidly. This causes the deposition of sediment because the mechanism for its transportation, that is, the flow, stops. Sediment is spread over a wide area, delivered primarily by **distributaries**, which are the numerous channels where the main stream breaks up (fig. 12.4). There may be multiple scales of distributary branches but all have similar morphology. They may be straight or meandering but they always have **natural levees** along the channel margins. These narrow, low ridges of sediment form as the result of flooding conditions that cause sediment to settle rapidly from suspension as bank overflow occurs. Some flood events result in the levees being breached and a fan-shaped **crevasse splay** is deposited adjacent to the levee (fig. 12.4).

Interdistributary areas contain shallow brackish bays, tidal flats, and marshes. These may vary in relative abundance depending on the stage of development of the delta. A predominance of open-water

Figure 12.4 Photograph of a distributary area on the Mississippi Delta showing natural levees and a crevasse splay.

bays signifies an immature delta, whereas dominance by marshes indicates maturity. This tends to parallel the situation in the estuarine system.

Most of the sediment deposited on the delta plain is mud, which accumulates due to flooding of the distributaries. Sand is generally restricted to the channel deposits in the distributaries.

Delta Front

The subtidal region seaward of the delta plain is called the delta front. It is a narrow, linear area that outlines the zone where river-mouth processes and marine processes come together. Among the marine processes are tidal currents, waves, and coastal currents, which may be wind or wave generated. At the mouth of the river or its distributary there is sudden loss of competence and capacity and the coarser sediment tends to accumulate. This sediment accumulation is affected by tides and especially by waves, which sort out fines and mold the sediment into a stable morphology. As a result, the delta front sediments are fairly homogenous and are dominated by moderate to well sorted sand in the shallowest areas, grading to interbedded sand and mud in deeper areas. This zone ranges in depth from slightly subtidal to a few tens of meters, with the outer limit controlled by wave activity; a higher energy wave climate will extend the delta front to deep water. In the Mississippi Delta, for example, the wave environment is one of low energy, except for occasional hurricanes, and the delta front extends to about 20 m of water.

Overall, the delta front deposits are the coarsest and best sorted of the delta. Elongate accumulations of sandy sediment (sandbars) predominate in the shallower areas (fig. 12.5). They tend to be oriented to the dominant process: tides generate sandbars that are perpendicular to the coast or parallel to the distributaries, and waves generate sandbars that are parallel to the coast. These processes also form sedimentary structures, such as ripples, horizontal layers, and scour and fill, where currents excavate sediment, then fill it back in. The relatively high-energy conditions keep the substrate somewhat mobile and restrict benthic organisms. Consequently, burrows or other biogenic features, including shell debris, are uncommon.

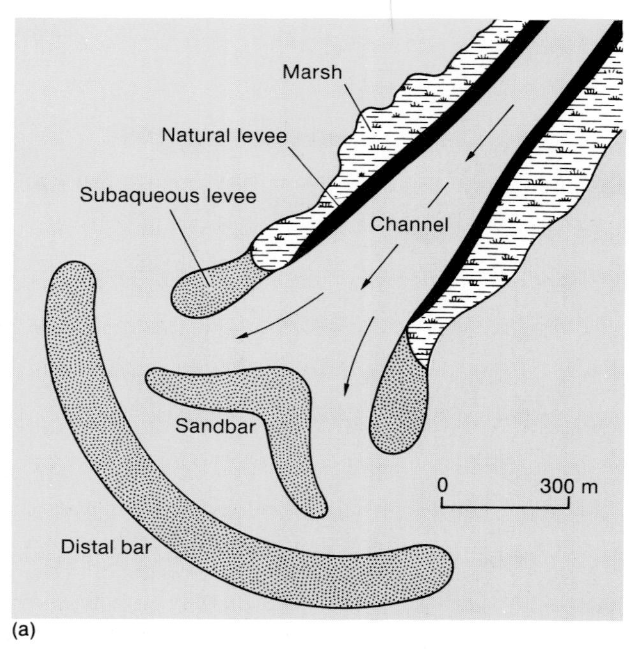

(a)

(b)

Figure 12.5 Diagram (a) and photograph (b) of the mouth of a distributary showing subaqueous natural levees and distributary mouth bars that are part of the delta front.

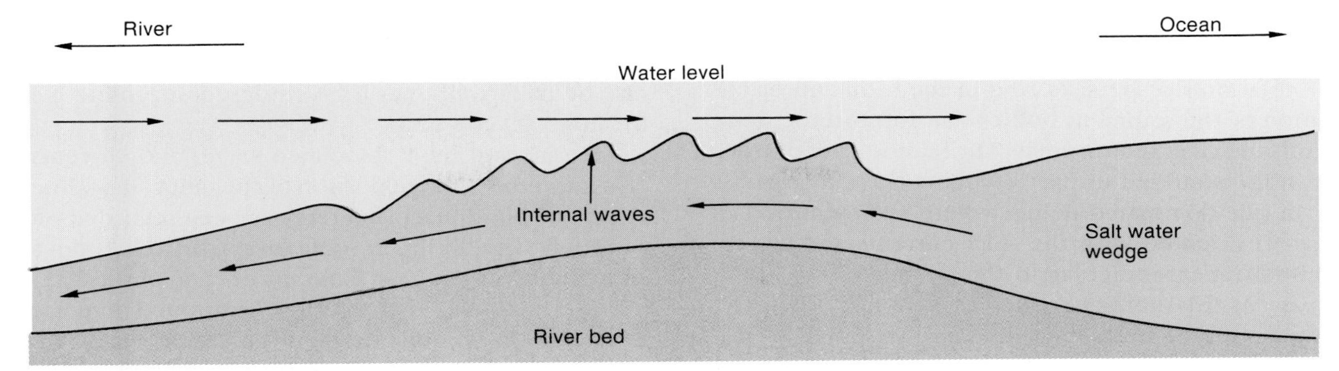

Figure 12.6 Schematic profile diagram showing the interaction of the saltwater wedge with the outflowing river water during flood tide.

Prodelta

This portion of the delta is the most homogenous and generally the most widespread. The prodelta is the seawardmost portion of the delta and is comprised mainly of mud, silt, and clay. There is a gradual decrease in grain size and also in rate of sediment accumulation in the seaward direction. Sedimentation is dominantly from suspension, thus producing thin, laterally continuous layers of mud.

The prodelta portion of the delta is actually prograding or growing seaward, thus its name. As a result, these sediments are deposited over continental shelf sediments. In a few deltas, this progradation has already covered the shelf and prodelta sediments move down the continental slope. This has happened on part of the active lobe of the Mississippi Delta.

The outer portion of the prodelta is a true marine environment and supports a marine community. Benthic organisms are fairly common in this area and show a gradual decrease in abundance toward the shallow part of the delta, reflecting the increase in sedimentation rate.

Fluvial-Marine Interactions

Processes on the delta, especially the outer delta plain and the delta front areas, represent a confrontation between riverine forces and marine forces. By definition, a delta represents a net accumulation of sediment produced by the river, an accumulation that is not carried away by marine processes. That in itself indicates the conflicts at work in this transition system. Once a delta is developed there is still conflict between the constructive forces, primarily the river, and the destructive ones, primarily the waves and currents of the marine environment. Virtually all deltas have constructive portions and destructive portions. It is also common for the nature and extent of these to change through time, in some cases even on a cyclic basis, such as seasonal. During the rainy season, runoff is great and river processes may dominate the delta environment. By contrast, during the dry season there would be much less runoff and discharge at the river mouth, thus permitting waves or tides to dominate deltaic processes.

Tidal Effects

The effects of tides in a deltaic system are much the same as in the estuarine system. Insofar as most of the processes are concerned, a delta is just a special type of estuary. They may display the same three styles of mixing or stratification as discussed for estuaries (see chapter 11, p. 188). The fresh but sediment-laden water debouched from the river mouth is buoyed up by the denser salt water. In some cases internal waves may develop along this interface because of the friction between the water masses (fig. 12.6).

The long, narrow configuration of the river and its distributaries tends to mask the Coriolis effect in a delta. There is no apparent rotational circulation and an amphidromic system is not present. The tilting of the water level is barely recognizable. The same is true for the interface between fresh and salt water.

The relative influence of tidal compared to riverine processes can be seen in the type of estuarine circulation that persists, and in the form and orientation of the sediment bodies that form in the delta front and river mouth areas. The least impact of tides is in the stratified estuary (type A) and the greatest is in tide-dominated deltas, where sediment bodies are all oriented with the tidal currents and are essentially perpendicular to the overall trend of the coast. As tidal influence increases, it first affects the interaction of fresh and saltwater masses, then proceeds to affect the morphology of the river mouth and its sediment bodies.

Tides can be both a constructive and destructive force on the delta. They may carry sediment to the river mouth and deposit it in the form of elongate sandbars. On the other hand, tidal currents can be strong and erode sediment from the river mouth and carry it away from the delta environment.

Storm Effects

Waves serve as the other primary oceanic process that influences deltas. Under some low-energy conditions, waves may be a constructive force on the delta, but usually waves are destructive. Of particular significance are the severe storms that affect coastal areas, including many of the deltas of the world. The most intense of these are hurricanes and typhoons, which generate huge waves and elevate tides up to several meters above normal. The Mississippi Delta in the Gulf of Mexico and the Ganges-Brahmaputra and Irrawaddy Deltas in the Bay of Bengal of the Indian Ocean are among those most influenced by severe storms. During the past few decades, hurricanes have cost many lives and many millions of dollars on the Mississippi Delta.

Because deltas tend to form on broad continental shelves, storm tides or surges associated with hurricanes or other intense storms are usually large due to the tendency for the storm to pile up water along the coast. This allows the large waves generated by such storms to strike not only at the delta front area and the outer limit of the delta plain, but also substantially inland on the delta plain due to the low relief, which allows extreme flooding. The delta plain is just about at normal sea level and an increase in water level of only 2–3 meters would be catastrophic for not only residents of the subaerial delta plain but also the environment and its fauna and flora. Certain communities, such as oyster reefs or marshes, may be destroyed or at least rendered unproductive for up to several years.

Storms and their associated waves and currents also may erode a large portion of the delta. Shoreline retreat of hundreds of meters has been recorded on the Mississippi Delta. In some cases, the tremendous amount of sediment lost from any portion of the delta may be redeposited at another location. There are also many situations in which the eroded sediment is lost to the delta system and carried away, both alongshore and offshore.

Classification of Deltas

The most commonly used classification of river deltas is based on morphology and is associated with the dominant process or the combination of processes that form the delta. This is a useful classification because it is simple and can be applied to all river deltas regardless of size or location.

The previous section pointed out that there are really three major processes or forces at work on the delta: the river, the tides, and the waves. These serve as end members in the delta classification, but there

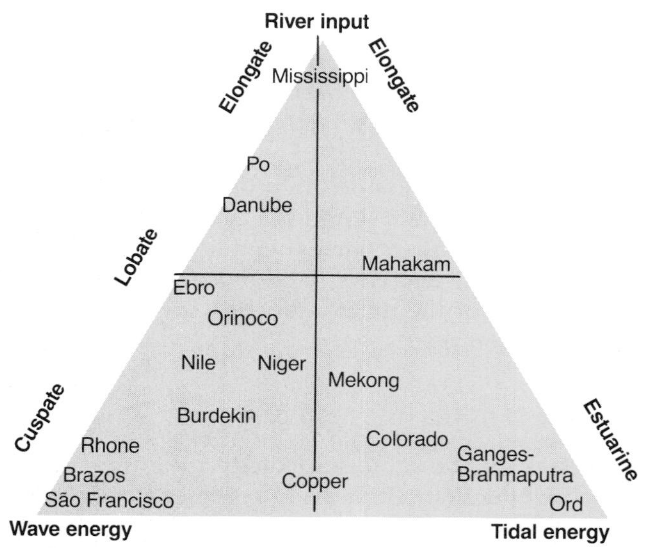

Figure 12.7 Diagram of the classification of river deltas based on the dominant process that controls the shape of the delta. The three end members are river processes, waves, and tidal currents.

is complete transition among all three (fig. 12.7). During the life of a given delta, the dominating process may change and therefore the morphology may change. The end members are therefore called river-dominated deltas, tide-dominated deltas, and wave-dominated deltas. The same environments are present in all of the morphologic types but they differ in shape and extent. For example, one type of delta may have extensive marshes, whereas in another this environment is almost absent.

River-dominated Deltas

The general morphology of a river-dominated delta is one of an irregular, digitate shoreline (fig. 12.8). Typically there are several distributaries with extensive interdistributary areas occupied by bays, marshes, and in some cases, tidal flats. These deltas tend to protrude far out onto the continental shelf. The modern Mississippi Delta (fig. 12.9) is a textbook example of a river-dominated delta. The Po and Danube deltas of Europe also fall into this category.

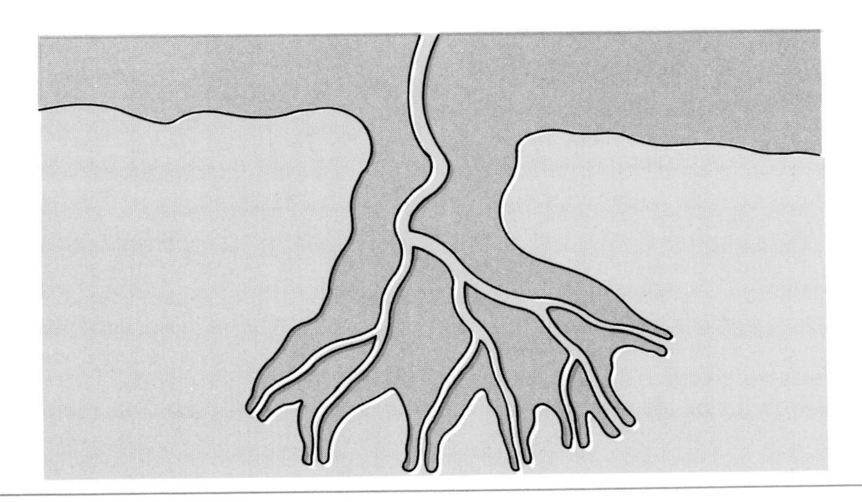

Figure 12.8 Outline map of a river-dominated delta showing the irregular shape and numerous distributaries.

Figure 12.9 Photograph of the present active lobe of the Mississippi Delta, a good example of a river-dominated delta.

In the river-dominated delta the river completely overpowers the marine processes and produces a rapidly prograding delta with a high rate of sediment accumulation. Deposition is concentrated at the distributary mouths with the interdistributary areas receiving sediment during times of flooding. This type of delta is primarily constructional. Destruction or erosion takes place only in areas where there are no active distributaries to supply sediment.

River-dominated deltas have extensive and diverse delta plain complexes, more than the other types. A typical delta will have bays, marshes, tidal flats, natural levees, crevasse splays, and channel deposits. The delta front is well defined but restricted in extent and thickness. Most of the volume of the river-dominated delta is contained within the prodelta muds.

BOX 12.1

The Mississippi Delta

The Mississippi Delta is one of the largest deltas in the world because it drains a large area of the United States and because it has developed in the Gulf of Mexico, a basin characterized by low wave energy and low tidal ranges. The combination of these conditions produces a huge river-dominated delta.

Although the Mississippi Delta complex is quite large, the present active lobe of the delta is relatively small. During the past several thousand years the active lobe of the delta has shifted over 100 km in its location. The present active part of the delta has achieved much of its appearance in only about the past 100 years. It has extended over 20 km in that time.

The reason for this rapid growth in the Mississippi Delta can be seen in the volume of water and sediment that is discharged annually. The mean annual rate of discharge is 15,300 m^3/sec, ranging from 2,800 m^3/sec during the dry season to 57,400 m^3/sec during the spring flood season. Simple arithmetic indicates that this is a mind-boggling number when considered over a year's time. The amount of sediment discharged annually is 5×10^{11} kg or 5.5×10^{8} tons. This is about 10 million dumptruck loads per year!

The delta plain area of the Mississippi Delta complex contains extensive marshes and interdistributary bays. These environments are very productive biologically. The river provides a great supply of nutrients and the shallow, warm, and fairly calm waters provide ideal conditions for development of extensive brackish water communities. This part of the delta is one of the primary nursery grounds for shrimp and fish, which migrate to the shelf and beyond when they mature. The delta itself is an important fishing resource both commercially and for sport. The oyster beds of the Mississippi Delta are among the most productive in North America.

Like most fragile coastal environments, the Mississippi Delta is not without its problems; these are partly due to human activity and partly to nature. The sediments that comprise the delta are mostly mud and accumulate very rapidly. As a consequence they are not well compacted. As time passes, compaction slowly takes place and the fragile marsh areas of the delta subside causing a localized rise in sea level. In the delta area the rise of sea level is 3–4 times the global rate or about 1 cm/yr. The result is that marshes are destroyed rapidly as the edge of the delta migrates landward due to the compaction and settling of sediment.

The more obvious problem on the delta is the impact of humans, especially that of the petroleum industry. The search for oil and gas is a necessary one and deltaic sediments are a prime location for extensive hydrocarbon resources. Much construction must be done in order to drill a well on the delta plain, especially in a marsh area. Typically a channel must be dug to provide access for equipment, then the site must be prepared. Drilling may last for months. If drilling is successful, permanent equipment must be installed and perhaps a pipeline as well. All of these activities have a negative impact on the marsh environment. It should be stated, however, that the petroleum industry has become more sensitive to problems of the environment and makes a concerted effort to minimize its impact and to follow environmental regulations. Furthermore, the delta plain environments are resilient and can recover quickly from storms or from the impact of man.

Wave-dominated Deltas

Deltas that develop along coasts with a high-energy wave environment display a distinct morphology. They are smoothly arcuate in outline, extend only a modest distance onto the shelf, and generally have only one or two channels (fig. 12.10). Typically, wave-dominated deltas are small due to the tendency for waves and wave-generated currents to carry sediment away from the river mouth. The delta plain area is limited in extent and is dominated by sand beaches and related ridges formed by waves. Interdistributary areas are nearly absent and marshes are restricted to low areas between ridges (fig. 12.11). Good examples of wave-dominated deltas are the São Francisco Delta in Brazil and the Senegal Delta on the west coast of Africa.

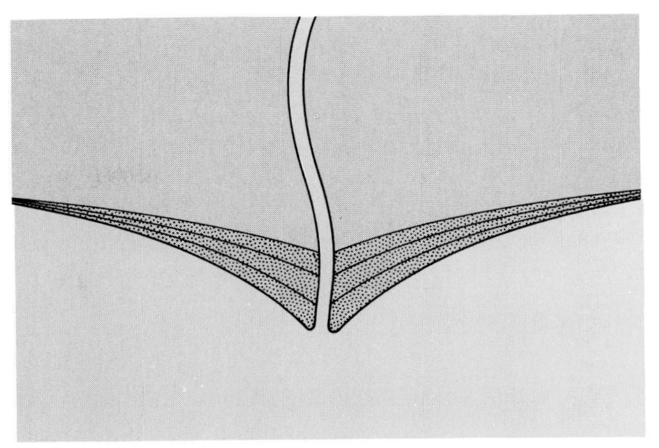

Figure 12.10 Outline map of a wave-dominated delta showing the smooth, cuspate outline and single major channel.

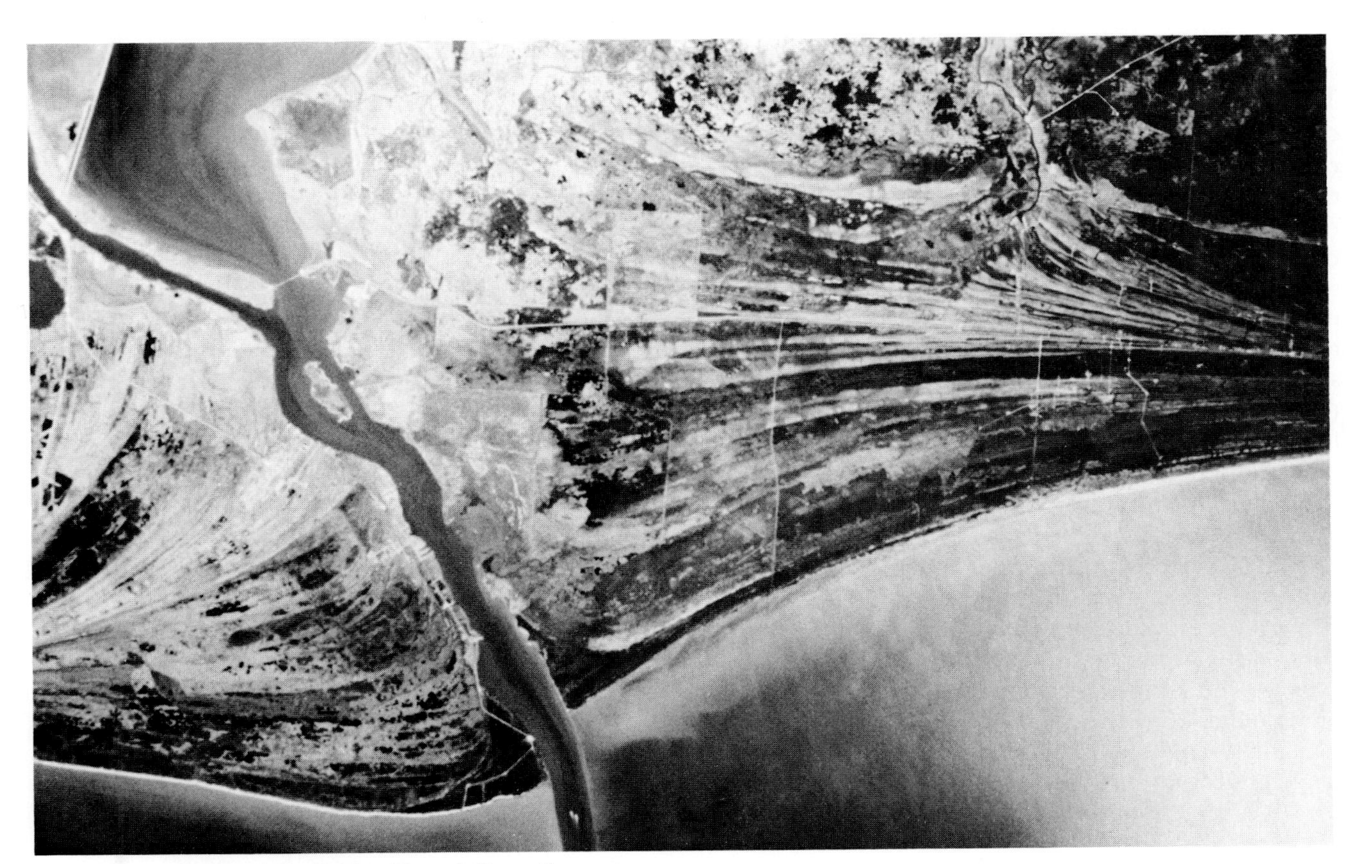

Figure 12.11 Photo of the Sabine River delta on the Louisiana-Texas border, a good example of a wave-dominated delta.

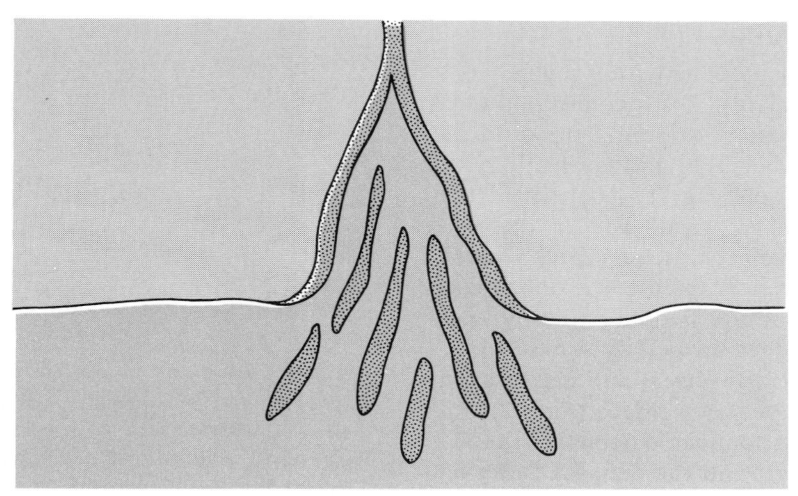

Figure 12.12 Outline map of a tide-dominated delta with its funnel shape and large sediment bodies that are perpendicular to the coast.

Tide-dominated Deltas

Coasts that have low wave energy and high tidal ranges develop tide-dominated deltas. Generally there is low to moderate sediment discharge from the river involved. The shape of such deltas is similar to tide-dominated estuaries. The overall configuration tends to be funnel-shaped and there are large sediment bodies oriented with the tidal currents at the mouth of the river. There is little protrusion onto the shelf and the outline of the delta is poorly defined (fig. 12.12). There is a transition with some overlap between tide-dominated estuaries and tide-dominated deltas. A general rule of thumb is that if a funnel-shaped, tide-dominated coastal embayment is served by several small rivers, it is best considered an estuary, whereas if a similar estuary is served by one large river, it is a delta. An additional consideration is that the delta will have more sediment accumulating at the mouth of the river than will the estuary. The Ganges-Brahmaputra, Amazon, and Colorado deltas (fig. 12.13) are tide-dominated.

Tidal ranges are generally in the macrotidal range. Currents of 1 m/sec and more are typical. These currents overwhelm any wave action and distribute the sediment so that accumulations are generally parallel to the tidal currents. It is common for the relief on these bars to be a few meters.

Tide-dominated deltas have little delta plain area and that which is present tends to be dominated by tidal flats with some marsh areas. Much of the delta is dominated by delta-front sands. In contrast to wave-dominated deltas, where delta-front sand bodies are parallel to the coast, in tide-dominated deltas these sand bodies are perpendicular to the trend of the coast. Little prodelta mud is deposited because tidal currents carry fine sediment beyond the delta.

Wave-dominated deltas are the most destructive of the three delta types. The high level of wave energy imparted on the coast removes all of the fine sediment and carries it from the delta. The waves smooth the outer edge of the delta plain and at the same time produce beach ridges, or storm ridges as they are sometimes called. These ridges may be 2–3 meters high and are composed of sand, which remains after removal of the fine sediment, and the shells of organisms. The sand that comprises the beach ridges may extend to relatively deep water due to large waves. The prodelta portion of the delta is small and is only about 30% or less of the total delta accumulation.

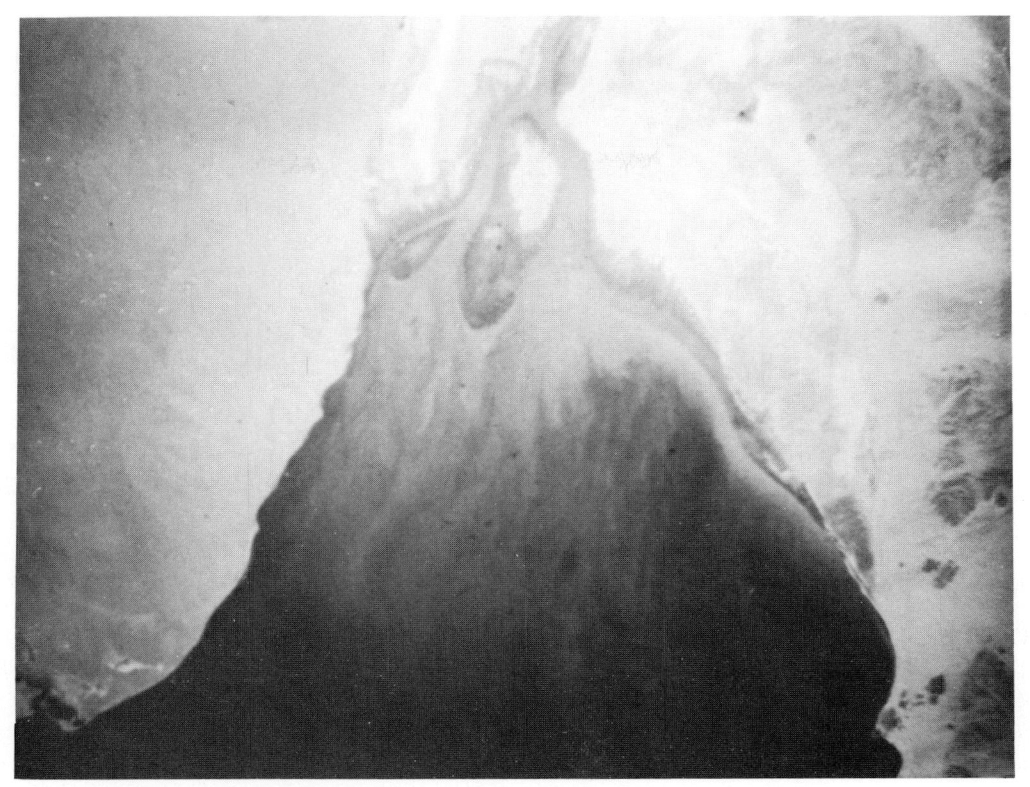

Figure 12.13 Photograph of the Colorado River delta in the Gulf of California, a good example of a tide-dominated delta.

Delta Communities

The organisms that inhabit the delta environment are essentially the same as those found in estuaries. Fresh to brackish water conditions persist on the delta itself and may extend out several kilometers beyond the delta plain, especially on river-dominated deltas. Just as is true for estuaries, deltas are areas of extremely high nutrient concentration. The discharge of the river coupled with nutrients from the highly productive delta plain serve as the basis for the ecosystem. The delta front and the landward portion of the prodelta are the areas where organisms are least abundant within the delta system.

The most productive areas of the delta are the interdistributary regions of the delta plain. Here the shallow bays, marshes, and tidal flats are home to diverse and extremely abundant life. In nearly all respects these communities mirror those of the low-energy estuaries. Photosynthesis is usually more restricted than in the estuary because of the extreme turbidity of the water being discharged from the river distributaries. Nevertheless, there is a great abundance of organisms and some of the most productive shellfishing is found in such areas. Oysters and shrimp are especially numerous, with the latter extending out beyond the edge of the delta plain. Fish are also abundant but because of the nature of the environment and the shallow water, large-scale commercial fishing is not widespread.

Planktonic, nektonic, and benthonic organisms abound in the interdistributary environments, with salinity being the primary controlling factor in distribution. The distributary channels are typically completely filled with fresh water or show some stratification. The distribution of organisms will reflect the nature of the water mass.

Figure 12.14 Flooding on the Mississippi Delta showing only natural levees and crevasse splays above water. A large crevasse splay is shown near the middle of the photo.

Because the delta plain is a fragile environment that is subject to change as the result of storms, floods, or, in some cases, the influence of humans, the communities may also show related and marked changes. A severe flood (fig. 12.14) can destroy certain species due to salinity changes or suspended sediment concentrations. Storms may remove extensive oyster beds or destroy marshes that are home to numerous species. Many deltas are areas of extensive exploration for petroleum and the impact of humans, such as channel dredging, pollution, and overall disruption of the environment may cause serious impact on delta plain communities.

Benthic organisms are limited on the delta front and the upper portion of the prodelta due largely to the mobility of the substrate and the rate of sediment accumulation. In the case of the delta front, wave action keeps the sediment substrate in near constant motion. The prodelta area is one with a high sedimentation rate and as a result the bottom is unstable, with slumping and other gravity-related phenomena being common. Planktonic and nektonic organisms in these areas are fairly abundant, but not nearly so as on the delta plain because of the lower nutrient levels over the prodelta.

The outer prodelta area is essentially a marine environment where benthic organisms are common. The fauna is much like that of the mid- to outer continental shelf, with mollusks, echinoderms, and worms being the most abundant.

There are some generalizations that can be made about delta communities with respect to the major categories of deltas. By far the most productive is the river-dominated type. Nutrients are in abundance and there are widespread and varied shallow environments for organisms to inhabit. The extensive interdistributary areas serve as excellent habitats with the right combination of nutrients, circulation, salinity, and protection.

Wave-dominated deltas are typically the least productive of the delta types. The nutrient material is rapidly dispersed, there are no shallow, protected habitats, and wave energy keeps the bottom mobile, thereby restricting benthic organisms. Tide-dominated deltas show a wide range in productivity but they fall in between the other two types in overall productivity. Some tide-dominated deltas, such as the Amazon, discharge great amounts of nutrients, but photosynthesis is severely limited by extreme turbidity. The mangrove islands near the mouth are similar to the interdistributary areas of river-dominated deltas. On the other hand, there are unproductive tide-dominated deltas such as the Ord Delta on the north coast of Australia. This river drains a desert and has tidal ranges of 8 m. The result is a delta that is nearly a biological desert due to the lack of nutrients, the small discharge, and the extreme tidal currents, which keep the substrate in motion.

Summary of Main Points

Deltas are areas of sediment accumulation at the mouths of rivers. They also form wedges of sediment that protrude beyond the nondeltaic coastline. They require an abundance of sediment and a limited amount of physical energy along the coast of the basin of accumulation. The interrelationships between three main processes, river discharge, waves, and tidal currents, control the morphology of the delta.

There are many similarities between deltas and estuaries. Both are influenced by freshwater runoff; there are many fluvial and related environments in both; and marshes are extensive in both systems. Deltas and estuaries are both sediment sinks, but whereas the lifetime of the estuary is limited to the time it takes to fill the coastal embayment with sediment, the delta protrudes into the basin of deposition and is not generally limited by spatial constraints, except the width of the continental shelf.

All deltas consist of three major elements: (1) the delta plain, which consists of several distinct environments and has fluvial affinities, (2) the delta front, which is the site of the greatest concentration of physical energy, and (3) the prodelta, which typically comprises most of the volume of the delta.

Delta classification is based on morphology and is the result of the three major physical processes to which the delta is subjected. River-dominated deltas are irregular in outline with numerous distributaries and a well-developed delta plain. Wave-dominated deltas have smooth, cuspate outlines with large amounts of sand in the delta front and as beach-dune complexes. Tidal deltas are also relatively high in physical energy, with large sand bodies oriented perpendicular to the coast by strong tidal currents.

Deltas are the site of a variety of coastal marine communities. The most diverse communities inhabit the delta plain, where marshes and interdistributary bays are widespread. The abundant nutrient material provided by the river and the generally shallow, well-oxygenated waters provide ideal conditions for the development of brackish fauna and flora.

Suggestions for Further Reading

Morgan, J.P. 1970. "Deltas, a resume," *Jour. Geol. Education,* v. 18, p. 107–17.

Morgan, J.P., and R.H. Shaver, eds., 1970. **Deltaic Sedimentation**, Soc. Econ. Paleontologists and Mineralogists, Spec. Publ. No. 15, Tulsa, Oklahoma.

Shirley, M.L., ed. 1966. **Deltas in their Geologic Framework.** Houston, Texas: Houston Geol. Soc.

Shepard, F.P., F.B. Phleger, and T.J. VanAndel, eds., 1960. **Recent Sediments of the Northwest Gulf of Mexico.** Tulsa, Oklahoma: Amer. Assoc. Petroleum Geologists.

13
Barrier Island Complexes

Chapter Outline

Barrier islands and related environments comprise one of the most complex and widespread coastal systems. These barrier systems contain long, sandy barrier islands that are generally parallel to the coast, are composed primarily of sand, and are interrupted by tidal inlets. Typically, barrier islands contain linear dune complexes landward of the beach that fronts the open marine environment. Extensive marsh and tidal flat environments as well as washover fans of sediment deposited by storms may be present on the landward side of the barrier, and estuaries or lagoons separate the barrier island from the mainland (fig. 13.1).

Barrier island systems extend along about 20% of the world's coastline and are the dominant coastal landform type on submergent passive margin coasts, such as the Atlantic coast of the United States. They may also be extensive along the coasts of mediterraneans like the Gulf of Mexico and the Arctic Sea. The present active barrier island systems are geologically young, having formed during the latter part of the **Holocene**, the recent period of sea level rise associated with the melting of the ice age glaciers, beginning about 18,000 years ago. Most of the modern active barrier islands are less than 7,000 years old.

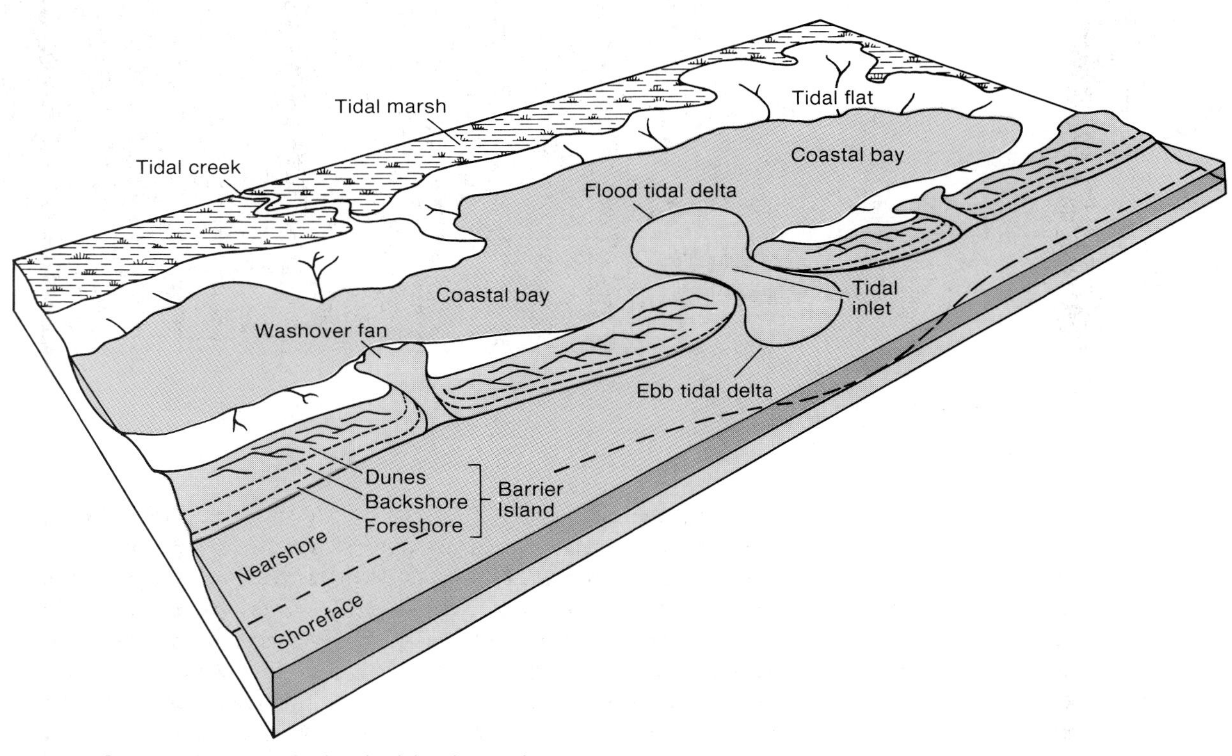

Figure 13.1 General diagram of a barrier island complex showing the location of the various component environments. (Blatt/Middleton/Murray, *Origin of Sedimentary Rocks,* © 1980, p. 660. Reprinted by permission of Prentice-Hall, Englewood Cliffs, New Jersey.)

Elements of the Barrier Island System

Because the barrier island systems are complex and because some of their major elements have not yet been discussed, it is appropriate to consider each of the environments separately before considering the system as a whole.

Beach and Nearshore Environment

The components of the seawardmost part of the barrier system are the beach and adjacent nearshore. These are actually separate environments but are discussed together because of their important interactions. The **beach** is the zone of unconsolidated sediment that extends from the low tide line on the seaward side to the next important change in topography or composition. This may be the base of dunes or bluffs of bedrock, or it may be man-made features such as seawalls. The **nearshore** is the zone from low tide across the sandbar and trough topography (fig. 13.2), or essentially equivalent to the surf zone during storm conditions. The beach and nearshore represent one of the most dynamic parts of the coastal zone. Here is where most of the wave energy reaches the coast and where wave-generated longshore currents develop.

The beach may be only a few meters wide (fig. 13.3), or it may be hundreds of meters wide (fig. 13.4). There are steep beaches (fig. 13.5) and there are those with gentle gradients (fig. 13.6). The beach

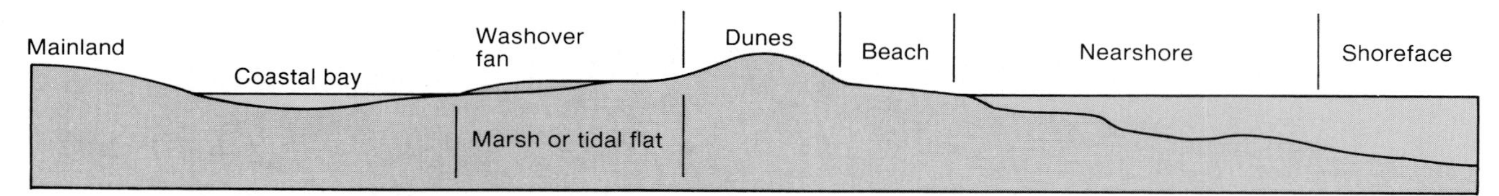

Figure 13.2 Profile diagram of a barrier island complex from the nearshore environment on the seaward side to the mainland.

Figure 13.3 Photograph of a narrow beach on the destructive portion of the Mississippi Delta.

Figure 13.4 Photograph of a wide beach on Padre Island, Texas.

Figure 13.5 Photograph of a steep beach on the high-energy Pacific coast near Carmel, California.

Figure 13.6 Photograph of a gently sloping beach on the low-energy Gulf coast on Mustang Island, Texas.

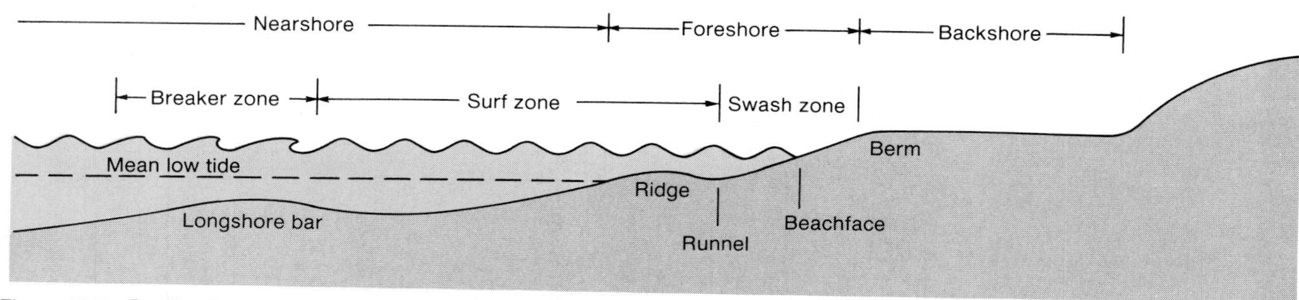

Figure 13.7 Profile diagram of the beach and nearshore showing the component subenvironments.

Figure 13.8 Photograph of a narrow and steep beach after a storm.

consists of two major elements: the **backshore** or backbeach and the **foreshore**. The foreshore is the seaward portion that is essentially equivalent to the intertidal portion of the beach. It is steep relative to the backshore (fig. 13.7), which varies from nearly horizontal to slightly landward sloping. The backshore is supratidal and is inundated only during storms or some spring tides.

This description is applicable to a beach that is accreting or adding sediment, such as would be the case during relatively low wave energy conditions. The other type of beach profile is the result of storm erosion from high levels of wave energy. Storm beach profiles tend to slope uniformly seaward with little or no backshore area; during these conditions the beach is essentially all foreshore (fig. 13.8).

The nearshore zone receives much wave energy and is also the region of longshore current. As waves enter shallow water and feel bottom, they steepen. As this process occurs, energy is transmitted to the sediment on the bottom, transporting grains in a back-and-forth motion due to the flattening of the orbital paths of the water particles. If the bars are shallow enough, the waves will break. They then re-form but are smaller, and continue moving shoreward. The bar and trough system may cause multiple breaking wave situations at any given time (fig. 13.9). The final break of the waves is always at the shoreline.

(a)

(b)

Figure 13.9 The longshore bar and trough (a) may be visible during calm waters or (b) the definition may be indicated by the linear breakers during high-energy conditions.

Refraction of waves in this zone causes longshore currents. These currents move more rapidly in the troughs than over the bars. The highest velocity longshore currents are commonly located between the shoreline and the first bar. As waves break, they cause a large amount of sediment to be thrown into suspension that is then carried alongshore by the longshore currents. Currents may travel at greater than 1 m/sec during storm conditions, thus carrying large quantities of sediment.

(a)

Figure 13.10 (a) Aerial photograph and (b) sea level photograph of a ridge and runnel system along a barrier beach, and (c) diagram of a ridge migrating shoreward.

The nearshore may include one or more **longshore bars,** or in some cases, it may not have any. Two bars are the most common but there may be more. The number of bars is limited by sediment availability and by the gradient of the nearshore bottom. The gentler the gradient, the more bars that are present. These longshore bars are essentially parallel to the shore and are subtidal. The depth of the bar crest and the distance of the bars from the beach varies from location to location and also at any given location.

An ephemeral intertidal sandbar termed **ridge and runnel** includes both the storm-created bar (ridge) and the trough (runnel) that separates it from the beach. The ridge is composed of some of the sediment eroded from the beach during the storm. During low wave-energy conditions between storms, the ridge slowly migrates back to the eroded beach by means of wave-generated currents and the rise and fall of the tide (fig. 13.10). As the tide rises and covers the ridge, waves carry sediment over its crest and deposit the particles on the landward side. As this occurs over and over, it moves the ridge landward. Eventually, it welds onto the previously eroded beach and repairs some of the damage done during the storm (fig. 13.11).

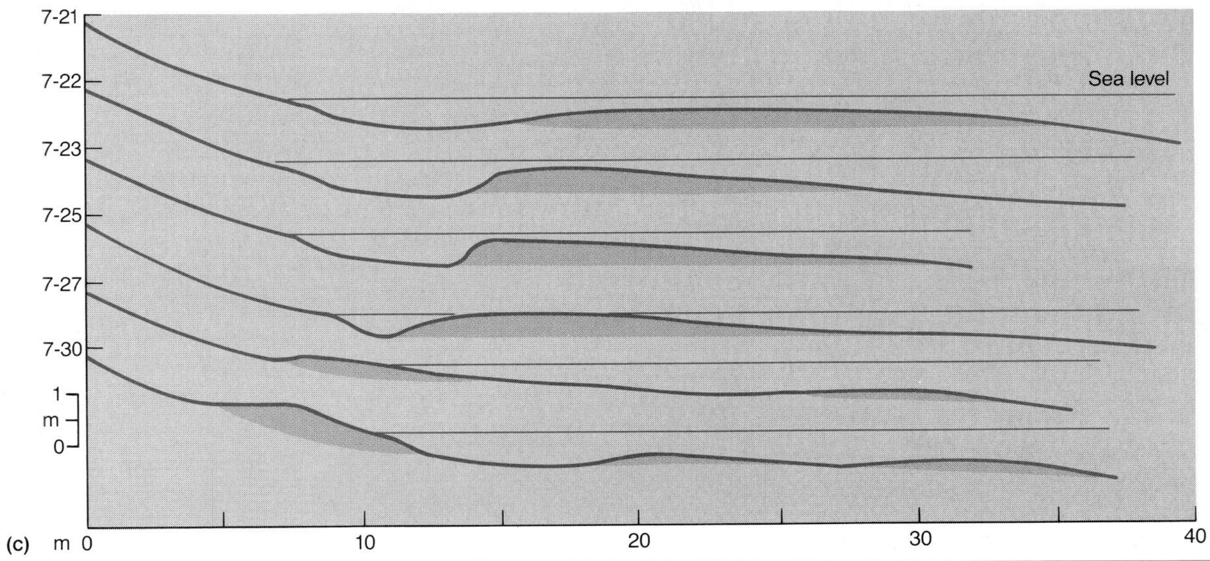

Figure 13.11 Ridge that has welded onto the beach thus returning much of the sand lost during the storm that eroded the beach and formed the ridge.

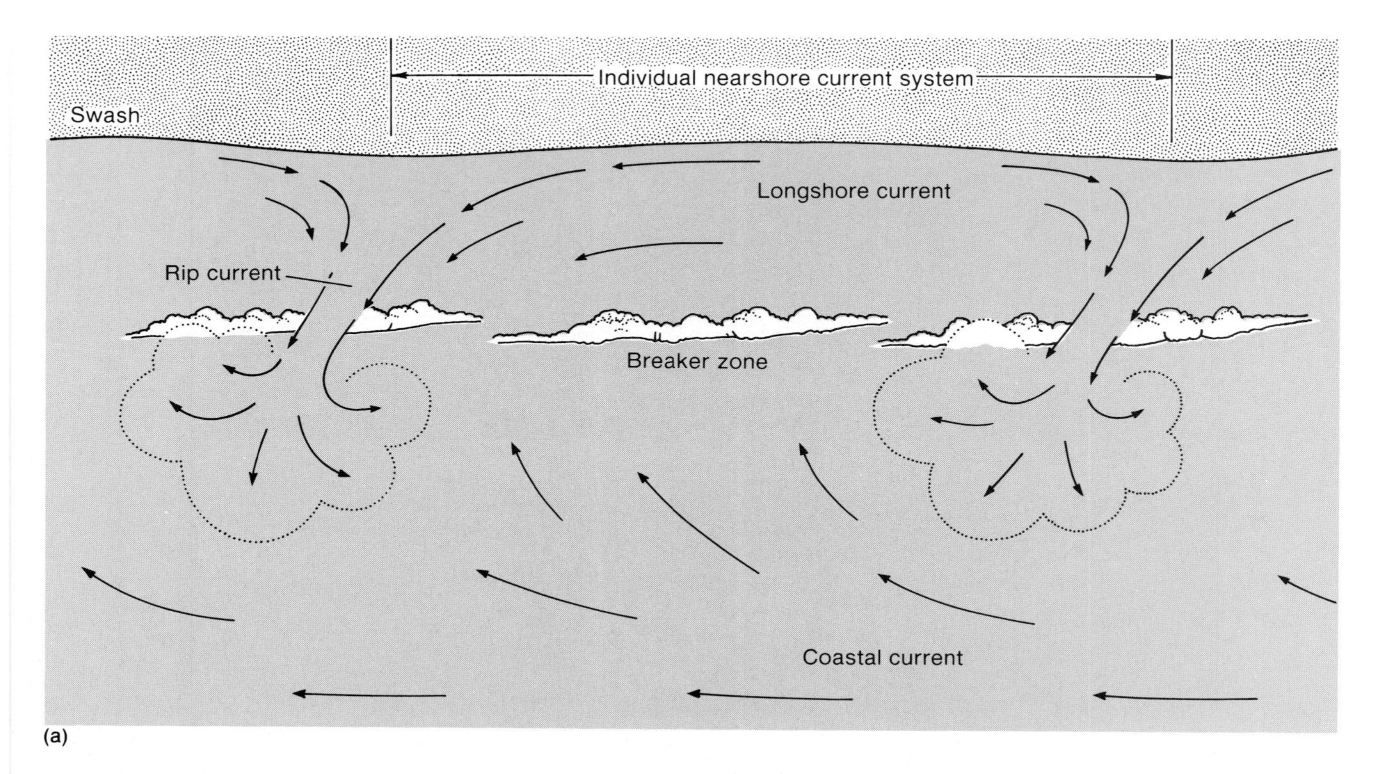

Figure 13.12 (a) Schematic diagram of nearshore area showing the nature and location of rip currents and (b) photograph of a rip current as shown by a plume of suspended sediments.

Figure 13.13 Aerial photo of a barrier with numerous linear and parallel dune ridges.

Both the subtidal longshore bars and the intertidal ridge are interrupted by rip channels, which carry rip currents. These are currents that form due to water piling up along the shore between the beach and the longshore sandbars. The water seeks a path of least resistance and returns seaward through saddles or low areas in the bars (fig. 13.12). These rip currents can transport sediment and they are a hazard to swimmers. In some areas the rip channels tend to persist for months or more, whereas in others, these channels move gradually or quickly, up to hundreds of meters in position in only a few days. Persistent rip channels are generally along beach and nearshore areas where long waves are refracted and approach the shore nearly parallel. These areas also have only slight longshore currents; it is these currents that move sediment alongshore and cause the rip channels to move as a result. In areas where waves approach the shore at a relatively high angle, there is generally a prominent longshore current that moves abundant sediment, and rip channels move readily.

Dunes

Large, mound-shaped, linear accumulations of sand called **dunes** are commonly formed immediately landward of the beach. These accumulations of sand-sized sediment result from persistent onshore winds that blow dry sediment from the backbeach into piles. Initial sand accumulation may be caused by a variety of obstructions, such as plants, pieces of wood, or any other object that interrupts the transport of the sediment by the wind.

Coastal dunes are commonly linear and may be in several rows (fig. 13.13). Those located immediately behind the beach are called **foredunes**. They are the youngest and usually the least vegetated. Most coastal dunes become somewhat stabilized by vegetation; however, intense storms or salt spray may remove vegetation, with a resulting mobility of the dunes.

Figure 13.14 Small incipient dunes forming on the back beach area on Padre Island National Seashore.

Coastal dune size covers a broad range. Some are so low as to barely be considered dunes (fig. 13.14), whereas others may rise tens of meters above the surrounding landscape (fig. 13.15). Sediment availability, strength and duration of onshore winds, and frequency and intensity of storms may be limiting factors in dune size.

Islands with numerous dune ridges are those that are building seaward due to the addition of large volumes of sediment. Barrier islands with no dunes or with one small dune ridge are commonly those that are migrating landward and decreasing or maintaining this width.

Inlets

Barrier islands are interrupted similar to longshore bars. In barriers, the breaks are larger than those in longshore bars, they are typically more permanent, are deeper, and their migration is slow relative to the interruptions of the bars. These breaks in the barrier are **tidal inlets,** so named because they serve as the pathway for tidal currents that transfer the rise and fall of the tides to the coastal bays that are landward of the barrier islands (fig. 13.1).

Sediment bodies called **tidal deltas** accumulate at each end of the inlet. They are formed by the sudden loss of speed and carrying power of tidal currents as they exit the inlet. In this respect these sediment bodies are much like river deltas, except that here the "river" is short and has two mouths at which the deltas form. The accumulation on the landward side is deposited largely by flood tidal currents and is called the **flood tidal delta,** whereas the one on the seaward side is the **ebb tidal delta** for similar reasons (fig. 13.16).

Inlets originate primarily in two fashions. They may develop after a longshore current deposits sediment in the form of a spit, also called a baymouth bar, across a coastal embayment. Because of tidal currents moving between the embayment and the

Figure 13.15 Large dunes, well covered with vegetation on St. Joseph Island, Florida.

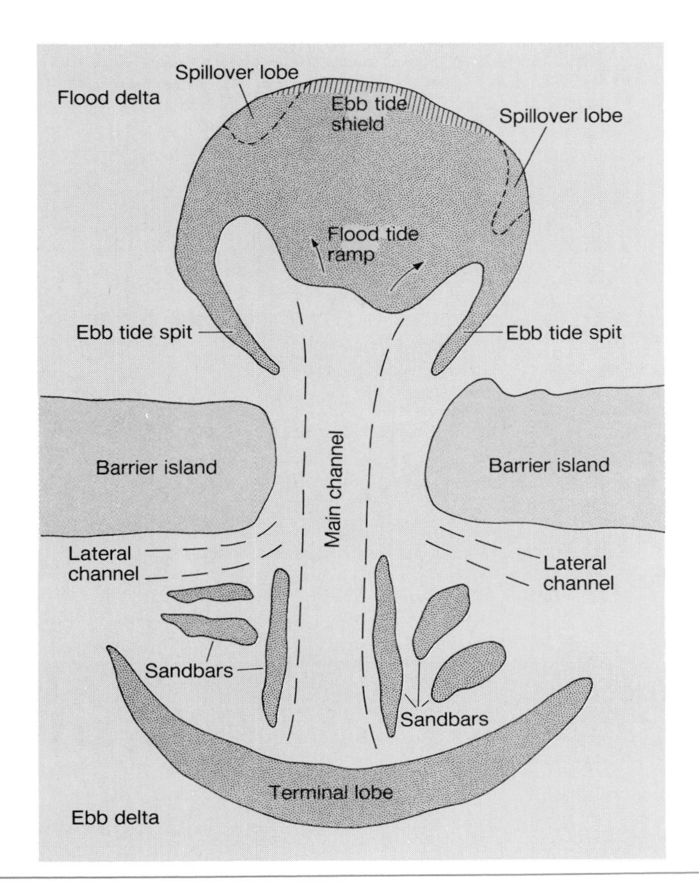

Figure 13.16 Generalized diagram of an inlet system showing the major components.

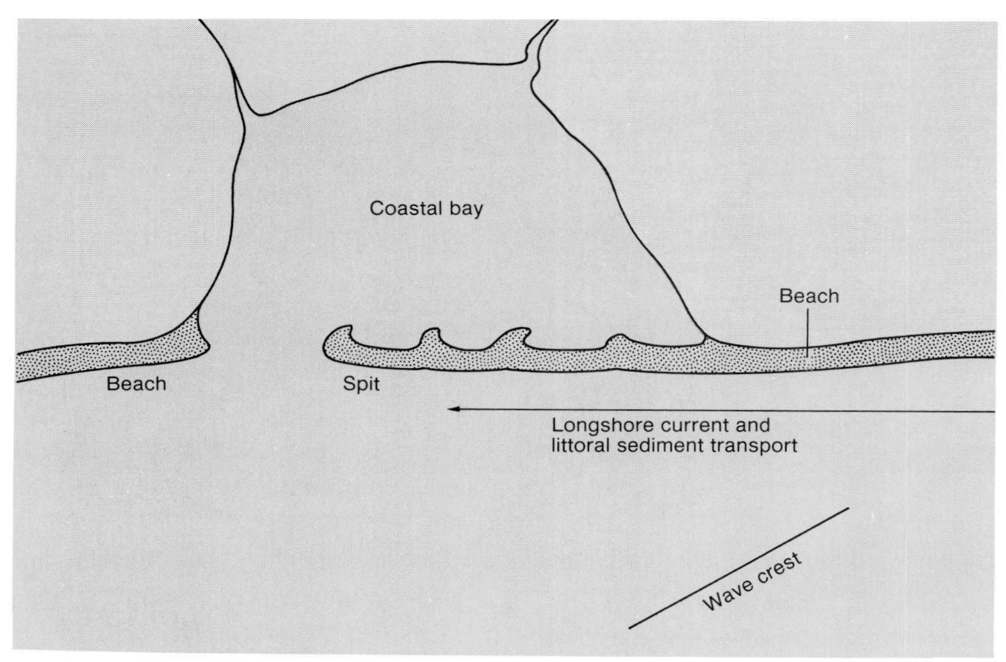

Figure 13.17 Tidal inlet formed by spit migration across a coastal bay.

Figure 13.18 Photograph of Corpus Christi Pass on Mustang Island, Texas, a small inlet with small but recognizable tidal deltas.

Figure 13.19 Large, well-developed flood tidal delta at Moriches Inlet, Long Island, New York.

open ocean, the spit is restricted from closing the embayment completely and a pathway for tidal exchange is maintained; this pathway is the tidal inlet (fig. 13.17). The other common mechanism for inlet formation is the breaching of a barrier island by intense storm conditions. The combination of large waves, high water, and tidal currents cuts a tidal inlet through a barrier (fig. 13.18). Storm-generated inlets, or **hurricane passes** as they are sometimes called, usually do not persist. Unless the inlet is fairly deep and tidal currents are strong, the longshore currents and resulting littoral drift will close storm-generated inlets in days or months.

Inlets represent important links between the highly productive coastal embayments and the open ocean. They carry nutrients, circulate oxygen and dissolved solids, and serve as pathways for the migration of multitudes of marine organisms. Because of this great value to the well-being of coastal and shelf environments, it is important to understand the dynamics of inlets and keep them "healthy."

Tidal inlets may be stable or unstable in terms of their size, shape, and position. Unstable inlets change size, usually by decreasing, and migrate along the barrier system. This is accomplished by longshore currents and littoral drift, which are wave-generated processes, predominating over tidal currents. The relatively stable inlets are ones where tidal currents are strong and cut deep inlet channels. In these, tidal currents predominate over wave-generated nearshore processes. Over a period of several years a single inlet may change from stable to unstable or vice versa.

Tidal deltas vary in size and shape depending on dominant processes. Flood deltas tend to be large and fan-shaped, with several lobes (fig. 13.19). They do not show any major modification due to waves because they are sheltered from large waves. In mesotidal areas the distal part of the flood tidal delta may be smoothed by ebbing tidal currents. Ebb tidal deltas are exposed to open water marine processes, especially waves and wave-generated currents. As a result, the ebb deltas range from being nearly absent on high wave-energy coasts (fig. 13.20) to being large and nearly perpendicular to the coast where tidal currents dominate over wave processes (fig. 13.21).

Figure 13.20 Small, wave-modified ebb tidal delta at Matanzas Inlet on the Atlantic coast of Florida.

Figure 13.21 Large, tide-dominated ebb tidal delta at Bunces Pass, Florida, on the Gulf Coast.

Figure 13.22 Thick and extensive washover deposits on Santa Rosa Island, Florida. These were deposited in association with Hurricane Frederick in 1979.

Washover Fans

Storms cause significant change to barrier islands, including erosion of the beach and sometimes the dunes, formation of tidal inlets, and overtopping the barrier, which typically produces **washover fans** (fig. 13.1). This feature is a thin, fan-shaped sediment accumulation that is deposited landward of the beach and dune region of the barrier island. It is formed by the combination of elevated water level and large waves carrying sediment from the beach and nearshore zone over or through low areas in coastal dunes.

Washover fans may be scattered along a barrier island, or, in the event of a major storm or low barrier island, they may be continuous (fig. 13.22). These are typically single event features that become vegetated and are not mobilized again, but there have been examples of large washovers that contain tidal channels and are mobilized regularly. Such a washover fan is present on San Jose Island along the central Texas coast.

Each washover event generally deposits a few tens of centimeters of sediment. The washover extends over the flat portion of the back barrier, which may include marshes and tidal flats. Numerous layers of this type of sediment deposit may be superimposed. In the case of narrow barrier islands, the washover may traverse the entire island and extend into the bay behind the island (fig. 13.23). Because individual fan layers are not thick, they rarely have severe impact on the vegetation of the marsh or back-island areas.

Marshes and Tidal Flats

The landward portion of the barrier island complex is a gently sloping intertidal area, typically containing both salt marsh and tidal flat environments. These environments are as discussed in chapter 11. Actually, the same environments are present throughout the coastal zone wherever intertidal conditions persist and provide the requisite conditions for their development. They may be on estuary margins, lagoon margins, on the landward side of the barriers, or even on intertidal portions of flood tidal deltas.

Figure 13.23 Narrow barrier island that has large washover fans that extend well into the coastal bay on the landward side of the island.

Marsh and tidal flat environments are typically not extensive on barrier islands that are developed on microtidal coastal settings. The attenuation of tidal range from the open marine environment, which in this case is the seaward side of the barrier, to the bay systems on the landward side of the barrier, is typically significant. The tidal range on the landward side of a barrier may be 30% to 50% less than on the open marine side because the barrier prevents some water from moving through the inlets and into the coastal bay. Tidal range is further decreased in related estuaries and lagoons that are more landward. A good example is Laguna Madre on the south Texas coast. The open marine side of the barrier here has a tidal range of nearly a meter, whereas in Laguna Madre there is a marked decrease in tidal range from both the north and the south, so that much of this lagoon has no significant lunar tides, again because of the inability of tidal currents to carry enough water in and out of the inlets during the allotted period of the tidal cycle. Similar patterns persist throughout barrier island systems.

As a result of this decrease in tidal range, the intertidal environment on the landward side of barrier islands on microtidal coasts is narrow. By contrast, mesotidal barrier coasts tend to have extensive salt marsh and tidal flat environments. This situation is caused by the combination of a larger tidal range and the fact that barriers on mesotidal coasts tend to be short with numerous inlets. As a result, the tidal range is reduced less on this coastal setting because the numerous, wide inlets allow tides to flood and ebb

Figure 13.24 Photograph of extensive marshes along the south Atlantic coast.

with little reduction in volume. The coastal complexes of Georgia and South Carolina serve as excellent examples of this type of barrier system with coastal marshes of several kilometers width between the barriers and the mainland (fig. 13.24).

Wind Tidal Flats

There is a special type of tidal flat that is common on, but not restricted to, the landward side of barrier islands. **Wind tidal flats** are broad continuations of the intertidal zone into the supratidal zone. They are flooded by storm or wind tides, thus their name. These supratidal flats are commonly flooded in association with the passage of frontal systems or severe storms. Because they are so flat and are near sea level, a storm tide of only 10–20 cm can flood at least part of the wind tidal flat.

This environment is covered with a combination of sand and mud that is a continuation of the adjacent tidal flat. It is common for blue-green algal mats to cover at least part of the flats because they can withstand prolonged periods of exposure. Mud cracks or other desiccation features are also present.

At many locations the wind tidal flats are much more extensive than the intertidal flats. This is especially true in microtidal areas where tidal range on the landward side of a barrier is low (fig. 13.25).

Coastal Bays

The protected aquatic environments between the barrier island and the mainland take on a variety of shapes and locations. Those immediately landward of the barrier tend to be elongate and parallel to the coast. Depending on a variety of circumstances, they may have differing chemistry and communities. More landward, there is commonly another type of coastal bay environment that is expressed as a distinct embayment in the adjacent mainland region: the estuary.

Figure 13.25 Extensive wind tidal flat environment on the landward side of Padre Island, Texas, (a) with low sand dunes and (b) without dune development.

In addition to estuaries, the other coastal bay environments that are associated with barrier islands are polyhaline bays and lagoons. A **polyhaline bay** is one that has a variable salinity, both above and below normal marine values of 35‰. These bays are influenced by estuaries and by circulation with the open marine environment through inlets. **Lagoons** are those coastal bays that have no appreciable runoff from land and have restricted tidal circulation. The consequence is hypersaline conditions and, under extreme circumstances, evaporite precipitation. All of these types of coastal bays are present within the coastal barrier complex of the Texas coast (fig. 13.26). The estuaries are irregular embayments in the mainland coastal plain that each have at least one significant stream entering them. Salinities range from essentially zero at the mouth of the stream up to the 20s of parts per thousand toward the seaward margin of the estuary. Immediately behind the barrier island are shore-parallel bays, some of which have polyhaline salinity conditions (Aransas Bay) due to seasonal variation in precipitation, and others of which are hypersaline (Laguna Madre). Salinity is a primary factor in controlling the nature of the biota present in each type.

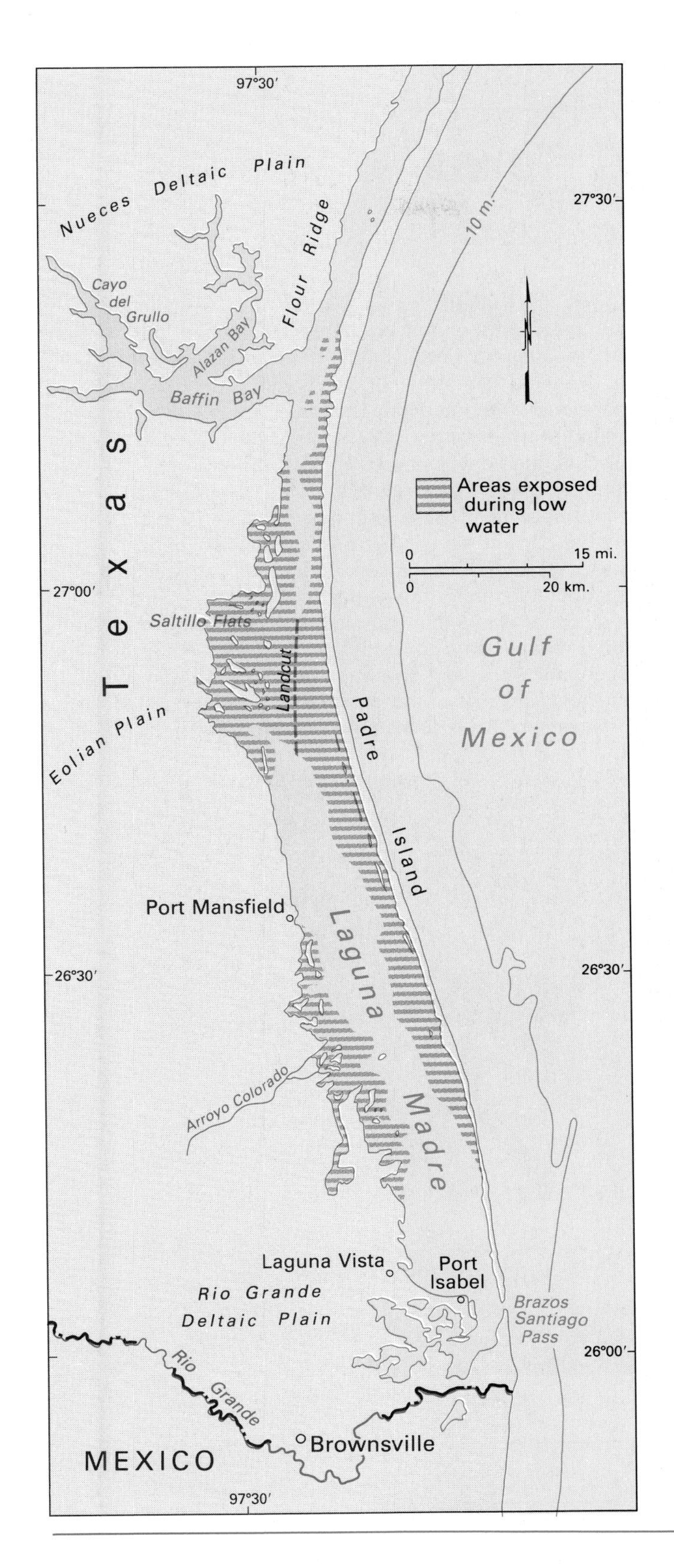

Figure 13.26 Map of the south Texas coast showing Padre Island and Laguna Madre, which comprise one of the longest barrier-lagoon complexes in the world.

BOX 13.1

Padre Island, Texas

Padre Island is located along the south coast of Texas in the northwestern portion of the Gulf of Mexico. This is one of the longest barrier islands in the world and by far the longest in North America. Because of its size, Padre Island contains nearly all of the environments and conditions that characterize barriers.

The island extends from near Corpus Christi to the Rio Grande Delta near the Texas-Mexico border. This distance of about 200 km is interrupted only by Mansfield Pass, a man-made inlet used for fishing boats. There have been natural storm-generated inlets in the past but they did not persist.

The south Texas coast is distinctly wave-dominated, with tidal ranges of 70–90 cm, and strong longshore drift that converges near the middle of the barrier island. Relief and width of the island range widely: the island is from a few to nearly 10 km wide, and relief is dependent on the development of dunes, which range from only about 1 m up to nearly 10 m above the adjacent barrier. The beach is generally well developed with foredunes landward of it. Most of the island shows a narrow dune complex that lacks distinct ridge development (fig. B13.1a). Washover fans are present throughout the island but are most widespread near the middle, where dunes are low.

The back portion of Padre Island consists of either low grassland areas that are used for cattle grazing or unvegetated, active dunes (fig. B13.1b). These dunes are concentrated near the middle of the island where longshore currents converge and the sand supply is abundant. Persistent onshore winds blow the sand to the landward side of the island where it accumulates. Adjacent to these dunes areas and in many areas, to

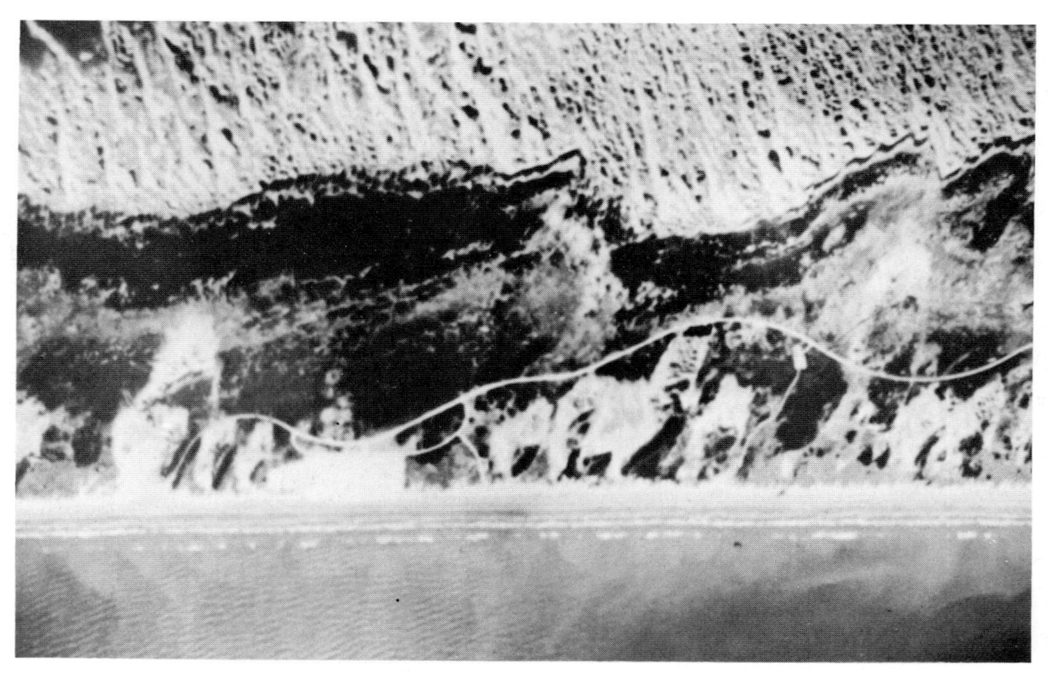

Figure B13.1a Aerial photograph of Padre Island showing low dunes and well-developed beach.

the grasslands as well, is an extensive wind tidal flat. This environment has essentially no relief and grades into the lagoon (Laguna Madre) that separates the barrier from the mainland. Slight storm or wind tides cause these flats to be inundated. Algal mats formed by blue-green algae dominate this environment.

Most of the coast of Padre Island appears to be reasonably stable. There is local erosion and local progradation but this varies in both space and time. The overall change in island configuration and shoreline position is modest but shows a trend toward landward movement. The abundant sediment supplied by the longshore current system tends to balance sea level rise and storm-generated erosion.

Because of the location of Padre Island, it is susceptible to hurricanes. Major changes to the island have been caused by these severe storms including abundant breaching of the island, destruction of dunes, and extensive beach erosion. Although these events cause major changes, the island is resilient and recovers in a matter of months. A recent hurricane cut more than 40 storm passes through the island, but after only a few months the longshore currents had transported sufficient sediment to close them off. The type of damage that takes the longest to repair is the erosion of dunes. In some cases it takes many years and in others, the dunes never return to their pre-storm configuration.

In general, Padre Island is stable, with no major long-term changes evident. It is mostly a national seashore and remains relatively pristine, for everyone to enjoy.

Figure B13.1b Active dune field on the landward side of Padre Island.

Origin and Development of Barrier Island Systems

Although barrier islands are widespread and their dynamics are well understood, there is still much to learn about their origin. Three basic concepts of the origin of barrier islands have been formulated and investigated over the past 125 years. In order of their proposal they are: (1) by the upward shoaling of subtidal sandbars due to wave action, (2) by the generation of long spits due to littoral drift and then breaching to form inlets, and (3) by the drowning of coastal ridges under conditions of rising sea level. For some time, supporters of each mechanism tried to prove that one concept was the origin of all barriers. It is now known that barriers form by different mechanisms, but there is still disagreement about the origin of many of the world's barriers.

Barriers that develop as the result of spit formation are obvious because of the coastal morphology necessary to produce such elongate spit complexes. There must be either a protrusion of the coast or an embayment from which the spit develops (fig. 13.27). A good example of a barrier that formed as a spit is Sand Key on the west-central coast of Florida (fig. 13.28).

Most of the controversy about the origin of barriers centers on the other two proposed mechanisms (fig. 13.29). It appears that both are possible but at this time there is no agreement on the relative importance of one over the other. Both of these origins produce a shore-parallel barrier island with water between it and the mainland. Also included would be tidal inlets with the number and size related to the amount of water that must pass from the open ocean into the coastal bay complex and back again during each tidal cycle. The important parameters that control this volume of water are the tidal range and the area of the bay complex landward of the barrier system.

Barrier Island Dynamics

Once formed, barrier islands are not static. They are dynamic systems with many variables influencing their change in size, shape, and location. Marine processes, such as waves, wave-generated currents, and tidal currents, are important to the dynamics of barrier islands. The setting, including the slope of the shelf and the shoreline configuration, must be considered and the availability of sediment is critical. Another important factor, especially at this point in time, is the change in sea level. It is presently

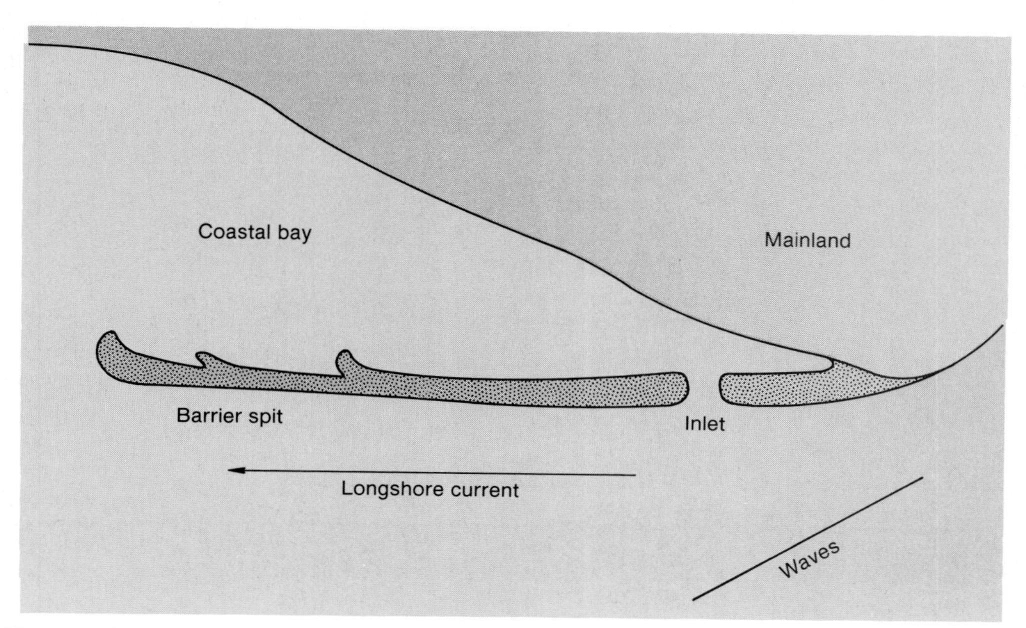

Figure 13.27 Diagram of barrier spit developed from a headland.

rising, with indications that the rate has increased during at least the past fifty years. The present rate of sea level rise is 3 mm per year or about one foot per century.

First of all, let us consider each of the variables individually to obtain a perspective on the possibilities that exist. Large waves erode beaches and therefore barriers, whereas small waves tend to build up these areas. The most important wave-generated current is the longshore current, which carries sediment along the beach and nearshore zone. Because this current is dependent on the orientation of approaching waves and therefore upon the wind direction, the distribution of wind controls the littoral drift along a coast. If winds blow from two distinctly different onshore directions, the longshore current and littoral drift will shift back and forth with little net transport in either direction along the barrier. On

Figure 13.28 Sand Key, Florida, an example of a barrier spit. The headland that served as the sediment source is in the background.

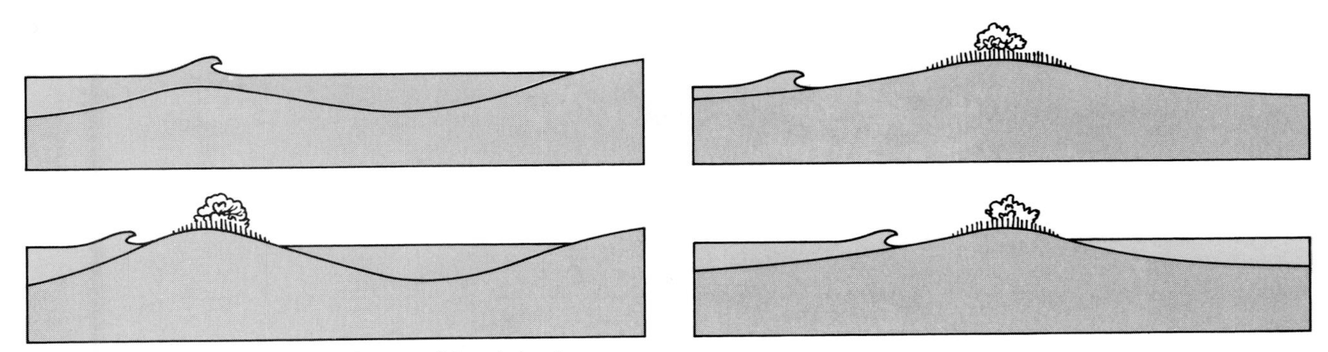

Figure 13.29 Diagram showing the possible origin of barriers from (a) upward shoaling of sand bars, and (b) drowning of beach ridges.

the other hand, if there is a dominant longshore current and therefore littoral drift, it will have a marked effect on the barrier island by transporting a large amount of sediment to an inlet or to the end of the barrier.

Tides themselves have a minor effect on the beach, however, they are dominant in the inlet environment. Strong tidal currents tend to stabilize the position of an inlet, whereas weak currents permit waves and longshore currents to dominate and cause the inlet to migrate in the direction of littoral drift (fig. 13.30).

The gradient of the inner shelf adjacent to the barrier is important for various reasons. First of all, a steep slope adjacent to the barrier will not attenuate any wave energy and thus wave erosion of the beach will be high compared to a gentle slope, where waves are dampened by the bottom before they reach the beach environment. Additionally, the development of a barrier requires a gentle gradient simply because this provides the appropriate setting on which the barrier can develop. Steep gradients do not provide a place for sediment to become stable.

Obviously, sediment is necessary in order to develop a barrier island regardless of any of the other variables. There are basically two places where sediments are obtained for the construction of a barrier island: (1) from the adjacent mainland and (2) from the inner shelf where the barrier develops. In the case of barriers that originate from spits, the mainland tends to be the dominant source of sediment (fig. 13.28). Those barriers that develop from the upward shoaling or drowning of beach ridges receive the bulk of their sediment from the reworking of inner shelf sediments. Changes in dominant processes or the direction of the dominant process may cause variation in the amount of sediment available for building the barrier island. Changes in the environment may make sediment available that previously was not or vice versa. A good example is the removal of vegetation that stabilized sediment. The result is that much sediment is suddenly available for incorporation in the barrier.

Sea level rise itself is extremely important and tends to affect most of the other variables. If all other variables are held constant, a rise in sea level tends to cause the barrier to migrate landward. It may also result in erosion and a decrease in the size of the barrier, perhaps even its complete destruction. A lowering of sea level under the same circumstances causes the opposite changes.

Some Modern Examples

Conditions along the present-day barrier island coasts include a rapidly rising sea level, little sand-sized sediment being supplied to the barrier systems from the mainland, a generally broad and low-gradient inner shelf, and a varying tidal and wave climate. Using these conditions it is possible to show different responses of barriers.

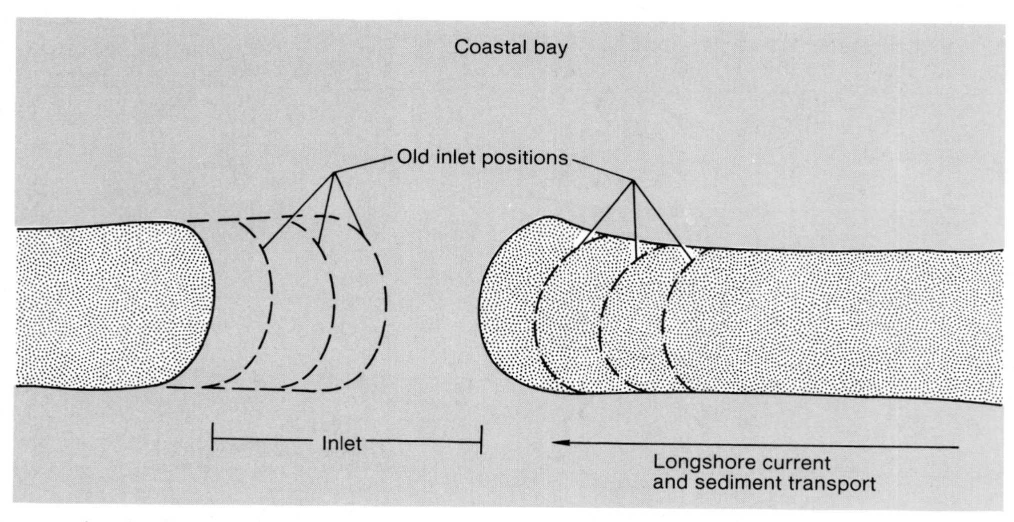

Figure 13.30 Diagram showing how spit migration along shore causes the related inlet to also migrate.

Figure 13.31 High altitude aerial photograph of the Outer Banks of North Carolina showing the barriers, inlets, and estuaries that comprise this system.

Outer Banks, North Carolina

Long, narrow barriers with only a few tidal inlets characterize most of the Outer Banks barrier island system of North Carolina. This barrier complex is separated from the mainland by large estuaries (fig. 13.31). Sediment is derived from the shelf being reworked and there is a marked north-to-south littoral drift. Some parts of the barriers have dunes that rise several meters above the beach, whereas others have dunes of only a meter or so in height. The coast is microtidal, with tides typically between 1.0 and 1.5 m. This coast receives a modest amount of wave energy and is subjected to intense storms from the northeast and hurricanes from the south.

The net result is that the Outer Banks barrier system has beach and dune erosion, abundant washover fan development, and occasional storm-generated inlets, which tend to be short-lived. The barrier island is moving landward, that is, it is **transgressing** over the adjacent estuary. The natural barrier system is such that transgression or landward migration is expected during times of sea level rise and limited sediment supply. The profile of the barrier island remains essentially the same, but it is displaced landward (fig. 13.32). Tidal inlets tend to migrate with the littoral drift or in some cases, they may close completely. Numerous inlets have closed along this coast during historical time. Parts of the Texas barrier island coast are responding in a fashion similar to the Outer Banks.

Sea Islands of South Carolina and Georgia

Farther down the Atlantic coast it is possible to see a striking contrast in the barrier island system of South Carolina and Georgia as compared to North

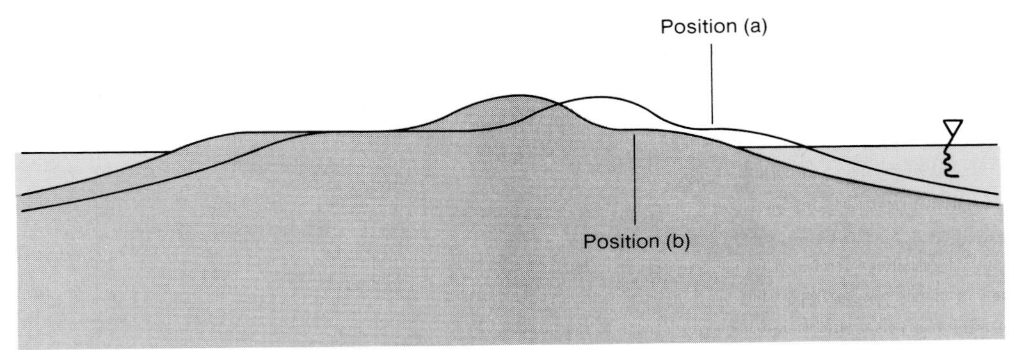

Figure 13.32 Diagram showing the landward migration of a barrier island due primarily to washover. Note that the configuration of the barrier profile does not change and that only the position of the island has changed.

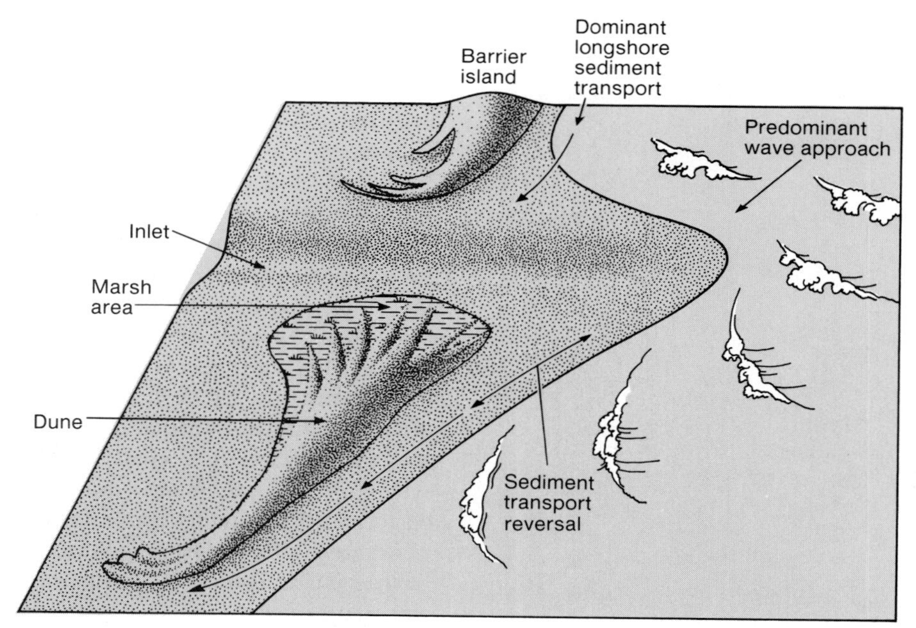

Figure 13.33 Sketch showing the process-response mechanism for the formation of a drumstick barrier island.

Carolina. This mesotidal coast is characterized by extensive marshes between the mainland and the barrier system. Barriers are short, generally wide, and with numerous broad, deep inlets. The inlets tend to be stable in position and have large ebb tidal deltas. The adjacent shelf provides the great bulk of the sediment for the barriers and the gradient is much less steep than off the North Carolina coast. This results in a low wave energy, although the same types of storms affect this coast as on North Carolina.

A close look at the barrier system on the South Carolina-Georgia coast indicates that there are additional differences, especially in terms of the shape and arrangement of the barriers. These islands are typically wide, but they display a marked difference in width from one end to the other. A glance at the overall coast shows that nearly all islands have the wide end toward the north. Additionally, the islands show a displacement of the shoreline on one side of the inlet as compared to the other; there is an offset. That is, the shoreline position is more landward or seaward on one side of the inlet as compared to the other side. The offset is in the southerly direction, which is the direction of the littoral drift system along the South Carolina-Georgia coast. The islands are called "drumstick" barriers because of their resemblance to the popular part of the chicken and their arrangement is called a **downdrift offset**.

The shape and arrangement is a response to the process-response mechanisms operating along this typical mesotidal coast. Sediment accumulations are concentrated in the downdrift shadow of the ebb tidal delta, producing numerous beach ridges and the fat part of the drumstick, while the downdrift part of the island receives only a small amount of sediment (fig. 13.33). The result is that the island tends to rotate seaward or prograde at one end and experience erosion or transgression at the other. This causes the offset in the shorelines.

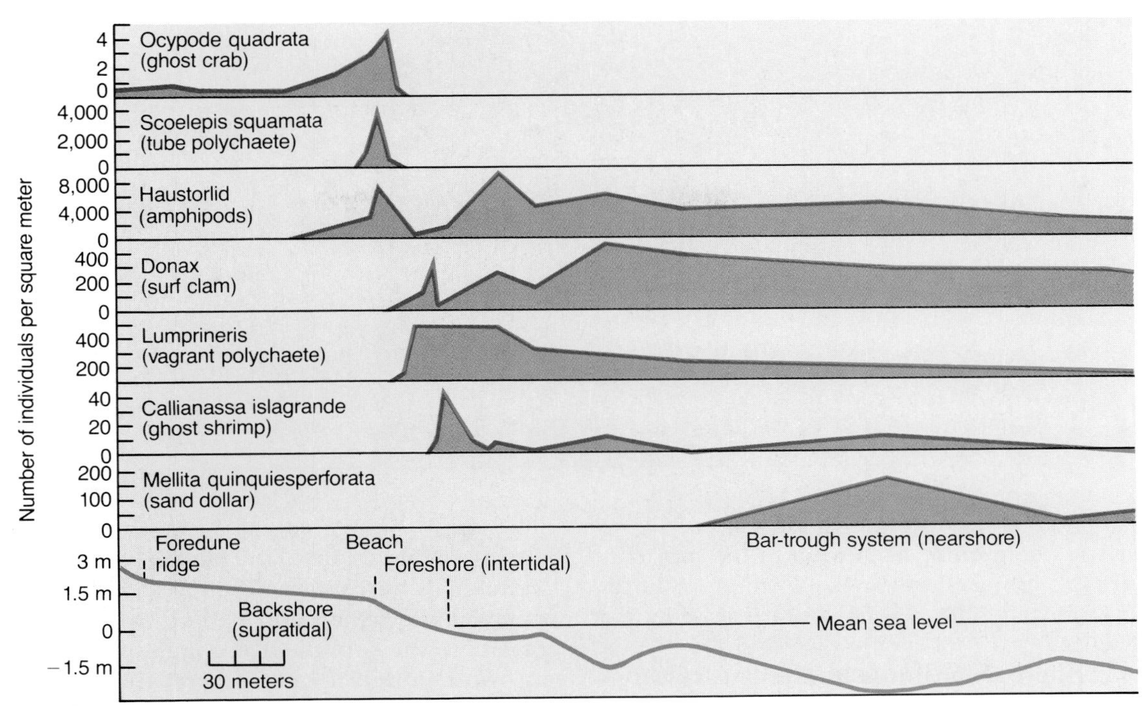

Figure 13.34 Diagram of beach and nearshore environment showing the location of dominant organisms.

Some coasts of this type also have migrating of tidal inlets, which compounds the nature of the barrier dynamics. Erosion on one side of the inlet and deposition on the other causes the barrier to migrate along the coast as well as to rotate. This is not the case on the South Carolina-Georgia coast because the deep inlets are relatively stable. It is common on the Florida Gulf Coast, where inlets are shallow and migrate rather rapidly.

Barrier Communities

The great diversity of environments that comprise the barrier island systems gives rise to an equally varied plant and animal community, which is really several communities because each element of the barrier systems tends to have its own characteristic organisms. Because the barrier includes subtidal, intertidal, and supratidal environments, the spectrum of organisms is broad. In the discussion that follows, each of the major elements is considered separately.

Beach and Nearshore Zone

The combination in this environment of sandy sediment and high energy from wave action tends to restrict organisms to a few types that are well adapted to this dynamic and rigorous environment. The wave action keeps the bottom in nearly constant motion, thus causing problems for most benthic organisms. Plankton are present but not in great quantity and not in differing types than over the adjacent shelf. Fish are common, especially some flatfish that prefer to be slightly covered by sand, such as the flounder.

Wave activity and sand-sized sediment tend to prevent organic detritus from accumulating in the substrate. If the waves do permit debris to settle to the bottom, it decays quickly in the well-oxygenated sediment.

The result is that the beach and nearshore area is occupied by a small number of species (fig. 13.34), each of which may be represented by numerous individuals. There are essentially three subdivisions of this environment, each having a

Figure 13.35 Shore birds feed along the beach, especially at low tide when many beach organisms are vulnerable.

characteristic community. They are (1) the subtidal bar and trough or nearshore area, (2) the foreshore or intertidal beach, and (3) the supratidal or backshore beach.

The nearshore is inhabited by a few scavenging detritus feeders and some filter feeders that live on plankton. Among the most abundant are sand dollars, small amphipod crustaceans commonly called mole crabs, and small clams, with some worms and burrowing shrimp also present. The mole crabs are rapid and efficient burrowers that expose their antennae to capture plankton. Small sand dollars are most abundant in the troughs between longshore bars, where they burrow just under the sand and are relatively protected from the waves. Clams that live in the surf are generally small and sturdy to withstand the high energy of waves and wave-generated currents. Detritus-feeding worms ingest sand grains and remove the digestable organic debris that clings to the grains. They then pass the grains through their system.

The foreshore portion of the beach is perhaps the most interesting part of this environment. There are but a few species present; however, each is represented by many individuals. Included are some of those species that occupy the nearshore, such as worms, amphipods, and the surf clam, *Donax variabilis*, which is common along the south Atlantic and Gulf coasts. This species is extremely abundant and colorful. The individual is about 2–3 cm across the long axis and comes in variegated colors. This efficient burrower moves up and down the beachface with the tide. It is exhumed by the uprush of the last breaking wave. During the backwash, *Donax* individuals burrow into the foreshore deep enough to cover their shell and still allow their siphon to extend to the surface. It is common for this species to occur in densities of several thousands of individuals per square meter.

Other clams are also present along the foreshore beach but they are not generally seen because they burrow well below the zone of sediment movement. There is generally a relationship between the wave energy imparted to the beach and the size and bulk of the clams that live there. On the Pacific coast for example, the clams are relatively large and have a thick shell, an adaptation to the rigorous wave energy that is typical along that coast.

Lugworms (*Arenicola*) and a group of amphipod crustaceans called sand hoppers also inhabit the foreshore beach. As is true for all of the benthic invertebrates in this environment, they too are burrowers. The sand hoppers also extend up onto the backbeach (fig. 13.34). The ghost shrimp *Callianasa*, common on tidal flats (fig. 11.13), is also present beneath the foreshore. This organism occurs out into the shallow subtidal area and up onto the backbeach. It is a deep burrower, with some reaching more than a meter below the surface. Ghost shrimp pump water through their burrow and filter out nutrient materials.

Many predators and scavengers rely upon the foreshore or intertidal beach for much of their food supply. At low tide, the beach is teeming with shore birds feeding on the worms, crustaceans, and small mollusks (fig. 13.35). Some crabs and scavenging insects also frequent this zone during low tide.

Figure 13.36 Ghost crab (*Ocypoda*), which is a common nocturnal inhabitant of the backbeach.

Figure 13.37 Burrow and characteristic excavation pile of the ghost crab.

The backshore zone has the sparsest and least diverse fauna of the beach and nearshore environment. The only common and fairly permanent residents are the ghost crabs (*Ocypoda*) (fig. 13.36) and small rove beetles. The ghost crabs are the most noticeable due to their numerous, easily recognizable burrows (fig. 13.37). The large holes and associated piles of excavated sand balls are characteristic of the backbeach area. Ghost crabs are nocturnal scavengers that are very fast on their feet.

Figure 13.38 Opportunistic vegetation on the backbeach area. This vegetation is instrumental in trapping sand, which eventually forms dunes.

The backbeach area is the most seaward zone in which land plants can exist. Various opportunistic plants that can withstand dry conditions and salt spray occupy this area. Most are either grasses, succulents, or spreading vines (fig. 13.38).

Dunes

Few species of organisms are characteristic of coastal dunes. The base of the foredunes may be occupied by scattered ghost crabs, but there are no animals that can be considered dune dwellers only. The most diagnostic life on the dunes is the vegetation. There are many coastal dunes that are mobile and do not contain vegetation. Those that do are dominated in the northern latitudes by American beach grass (*Ammophila breviligulata*) and, in the southern latitudes, by sea oats (*Uniola paniculata*) (fig. 13.39) and salt meadow grass (*Spartina patens*), a close relative to the common salt marsh grass.

Dune vegetation is an important aspect of barrier ecology because of the stabilizing effect it has on the dunes. There are many areas where the activities of man have severely affected dune vegetation and as a result there has been considerable erosion of the coastal dunes (fig. 13.40). Footpaths, vehicular trails, and any other activity that destroys vegetation can contribute to this problem.

Back-island Environments

Landward of the coastal dunes is a complex that includes several different environments. These include the washover fans, back-island flats, marshes, tidal flats, and the estuaries. These have either purely terrestrial communities or they have been discussed in previous chapters and will not be repeated.

Coastal Lagoons

An important coastal aquatic environment that is associated with the barrier island complex and has not been discussed previously is the coastal lagoon. These shallow, generally elongate water bodies have no appreciable freshwater runoff from land and they have restricted tidal circulation. The result is a hypersaline environment with a nominal tidal range. Because the chemical environment is unusual, there is a special community that is adapted to this environment. The general rule for hypersaline environments is that the variety of organisms is limited but the number of individuals is high.

Figure 13.39 Sea oats (*Uniola paniculota*), which are important vegetation for dune stabilization.

Figure 13.40 Erosion of dunes can be caused by human activities, including footpaths and trails made by off-road vehicles.

BOX 13.2

Laguna Madre, Texas

As Padre Island is the largest barrier island in North America, Laguna Madre is the longest coastal lagoon. It is the water body that separates Padre Island from the south Texas mainland. It too exhibits variety in its environment, both geographically and chronologically. The lagoon is typically a few kilometers wide but narrows to nothing in the landcut area where wind-blown sand has filled in completely (fig. 13.26). Water depth is only a meter or so except in the Intracoastal Waterway, which runs the entire length.

The salinity of Laguna Madre is one of its unique features. It ranges from essentially normal marine concentrations of 35‰ at either end up to at least twice that level near the middle (fig. B13.2). Actually, the gradient is greater on the south end than on the north end. Adjacent to Laguna Madre on the landward side is Baffin Bay, which is a drowned drainage system that was an estuary but has become a hypersaline lagoon.

Laguna Madre owes its nature to the general absence of tidal circulation with the open ocean and to the climate of the adjacent land area of south Texas. There is a general trend of decreasing rainfall toward the south so that most of this area is arid. The total rainfall is low, and it comes during the wet season, when salinities become low, even below normal marine levels locally. The long dry season with its evaporative conditions reverses the trend and salinities become high. Large amounts of rainfall from tropical storms also cause the same situation.

Most bottom sediments in Laguna Madre are sand, without much mud. Biogenic shell debris is also a common constituent. Although the environment is hypersaline, there is an abundant but not very diverse community that lives there. Various worms, mollusks, and fish are common. Productivity is moderately high with eel grass and various types of algae.

Local embayments and isolated ponds experience very high salinity concentrations, eventually resulting in the precipitation of evaporite minerals, such as gypsum ($CaSO_4$) and halite (NaCl) or common salt.

Laguna Madre is a large, unique coastal lagoon with extreme conditions, but it supports an abundant community. It is one of the better places for sport fishing on the entire Texas coast.

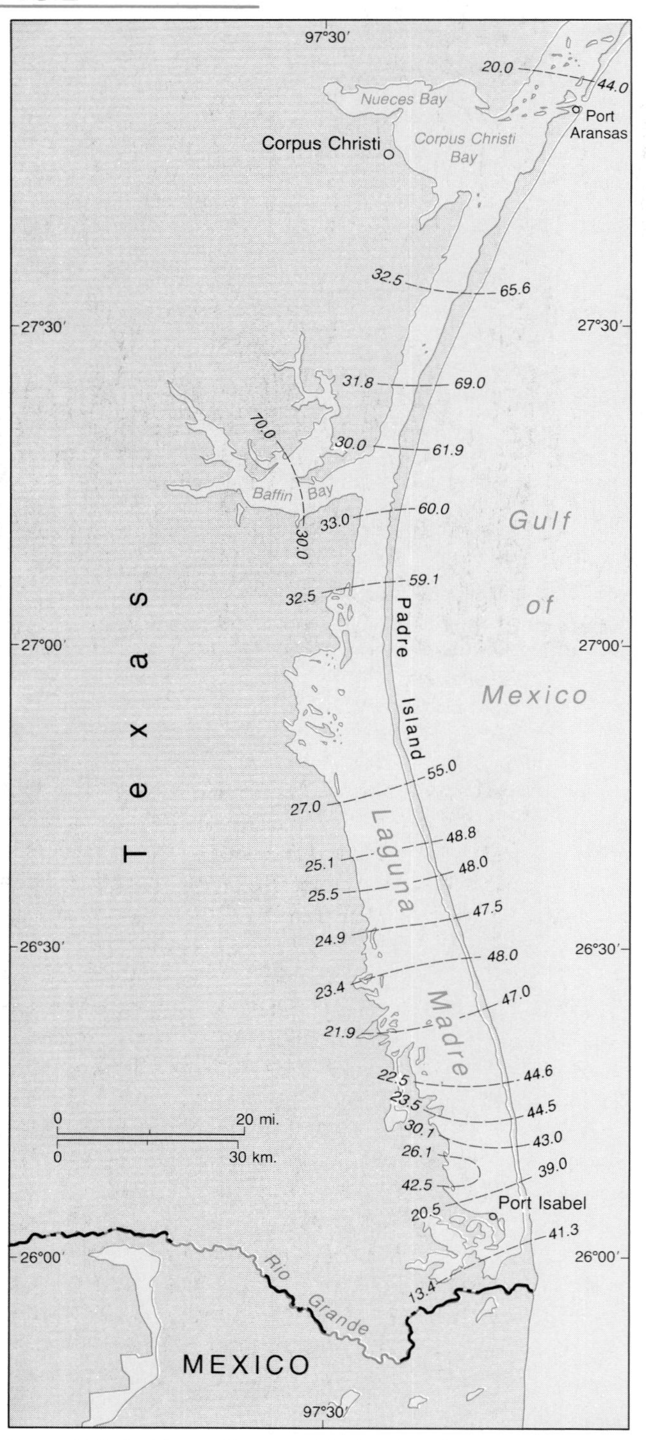

Figure B13.2 Map of Laguna Madre showing salinity ranges and annual variations in salinity.

Figure 13.42 Photograph of Lake Reeve, a polyhaline coastal lagoon in Victoria, Australia.

Depending on the nature of the lagoon and the climate of the surrounding area, there may be varying characteristics to the environment. Some are hypersaline throughout, without variation in salinity at any place within the lagoon, but with a range in salinity from one location to another. Such is the case in Laguna Madre, Texas (fig. B13.2). It is one of the largest hypersaline lagoons in the world and has salinities from near normal marine levels to nearly 100‰, generally increasing away from the areas of interchange with open marine waters. Local embayments may have salinities near 200‰ in the summer. This is common for lagoons in arid climates.

Another type of coastal lagoon is polyhaline in nature, with the major difference caused by climatic effects. This type of salinity pattern is displayed by coastal water bodies that have distinct wet and dry seasons. A good example is Lake Reeve on the southeastern coast of Australia (fig. 13.41). This lagoon receives a fair amount of rainfall and some runoff during the wet season, causing the salinity to fall below normal marine levels of 35‰. During the dry season the lagoon experiences considerable evaporation with salinities elevated considerably above normal marine levels. Some of the shallow portions become salt pans in which halite precipitates; they eventually dry up before the next wet season.

The fauna of hypersaline and polyhaline lagoons reflects the salinity conditions. Some organisms can tolerate both situations but others cannot. Polyhaline conditions are characterized by euryhaline organisms, or those that can withstand wide ranges in salinity. There are also certain organisms that can only tolerate hypersaline conditions. They are generally stenohaline in that they cannot handle salinities below that of normal marine conditions, but are able to tolerate elevated salinity.

Hypersaline coastal lagoons tend to be shallow, with water temperature high and circulation modest at best. The result is that oxygen is limited, unless there is a good population of marine grasses. Fortunately most hypersaline lagoons do have such photosynthesizers. Benthic animals are restricted to a few species of mollusks, foraminifera, and worms. In some lagoons only three to six species are present in large numbers. There may be more than 1,000/m^2. This contrasts with most estuaries, where dozens of species may be abundant.

There are some fish, such as the redfish, that thrive in elevated salinities. A particularly well-adapted fish is the killifish (*Funulus*), which has amazing abilities to osmoregulate, that is, to control the salinity of its body fluids regardless of the surrounding environment. Consequently, this fish can adapt to fresh water and to salinities of well over 100‰. Such adaptability enables the killifish to live in nearly any salinity that might occur along coastal bays.

Polyhaline bays show similar patterns overall; however, the fauna is typically different. Some species are present here as well as in hypersaline environments and some are different. Typically the fauna in the polyhaline environment is also low in diversity but high in number of individuals. A few species of clams and snails, both detritus and filter feeders, are present along with worms, foraminifera, and fish. These bays tend to be high in nutrients and serve as nursery grounds for many open marine species, much as estuaries do.

Summary of Main Points

Barrier island systems are extremely complex in that they contain a broad spectrum of environments from fresh water to hypersaline, from low-energy bays to dynamic beaches, and from subtidal to supratidal. Life is generally abundant and diverse. Barrier systems are important to the open marine ecosystem in that the coastal bays behind barrier islands serve as a food warehouse for numerous species that hatch and grow to maturity prior to migrating to the open waters.

Barrier islands are also important to humans because of their recreational value, and because of the protection they afford the coast from hurricanes and other severe storms. Large-scale commercial development has occurred on barrier islands, with extreme examples including Miami Beach, Florida, Atlantic City, New Jersey, and Ocean City, Maryland. There is much concern about such development because of its impact on the natural processes that characterize the barrier islands. It is likely that future barrier island development will be severely controlled.

The beach and nearshore environments of the barrier island system are wave-dominated and tend to be characterized by sandy sediment. Cyclic changes characterize this area due to seasonal changes in wave energy. The fauna here is abundant but not diverse, due to special adaptations necessary to survive on a mobile substrate.

The supratidal part of the barrier includes the dunes, washovers, marshes, and tidal flats. The size and abundance of dunes is a direct function of sediment supply: if it is great, the dunes build up and the island progrades, whereas if the sediment supply is limited, the island tends to show an erosional trend. Washover fans are a primary mechanism that allows the barrier to migrate landward. These storm-generated features are especially widespread on barriers with small dunes.

Inlets are the tide-dominated part of the barrier system. They tend to vary: some are short-lived due to wave-dominated conditions, and closure takes place as the result of longshore transport of sediments. The size and stability of an inlet tends to be a function of the tidal prism or water budget that flows through it.

The overall barrier island system is a complex and dynamic one that tends to show an equilibrium between the various environments. If there is a significant change imposed on one of the environments, this causes an adjustment for most or all of the others.

Suggestions for Further Reading

Gosner, K.L. 1979. **A Field Guide to the Atlantic Seashore.** Boston, Massachusetts: Houghton Mifflin Co.

Leatherman, S.P. 1979. **Barrier Island Handbook.** Amherst, Massachusetts: Univ. of Massachusetts Press.

Kaufman, W., and O. Pilkey. 1979. **The Beaches are Moving.** Garden City, New York: Anchor Press/ Doubleday.

Schwartz, M.L. 1973. **Barrier Islands.** Benchmark Papers in Geology. Stroudsburg, Pennsylvania: Dowden, Hutchinson and Ross.

14
Rocky Coasts

Chapter Outline

The coastal environment that is characterized by rugged cliffs and crashing waves (fig. 14.1) differs greatly from the placid, broad, sandy beach with waves gently breaking across the surf zone. Rocky coasts are erosional coasts, whereas the sandy coasts are depositional. Rocky coasts are widespread and present a marked contrast to sandy barrier island coasts primarily in their physical energy and in the type and distribution of organisms that occur there. In the United States, rocky coasts are widespread in northern New England, especially along the coast of Maine. They are absent throughout the rest of the Atlantic coast and along the entire Gulf coast, except for the Florida Keys area. Much of the Pacific coast is rocky, and both Alaska and Hawaii have extensive rocky coasts.

Rocky coasts include a broad spectrum of environmental conditions, such as tidal range, wave climate, and water temperature. There is also a range of coastal morphology. In general, however, the rocky coasts of the United States are characterized by at least a modest tidal range, high wave energy, rugged, high-relief coastal terrain, and a narrow, relatively steep adjacent shelf. Estuaries, small barriers, and even small deltas are also associated with rocky coasts.

Figure 14.1 Photograph of rugged, rocky coast at Acadia National Park on Mount Desert Island, Maine, with waves breaking along the shore.

Most rocky coasts are associated with specific tectonic situations. Nearly all of the western coastal areas of both North and South America are rugged rocky coasts adjacent to narrow or nonexistent shelves. This is directly related to the fact that these coasts are associated with the emergent leading edge of crustal plates, a zone of compressional tectonics. Other rocky coasts may be associated with certain climates, such as glacial coasts on the coast of Alaska and reef development in the Florida Keys.

Sea level change is an especially important influence on rocky coasts. This includes not only the worldwide sea level changes associated with recent glaciation, but also those of a more local or regional nature that are related to tectonics or to postglacial uplift. For example, there are numerous places, such as near Anchorage, Alaska, and many places in southern California, where uplift has raised old shorelines many meters above present sea level (fig. 14.2).

The organisms that inhabit rocky coasts, both subtidal and intertidal, present unique marine communities. There are narrow, distinct zones occupied by various species controlled by such environmental factors as physical energy, time of exposure, and food supply.

Coastal Morphology

The coastal morphology of a rocky coast differs greatly from a sandy, relatively low-energy coast both in the profile of the coast perpendicular to the shoreline and also in the configuration of the shoreline itself. There is commonly variation in the coastal profile of even a small area; however, some generalizations can be made. It is also appropriate to compare this profile to that of the typical sandy coastal area.

Coastal Profile

The obvious and readily visible portion of the rocky coast profile is that part above sea level. Typically this is cliffed bedrock with a steep face that rises at least several meters above the water (fig. 14.3). Some areas may have compacted, resistant sediment that is not bedrock but can hold up a steep slope. Such conditions are common along some of the southern California coast and also along glaciated coasts.

The base of the cliffed coast commonly abuts against a beach, although there are some rocky coasts where beaches are absent. The presence of a beach is dictated by two primary criteria: (1) the availability of unconsolidated sediment from which to construct the beach and (2) a place for this sediment to accumulate. There are some locations where beaches do not form because the steep cliffs continue well below sea level. This is most common in fjords or other protected, cliffed coasts. Most rocky coasts do have at least narrow platforms where waves have cut the rock over time and where sediment may accumulate to form a beach. These **wave-cut platforms** (also called wave-cut benches or wave-cut cliffs) may be hundreds of meters wide or they may

Figure 14.2 Photograph of uplifted terrace on the coast of California.

Figure 14.3 Cliffed coast in central California on the Big Sur Coast.

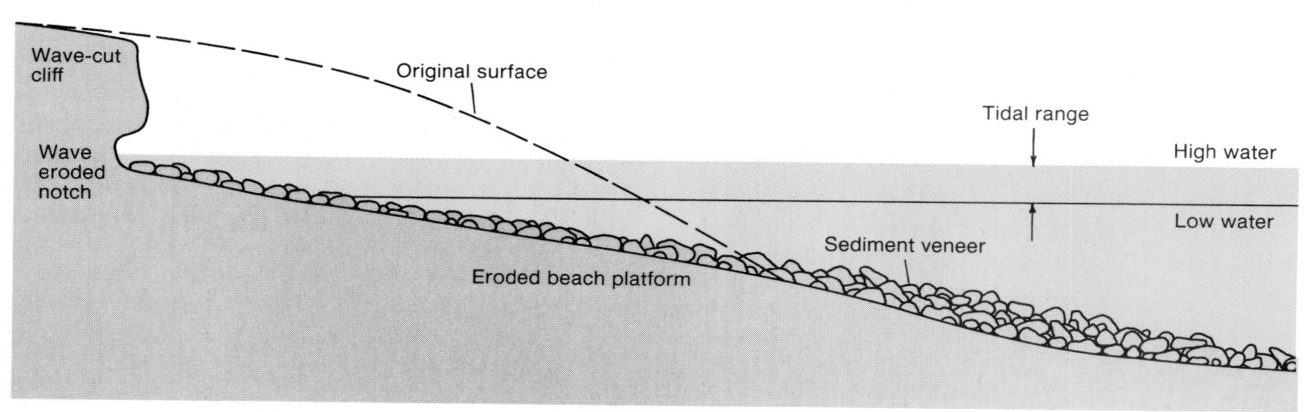

Figure 14.4 Diagram showing rocky coast with wave-cut platform.

be narrow (fig. 14.4). Because of the wave energy, sediment may be swept clear of these platforms causing beaches to be absent. There are, however, beaches that persist in this environment. Generally they are comprised of very coarse sediment, typically gravel. Some of these beaches have particles the size of boulders (fig. 14.5).

Beaches on rocky coasts consist of a thin wedge of unconsolidated sediment resting on a wave-cut platform cut in the bedrock. Because of the small amount of sediment and the high wave energy, the beach commonly is removed, at least in part, during the storm season, usually the winter. This is especially true of sand beaches along rocky coasts. A particularly good example is the section of the Oregon coast near Ona Beach State Park. The coast has a spring tidal range near 4 m, thus producing a wide beach that is dominated by sand (fig. 14.6). This photo was taken during July. The same location

Figure 14.5 Photograph of a pocket boulder beach between rocky headlands at Cape Ann, Massachusetts. The boulders are about 30–40 cm in diameter.

Figure 14.6 Sandy beach near Ona Beach State Park in the central Oregon coast.

Figure 14.7 Wave-cut platform at Ona Beach State Park during the winter when all of the beach sediment has been washed offshore by high wave energy.

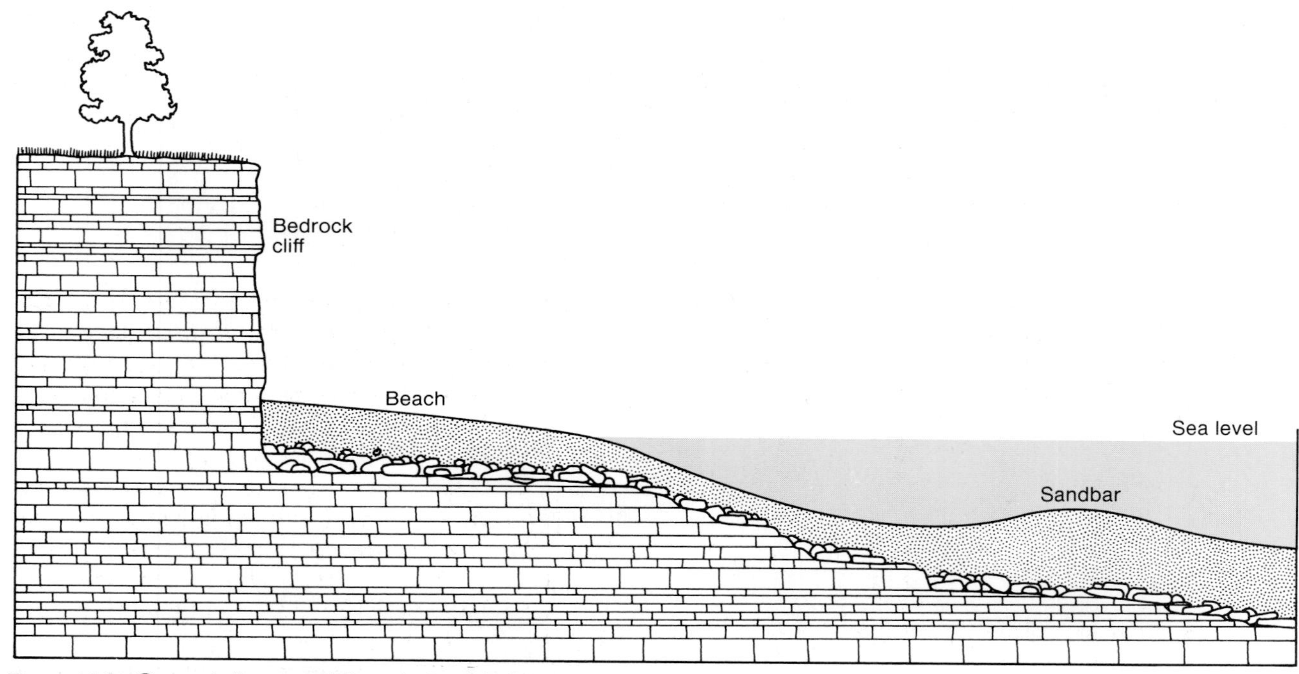

Figure 14.8 General diagram of the relationship between a sandy beach and nearshore area, and the underlying bedrock along a rocky coast.

Figure 14.9 Irregular rocky coast with headlands and embayments along the central Oregon coast.

during February shows almost no sand (fig. 14.7), but instead only a wave-cut platform. Large storm waves have eroded the beach sediment and carried it offshore. During the relatively low-energy summer period, the smaller waves and currents move the sand back onto the platform.

The shallow subtidal zone seaward of the beach along a rocky coast reflects the nature of the beach. Sediment may be present or it may be absent depending on the profile configuration and the wave energy. This nearshore zone is typically fairly steep with little or no unconsolidated sediment present.

A comparison of the profile of a cliffed rocky coast with a sandy low-energy coast shows several differences (fig. 14.8). In general, the rocky coast has more relief, steeper slopes, narrower beaches, and coarser sediment than the sandy low-energy coast. There is much variation on this profile, but the overall shape is common.

Longshore Variation

There is great variety in morphology along a rocky coast. This includes the profile previously discussed, but is also related to wave energy, rock type, structure and weathering characteristics, and the effects of humans. Whereas sandy, low-energy coasts commonly display a smooth and uniform configuration, rocky coasts are generally irregular with numerous embayments and protruding headlands (fig. 14.9). Most of the beaches along high-energy rocky coasts develop in the embayments where wave energy is low and where sediment can be trapped (fig. 14.10).

Headlands are the zones of highest wave energy and therefore are the places where erosion is most rapid, relative to the embayed areas (see chapter 6). As such, headlands serve as source areas for sediment, which may make its way to the shore and eventually become incorporated into the beach. The headland areas also provide sediment for transport

Figure 14.10 Sandy pocket beach between headlands on the New South Wales coast of Australia.

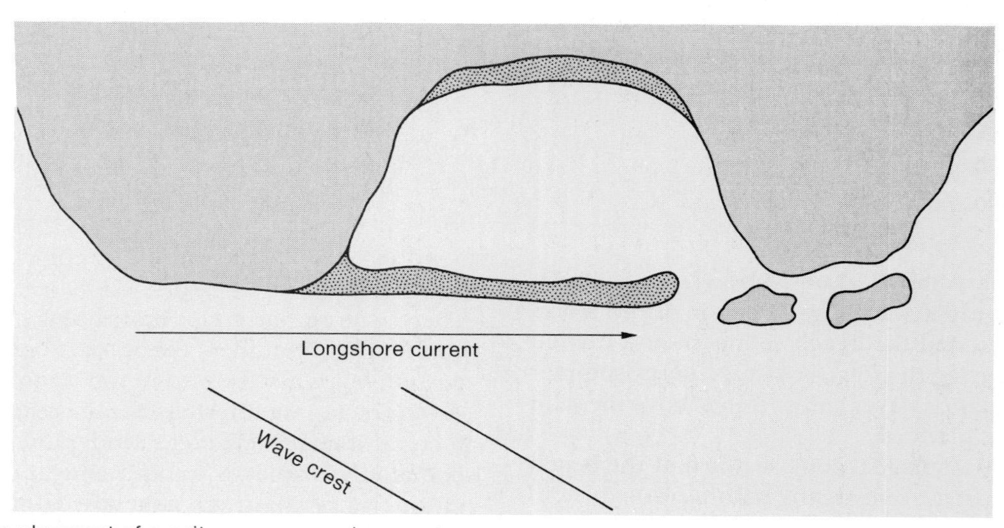

Figure 14.11 Development of a spit across an embayment between rocky headlands due to wave-generated longshore transport of sediment.

by longshore currents, which may cause the development of spits or baymouth bars (fig. 14.11). These depositional features can form barriers protecting the coastal bays on their landward side. In doing so, deltas and low-energy estuaries may develop in the embayments, such as the Columbia River mouth and Willapa Bay along the northwest coast of the United States. Inlets and flood tidal deltas are also common in this coastal setting. Ebb tidal deltas are typically absent because the high wave energy in this environment carries all of the sand away on the seaward end of the inlet, thereby prohibiting their development.

Figure 14.12 Rocky coast near Point Reyes, California, with reach of straight beach along which a longshore current can be developed.

Coastal Processes on Rocky Coasts

The same processes operate on rocky coasts as along sandy depositional coasts. The major differences between the two are the rate at which they operate and the relative importance of specific processes. Rocky coasts are distinctly wave-dominated. Even in areas where tidal range may be high, the wave energy still dominates and controls the morphology of the coast. Another significant difference is that because of the resistance of bedrock to coastal processes, the rate of change, that is, erosion, is slow on rocky coasts, even though the waves are large and the coast is subjected to high-energy conditions.

The relatively steep shelf and nearshore gradient of rocky coasts permit waves to move to the shore with little or no interference by the bottom. They are neither slowed nor dampened, thus they strike the coast with undissipated energy. Some rocky coasts reflect the waves with little absorption of energy. This is especially true if there is no beach and the waves strike directly on the cliff.

Erosion of the rocky coast is brought about primarily by waves striking the cliffs. This permits water to make its way into cracks and crevasses, causing erosion. Large rocks are thrown at the cliffs by the waves and some runoff from land also contributes to the erosion process.

Longshore currents are strong and persistent on some rocky coasts and absent on others. In order for these currents to develop, it is necessary to have waves approach the coast at an angle and for water depth to be shallow enough for refraction to occur. There must also be a reasonably straight reach of shoreline. Some rocky coasts provide these conditions (fig. 14.12). There are others, however, where waves approach parallel to the coast or where the nearshore bottom gradient is too steep to refract the waves. The result is no significant longshore current.

BOX 14.1

Erosion Along the California Coast

The coast of California is one of high relief, irregular bluffs and cliffs, and high wave energy: all the conditions that characterize an eroding rocky coast. These conditions tend to be common along all of the west coast of the United States. Some areas, however, also have a compounding problem, in that they are heavily developed and populated. The coast of southern California is such a place.

This area is not only one of high coastal relief, high wave energy, and many people, but is also an area where the earth's crust has been greatly disturbed by tectonic activity. This tectonism is partially responsible for the high relief along the coast and into the ocean, where complicated high-relief topography duplicates that on land.

Like most rocky coasts, the southern California area has been subjected to development of wave-cut platforms or benches. Sediment has accumulated on these benches and formed beaches, some big and others not so big. One of the important aspects of this beach sediment is its cyclic nature. During the summer, waves are relatively small and sand accumulates on the beach, but in wintertime the waves are large, so that most, or in some places, all of the beach sand is eroded away, leaving bare rock or coarse gravel. The following summer the sand returns during low wave-energy conditions.

Another factor limiting the amount of sand on the coast of southern California is the presence of numerous dams on many rivers that empty into the sea. These rivers carry large amounts of sediment, but the dams prevent it from reaching the sea and being incorporated into the beaches. In many places the sediment impounded by the dam is being dredged and placed seaward of the dam to allow it to proceed to the coast.

A third important factor in limiting beach development along this coast is the steep nearshore area and related submarine canyons, which carry sediment away from the coast, out to deep water. This topography makes it difficult for beach sediment to accumulate.

Beaches and sediments are important to the discussion of rocky coasts because beaches afford the best protection possible for cliffs and bluffs. Erosion along coasts, especially rocky coasts, is the result of direct wave attack. The beach provides a buffer between the waves and the cliffs and bluffs, thus protecting them from erosion. Loss of this beach results in accelerated erosion. The beach is therefore an insurance policy against erosion.

Another situation that impedes the formation of longshore currents is an irregular coastline, a common feature of rocky coasts (fig. 14.13). Such coasts do not allow wave refraction and they do not have a straight reach of shore.

An important effect of tides on rocky coasts is to make it possible for waves to attack a vertical zone of the coast. Most rocky coasts have steep gradients near the shore and are backed by cliffs. A small tidal range therefore concentrates the wave energy in a narrower zone than a large tidal range. For the same reason, the wave-cut platform generally will be wider on a coast with a large tidal range than on one with a small tidal range.

Rocky Coast Communities

The hard substrate and high-energy conditions of the rocky coast provide conditions conducive to a broad range of benthic communities, both subtidal and intertidal. These specially adapted organisms have been the subject of numerous investigations, books, and documentary movies. The intertidal zone and the tidal pools of rocky coasts provide fascinating viewing, even for the casual observer.

There are three natural niches in a rocky coast environment that can be considered separately: (1) the subtidal environment, (2) the intertidal environment, and (3) the tidal pool environment. Be-

Figure 14.13 Irregular rocky coast in northern California where longshore currents would not develop along the shore.

cause there are ecological factors that are common to all of these, they share common organisms. Each niche has its own special characteristics and its own communities, however.

Ecological Factors of Rocky Coasts

Waves are by far the dominant factor in the rocky coastal environment. All organisms that inhabit this type of coast must be able to cope with continually high physical stress caused by wave surge. Consequently, sessile benthos dominate the communities, with some specially adapted vagrant or migrant benthos present as well. All organisms in this environment are equipped with some mechanism with which to firmly grip the rocky substrate.

Most other ecological factors, however, show no special adaptations or variations to the rocky coast. Salinity is usually normal, although locally it may be influenced by freshwater runoff. Temperature is dependent upon latitude and oceanic circulation patterns, and light penetration is generally not a limiting factor in the distribution of organisms on the rocky coast.

Availability of food along rocky coasts is directly related to wave activity. The waves carry plankton throughout the environment and make it available to the numerous filter feeders attached to the rocky substrate. Many vagrant benthos feed upon the attached forms of algae and sea grass. Food is more readily available to filter feeders along rocky coasts than on sandy coasts because of the high wave energy, which keeps detritus in suspension.

Another factor that is important in controlling the distribution of life on rocky substrates is the resiliency of the organism itself or of its shell. The crashing waves or their surge cause considerable stress on algae, starfish, snails, or whatever organisms happen to be in this zone. Only those individuals that can withstand the physical rigors of this wave impact can persist on the rocky coast.

(a)

(b)

Figure 14.14 Examples of kelp, a brown algae that attaches itself to rocky substrates in areas of cool water.

Figure 14.15 Encrusting barnacles with snails and algae on a rocky intertidal surface.

Subtidal Community

Compared to the intertidal zone, there is considerably less information on the subtidal environment of the rocky coast. This is due largely to the fact that the shallow subtidal environment on a high-energy, wave-dominated coast is difficult to observe firsthand. There is much danger for the SCUBA diver in such a dynamic zone.

In many respects the community on the subtidal portion of the rocky coast reflects that of the lower intertidal environment. As depth increases and the wave energy decreases, the variety of organisms increases. That is, the rigors of the intense wave energy in the shallow subtidal zone restrict the community to a few specially adapted species. With increasing depth, the need for these special adaptations diminishes and diversity increases as a result.

Practically speaking, the shallow subtidal and the lower intertidal communities are the same. Virtually all of the organisms that occupy the subtidal rocky coast can be exposed for short periods of time during some of the daily tidal cycle. The subtidal rocky coast community ranges up to neap low tide or slightly above and includes many species of sessile algae, grasses, and invertebrates, as well as some slow-moving vagrant benthic forms.

Kelp, which is brown algae, is common in this environment. Its holdfast systems keep this long organism fixed to the rocky substrate (fig. 14.14). These individuals are as much as 10 m in length and, along with phytoplankton, serve as the producers in this environment. There may be more than a dozen species of marine algae at any given location, with a general tendency for the species with the largest individuals to be in the subtidal and lower intertidal zone. Brown algae dominate this region.

Several varieties of invertebrates occupy the subtidal environment, including filter feeders, scavengers, and predators. Filter feeders are sessile benthos that are attached to the rocks either by threadlike structures, such as those of the mussels, or by cement, such as that used by encrusting barnacles (fig. 14.15). Some invertebrates bore into the hard substrate by either chemical means or physical abrasion. In addition to the aforementioned groups,

Figure 14.16 Chitons on a rocky intertidal surface.

anemones, goose-necked barnacles, encrusting oysters, sponges, and various worms also are common sessile organisms on the subtidal rocky coast.

Barnacles are among the dominant organisms in the intertidal zone and are also present in the shallow subtidal environment. Studies have shown that these filter feeders grow more rapidly in the subtidal environment, where feeding can take place continuously, than in the intertidal environment. Their dominance in the intertidal environment is the result of their adaptation to this niche in which many organisms cannot survive.

Scavengers and grazers move slowly across the rocky substrate, scraping algae and lichens from the surface. They possess large, muscular organs that both grip the substrate and provide locomotion. Chitons (fig. 14.16) and limpets are good examples. Both have external shells that encase and protect their soft organs, low-profile shapes that present little resistance to waves, and scraping mechanisms for gathering nourishment from the algae on the rocky surface. Various snails also occupy this environment, but they have more bulbous shapes and are, therefore, more susceptible to removal by waves.

The predators in this environment move relatively rapidly, compared to the scavengers. These organisms also must withstand the constant rigors of waves and therefore must be able to grip the substrate. Such animals as starfish (fig. 14.17), sea urchins, and octopi are among the more common predators. They have sucker-like feet that grip the substrate and, in the case of the starfish and octopus, also are used to grip prey. There are a few sessile predators, such as some anemones, and also slow-moving vagrant types, such as the carnivorous snails.

Figure 14.17 Small starfish and limpets on a rocky intertidal surface.

Intertidal Community

Most of our knowledge of rocky coast communities is limited to that of the intertidal zone. It is here that one can see distinct zonations of species of both plants and animals (fig. 14.18). All of these organisms must have the same general adaptations as those in the subtidal zone because they are subjected to rigorous wave attack. Additionally, intertidal organisms must be adapted to repeated exposure to the atmosphere for at least short periods of time. This adaptation typically is some ability to retain moisture during the time of exposure.

Two zones that can be treated separately, although both are intertidal, are the below-midtide area and the above-midtide area. Below midtide, organisms are exposed subaerially for only a few hours, with the absolute time dependent upon semidiurnal or diurnal tidal conditions. Above midtide, organisms are exposed for a relatively long time; the area between neap high tide and spring high tide is a special environment with organisms exposed for several days at a time.

Lower Intertidal Zone

The ranges of most of the animals in the lower intertidal zone also extend into the subtidal zone. Various species of brown algae, such as the kelp, and green algae are common in this environment. In addition, there are both sessile and vagrant benthos that make their living in the lower intertidal zone. Those organisms that photosynthesize must be able to keep moist for a few hours during exposure. This is not a problem for the large kelp or the green algae because their fleshy tissues tend to retain water. Photosynthesis may proceed at an increased rate in some species during exposure due to the increase in sunlight.

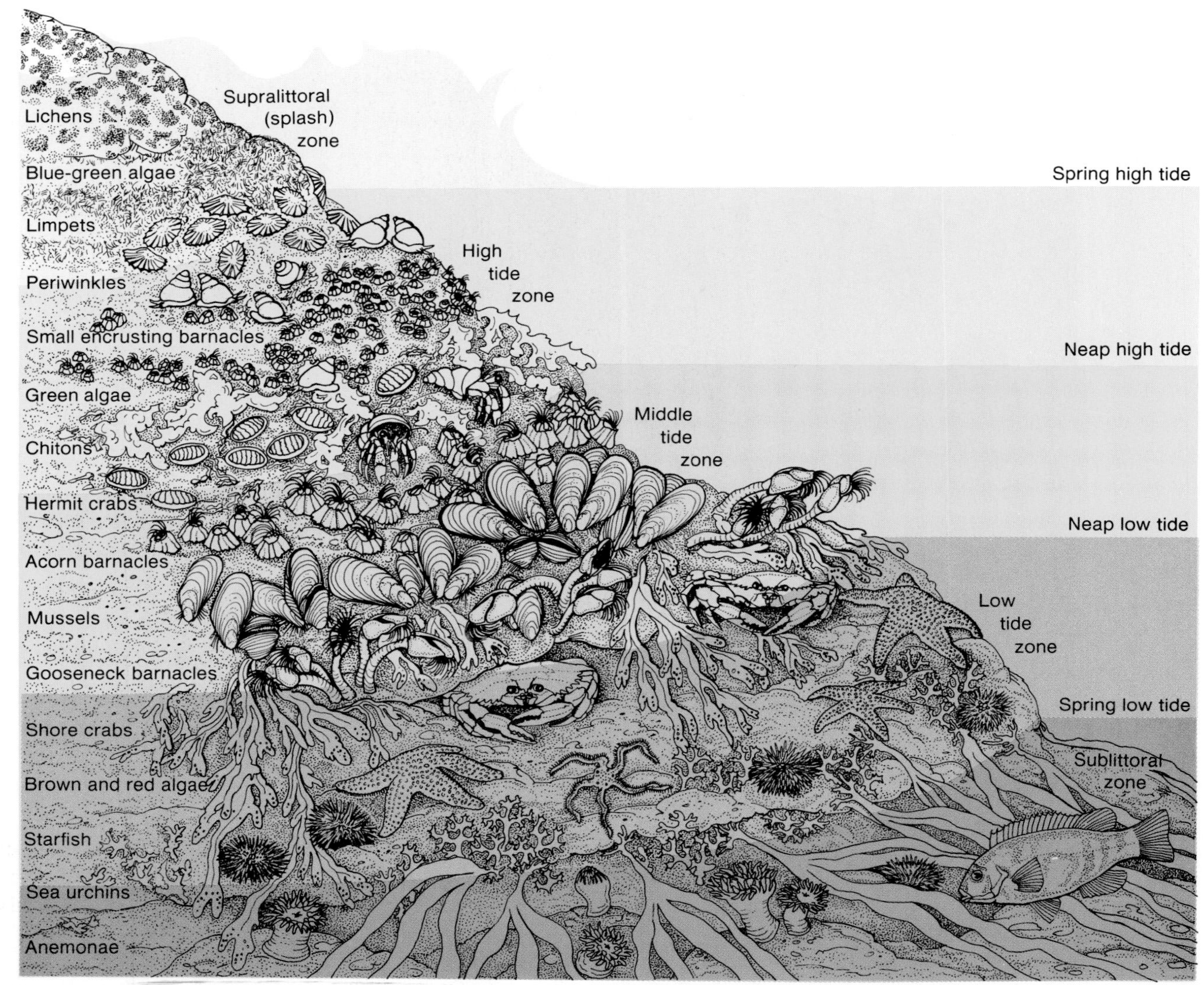

Figure 14.18 Zonation of common organisms on the intertidal rocky coastal environment.

Sessile benthos such as anemones, mussels, barnacles, oysters, and some tube worms do very well in this environment. They are filter feeders and most of the time are submerged, with a good food supply of plankton and detritus. The anemones are fleshy and contain much water in their tissues. The others have external shells that can be effectively sealed during exposure. These features allow the organisms to experience exposure to air without any detrimental effects. The one important activity that such filter-feeding organisms cannot do while exposed is take in food because their food supply is contained in the water.

The vagrant benthos of this environment fall into the same two categories that were mentioned in the discussion of the subtidal rocky coast: those that move slowly and those that can move relatively rapidly. Each has special adaptations that operate to the

Figure 14.19 Upper part of the intertidal zone with algae and lichens.

advantage of the species. For example, the slow-moving types, such as limpets, chitons, and abalone, that crawl slowly across the substrate scraping rocks for food, all have shells that protect the fleshy part of the organism from exposure. These are the same low-profile shells that are wave-resistant. Both of these features are beneficial for life in the high-energy, intertidal zone.

The faster-moving benthos, such as crabs and some starfish, have advantages of their own. These organisms do not have shapes that are well designed to resist wave attack, but they can move easily to places of protection. This same capability of movement enables these organisms to move to sources of food. During low tide, when the substrate is exposed, crabs can quickly move to prey or detritus. Some crabs also have a special adaptation in their gill structure that enables them to live subaerially for some time. They can feed regardless of the tidal stage, whereas the sessile types are restricted to feeding while submerged because their feeding organs or structures operate only underwater and their food source, whether detritus or living organisms, is only available in the water.

Upper Intertidal Zone

There is a somewhat different trend in the organisms present above mean tide. In general, there are fewer plants relative to animals and there are fewer species overall. The individuals tend to be somewhat smaller toward the upper portion of the intertidal zone because of the decrease in food supply, especially for filter feeders.

Blue-green algae and lichens are abundant in the uppermost intertidal zone along with barnacles (fig. 14.19). Algae occupying this zone have thick cell walls that help to retain moisture and thus withstand the relatively long exposure to the atmosphere. Some

BOX 14.2

A Rocky Coast Community

The community that occupies a rocky coast includes the complete spectrum of trophic levels, from the algae that are primary producers to mammals, such as sea lions, that are the top carnivores. Although the environment requires special adaptations in order for organisms to survive, it is one of both great diversity and abundant numbers. This is in part due to the opportunistic nature of the specially adapted organisms and the lack of competition due to this required specialization. It is also due to the good circulation, which makes nutrients and oxygen readily available.

Waters of the rocky coast tend to be clear and free of suspended sediment because of the general absence of fine sediment along rocky coasts. Consequently, light penetration is good, inhibited only by reflection from the extremely irregular surface of waves. The net result is that light is not a limiting factor in photosynthesis. All forms and types of algae are present, including blue-green (found near the spring tide or supratidal zone), green, red, and brown. Green and brown algae are common in the intertidal zone with the brown being generally lower. Red and brown algae are subtidal and may extend to depths of several tens of meters, especially the kelp.

The primary consumers of the intertidal community include numerous filter-feeding sessile species, such as barnacles, mussels, and worms. There are also several types of vagrant scavengers, such as the chitons, limpets, and periwinkle snails. First-level carnivores are typically invertebrates that feed on worms, snails, and crustaceans. Most of these are vagrant, such as crabs and echinoderms; however, the sessile anemones also are in this category.

The second-level and top carnivores are the primary predators that occupy this environment. Various fish, octopi, and mammals are in this category. The sea lions are the most voracious predators, living almost exclusively on fish.

limpets and chitons also live here, along with small detritus-feeding snails. Crabs also venture into the high intertidal zone on occasion, but the food supply there is limited. The two primary adaptations required of organisms that inhabit this zone are the ability to withstand prolonged subaerial exposure and the requirement of only a modest food supply.

The rocky coast produces a zone that is not common along other coasts: the spray zone. This is above high tide but it is not the same as the supratidal zone because it is sprayed or wetted regularly, in some places continually. This zone is caused by spray from waves breaking on the rocky coast. As a result, colonization by land plants is prevented. Because it provides for an extension of the normal habitat of certain organisms, this unique environment allows some organisms, such as chitons, limpets, algae, and lichens, to extend their habitat above the high-tide line.

Tide Pools

The irregular surface that is typical of intertidal rocky coasts produces small basins that remain filled with water when the tide falls. These small pools of seawater present a distinct niche in which coastal organisms can exist during the tidal cycle (fig. 14.20). Tide pools present some benefits and some disadvantages to the organisms that become trapped in them. Depending on its location, there may be some marked changes that occur within the pool. These tend to be more prevalent in the uppermost part of the intertidal zone. Some pools may be between neap and spring high tide, in which case they may be outside of the tidal influence for several days. Various things may happen during this period: rain may fall and lower the salinity, evaporation may cause the salinity to be elevated, or the sun may significantly elevate the temperature. Any of these conditions can be fatal to the normally stenohaline marine community that inhabits the open rocky coast.

Figure 14.20 Tide pool, where water is trapped during low tide, making this a unique intertidal environment on the rocky coast.

The organisms that inhabit tide pools are those typical of that part of the intertidal zone in which the pool develops. There may, however, be some individuals that normally move with the tide and others that are nektonic but cannot escape. It is common to find fish, an octopus, or other subtidal individuals in tide pools. This may be advantageous in that the fish or other carnivore has a captive population on which to feed. But it can also be a problem if there is little or nothing to eat or if the oxygen supply becomes depleted and rising waters do not alleviate the situation. Crabs seem to be among the best adapted organisms to deal with the tide pool situation. They can, in effect, move from one pool to the other in search of food. If the salinity changes, the crabs can leave. Virtually all other organisms are trapped except for some of the more mobile echinoderms and mollusks.

Summary of Main Points

Rocky coasts are characteristic of active margin areas, although they also occur in other areas. A rocky coast is a coast of high wave energy and is typically characterized by erosion. Relief is high and only thin veneers of sediment are present locally on the bedrock surface. Typical coastal processes, such as longshore currents and rip currents, tend to be either absent or just local.

Organisms that occupy the rocky coast are specially adapted to life on a hard substrate in a physically rigorous environment. They must be able to withstand pounding by waves and they must have the ability to attach themselves firmly to the rocky substrate. Generally this is accomplished by a low morphology protected by a shell and either a sessile mode of life or a slow-moving lifestyle that includes extensive attachment by large muscular structures. Fish and some mobile benthic invertebrates, such as crabs, are also common on the rocky coast.

The great circulation provided by waves results in an abundant food supply that tends to favor filter-feeding organisms. The encrusting lichens and algae also provide a source of nourishment for scraping scavengers.

In addition to these adaptations, the organisms that live in the intertidal zone on the rocky coast have adapted to intermittent exposure. Typically this is in the form of protection from desiccation. Those that occupy the spray zone have adapted to survive on minimal moisture.

Suggestions for Further Reading

Amos, W.H. 1966. **Life on the Seashore.** New York: McGraw-Hill, Inc.

Carefoot, T. 1977. **Pacific Seashores.** Seattle: University of Washington Press.

Newell, R.C. 1979. **Biology of Intertidal Animals.** New York: American Elsevier Publishing Co.

Stephenson, T.A., and A. Stephenson. 1972. **Life Between the Tidemarks.** San Francisco: W.H. Freeman Co.

Zottoli, R. 1976. **Introduction to Marine Environments.** St. Louis: C.V. Mosby.

PART FOUR

Open Marine Environments

15
Reefs

Chapter Outline

- Noncoral Reefs
  - Oyster Reefs
  - Worm Reefs
  - Other Reef Types
- Coral Reefs
  - Reef Classification
  - Reef Environments
    - Reef Slope
    - Reef Surface
    - Lagoon
  - Reef Sediments
  - Reef Morphodynamics
- Summary of Main Points
- Suggestions for Further Reading

Reefs have long been studied by scientists. They have also long been popular with tourists vacationing in exotic low-latitude regions, and they are of great commercial importance in the present world ocean because of the organisms living on them. Actually, the term reef means different things to different people, depending largely on the nature of their livelihood. The sea captain considers any shallow navigation hazard a reef; fishermen commonly use the term for submarine features where nets are snagged and fish abound, or for shallow-to-supratidal rock structures that cause waves to break. The scientist reserves the term **reef** for wave-resistant, organic-framework structures. In other words, the structure must be dominated by a rigid framework of organisms. Framework organisms include a variety of invertebrates and algae such as certain corals, coralline algae, oysters, and some worms. During the geologic past other groups of organisms now extinct served as framework builders for organic reefs.

The term reef is typically applied to coral reefs today. These are certainly the largest, the most widespread, and the most beautiful of modern reefs. They should, however, be called coral-algal reefs because red, coralline algae are an important framework constituent of modern reefs. The bulk of this chapter is devoted to coral reefs; however, most of the basic principles and generalizations made about this environment can be applied to reefs in general, regardless of the dominant organisms.

Noncoral Reefs

Oyster Reefs

Other than corals, oysters are probably the most familiar and widely distributed framework-building organisms. They are attached to rocks, pier pilings, and seawalls, and they may be found scattered over the substrate of shallow coastal water bodies. Oysters prefer a firm substrate; as a result they develop clusters of individuals, with the living ones attached to underlying expired ones. The oyster shell is the preferred substrate for a larval form to come to rest and mature to a reproducing adult.

Oysters are most common in low-energy estuaries with salinities between 10 and 20 parts per thousand. Salinities below about 10‰ make it difficult for young oysters to mature, whereas above 20‰ there tend to be problems with marine predators such as the oyster drill, a snail. Shallow, muddy estuaries such as those along the Atlantic and Gulf coasts of the United States serve as ideal oyster habitats. It is also common for oysters to live in the intertidal zone (fig. 15.1) up as high as about neap high tide, where the marsh grass begins.

Figure 15.1 Oyster reefs at the tidal flat-marsh boundary in a South Carolina estuary.

Because of their sessile lifestyle, oysters rely on currents to carry suspended organic detritus and phytoplankton, which serve as the main food supply. Tidal currents in the estuaries serve as the primary transport mechanism for this food supply. As a result, the oyster reef tends to develop a characteristic linear morphology with the currents moving perpendicularly across the length of the reef. Such a combination of reef shape and current direction provides for the maximum exposure of the largest number of individuals to the food supply. These reefs tend to be located where currents are relatively swift and nutrient supply is greatest in a given estuary (fig. 15.2).

There are also places where oyster reefs develop offshore from the coast. A good example is the northwest coast of the Florida peninsula. Here the shallow shelf is one of low wave energy and the environment is essentially that of an estuary. Embayed river mouths empty into the Gulf of Mexico and salinity is below normal several kilometers from the coast. Long oyster reefs parallel the coast with the tidal currents that move across them carrying their food supply (fig. 15.3).

Figure 15.2 Photograph of an estuary at low tide showing numerous exposed commercial oyster reefs.

Figure 15.3 Elongate oyster reef along the Gulf Coast of Florida. These reefs develop a linear shape perpendicular to tidal currents in order to take advantage of the food supply provided by the currents.

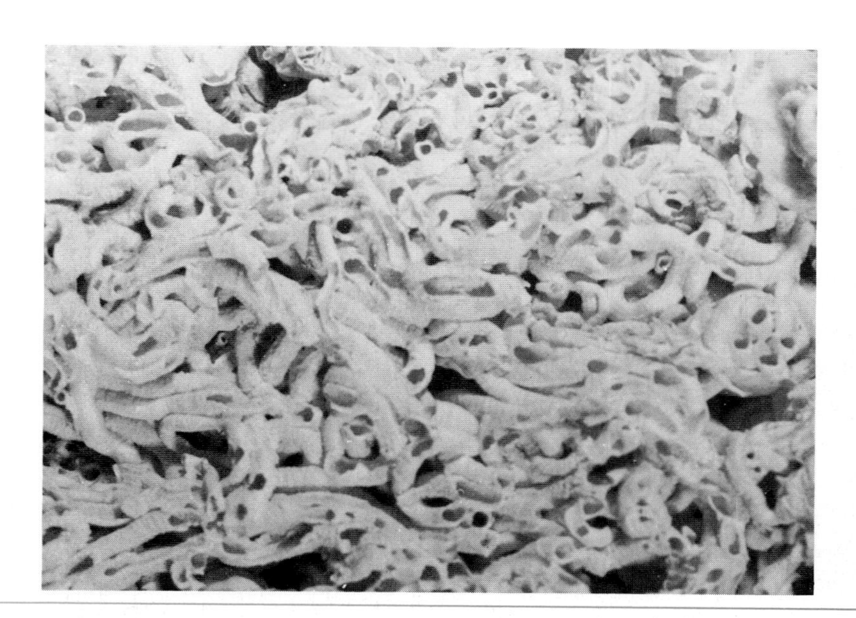

Figure 15.4 Close-up photograph of the calcareous serpulid worm tubes that develop reefs.

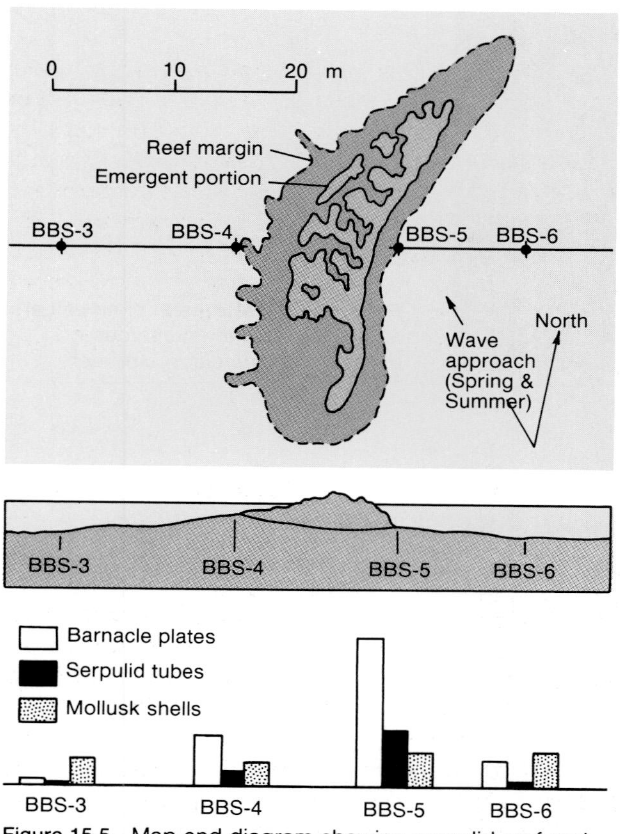

Figure 15.5 Map and diagram showing serpulid reef and adjacent area from Baffin Bay, Texas.

Although oysters are the basis for one of the most economically important shellfish industries of the marine environment, they are currently being subjected to ever-increasing pressures from pollution due to the activities of humans. Numerous estuaries have experienced at least temporary bans on harvesting due to contamination by toxic chemicals or bacteria. Oysters are not capable of selecting what they consume during the filtering process and consequently they are susceptible to a variety of potentially hazardous materials that are in suspension.

Worm Reefs

Some worms secrete tubes of calcium carbonate that, when formed into large clusters, make a wave-resistant, organic framework, that is, a reef. Among the most prominent of these are the serpulid worms (family—Serpulidae), a type of polychaete annelid. The serpulids are a family of roundworms that have tubes of a few millimeters to a centimeter or so in diameter and are several centimeters long (fig. 15.4). Like most colonial organisms, these worms are continually growing and expiring, with the framework mass increasing as the living individuals occupy the outermost tubes.

Although these worm reefs are small, with most confined to areas on the order of tens of square meters (fig. 15.5), they may occasionally reach hundreds of square meters. The reef framework rises a meter or so above the surrounding substrate.

Figure 15.6 Algal stromatolites that form small reef structures in Shark Bay, Western Australia.

Serpulid worm reefs presently occupy a range of physical energy conditions, from the relatively low-energy environment of Baffin Bay on the south Texas coast, to the wave-dominated, high-energy environment of the Atlantic coast of Florida. Like the oysters, these tube worms are also filter feeders. In an area such as Baffin Bay, the reefs rely on tidal currents and wave-generated currents to carry their food supply, whereas off the east coast of Florida, waves are the primary energy source. The latter is more of a typical reef-type environment, with waves breaking on the worm reefs, which rise to about a meter or so below low tide at a distance of 30 to 50 m offshore.

Individual worm tubes are arranged in two distinctly different habits. One has randomly arranged tubes and the other has an obvious fabric with tubes oriented essentially parallel to one another. Studies indicate that the randomly arranged tubes tend to be constructed under low-energy situations, whereas oriented tubes are associated with locations where strong currents persist.

Other Reef Types

The term reef has also been applied to mangrove root systems and their related sediment accumulations. The complex prop root structures form the organic framework. There is at least one fossil example of such a reef located on Key Biscayne in Florida.

Some blue-green algae form small reefs. These algae live in the intertidal environment and collect sediment, which lithifies rapidly. The form taken by these wave-resistant, organic frameworks is that of a large cabbage head (fig. 15.6). These structures are called **algal stromatolites** and they are very important in the fossil record. Stromatolites reach nearly a meter in height and are scattered throughout the current-dominated intertidal zone. The best location for modern examples is in Shark Bay, Western Australia.

Coral Reefs

Although most people generally associate corals with reefs, there are many corals that do not live in the reef environment. Some are solitary rather than colonial and live in deep water. Corals of the reef-building type are called **hermatypic corals**, and contain the symbiotic algae **zooxanthellae**. These are microscopic brown algae that live within the coral animal. Both organisms benefit from the arrangement because the algae receive shelter and waste products containing nutrient materials, the coral receive oxygen and certain organic compounds from the algae, and the algae itself serves as food.

Hermatypic corals have several ecological restrictions that limit their distribution. They are limited to shallow, warm water, in which zooxanthellae

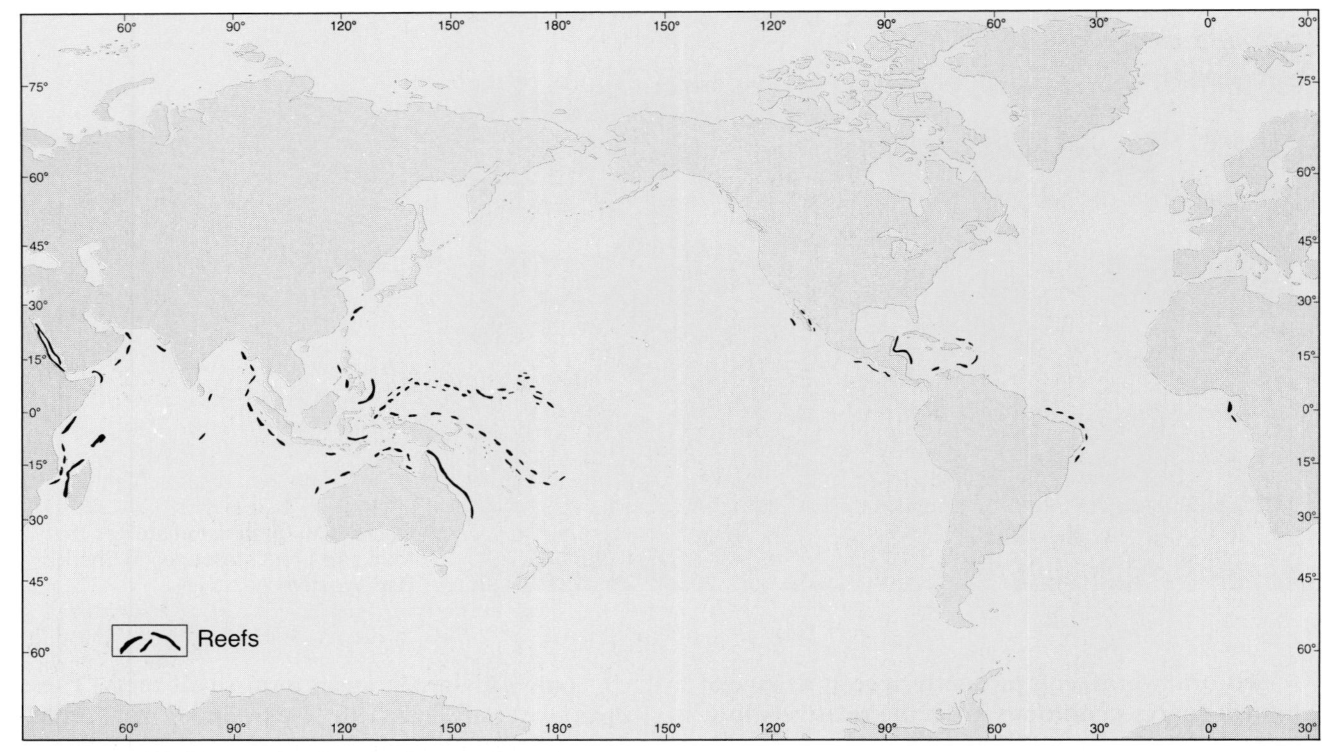

Figure 15.7 World map showing distribution of coral reefs.

can exist. The minimum temperature at which coral reefs can survive is 18°C. This restricts their distribution to about the thirtieth parallel of latitude in both the northern and southern hemispheres (fig. 15.7). Reef corals are slightly stenohaline, preferring normal salinity, but tolerating values a bit lower. Waters in which the corals live must not be turbid, partly because of the interference of light penetration, but also because the filter-feeding coral animal cannot tolerate much suspended sediment.

Reef Classification

The shape of the reef and its relationship to adjacent landmasses served as the basis for the original reef classification by Darwin, which is still widely used. During the voyage of the *HMS Beagle,* Darwin made extensive notes on reefs as well as many other natural phenomena. Based on his observations he found that reefs can be placed into one of three categories: (1) fringing reefs, (2) barrier reefs, and (3) atolls. This classification is based solely on morphology and location. It does not consider the types of organisms present or their distribution on the reef.

Fringing reefs are, as the name suggests, next to the landmass. The reef framework develops along the shore and extends seaward for distances of hundreds of meters or more. There is no lagoon or non-reef environment between the reef and the adjacent landmass. This type of reef provides natural protection from erosion of the beach and nearby coastal environments because it bears the brunt of the waves that impinge on the shore. It is also the type of reef that causes the waves to develop large breakers and cause the potential dangers in the great surfing areas of Hawaii.

Some reefs parallel the coast but are located at some distance from the shore. A lagoon that lacks widespread reef framework, although it may contain hermatypic corals between the reef and the adjacent landmass, is called a **barrier reef** and occupies a position much like that of a barrier island (chapter 13). Most barrier reefs also have passes or inlets forming breaks in the reef. These are similar in function to the inlets of the barrier island system. Barrier reefs are commonly hundreds of meters to a few kilometers from the landmass; however, the Great Barrier Reef is at least 50 km from the coast. It is

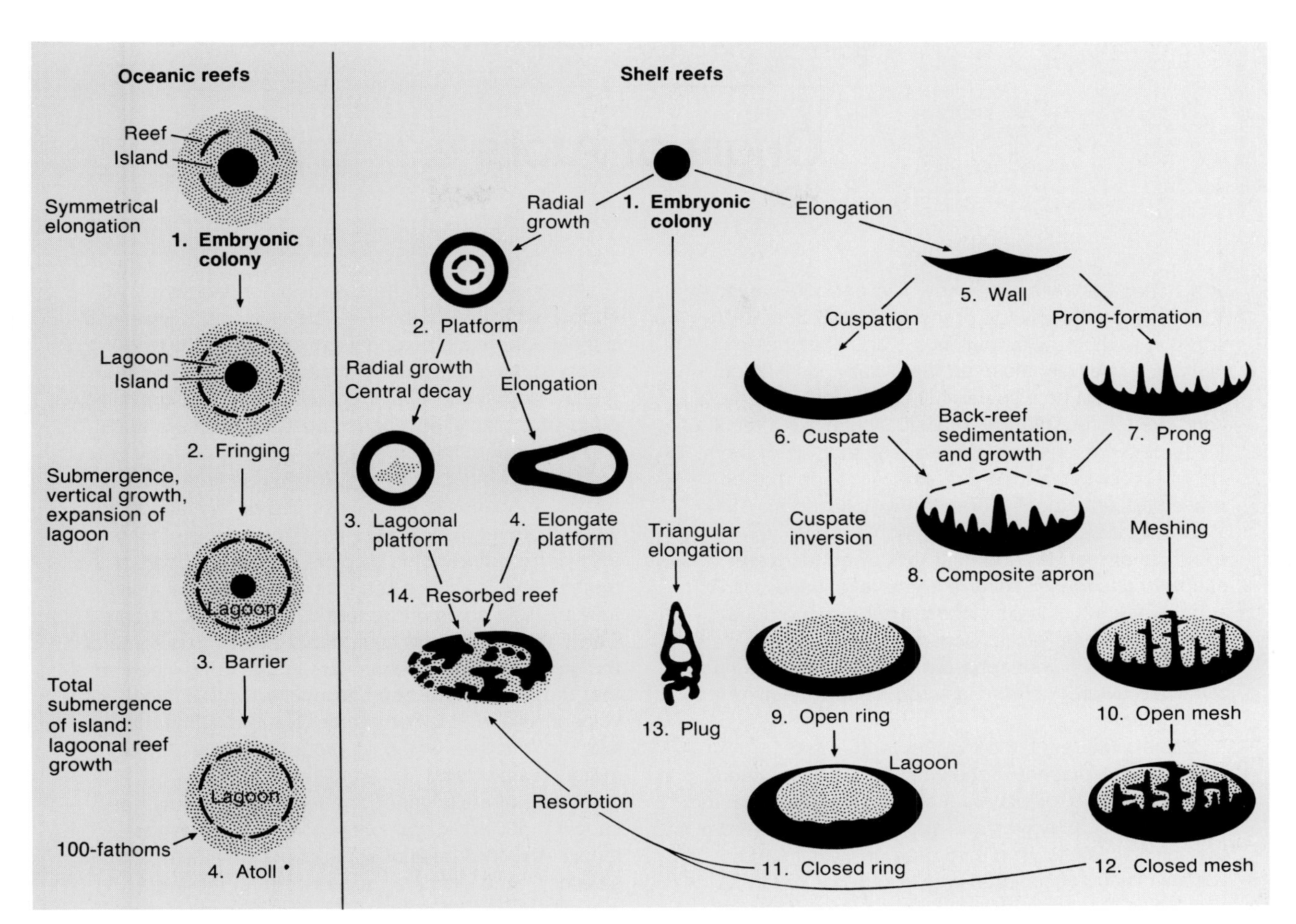

Figure 15.8 Classification of coral reefs according to W.G.H. Maxwell.

hundreds of kilometers wide and over 1,500 km long, the largest reef complex in the world. It is composed of thousands of reefs and is not one gigantic reef.

Atolls are circular reefs that do not have an adjacent landmass exposed above the ocean surface, unlike the previously discussed types. There is a lagoon in the center of this circular complex and channels or tidal passes provide circulation into and out of the lagoon in much the same way as the barrier reef. This type of reef may rise thousands of meters above the seafloor or it may be located on relatively shallow shelf environments. Atolls in deep water have oceanic islands beneath the reef framework. The origin of atolls was debated for many years, but has now been solved as described in Box 15.1, "Origin of Atolls."

Many other types of reefs have also been proposed to supplement the three major types listed previously. Included are patch reefs, table reefs, and pinnacle reefs. These are all small reefs, relative to the other three, and typically are associated with one of them.

Examination of reefs all over the world led scientists to realize that the classification described in Box 15.1 was not comprehensive. This led to a more recent classification by W.G.H. Maxwell, which considers two major categories: (1) **oceanic reefs** and (2) **shelf reefs**. Oceanic reefs are those that rise at least 100 m above the seafloor, typically beyond the continental shelf. Shelf reefs develop on the continental shelf, where water depth is relatively shallow. Although the complete classification looks complicated (fig. 15.8), it is not. Each of the two major cat-

BOX 15.1

Origin of Atolls

Charles Darwin made many different observations while on the voyage of the ship HMS *Beagle.* In addition to the now famous notes and sketches on animals, especially those on the Galapagos Islands and on finches, he also carefully observed numerous coral reefs throughout the Pacific Ocean. As a result of these observations and subsequent analysis of them, Darwin proposed a theory of atoll formation that he presented in his book *Coral Reefs,* published in 1842.

Darwin proposed that atolls resulted from an evolution of reef development that owed its overall shape to changes in relative sea level caused by subsidence of volcanic islands. After viewing coral reefs in the open Pacific Ocean in various configurations, it was apparent to Darwin that he had observed different states in a sequence of reef development.

Some coral reefs, the fringing reefs, were immediately adjacent to the volcanic islands. Others, the barrier reefs, displayed a distinct lagoon protected from open-ocean waves and separating the reef from the volcanic island. The third type, the atoll, had no volcanic island associated with it, but was rather a somewhat circular coral reef with a central lagoon (fig. B15.1).

Darwin felt that subsidence of the volcano and continued upward and outward growth of the reef produced the sequence of reef development that he observed. This mechanism can easily be envisioned for the fringing and barrier reef types, but the atoll would require complete subsidence of the volcanic island below the floor of the reef lagoon. At the time of Darwin there were no data that could support this contention.

Although much of the scientific community supported Darwin's theory, there were also those who did not. One such person was R.A. Daly, who had a theory of his own on atoll formation. He believed that the lowered sea level and cooler waters that prevailed during the glacial advances caused coral reefs to die and that during this time of low sea level, waves eroded the volcanic islands. The flat surface produced by this truncation eventually served as the lagoon floor when reefs recolonized the area and sea level rose following melting of the glaciers. Upward growth of the reefs during the rising sea level produced the atolls that are now common throughout the Pacific Ocean. Daly's theory seemed reasonable except that some of the atolls are tens of kilometers in diameter, meaning that they must have been associated with a large volcanic island. It is not reasonable to assume that an island of this size would be eroded away in only a million years or so.

Drilling of atolls during the twentieth century has shown that Darwin was correct in his subsidence theory. The thickness of some of the reefs is much greater than a kilometer and the age of the reefs extends to tens of millions of years. Neither of these parameters could be accounted for by Daly's glacial control theory of atoll formation. It is now well established that volcanic islands subside due to their mass on the thin oceanic crust.

BOX 15.1

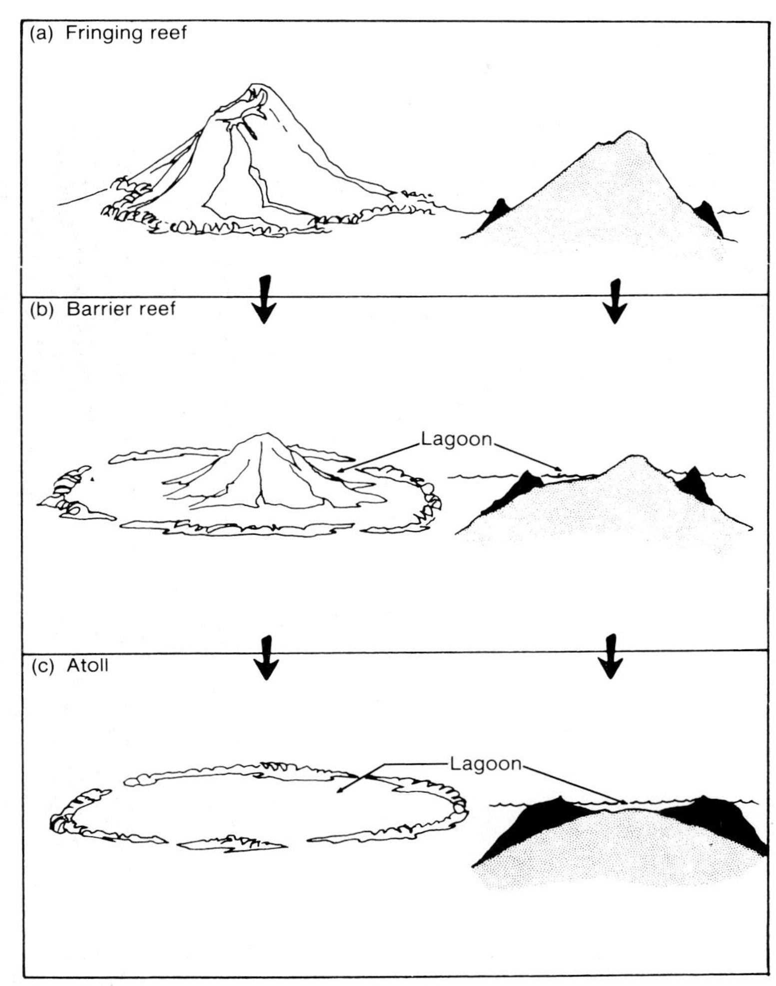

Figure B15.1 Diagram showing the three types of reefs in the development of an atoll according to Darwin.

egories culminates in a morphology that is essentially circular, that of an atoll. In the case of the oceanic reefs, they go through stages of development, from fringing through barrier to atoll. The shelf reefs achieve the atoll-like configuration through a different route because they have no adjacent landmass. Whether the reef originates as a circular platform reef or a linear wall reef (fig. 15.8), the end product displays an elliptical to circular configuration because of wave refraction. The framework, however, is or may be asymmetrical. The more resistant area is facing the prevailing wind (windward side) and therefore the large waves, whereas the less resistant part faces the leeward side of the reef.

Reef Environments

Coral reefs are comprised of a few distinct environments, each with a characteristic fauna and flora, physical characteristics such as waves and currents, and its own morphology. These environments tend to be present in the same relationships to each other regardless of the size and overall morphology of the reef. Because of this, reef ecology is rather straightforward and predictable. Oceanic reefs or shelf reefs, Pacific Ocean or Indian Ocean, the overall reef system is fairly consistent.

The coral reef system can be considered as being composed of three broad environments: (1) the reef slope, (2) the reef surface, and (3) the lagoon. Each of these broad environments contains multiple environments, each with its own characteristics. Virtually all reefs, however, have at least these three environments in common.

Reef Slope

The reef slope is a transition environment between the relatively deep, non-reef environment that surrounds the reef and the shallow part of the reef system. At least part, if not all, of the slope consists of a base of organic framework, much of which is not living at the present. The reef slope is typically steep, with gradients of 30° being common. In shelf reefs, this slope extends for up to 100 m, but in oceanic reefs it may be thousands of meters, with much of it not of an organic-framework origin.

Figure 15.9 Underwater photograph of a steep face of Heron Reef, Australia. Note the dominance of massive corals and absence of branching types in this high-energy environment.

The lower portion of the reef slope is composed of debris derived from the breakdown of the organic framework. This debris material is dumped at the base of the slope, resulting in a thick sediment accumulation of angular and poorly sorted particles of varying size, up to boulders. Below about 45 m only a few hermatypic corals are present and are mixed with the rubble from above. These corals are generally small and delicate by coral reef standards. They may extend to depths of more than 100 m, depending upon light penetration. This part of the reef is below the zone of wave influence; circulation is from oceanic or tidal currents that transport food for filter feeders.

Above this zone, between about 40 m and the wave base (10–15 m), is the upper part of the reef slope. It is characterized by more typical reef-type corals including branching and foliate types, such as *Millepora, Porites,* and *Heliopora.* Light penetration is good and food supply is typically abundant, giving rise to a diverse, dense population of organisms. In addition to corals there are various algae, bryozoa, encrusting sponges, and other sessile invertebrates (fig. 15.9). The lower energy slopes on the lagoon side of the reef contain a similar community.

Many reefs, especially those in high wave climate environments, such as the Pacific, have a terrace at about 10–15 m that is the wave base during non-storm conditions. This terrace is horizontal and may range in width up to several tens of meters. It is typically covered with massive, sturdy corals and other invertebrates (fig. 15.10). Waves provide ample food supply and also keep the terrace swept clean of unconsolidated debris. There is virtually no sediment on its surface.

The uppermost portion of the reef slope is characterized by various species of the branching coral *Acropora.* In areas of low to moderate energy in Caribbean reefs, the staghorn coral, *A. cervicornis,* dominates and in places where wave energy is intense, the moosehorn coral, *A. palmata,* is dominant (fig. 15.11). Other coral types are also present, including massive or brain corals, but the *Acropora* characterize this zone.

Figure 15.10 Underwater photograph of wave-formed terrace on Heron Reef, Australia, at a depth of about 8 m.

Figure 15.11 Photograph of the common branching reef-forming coral *Acropora palmata,* the moosehorn coral.

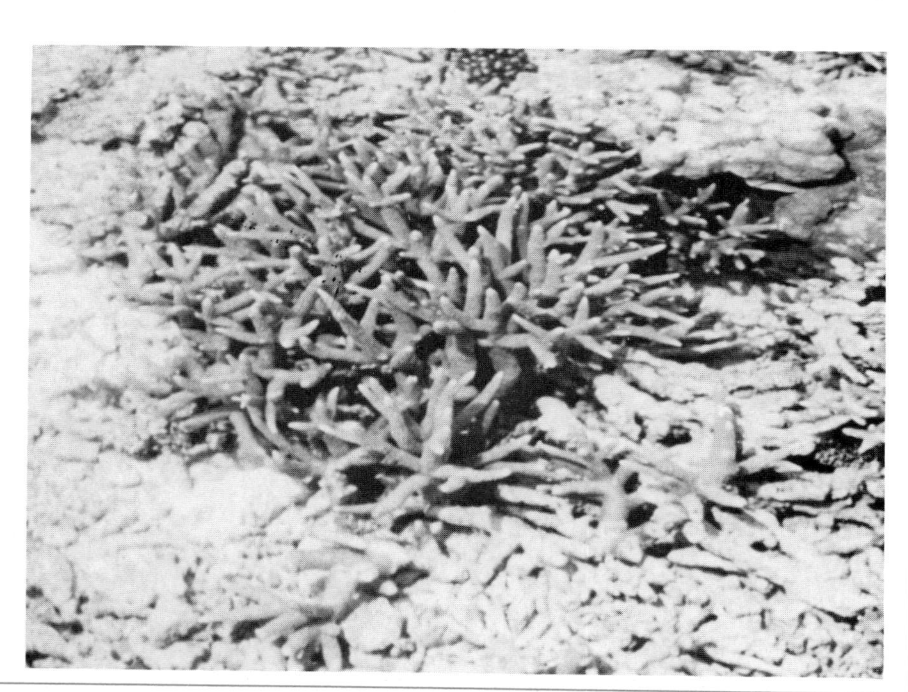

Figure 15.12 Short, stubby branching variety of the staghorn coral *Acropora cervicornis.*

Waves have considerable influence on the morphology of the corals as well as the species that are present. For example, the staghorn corals are long and delicately branching in areas of low wave energy, whereas in areas of higher wave energy the branches tend to be short and stubby (fig. 15.12). The moosehorn coral shows a similar morphologic adaptation with short, more massive branches typifying very high wave-energy conditions.

The upper slope of the windward reef typically develops a **spur and groove** morphology (fig. 15.13) due to the energy and surge associated with large waves. This morphology resembles the but-

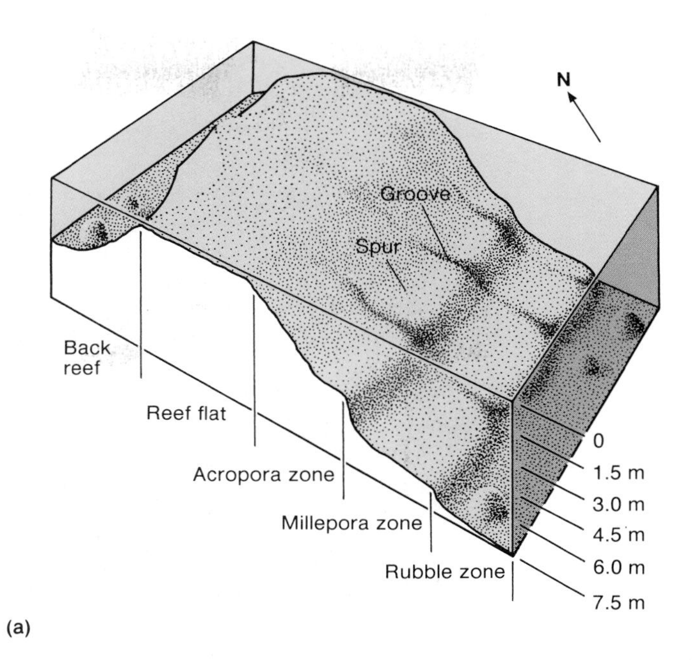

Figure 15.13 (a) Diagram of the spur and groove structure on a coral reef and (b) a low altitude aerial photograph of the same feature in the Florida Keys.

Figure 15.14 Intertidal reef surface exposed during low tide at Heron Reef, Australia.

tress system of Gothic cathedrals. The spurs are flat-topped ridges of organic framework composed primarily of *Acropora*, *Millipora,* and the red coralline algae *Lithothamnion,* which encrusts the coral and binds the framework. The grooves are a few meters deep and are floored by unconsolidated reef debris. This morphology may extend up to 100 m across the windward upper slope.

Reef Surface

The reef surface is relatively low in relief and contains multiple environments: (1) the algal ridge, (2) the rubble zone, (3) islands, if any, and (4) the reef flat. The reef surface generally rises to within 2–3 m of sea level. Some reefs are in regions of high tidal range and their surface is intertidal (fig. 15.14), such as the Great Barrier Reef. The Caribbean Sea is microtidal and its reefs are generally slightly subtidal. Width of the reef surface ranges widely, from tens of meters to kilometers.

The **algal ridge**, also called the *Lithothamnion* ridge, is the seaward zone of the reef surface. It is a smooth, almost concrete-like pavement that is formed by the dense encrustation of corals by the red coralline algae (fig. 15.15). This ridge is only a few to perhaps 100 meters in width and is best developed in the high-energy reefs of the Pacific.

Just behind the algal ridge is a **rubble zone**, sometimes called the boulder rampart, which is characterized by scattered debris thrown up onto the reef surface by storm waves. This debris ranges in size up to large boulders and is unsorted and angular (fig. 15.16).

Figure 15.15 The red coralline algae (*Lithothamnion*) ridge on a reef at Fiji.

Figure 15.16 Rubble zone showing various pieces of reef debris thrown onto the outer surface during high wave-energy conditions. Heron Reef, Australia.

Figure 15.17 Heron Island, an example of a small stabilized reef island forming from wave and current accumulation of reef-derived sediment.

Some reefs contain islands comprised primarily of biogenic sand derived from the skeletal remains of various reef organisms. These islands are typically no more than a few square kilometers in area. Some are completely unvegetated, whereas others have mature vegetation and may be occupied by human settlements (fig. 15.17). The islands are located wherever waves and currents accumulate sediment that remains stable over at least several years. Commonly a severe storm will destroy such an island. Many reef islands are somewhat protected by **beachrock**. This is, as the name suggests, a type of rock that forms in the beach zone. Biogenic sediment composed of calcium carbonate cements together rapidly in the tropics and forms what is similar to a natural seawall (fig. 15.18).

The portion of the reef surface that does not fall within the previous three environments is the **reef flat**. It is commonly the most widespread of the reef surface environments and is located in the relatively protected region behind the algal ridge and rubble zone. Islands tend to form over the reef flat. Here a diverse community thrives, comprised of both the framework builders and the other organisms. Many types of corals, red algae, green calcareous algae, mollusks, echinoderms, and others are present. It is not unusual for this environment to contain hundreds of different species of benthos. Fish are also abundant and diverse.

The upper surface of the reef flat is controlled by wave activity in subtidal reefs and by tidal level in intertidal reefs. There is a relationship between the level of physical energy and the nature of the surface of the reef to about the low-tide mark. In places where changes in tidal conditions have occurred, much of the framework of the upper surface has expired.
conditions have fairly uniform upper surfaces.

Corals, as well as other reef organisms, cannot tolerate much exposure. This tends to limit the upper surface of the reef to about the low tide mark. In places where changes in tidal conditions have occurred, much of the framework of the upper surface has expired.

Lagoon

Most reef types have a central area that is deep, relative to the upper reef surface. These low-energy areas are called lagoons and they take on a range of shapes, sizes, and depths. Some are so deep that photosynthesis cannot occur at the bottom and therefore

Figure 15.18 Beachrock, which is rapidly cemented beach sediment that forms in tropical environments and helps to stabilize the reef islands.

Figure 15.19 Shallow lagoon in a shelf atoll showing patch reefs with branching corals.

hermatypic corals are absent. Most, however, do contain a fauna and flora similar to that of the reef slope and surface. Most of the difference is in the dominant species and in the form taken by various species in this low-energy environment.

Lagoons receive little in the way of wave energy and must rely on either tidal circulation to provide nutrients and food or they must produce these in the lagoon itself. Except for the relatively low physical energy, the lagoon environment has essentially the same characteristics as other reef environments. Commonly there are small framework structures within the lagoon (fig. 15.19). These have various names including patch reefs, pinnacle reefs, or faros. They are tens to hundreds of square meters in area and rise at least a few meters above the lagoon floor. Framework structures are scattered in the lagoon with much of the floor covered by unconsolidated biogenic sediment and various nonframework organisms. The protection afforded by the reef permits

Figure 15.20 Delicate sea whips and sea fans (soft corals) in a Caribbean reef lagoon.

Figure 15.21 Sandy beach formed from biogenic reef debris sorted and concentrated by waves and currents.

a variety of vagrant benthos to live there that cannot withstand the rigors of the reef slope or even the upper reef. Lagoons also tend to support growth of delicate sessile organisms such as sea fans, sea whips, and other **alcyonarian corals** (fig. 15.20). These soft-bodied organisms only contribute tiny spicules to the sediment, much the same as sponges.

Reef Sediments

Sediments on a coral reef are biogenic, calcium carbonate derived from the skeletal remains of the fauna and flora that comprise the framework as well as from the other organisms that live on the reef (fig. 15.21). The only exception to this is in situations where there are fringing reefs that are close to land, such as volcanic islands. These landmasses may provide a minor contribution to the sediment that accumulates in the reef environment.

BOX 15.2

Heron Reef, Australia

The Great Barrier Reef is a complex of thousands of shelf reefs off the Queensland coast of northeast Australia. This reef complex extends for 1,500 km and covers 200,000 km^2. The Capricorn Group of several small shelf atolls is on the southernmost part of this complex. This group of reefs, including Heron Reef, is well-known because of its proximity to land (about 65 km) and the presence of a research station on Heron Island.

Heron Reef is about 10 km long, with a maximum width of 6 km. There is a small island on the western end of the reef. The reef itself has distinct windward and leeward sides with a shallow central lagoon (fig. B15.2). The windward reef faces the intense wave energy of the south Pacific Ocean and as a result shows a smooth coast, whereas the protected lagoon exhibits an irregular outline. The lagoon contains depths of less than 10 m but displays the usual patch reefs common to these protected environments.

Heron Island is a small accumulation of reef-derived sediment that is about a kilometer long and a few hundred meters wide. It is stabilized by extensive development of beachrock along both of the elongate shores. The island houses a small research station operated by the Great Barrier Reef Committee and also a small fishing resort. This reef is probably visited by more people than any other in the Great Barrier Reef complex.

The reef provides good examples of the tremendous fauna and flora that abound in the Great Barrier Reef system. There are more than 200 species of algae alone and several hundred species of invertebrates. Both the diversity and abundance of organisms is great.

Tidal range is near 3 m and, like most reefs of this area, the reef surface is exposed at low tide. This enables one to walk over the reef surface and examine it thoroughly and completely. Because the environment is one of high energy, relative to Caribbean reefs, there are obvious differences in the overall nature of the forms taken by various framework groups. For example, branching corals have short, stubby branches to withstand high wave energy and the encrusting red, coralline algae that provide strength to the framework are abundant, relative to Caribbean reefs. Another interesting contrast is the soft corals: on Heron Reef they are massive, whereas in the Caribbean they are branching and delicate.

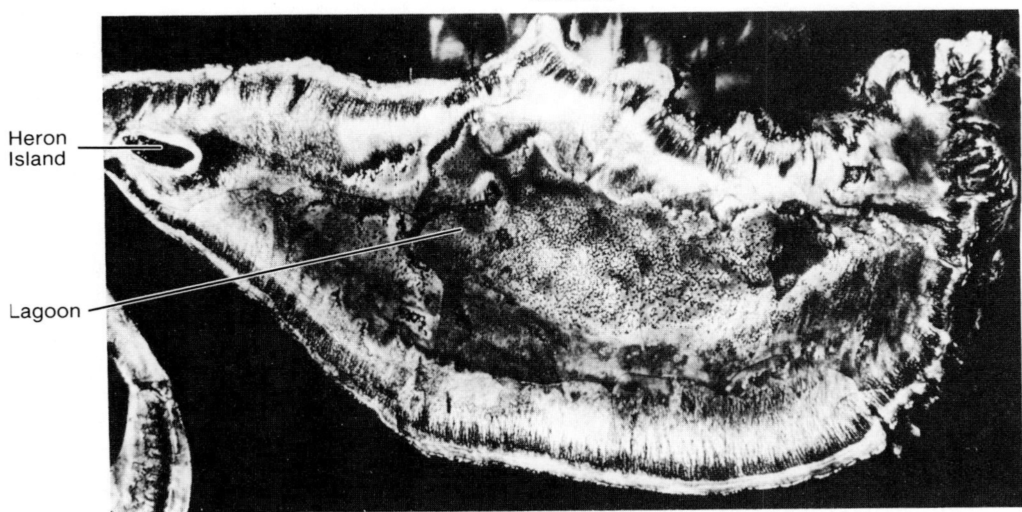

Figure B15.2 Aerial photograph of Heron Reef, Australia, showing the major environments.

Figure 15.22 (a) Parrotfish feeding on the framework of the reef. (b) Teethmarks of the parrotfish, and (c) parrotfish jaws.

Reef sediments range widely in texture and origin. In most reef environments grain size ranges from mud to boulders and sorting is poor. Both physical and biological processes produce reef sediment. Waves are an obvious and major means of breakdown for both framework and nonframework organisms. Another important source of sediment is **bioerosion:** during various feeding processes organisms break down skeletal material. The natural disarticulation of organisms upon expiration also produces much sediment.

Waves are important producers of sediment on the outer reef slope and to a lesser extent on the reef surface. Framework structure is broken up during storms, producing large particles and in some cases uprooting large coral colonies. High energy levels in these wave-dominated environments sweep them clean of sediment, sending the material down slope by means of gravity to accumulate on the lower reef slope. The rubble zone is also a product of intense

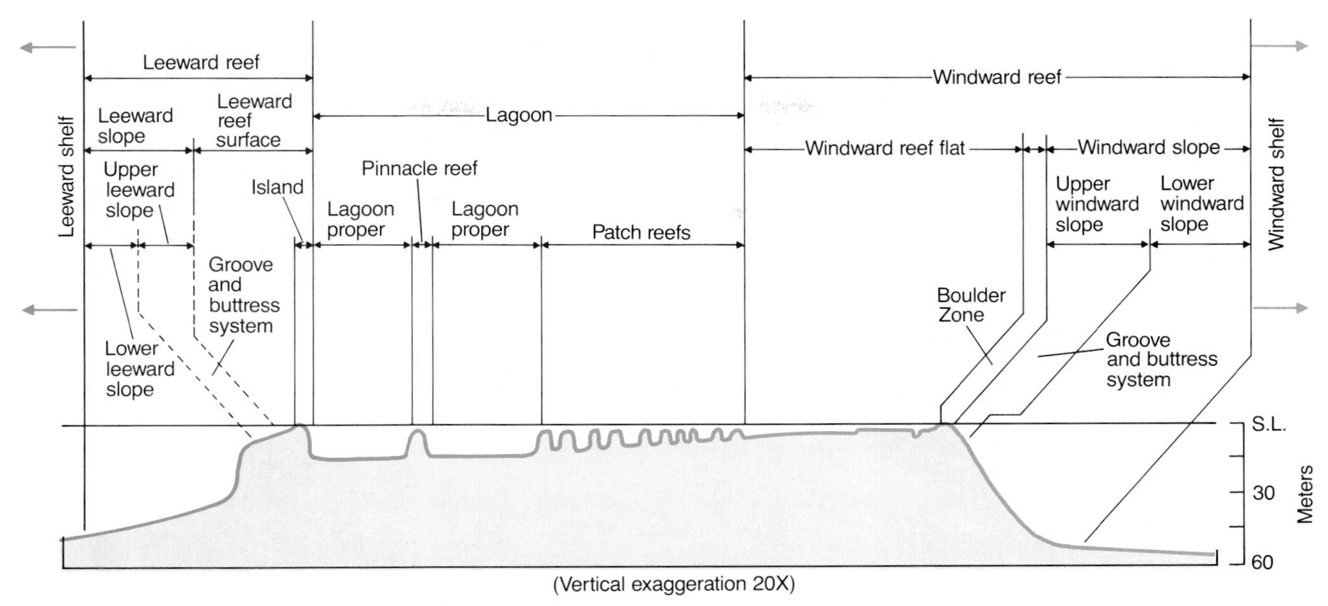

Figure 15.23 Diagram showing zonation of environments on the windward and leeward sides of a reef.

wave energy and is likewise comprised of unsorted, angular, coarse particles. Storm waves throw the broken framework fragments onto the reef surface. They are too large for nonstorm waves or tidal currents to sort them.

The natural disarticulation of calcareous skeletal material is widespread on the reef. Primary contributors are green algae, red coralline algae, mollusks, echinoderms, and the whole tests of foraminifera. Most of this sediment is the grain size of sand or fine gravel and may be sorted, such as on island beaches or shoals, or unsorted, such as that which accumulates in the framework on the reef surface and in the lagoon.

Bioerosion provides much sediment throughout the reef environment. This sediment generally ranges from mud to sand and is produced primarily by the feeding activities of reef organisms. The diver is quickly aware of a continuous rasping sound throughout much of the reef. It is produced by parrotfish feeding on corals. These beautiful fish are very abundant and feed exclusively by chewing the living coral from which they extract nutrient material. They also cause large quantities of sediment to fall on the reef floor. Parrotfish have large, massive jaws that permit this type of activity and produce teeth marks on corals (fig. 15.22). They also ingest sediment in the feeding process and eventually excrete it in the form of pellets, which quickly disaggregate, producing sediment.

Numerous other organisms are also involved in the sediment-producing process in the reef environment. Echinoderms, especially sea urchins and sea cucumbers, ingest sediment in their feeding process and produce pellets that disaggregate. There are several groups that bore into the reef substrate and thereby cause bioerosion, including sponges, gastropods, pelecypods, algae, and bacteria.

Reef Morphodynamics

The size and shape of a reef is the result of many interacting ecological factors, such as water quality, food supply, depth, climate, and others. If all of these are present in appropriate levels, it is the physical processes, especially the waves, that control the morphology of the reef. Reefs thrive in a high wave-energy environment.

The most basic aspect of the impact of waves on reef morphology is the difference between the windward reef and the leeward reef. Windward reefs experience the highest wave energy and also show the best framework development (fig. 15.23). Typically the windward slope is steeper than the leeward and it is about twice as wide on the reef surface.

Figure 15.24 Aerial photograph of a Pacific atoll showing asymmetry with widest part of the reef being the windward side and the narrowest being the leeward side.

Spur and groove development is more prominent on the windward reef. In line with this is the better development of the algal ridge on the windward side. The upper windward slope displays a smooth, well-defined shape, whereas the leeward reef is more irregular in outline.

In keeping with the generalizations mentioned previously, the atoll or circular-shaped reefs (fig. 15.24) show an asymmetry that is a measure of the overall response of the reef to relative wave energies on the windward and leeward sides. Another look at the reef classification by Maxwell (fig. 15.8) shows several stages in the morphologic history of reefs. Shelf reefs tend to have a distinctly dominant windward side and therefore display great asymmetry. Note the development of the elongate reef to the closed ring type, which is a good example of windward versus leeward morphology. Oceanic reefs tend to display less asymmetry because they are affected by large waves from various directions. Even though there is a direction that receives most of the wave energy, refraction around the island tends to distribute wave energy more uniformly than on shelf reefs.

Summary of Main Points

Reefs occur in a great variety of sizes and shapes, and can be formed by various organisms, both plant and animal. Oysters, blue-green algae, and worms are the framework organisms of modern reefs, in addition to the coral-algal reefs that typically come to mind when the subject is mentioned. The coral-algal reefs, or coral reefs as they are more commonly known, are restricted in their development to shallow warm waters and are thus low-latitude features.

Although the size, shape, and location of coral reefs may vary, all have similar elements, such as reef slope, algal ridge, reef surface, and lagoon. There is a windward side and a leeward side, with the windward side being the best developed due to the reef's preference for high-energy conditions.

Reefs may grow adjacent to landmasses, they may be separated from land by protected reef lagoons, or they may be isolated from landmasses. The organisms that comprise the reef community are opportunistic in that they establish a framework wherever conditions are favorable. These organisms are adapted to a high-energy environment and thrive there in tremendous numbers.

Suggestions for Further Reading

Darwin, Charles. 1842. **The Structure and Distribution of Coral Reefs.** Berkeley, California: Univ. of California Press (a 1962 reprint of the original).

Hopley, D. 1982. **The Geomorphology of the Great Barrier Reef.** New York: Wiley Interscience.

Maxwell, W.G.H. 1968. **Atlas of the Great Barrier Reef.** New York: Elsevier Publishing Co.

Newell, N.D. 1972. "The evolution of reefs," *Scientific American,* 226:54–64.

Stoddart, D.R. 1969. "Ecology and morphology of recent coral reefs," *Biology Review,* 44:433–98.

Wells, J.W. 1967. "Coral reefs," in Hedgepeth, J.W., ed., **Treatise on Marine Ecology and Paleoecology**, Geol. Soc. Amer. Memoir no. 67, v. 1, p. 609–32.

16
Continental Shelf

Chapter Outline

The continental shelf is the broad area adjacent to the continents that is really the drowned portion of the continental mass. This typically broad, gently sloping, subtidal extension of the landmass begins at the shoreline and extends seaward to the distinct change in gradient at the shelf-slope break. The discussion here considers the shelf to begin seaward of the surf zone or the bar and trough topography that typically characterizes this nearshore zone. This environment has already been discussed in detail in chapter 13.

Although it is possible to provide some generalizations on the general morphologic character of the continental shelf, these generalizations have little meaning because of the great variation in shelf characteristics throughout the world. The average width of the shelf is 75 km; however, the range is from just a few kilometers, such as along the southeast coast of Florida and the coast of California, to several hundreds of kilometers, such as the western Florida shelf and the Grand Banks off the coast of Newfoundland in the North Atlantic. In general the shelf width around the Pacific Ocean is narrow, whereas along the Atlantic Ocean it is wide. This corresponds to the active margin situations around the Pacific and the passive margins on the Atlantic Ocean.

Although the average depth at the shelf-slope break is 130 m, the range is from about 35 m to nearly 350 m. The depth at this break in gradient is largely controlled by the origin of the shelf in a particular area, but is also affected by recent glaciation and tectonic activity. The gradient of the shelf is low, about 2 m/km or 1:500. This is such a gentle gradient that if you were to walk across it you would probably not be able to tell if you were going up or down the slope. This parameter varies, but usually the inner shelf is steeper than the outer shelf.

The surface of the continental shelf is not the flat, featureless plain that early workers surmised. Instead there are numerous positive and negative features that provide at least tens of meters of relief. Many of these are older coastal elements, such as barrier islands, which have subsequently been submerged by a rise in sea level. Various other sand bodies, living and extinct reefs, and volcanoes are also present on the shelf. Negative features include drowned river valleys, heads of submarine canyons, fault valleys, and glaciated valleys.

Some continental margins contain complex fault systems with numerous small basins. These narrow margins are called **continental borderlands** and

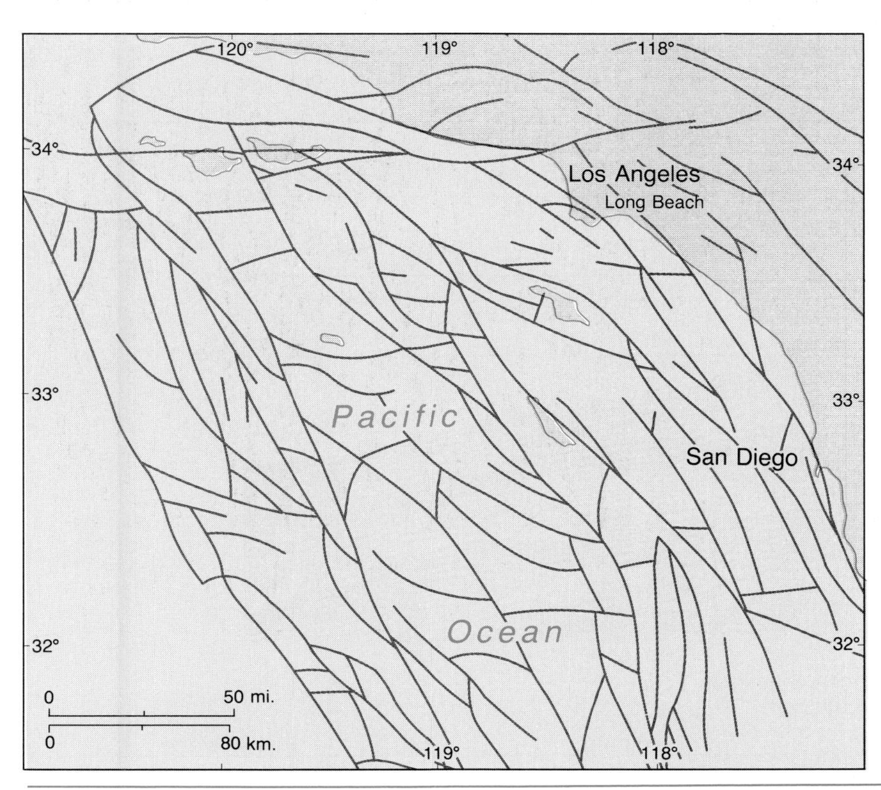

Figure 16.1 Map of California Borderland showing numerous fractures in the continental margin between Los Angeles and San Diego.

have a geologic structure characterized by numerous high-angle normal faults that bound small shallow basins (fig. 16.1).

Deep terraces called **marginal plateaus** occur adjacent to some continental shelves. These plateaus are distinctly deeper than the shelf. A good example is the Blake Plateau (fig. 16.2), which is located off the southeastern coast of the United States between Florida and North Carolina.

Origin and Distribution

In order to discuss the origin of the continental shelf, it is necessary to include the continental slope because they cannot be genetically separated. Because of the availability of geophysics, especially seismology, it is possible to see the internal geometry of the earth's crust that comprises the continental terrace (fig. 16.3). Prior to these kinds of data, only the

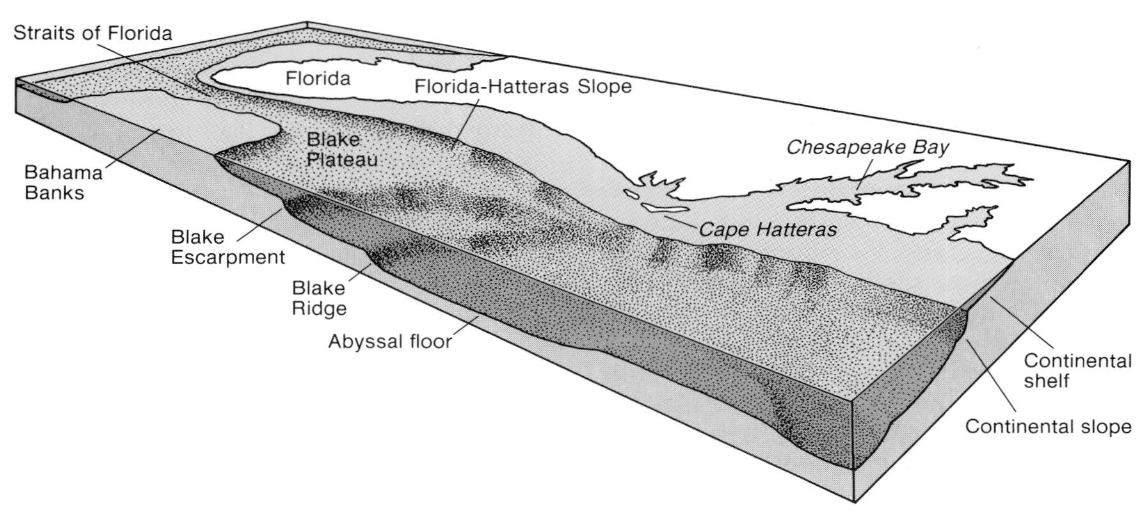

Figure 16.2 Map of Blake Plateau off the southeastern United States, which is an example of subsidence of the continental margin.

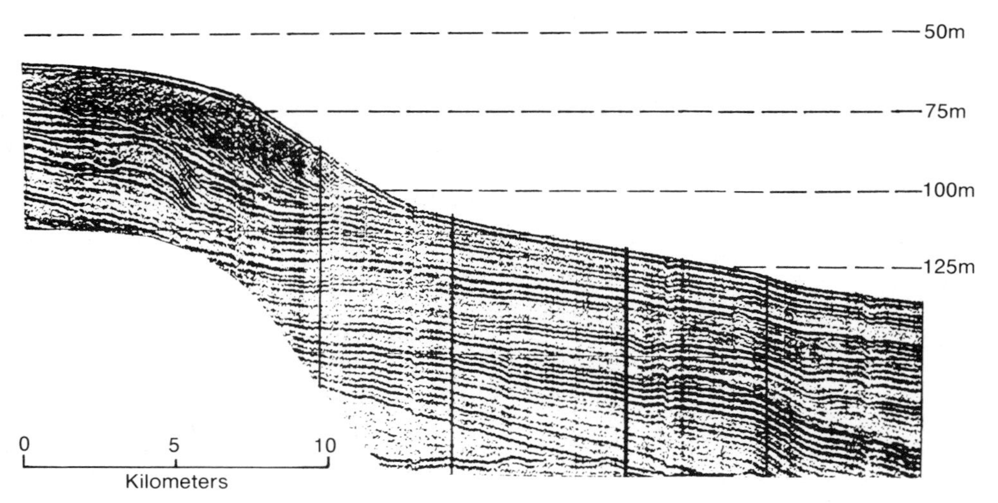

Figure 16.3 Seismic reflection profile across the continental shelf showing layering of sediments and truncation of layers.

surface configuration of the terrace was known and this looked similar in most places. It is now known that the continental terrace may have a wide variety of geologic origins. Many factors, including plate tectonics, latitude, and sediment supply, have strong influences on the geologic development of the terrace.

Faulting of the shallow crust along the slope (fig. 16.4a and b) is a common origin. In some situations of this type, the faulting results in blocks of crust forming dams with sediment ponded on the landward side. Faulted margins may be modified by slumping, erosion, and deposition of sediment so that the present surface morphology masks the original fault block surfaces. Such faulted slopes may develop on margins of continental blocks that have already been deformed (fig. 16.4a) or on those built by simple aggradation (fig. 16.4b).

The example shown in figure 16.4c is one of seaward accumulation of sediment essentially as a wave-built terrace, originally thought to be the dominant origin for terraces throughout the world. A somewhat related type is that formed as the result of the combination of progradation and subsidence (fig. 16.4d), which produces a simple, layered sequence of strata with a downward flexure.

There are numerous styles of continental margins that are basically similar: they develop as the result of some type of major damming feature. These geologic features cause the ponding of sediment between them and the continent, thus creating the shelf. The outer slope of this dam or some modification thereof forms the continental slope (fig. 16.4e–h). Erosion or faulting may cause the development of a topographically high feature separate from the main continental landmass. Filling of the

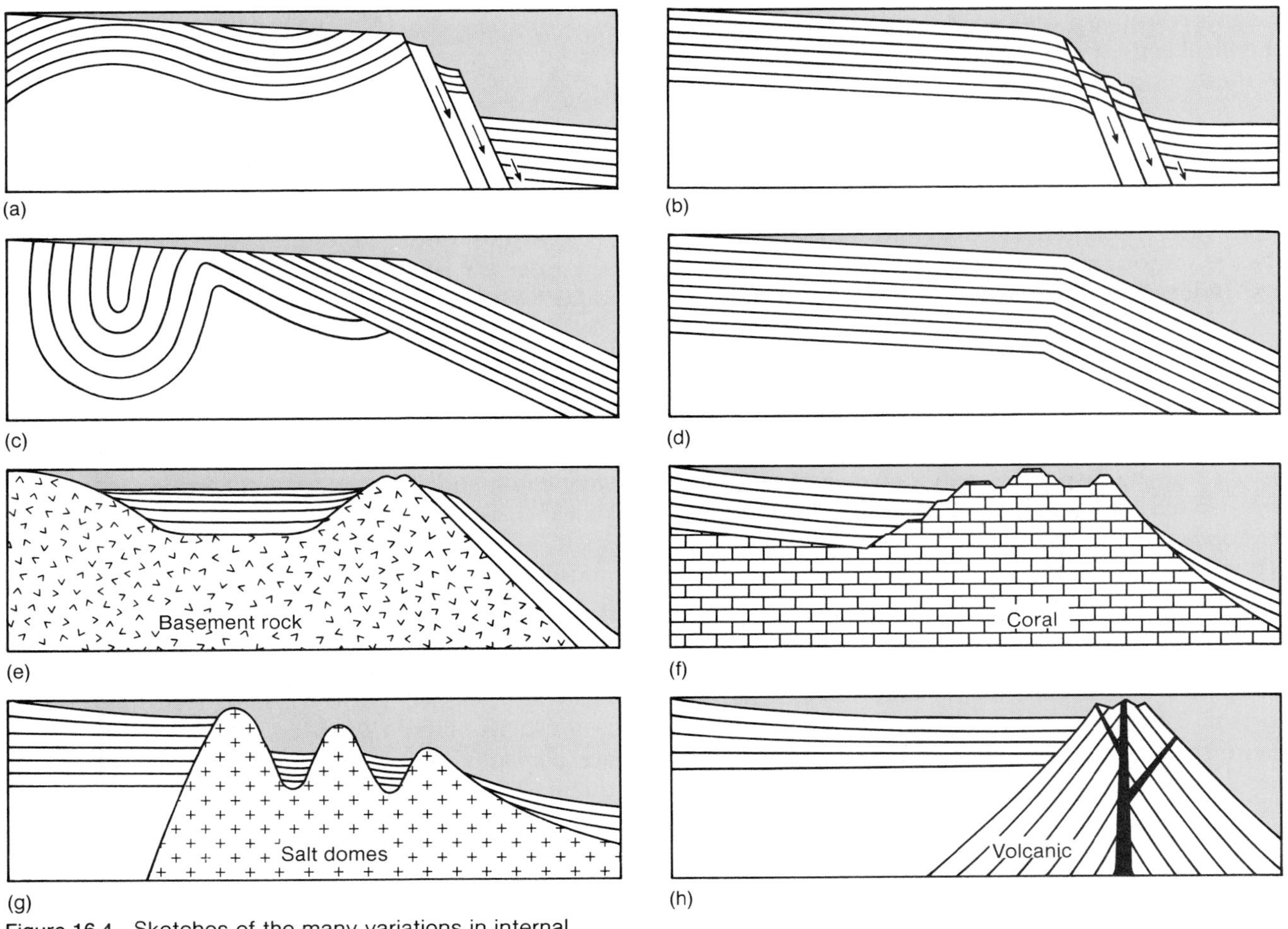

Figure 16.4 Sketches of the many variations in internal structure of the continental terrace.

basin causes formation of a continental terrace (fig. 16.4e). A similar situation can be caused by a large reef acting as the dam (fig. 16.4f). Some places, especially in the Gulf of Mexico, have salt domes that form the dam and have sediment filling in and around the intruded salt mass (fig. 16.4g). Although not common, volcanic complexes may also form continental terraces in the same manner (fig. 16.4h).

Processes

A variety of physical, chemical, and biological processes act on the continental shelf, with the physical processes having the greatest impact. Because of the continental shelf's location between the continental landmass, with its varied coastal environments, and the deep ocean basin, the continental shelf is influenced by both of these environments. Additionally, the shallow depths, combined with an abundant food supply, result in a diverse and abundant fauna. The chemical activity on the shelf takes a subordinate role to these other processes.

Physical Processes

Waves and tidal currents are the dominant physical processes on the shelf, however, various currents other than those generated by these processes may also be important. The degree to which these physical processes influence the shelf varies widely, so much so that the shelf is commonly considered to be either **wave-dominated** or **tide-dominated**. The term **storm-dominated** is generally applied to the former. Most of the shelves of the world are storm-dominated, with large waves generated by these storms causing suspension and transport of sediment and also having a great impact on the benthic fauna. It is only during the storms that wave motion interacts with the bottom except in the shallow, nearshore zone adjacent to the coast. Under storm conditions waves may disturb the bottom to depths of 100 m. Such storms cause not only waves, but they also generate currents. These currents can be important factors in transporting sediment and nutrients throughout the continental shelf and to other marine provinces.

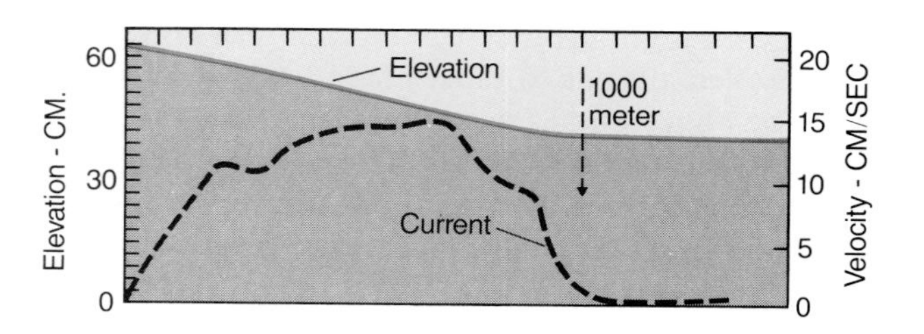

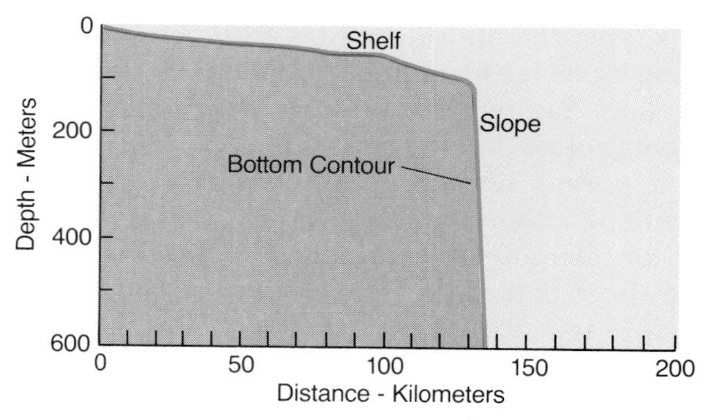

Figure 16.5 Generalized profile of the continental shelf and related tidal currents and elevation of the tidal wave of the New Jersey coast. The maximum velocity of tidal currents is about 15 cm/sec and the tidal wave is only about 15 cm high on the shelf.

On most of the continental shelves of the world, the tides are not a major physical process until the tidal wave reaches the coastal zone. Over the open shelf, tidal currents are typically only about 10–15 cm/sec (fig. 16.5). Such currents have little impact on sediments or organisms on the floor of the shelf. There are, however, some continental shelf areas where tides are the dominant process. Tidal currents may reach speeds in excess of 1 m/sec in extreme cases with most of the sediment transport on such a shelf being generated by tidal currents. The best example of such a tide-dominated shelf is the North Sea between the British Isles and northwestern Europe. While the tremendous storms that peril the North Sea and destroy drilling platforms have made headlines, the work of the tidal currents in this area is not such common knowledge. These tidal currents are continually at work and the storms are infrequent. The currents form and maintain large-scale linear sand bodies on the floor of this tide-dominated shelf (fig. 16.6). These features may be up to 30 m above the surrounding shelf floor, with lengths of several kilometers.

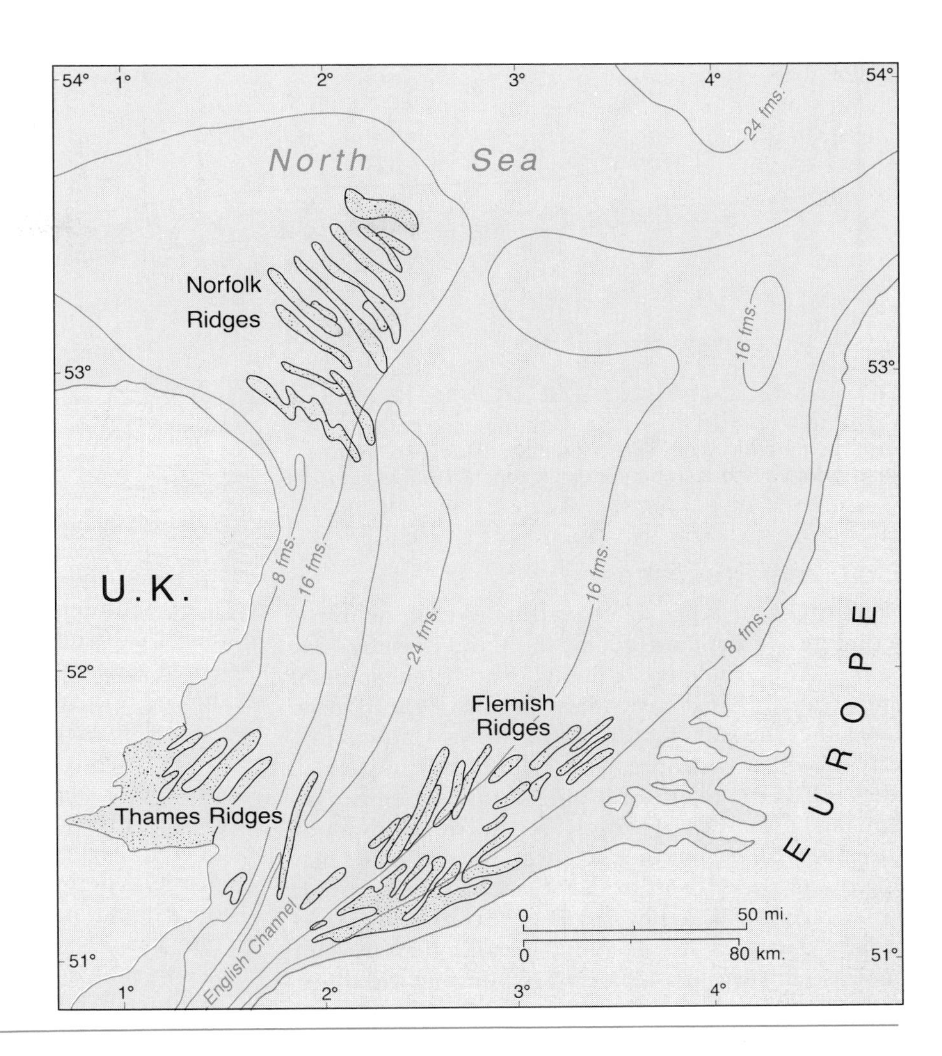

Figure 16.6 General map of the southern portion of the North Sea showing numerous linear sand ridges that are generated by strong tidal currents flowing between Great Britain and Europe.

Various other types of currents are present on the continental shelves. These may be generated by waves or by wind. Wind-generated currents are generally the most common ones on the inner one-half to two-thirds of the shelf. There are also, however, currents caused by density gradients, topography, and waves. These shelf currents are sluggish, with speeds typically in the range of 5–10 cm/sec, although they may reach 30 cm/sec. On the outer shelf near the shelf-slope break, the upwelling process may be a dominating factor, such as is the case along the west coasts of the United States and Peru. Along shelves that are removed from the area of upwelling or that are located on the west side of ocean basins, outer shelf circulation is controlled by the major circulation cells of the open ocean.

A variety of other physical processes may be present on the continental shelf, such as internal waves, tsunamis, and ice sheets or icebergs that drag over the bottom. These phenomena are restricted in their distribution and/or their occurrence but may be important locally. Internal waves may intersect the shelf and cause sediment to be placed in suspension well beyond the wave base of storms. Tsunamis are extremely long period waves that cause sediment motion throughout the entire continental shelf. Ice sheets from glaciers that move out onto the shelf may act as large plows (fig. 16.7), excavating and rearranging great quantities of sediment.

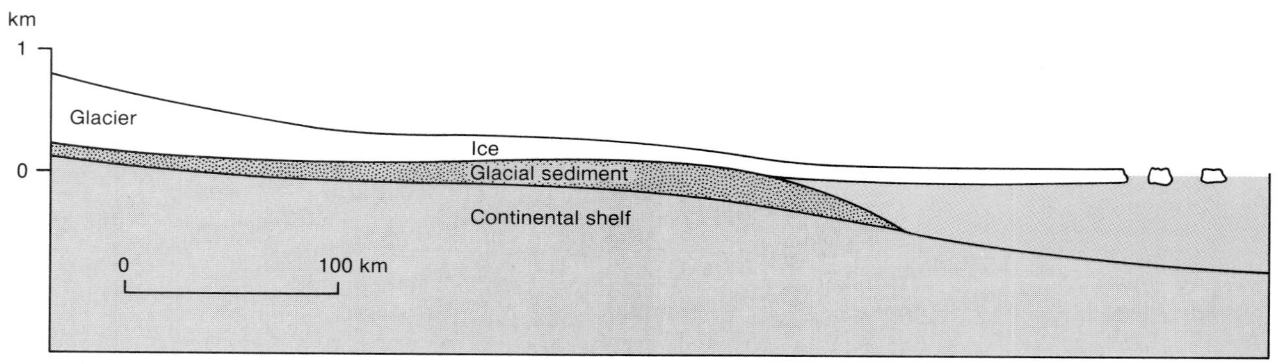

Figure 16.7 Sketch showing a grounded ice sheet on the inner continental shelf. These ice masses act like large road graders and also provide sediment when melting occurs.

Biological Processes

Many different types of organisms live on or in the sediments of the continental shelf and thereby have an impact on the environment. In addition, some of the organisms that float or swim above the bottom also affect the substrate in various ways. Unquestionably the most widespread and significant impact on the shelf is the result of bioturbation of benthic organisms, nearly all of which are invertebrates. Such animals as clams, worms, sea urchins, and snails may burrow in order to protect themselves and provide a place from which they can filter seawater to extract minute particles of suspended organic matter. They may also burrow as a means of facilitating the ingestion of sediment from which they extract organic debris. Both of these types of burrowing organisms are infaunal benthos. There are also organisms, such as some snails, worms, and crustaceans, that graze on the surface in search of food and although they do not process as much sediment as the infaunal grazers do, these animals move sediment, produce pellets, and typically leave tracks or trails.

All of these organisms take in sediment and pass the inorganic portion through their systems, producing fecal pellets. It has been estimated that all of the sediment on a muddy shelf is passed through organisms at least once each year. Some fecal pellets disaggregate easily and rapidly after formation, but others remain as pellets indefinitely. On most of the shelf areas where mud is abundant, pellets comprise a significant percentage of the total sediment present. In some low-latitude areas where the continental shelf sediment is dominantly carbonate mud, pellets may be the dominant constituent. Such is the case in some areas off the Yucatan Peninsula of Mexico and on the northwestern shelf of Australia.

In addition to bottom-dwelling organisms, some of the planktonic and nektonic animals also are pellet producers, although the volume and durability of the pellets produced by these organisms are low. Some of the herbivorous and carnivorous zooplankton produce pellets, as do many types of fish. Generally, the fish pellets disaggregate almost immediately after they are formed.

In the process of consuming sediment or other organisms, animals break up skeletal material and produce sediment. Although the highest rate of occurrence of this activity is in the reef environment (see chapter 15), it also occurs on the shelf and in most other marine environments. Similarly, organisms scrape material on rocky substrates or on shells and thereby cause bioerosion.

The activities of some organisms may interact with the shelf floor. Large organisms or large numbers of organisms, such as schools of fish, can cause the suspension of much sediment, which is then transported by waves or currents.

Chemical Processes

It is hard to perceive the chemical processes that take place on the shelf. Certainly the most widespread and significant is the precipitation of minerals. Because it is nearly impossible to achieve extremely high salinities on the continental shelf, evaporite minerals

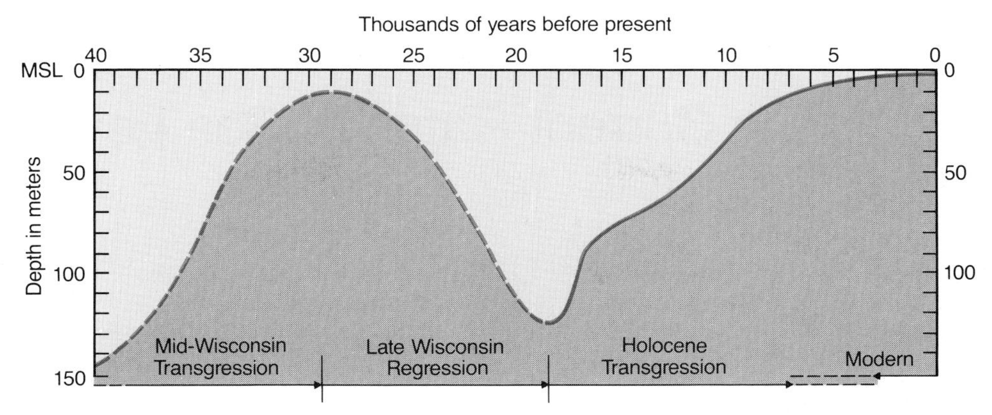

Figure 16.8 A simplified sea level curve representing the past 40,000 years. Notice that the Holocene rise in sea level shows distinct breaks in the rate of rise, with each segment becoming progressively slower.

do not form in this environment; however, they may form in enclosed, shelf-like basins. Probably the best example of this is the Gulf of Kara Bogaz in the Caspian Sea (which is really a lake connected to the marine environment), but less abundant occurrences also are present in the Gulf of California, the Persian Gulf, and the Red Sea. Carbonate minerals also may precipitate directly from seawater, although the volume that forms in this manner is small on most continental shelves.

Other direct precipitates are the phosphate deposits that occur adjacent to the areas of upwelling along the outer shelf near the western side of continental blocks. Similar material also is present on the outer shelf off of North Carolina.

Sediments

The continental shelf contains a sediment suite that is extremely diverse in its age, thickness, distribution, origin, and rate of accumulation. Much of this variety can be attributed to the **Holocene transgression** caused by melting of the last ice age glaciers. In order to understand the nature of sediment distribution on the continental shelf it is necessary to have a general background about this major period of sea level change (fig. 16.8).

Holocene Transgression

This rise in sea level began about 18,000 years ago and is continuing today. The total rise relative to the continents varies throughout the world. For North America it is about 130 m, although various researchers believe the change to be more or less. That means that during the low stand of sea level prior to the Holocene transgression, most of the present continental shelves of the world were exposed above sea level.

From the start of the sea level rise to about 7,000–8,000 years ago, sea level rose very fast, an average of almost 1 cm/year. During that period of time the total rise in level was over 100 m. Sea level was at a position about 10 m below its present level about 7,000 B.P., and from that time to the present the rate of rise was a little more than 1 mm/year (fig. 16.8). Sea level rise also slowed about 3,000–4,000 years ago, with disagreement about what has happened to sea level since that time. There are basically three theories: (1) sea level reached essentially its present position at 3,000–4,000 years ago; (2) sea level has moved above and below present sea level since that time; and (3) sea level has gradually risen from 3,000–4,000 B.P. to the present. Recent studies have shown that over the past several decades there has been a marked increase in the rate of sea level rise with some estimates of near 1 cm/year. Although there is not universal agreement about why the earth's climate is warming, it is the cause of increased melting of ice sheets and the resulting increase in sea level rise.

BOX 16.1

Some Ideas on Current Rates of Sea Level Rise

For the past several years researchers have been devoting much effort to determining and trying to understand the present rate of sea level rise. They are also trying to predict what the rate of change will be over the next century or so. Much of this effort has been undertaken in an extensive study by the United States Environmental Protection Agency (EPA).

Since civilization began, people knew that sea level was changing. In the eastern Mediterranean area much of the change is due to the fact that this is an area that is tectonically active, especially volcanically. The worldwide or eustatic changes in sea level tend to be somewhat slower and therefore more difficult to recognize and measure. They result primarily from melting (or expansion) of the large ice sheets and expansion of ocean water. Such changes in the ice sheets are caused by climatic changes. These changes in climate need not be severe but can be only on the order of a degree or two change in the mean annual temperature.

The primary culprit in the elevation of the mean annual temperature over the past decades is something known as the **greenhouse effect**. This is the containing of heat within the earth's atmosphere because it is being trapped and not allowed to escape. The analogy is with the greenhouse or the familiar situation of heat trapped in a closed automobile during the summertime. Since the industrial revolution there has been a great increase in the amount of carbon dioxide released into the atmosphere, especially by the automobile. This carbon dioxide acts as insulation in the upper atmosphere and keeps heat trapped near the earth. The result has been an increase in mean annual temperature,

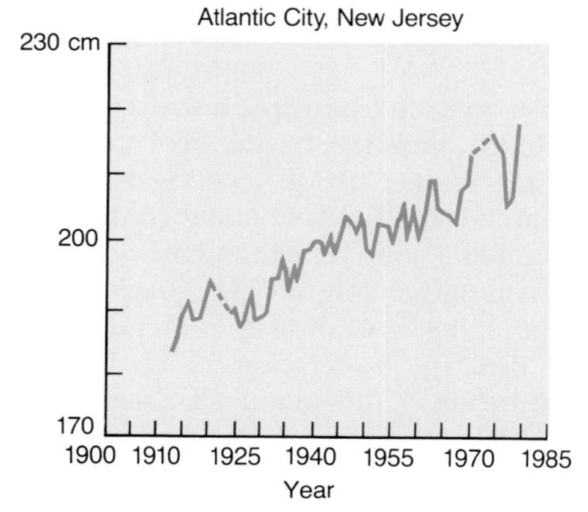

Figure B16.1 Records of tide gauges for the past several decades at various locations in the United States showing the general increase in sea level.

Sediment Types

The wide variety of sediments on the continental shelf includes terrigenous, authigenic, carbonate, volcanic, residual, and relict sediments. On a worldwide basis, the detrital sediments derived from land are most abundant, although some shelves are dominated by carbonate sediments. Volcanic, authigenic, and residual sediments are minor constituents of most shelves. **Relict sediments** actually comprise the bulk of surface sediments on the continental shelves of the world. These are a special type because this category includes sediments of all origins and compositions. These relict sediments were deposited on the shelf, generally in coastal or nearshore environments, as the rapid sea level rise occurred during the first portion of the Holocene (fig. 16.8) or in some cases, during the lowering of sea level that preceded the transgression. These sediments are considered relict because although they are on the shelf surface, they were deposited in an environment much different than the one that they now occupy: these sediments are out of equilibrium with their present environment. Examples of relict

BOX 16.1

increased melting of the ice sheets, and an increase in the rate of sea level rise.

Tide gauges have been installed at various locations around the country for several decades. Records from these gauges show a marked increase in sea level over this period of time (fig. B16.1), much greater than the average for the past 3,000–4,000 years. The present rate of sea level rise is about 3 mm/year or about 30 cm (1 ft) per century. This rate may continue or under extreme conditions, it could increase to 1 cm/year. Such an increase would be disastrous to most of the Atlantic and Gulf coastal areas of the United States and similar coastal areas of the world.

The greenhouse effect makes it apparent that pollution is not only directly harmful for humans and most other living things, but it can also cause other types of changes that have worldwide impact.

Miami Beach, Florida
130
100 cm
1900 1910 1925 1940 1955 1970 1985
Year

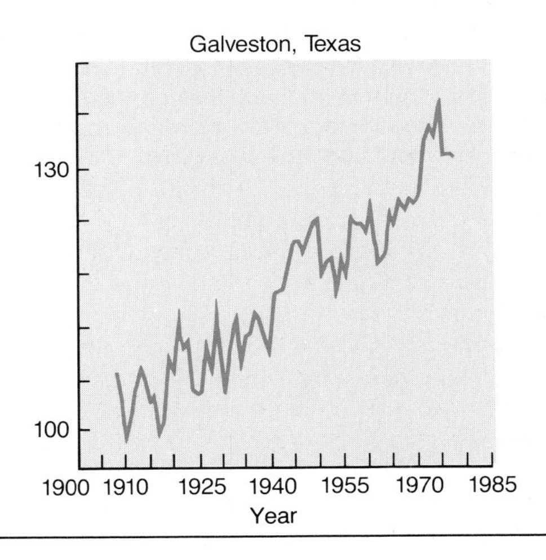

sediments are oysters, oolites, mastodon bones, coarse sand and gravel, peat, and other shallow, nearshore varieties that were deposited when the shoreline was out on what is presently the continental shelf. Additionally, the morphology of the area where relict sediments occur may contain old beaches and coastal ridges.

Detrital sediment presently being derived from the land and supplied to the shelf consists of nearly any mineralogy possible, but is dominated by quartz and clay minerals, with lesser amounts of feldspar and rock fragments. Although much sediment is trapped in deltas and estuaries, a significant amount also reaches the shelf.

Authigenic sediments are those that precipitate directly from the seawater. They comprise only a small percentage of shelf sediments and include primarily phosphorite, glauconite, and carbonates. Volcanic sediments vary greatly in abundance depending on the availability of a source.

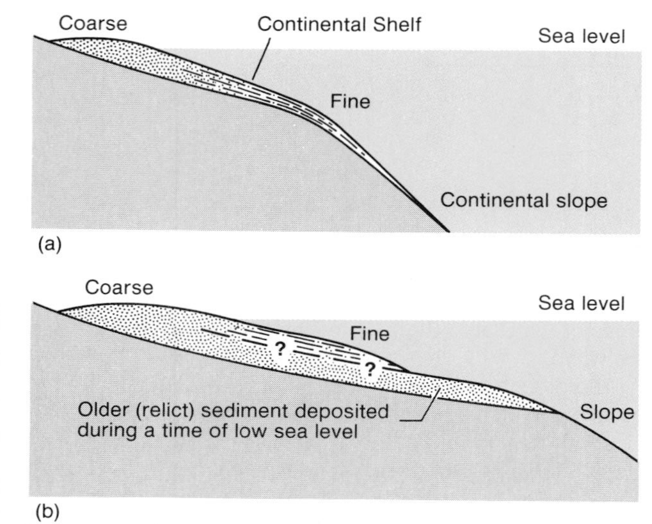

Figure 16.9 A general diagram showing the distribution of sediments across the continental shelf. On a few narrow shelves (a) the trend is from coarse near the shore to fine on the outer shelf, but (b) on most shelves the modern sediments only extend over part of the shelf and old, coarse, relict sediments are at the surface.

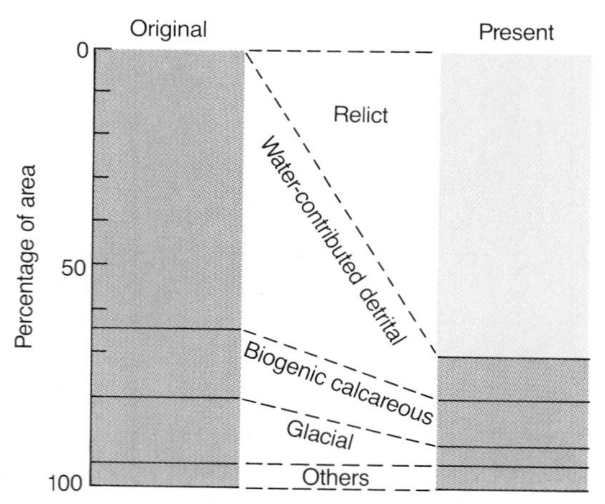

Figure 16.10 A general graphic depiction of the composition of continental shelf sediments showing original sediments deposited in this environment and the present sediments that occupy the shelf. About two-thirds of the present shelf sediments are relict from earlier periods of deposition.

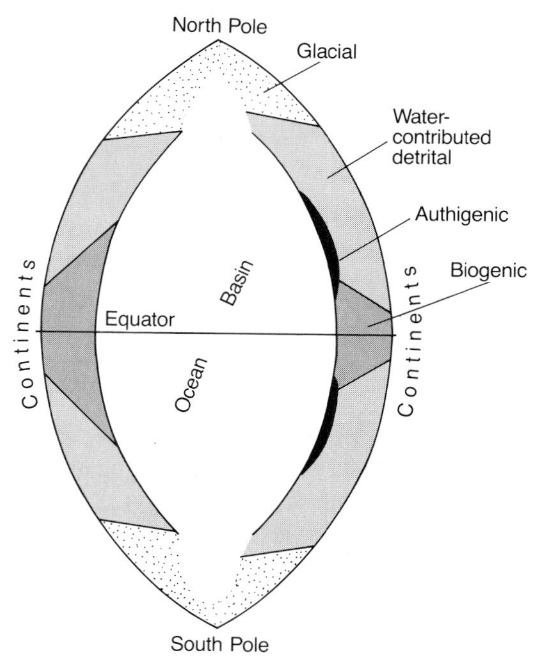

Figure 16.11 An idealized diagram of an ocean basin showing general depositional patterns on the continental shelf. The high latitudes are characterized by glacial sediments, the low latitudes by biogenic sediments (largely coral reef material), and the mid-latitudes are characterized by detrital sediments from land. Notice that the biogenic sediments extend further on the west side than on the east, due to warm currents, and that authigenic sediments (phosphorite) are common on the east side of the basin, due to upwelling.

Distribution of Sediments

Until about the time of World War II, when our knowledge of the ocean floor and particularly the shelf increased greatly, it was thought that the shelf was covered with a blanket of sand and mud derived from land, displaying a general decrease in grain size across the shelf, from sand at the coast to mud out on the outer portion of the shelf (fig. 16.9). Sampling has shown that the relict sediments are widespread across the outer shelf. Because relict sediments were deposited at or near the shoreline, they are coarser than shelf sediments and do not conform to this pattern. About two-thirds of the present continental shelves of the world are covered with relict sediments (fig. 16.10). A comparison of the originally deposited sediments with those on the

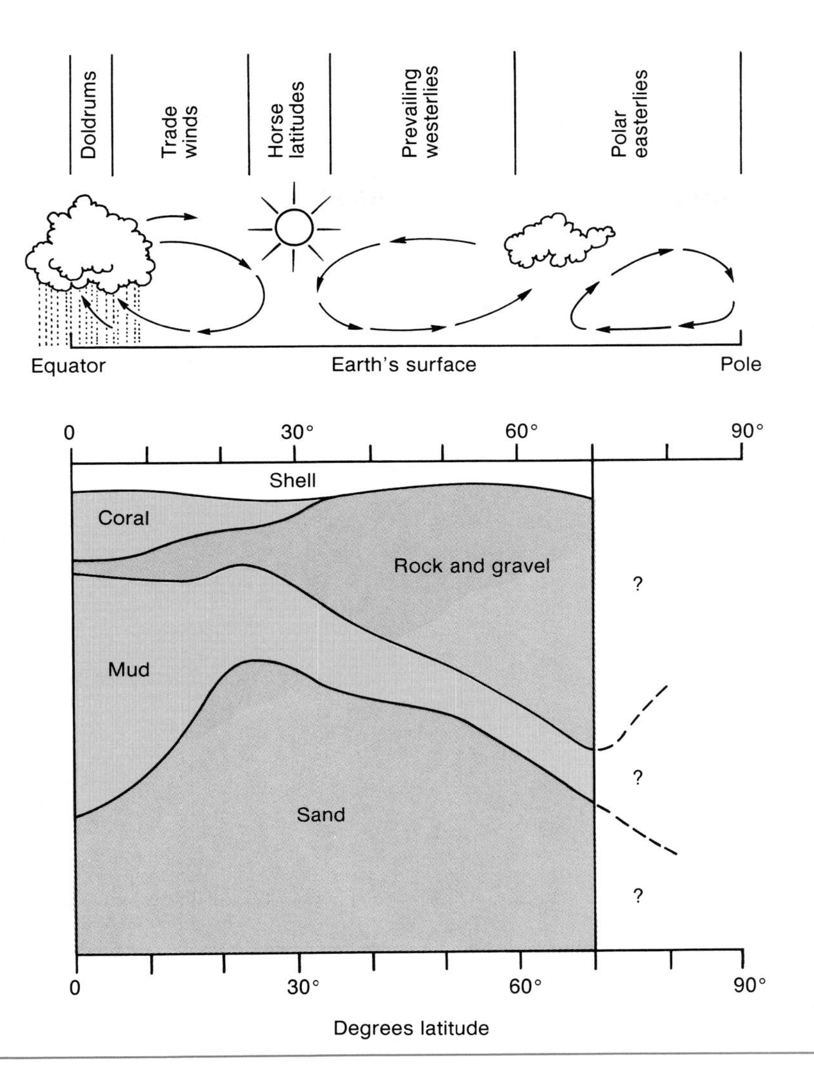

Figure 16.12 General diagram of how shelf sediments are related to latitude. The highest latitudes are not included because of absence of sufficient data.

present shelf shows that the vast majority of relict sediments are detrital. Biogenic sediments are about the same and glacial sediments were originally more common than now because of the effect of glaciers prior to the Holocene.

Although there is great variation in shelf sediments, there are some distinct and important patterns that can be seen within ocean basins. These patterns of sediment distribution are greatly affected by their latitudinal position on the globe. An idealized schematic of an ocean basin with its related shelf complex shows that high latitudes are dominated by glacial sediments (fig. 16.11). Biogenic sediments are adjacent to the equator, with greater distribution on the west side of the ocean basin due to the warming effects of major circulation systems (fig. 5.14). Small bands of authigenic minerals are present on the outer shelf of the east side of the ocean basin in the mid-latitudes (fig. 16.11). These represent the phosphorite that forms in the areas of upwelling. The remainder of the shelf is detrital sediment from land.

A summary of the inner shelf, or modern sediments, by latitude also shows some important generalizations (fig. 16.12). Note that the patterns stop at the high latitudes due to a lack of sufficient data. Biogenic (shell) material is uniform throughout, but coral shows a significantly higher value in the low latitudes, due to the restricted distribution of coral reefs. Rock and gravel sediments increase greatly in abundance from about 30° to the high latitudes thus reflecting glacial sediments on the shelf. Mud shows almost the reverse. It is dominant in the equatorial latitudes and decreases markedly to about 20° (fig. 16.12). The reason for this distribution is the extensive chemical weathering of rocks in the tropical climates and the large drainage systems that are present to carry these weathering products to the sea.

Carbonate Shelf Sediments

Some shelves are dominated by carbonate sediments and others are only a few percent, but carbonate sediments are present throughout the world. They are the only significantly abundant shelf sediment that is produced on the shelf itself. Nearly all of this carbonate material is of a biogenic origin, although some of a direct physicochemical origin may be present locally. Mollusks comprise the largest group of contributors, with coral, echinoids, and foraminifera also being important. The latter may be benthic or planktonic, with benthic types most common in the inner shelf and planktonic varieties dominating on the outer shelf. Barnacles and coralline algae may also be present.

Overall, the inner shelf contains less biogenic carbonate than the outer shelf. This is not because of a difference in abundance of organisms, but is because the inner shelf carbonates are greatly diluted by the relatively large amount of detrital sediment that is introduced. Currents on the outer shelf also prevent some fine detrital sediment from settling out and these currents further enhance the abundance of carbonate detritus. An example of carbonate distribution from the North Carolina shelf shows mollusks dominating the inner shelf and pelagic foraminifera dominating the outer shelf (fig. 16.13). Various other varieties are scattered throughout the shelf.

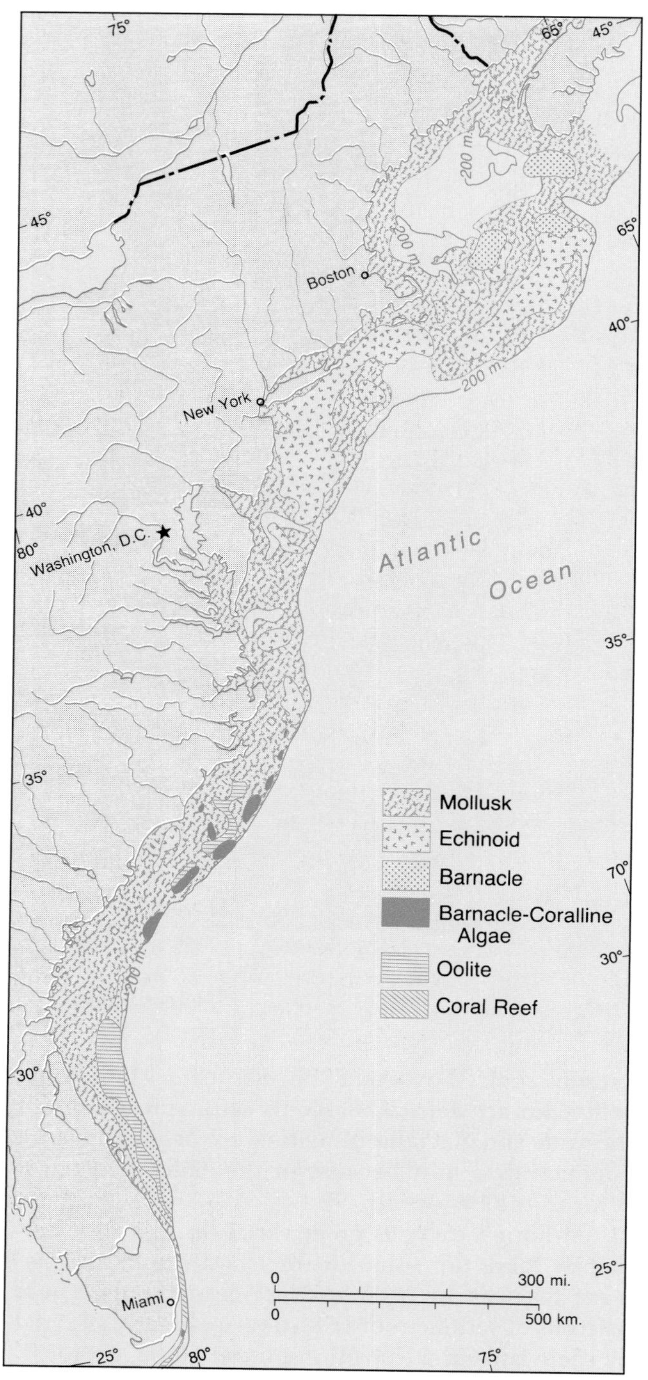

Figure 16.13 Map of shelf carbonate sediments along the entire Atlantic shelf of the United States.

Environmental Conditions

The proximity of land and the shallow depths make the continental shelf environment variable in terms of the conditions that support or limit life. Chemically the shelf is nearly ideal for supporting organisms in that there is good circulation to provide oxygen and to transport nutrient materials and organic debris. The salinity is generally constant at normal marine levels (35‰), however, it may be reduced near areas of high freshwater discharge, such as deltas and estuaries. The temperature is also fairly uniform, except that in the mid-latitudes there are seasonal changes. Light penetration is generally good but is less than the open ocean due to the influence of runoff of suspended sediment. As a result, there is typically a significant reduction in light penetration toward the shore.

Wave activity is one of the most important environmental factors in terms of its impact on marine life. Firstly, waves cause suspension of sediment, which can inhibit light penetration and therefore, photosynthesis. The wave climate in a particular area can also have a marked influence on the type of sediment that accumulates on the bottom. For example, in a high-energy area, much of the fine sediment will be winnowed away by the waves and the bottom will be dominated by sand. A low-energy wave climate tends to favor the opposite situation. Areas where unconsolidated sediment is limited and where waves are large are generally characterized by a bedrock shelf floor, with the sediment swept away. Any of the above scenarios has a marked effect on the nature of the bottom community that is present.

Another effect of large waves is to keep the unconsolidated bottom in a state of flux. This is a limiting factor to many benthic organisms, especially those that are infaunal. Organisms such as worms, clams, or snails that live just below the sediment surface are continually being uncovered by waves. Much suspended sediment caused by wave action is also a problem to filter-feeding organisms. For example, the same worms and clams mentioned previously rely upon particles of organic debris suspended in the water for their food supply. If there is also much sediment in suspension, then the filtering organs of these animals become clogged with sediment and the organisms expire.

Organisms

The continental shelf supports diverse fauna and flora. It is the most populated of the open marine environments except for coral reefs. The combination of abundant food supply, shallow water, good circulation, and diverse habitats results in a broad spectrum of benthic, nektonic, and planktonic organisms. Food supply is great because of proximity to the land from which much of the nutrient material comes, and also from recycling of organic debris and nutrient material within the marine environment. Currents carry nutrients from the coast out onto the shelf and the shelf organisms themselves provide nutrients when they expire and their tissues decay.

Flora

The photosynthetic organisms of the shelf environment include phytoplankton, benthic algae, and sea grasses. Typically phytoplankton are very abundant except in areas where there is much suspended sediment in the water column as the result of high discharge from deltas or estuaries. In some areas this limiting factor may be seasonal due to high discharges associated with rainy seasons or spring meltwater runoff.

Phytoplankton in the shelf environment show abundances that vary with latitude and with the seasons. This variation is largely in response to sun angle, day length, upwelling, and other factors that affect productivity. In clear tropical waters these organisms may occur throughout the water column above the shelf because of deep light penetration. In some mid-latitude areas there are periodic blooms with tremendous concentrations of phytoplankton. This phenomenon is due to a combination of ideal temperature, high nutrient levels, and good light penetration.

Sea grasses are few in number of species but in some shallow mid- and low-latitude areas they may form a dense carpet on the shelf floor. Sea grasses may extend to depths of 50 m but typically they are in depths less than one-half of that. They occur on all types of substrates but are most common in soft muds and sands. In addition to acting as a primary producer in the food chain, sea grasses also are efficient stabilizers of the sea bottom. The blades act

(a)

(b)

Figure 16.14 Underwater photographs of (a) eel grass (*Zostera*) and (b) turtle grass (*Thalassia*), the most common grasses on the shallow shelf.

Figure 16.15 Underwater photograph of shallow marine environment with calcareous algae (*Penecillus*) on a rippled surface.

to protect bottom sediment from erosion by waves and currents by preventing or dampening currents that may transport the sediment. They also act as traps for sediment that is carried through grassy areas.

There are only a few abundant and widespread types of sea grasses. *Thalassia* (turtle grass) is one that is dominant in the low latitudes and *Zostera* (eel grass) is abundant in temperate waters (fig. 16.14). Sea grass also hosts numerous encrusting and vagrant invertebrate benthos that may feed on the blades or simply use them for shelter or a substrate on which to attach.

There are also numerous benthic algae that live in the shallow marine environment. These include brown, green, and red algae, some of which are calcareous. The calcareous algae include greens (*Halimeda, Penecillus*) and coralline reds (*Goniolithon*). They occupy essentially the same niche as sea grasses (fig. 16.15) but are found in the low-latitude shelf areas.

Fauna

A great variety of planktonic, nektonic, and benthic animals live on the continental shelf environment. Much of our commercial fishery catch comes from this area. The abundance of food and the almost uniformly optimal chemical and physical environment make this an area of abundant and diverse life.

Pelagic Animals

Included in this group are the zooplankton and the nekton. Currents dictate the movement of the zooplankton and as a result there are seasonal patterns and geographic variations to their distribution. This group of organisms serves as a fundamental link in the food chain of the shelf environment. Areas where phytoplankton are abundant will also support a large zooplankton population because of the dependence of the latter on the former as a food supply. Consequently the shelf areas where light penetration, nutrient supply, and physical parameters are favorable will also be sites of abundant zooplankton.

BOX 16.2

Box Coring on the Continental Shelf

Research on the continental shelf substrate, as well as that on most other marine environments, requires that samples be taken. A major area of marine research has been the study of the interrelationships between benthic organisms, especially the infaunal organisms, and the substrate. The typical sediment sample is taken from a core of a few centimeters diameter or from a grab sampler, which is a type of clamshell sampling apparatus much like a bucket on a construction dragline. Generally, biological samples are taken by a dredge or a trawl that is dragged along the shelf floor. These techniques result in either a small, undisturbed sample (core) or one that is large, but disturbed (all the others).

A sampling device called a box corer has been developed and is used extensively to solve these problems. The corer consists of a frame that rests on the substrate. It contains a rectangular core that is about 0.1 m in area and that penetrates at least 30 cm. The apparatus is lowered over the side of the ship (fig. B16.2a) and penetrates the bottom. As retrieval takes place, the spade lever moves into the sediment and acts as a base to the box, which has penetrated the substrate (fig. B16.2b). In this manner a relatively large and undisturbed volume of shelf substrate can be collected, including not only the sediment but also its contained organisms.

The block of sediment is then prepared for sampling and study. Typically it is also x-rayed to examine the layering of sediment and burrows that may be present (fig. B16.3).

(a)

(b)

Figure B16.2 Photographs of (a) a box corer, cocked and being lowered to the floor of the shelf, and (b) the corer being retreived, along with its contained sample.

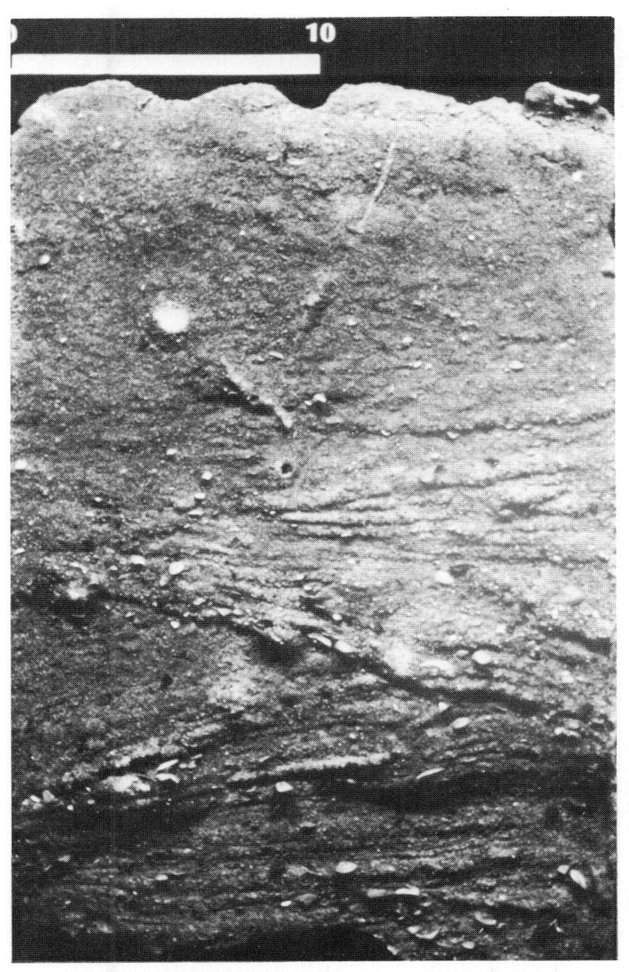

Figure B16.3 Photograph of a box corer sample from the shelf.

There are some shelf organisms that are on the borderline between plankton and nekton. That is, they have swimming abilities, but under intense current or wave energy conditions, they are not able to overcome the water motion. Among these are some important economic species such as the shrimp. These animals are one of the most valuable resources of the shelf environment. Like most invertebrates, shrimp have planktonic larvae. They spawn in late winter and early spring in the mid-latitudes. These organisms reach maturity within the first year and have a life span of only two or three years. This is ideal for the shrimping industry, which is concentrated along the Atlantic and Gulf coasts of the United States. Other broad mid-latitude shelves also host large shrimp populations.

Fish comprise the bulk of the pelagic nekton on the continental shelf, although squid, sea turtles, porpoises, and some types of whales are also in this diverse group. The great majority of commercial fishing in the world is on the shelf and this is one of the reasons that many countries have extended their territorial jurisdictions to a 200-mile limit (further discussion on this topic is in chapter 20). The shelf supports great populations of desirable fish, they are readily accessible, and the areas are fairly proximal to ports. Unfortunately, many of the productive shelf areas are being fished at or above their maximum so that future increase in the world catch is unlikely. In fact, there is serious concern about overfishing and the populations may have already begun to diminish.

Benthic Animals

Many thousands of species of invertebrates including mollusks, echinoderms, arthropods, coelenterates, and polychaete worms comprise the benthic fauna of the continental shelf. They occur as both infaunal and epifaunal varieties and numerous examples of vagrant and sessile forms occupy this environment. In short, there is nearly every type of invertebrate on the shelf. The discussion that follows is organized around the nature of the substrate and the life habits of its occupants.

Hard, rocky bottoms occur on the continental shelf, although this is a minor percentage of the total. A few borers, such as sponges or boring clams, may live on such a surface, but the dominant animals are the epifaunal types. There are vagrant scavengers such as crabs, lobsters, sea urchins, and snails. The most abundant are the sessile types, such as barnacles, sea fans, and sea lilies. Although some food is available for bottom detritus feeders, the filter feeders tend to do better in such an environment. This is because the absence of unconsolidated sediment is generally an indication of high-energy wave or current conditions, which also implies that bottom organic detritus will not be available as a food supply. If sediment is prevented from accumulating on the shelf, then the same holds true for organic detritus. The only abundant food source is from suspended material.

The organisms that live on and in unconsolidated substrates of the continental shelf are diverse and numerous. Epifaunal organisms include both filter feeders and deposit feeders. These organisms either graze over the sand-and-mud bottom or they are sessile and rely on filtering particulate organic matter from the water. Mollusks, crustaceans, echinoderms, and foraminifera are among the most abundant. Generally there is an abundance of food and other environmental factors are favorable, so that the most important limiting factors tend to be the nature of the substrate and the presence of predators.

Infaunal organisms have different problems with the substrate than do epifaunal organisms. In most cases the organisms burrow for a means of protection from predators and from physical processes such as strong currents or wave action. Bivalves, polychaetes, crustaceans, and snails have many infaunal sessile species. These filter-feeding organisms rely on currents to carry particles of organic detritus and nutrient elements to them in order to sustain themselves.

Other infaunal animals move through the substrate ingesting sediment and extracting the nutrients and organic detritus from it. These organisms move like plows or bulldozers through the sediment, churning it up. Included are numerous varieties of echinoderms, such as some of the sea urchins, starfish, and sand dollars, some bivalves and gastropods, and many types of worms. The distribution of these grazing organisms is also dependent on certain types of substrate. Not only is it important that the sediment contain the appropriate food supply, but it also must be the type of sediment that the animal can burrow into or move through.

Summary of Main Points

The continental shelf is among the most varied of the open marine environments in terms of processes, sediments, organisms, and morphology. There is also modest variation in the chemistry of the shelf, although this aspect is relatively homogenous and predictable away from the influence of land. The shelf receives deep oxygenated and nutrient-rich waters from the outer margin through upwelling.

Although there is a broad range in the characteristics of shelves on a global scale, there are appropriate generalizations that can also be applied. Classifying shelves according to their dominant process serves as a good example of this. Tide-dominated shelves are much less widespread than storm-dominated shelves. There are certain broad generalizations that can be made about each of these types. Those dominated by tides are generally high-energy environments. Rapid currents persist nearly continuously, and as a result, the substrate and the organisms that inhabit it reflect this feature. The substrate is typically free of fine-grained sediment, with sand-and-shell or glacial gravels most abundant. The water is well circulated, thus providing abundant oxygen and also nutrient material and other food for organisms inhabiting this type of shelf. In fact however, tide-dominated shelves do not contain broadly diverse and abundant benthic fauna and flora, although the pelagic communities are well populated. The currents tend to keep much of the fine sediment in suspension, which inhibits light penetration and thereby restricts photosynthesis. This affects the phytoplankton. The benthic photosynthesizers are additionally restricted by the mobility of the bottom. This latter aspect also is a serious problem for infaunal animals and for sessile forms. About the only benthic organisms that can do well in such an environment are the vagrant benthos or those sessile forms that can tolerate suspended sediment.

Wave-dominated shelves predominate around the globe. This is fortunate in many respects, the most important being that such conditions tend to support a larger and more diverse fauna and flora than is the case for tide-dominated shelves. Under normal conditions the wave-dominated shelf is a quiescent environment, except in the inner shelf where there is nearly continuous wave activity and even this varies in both time and space. There generally is adequate circulation from tidal, wind, and density currents to distribute oxygen and food throughout the shelf. The relatively low energy level allows mud to accumulate, as well as sand, and means that suspended sediment concentrations are not high, except near major sources of fines, such as deltas or estuaries. This all leads to a highly productive environment, beginning with light penetration and photosynthesis and continuing to the high biomass of fish, including many that are commercially desirable.

Organisms of the pelagic shelf environment are generally not restricted by light penetration or food supply and benthic organisms have the complete spectrum of bottom conditions from which to choose. Wave-dominated shelves have a broad cross section of sessile and vagrant epifauna and infauna as well as benthic grasses and algae.

This set of conditions is temporarily interrupted during and shortly after a major storm. This may be a tropical storm, such as a hurricane or typhoon, or it may be due to severe winter storms, such as those that occur along the Pacific Northwest or the New England coast of North America. The high physical energy created by storm waves and associated currents can wipe out nearly the entire benthic population. The environment, however, recovers quickly and in a matter of days various species return, with prestorm conditions achieved in a few months in most cases.

Suggestions for Further Reading

Emery, K.O. 1969. "The continental shelves," *Sci. Amer.,* 221:39–52.

Kennett, J.P. 1981. **Marine Geology.** Englewood Cliffs, New Jersey: Prentice-Hall, Inc.

Levinton, J.S. 1982. **Marine Ecology.** Englewood Cliffs, New Jersey: Prentice-Hall Inc.

McConnaughey, B.H., and R. Zottoli. 1983. **Introduction to Marine Biology.** St. Louis: The C.V. Mosby Company.

Shepard, F.P. 1977. **Geological Oceanography: Evolution of Coasts, Continental Margins and the Deep Sea Floor.** New York: Crane, Russak.

17
Outer Continental Margin

Chapter Outline

The outer continental margin consists of the continental slope and the continental rise. It represents the major transition zone between the continental and oceanic portions of the earth's crust. The continental landmasses, coastal environments, and the continental shelves represent distinct geologic and physiographic entities that are continental in nature. These continental blocks contrast markedly with the oceanic portion of the crust in terms of composition, physiography, and overall geology.

The continental slope is the major physiographic boundary zone of the earth's crust in that it connects the shallow shelf with the deep oceanic basins. It is distinctly continental in composition and is the side of the continental block of crust that is floating on the oceanic crust. The continental rise is continental in composition, in that the sediment of which it is comprised was derived from the continental landmass, but it is more like the deep sea in its physiographic character. It gradually merges with the ocean floor at its seaward margin.

In terms of the pelagic environment, all of the outer continental margin is oceanic in nature. The circulation, water mass properties, biota, and other characteristics are all those of the open ocean system.

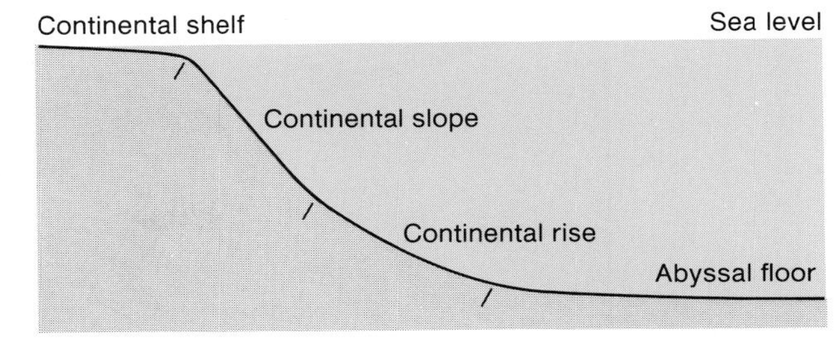

Figure 17.1 General diagram showing continental slope and rise environment relative to other major oceanic provinces.

Continental Slope

The continental slope extends from the shelf break to the continental rise, where the gradient becomes less than 1:40. In some basins the slope is continuous with a deep-sea trench or may terminate with a marginal plateau, such as the Blake Plateau off the southeastern margin of the United States (fig. 17.1). The gradient of the continental slope ranges widely, depending on the geologic nature of the edge of the continental block. The surface of the slope may be smooth, irregular, terraced, or cut by faults and submarine canyons. The width of the slope may be up to 100 km. Water depth is about 100–200 meters at the upper limit and is typically 1,500–3,000 meters at the seaward limit.

For a long time it was thought that the continental slope was featureless except for submarine canyons, and that the slope was a broad zone of sediment bypass that served as a transport environment for sediment from the shelf to the rise and ocean floor. Technological advances, which have permitted penetration of slope deposits by such geophysical techniques as high-resolution seismic profiling and core drilling, have shown the detailed geology of the slope. Many of the details of a given continental slope area are related to the plate-tectonic history of that particular location. It is now apparent that there is great physiographic variety on the continental slope and that some slopes have accumulated thick sequences of sediment.

Origin of the Continental Slope

This relatively steep outer margin of the continental block acquires its character from a variety of geologic situations. Some are the result of tectonic activities, such as faulting or folding, which may be accompanied by extensive erosion (fig. 17.2a). The upbuilding of sediments associated with subsidence and the outbuilding of sediments added to the continent (fig. 17.2b) tend to produce somewhat gentler slopes. The combination of tectonics and sediment accumulation also occurs (fig. 17.2c).

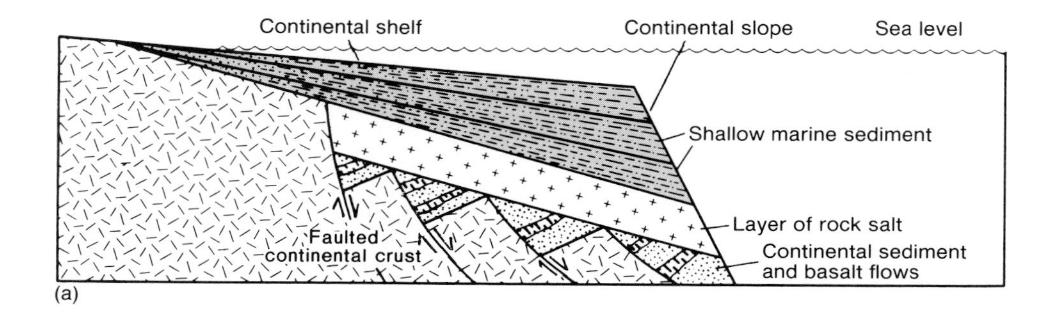

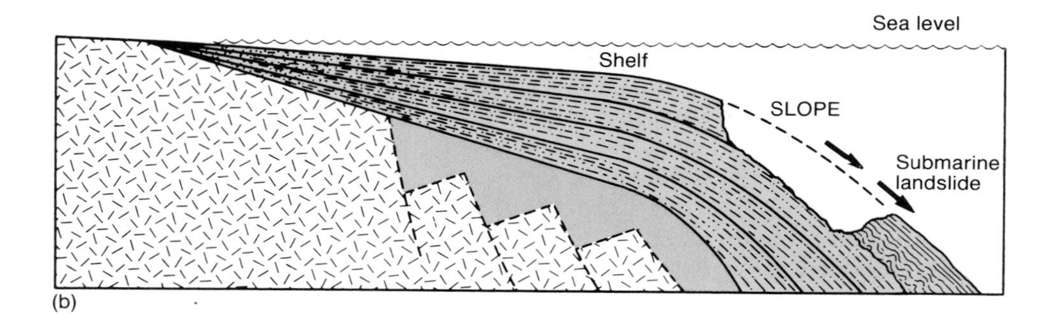

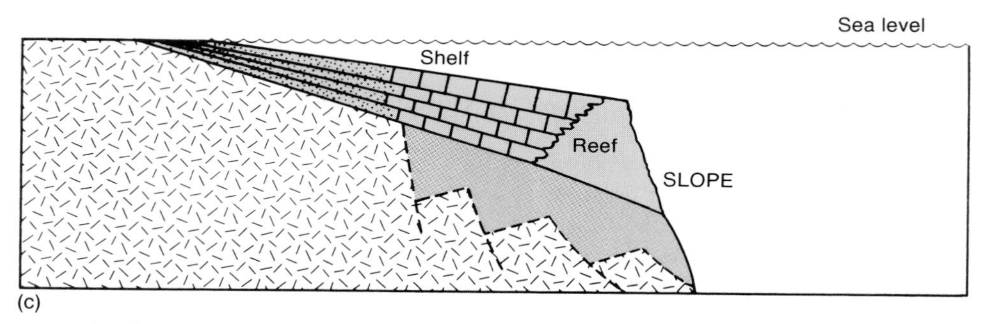

Figure 17.2 Diagrams showing some of the various possible origins for the continental slope.

Large reefs may act as dams and cause sediment to fill in behind them, thus producing the slope. A special type of continental margin called a **continental borderland** consists of considerable faulting, with upthrown blocks acting like dams and producing a stair-stepped slope. This situation is present off southern California (fig. 17.3).

Before the development of deep seismic profiling it was only possible to speculate about the internal structure of the outer margin from a knowledge of the surface configuration. Now that the internal character can be determined, it can be demonstrated that many different subsurface geologic situations can produce a similar profile (fig. 17.2).

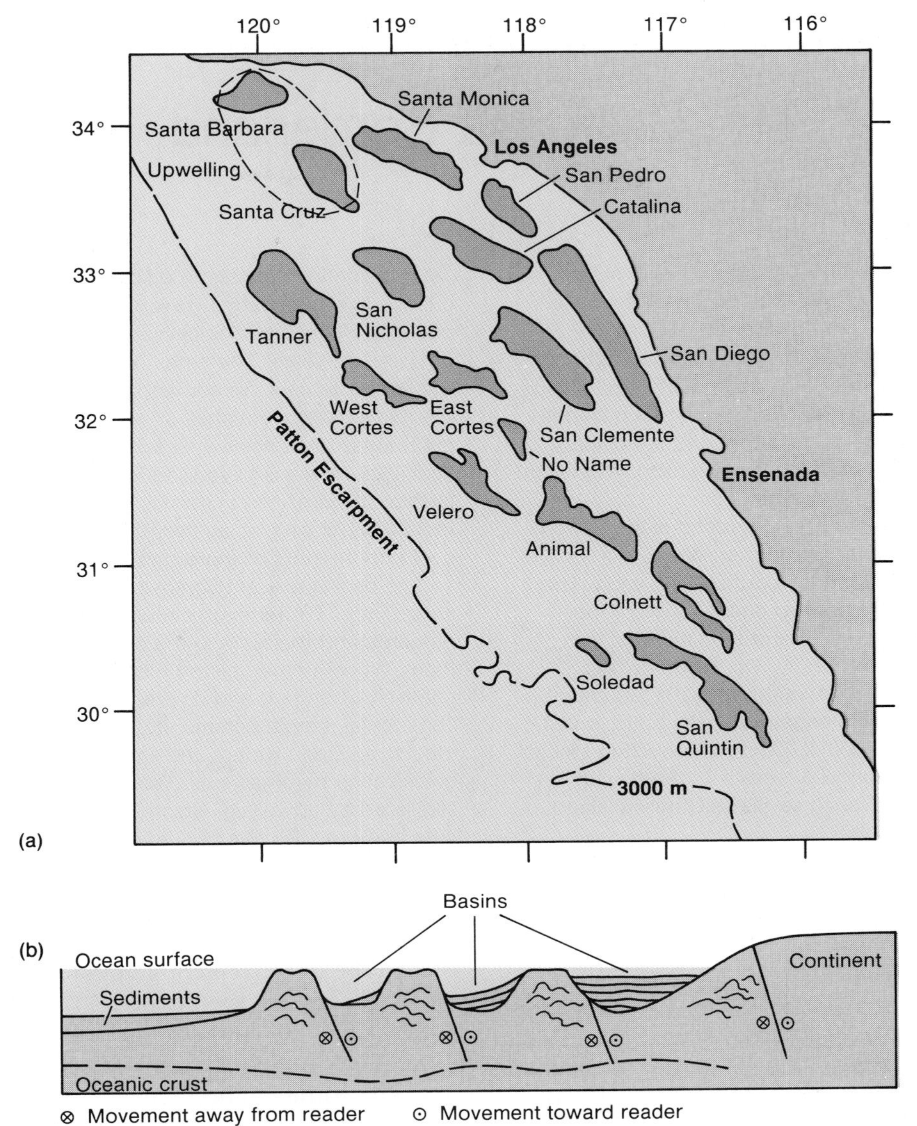

Figure 17.3 (a) Map of basins in the California Borderland and (b) a general cross section of this area showing uplifted and down-dropped blocks of the continental margin.

BOX 17.1

The California Borderland

The continental margin off southern California is unique in North America. The continental shelf is absent or very narrow. The bulk of the continental margin consists of a series of crustal blocks that have moved almost vertically relative to one another, giving a topography of basins and blocks with steep slopes in between. This terrain mimics that on the adjacent landmass and results from the faulting associated with the plate boundaries in this area.

The area of the California Borderland extends from Santa Barbara, California, to northern Mexico, a distance of 1,000 km, and is about 200 km wide. The region is essentially a seaward continuation of the crustal configuration on land and terminates in the continental slope. It has nearly two dozen basins, ranging from 50–200 km in length and 20–100 km in width. They tend to be elongate, parallel to the coast, and have depths of 1,000–2,000 m. The uplifted blocks are less than a few hundred meters in depth, with some forming islands, such as Santa Catalina Island and San Clemente Island.

This continental margin displays considerable relief relative to others, largely due to the fact that the relief originated only a geologically short time ago: about 5–10 million years. There has not been enough time for the basins to have filled in and built the margin up into the relatively smooth shelf and slope that is present in most areas. It is, however, apparent that the basins near the landmass have filled in much more than those further offshore. The system is filling in those closest to the sediment source; as they fill in, sediment spills over into the next offshore basin.

The basins present somewhat isolated environments in terms of circulation of water and nutrients and therefore in the organisms that inhabit them. Whereas the uplifted blocks tend to support a relatively abundant and diverse shallow marine continental margin community, the basins tend to be otherwise. Deep waters and limited circulation inhibit colonization by organisms. This condition is somewhat alleviated by upwelling, which is common along this area and provides a supply of nutrients and well-oxygenated water.

Slope Processes

The presence of large-scale downslope processes attributable to gravity has been known and studied since the early days of marine geology in the 1940s and 1950s. It was not until somewhat later that the presence and importance of processes that move parallel to bathymetric contours, or essentially horizontally, were recognized.

Density Currents

Gravity processes that carry large volumes of sediment down the continental slope are widespread and varied. They include **debris flow**, a mixture of sediment and water, **slump**, where large blocks of sediment move downslope similar to that on land, and **turbidity currents** (fig. 17.4), which are density currents of sediment-laden water that transport large volumes of sediment downslope. A density current is any type of movement generated by a density gradient (for review see chapter 5). In the case of turbidity currents this gradient is caused by the difference in density between a water with suspended sediment and the surrounding water.

There are various mechanisms that may disrupt the seafloor, cause sediment to become suspended in the water column, and thus create a density gradient in the form of one of the gravity processes. Among the most prominent mechanisms are: (1) earthquakes, (2) volcanic eruptions, (3) large influxes of sediment, and (4) large waves interacting with the ocean floor. The resulting gravity processes, such as turbidity currents, erode as they pass over the continental slope and deposit sediment at its base. Much of the continental rise, which is discussed in the following section, is comprised of **turbidites**, those sediments deposited by turbidity currents.

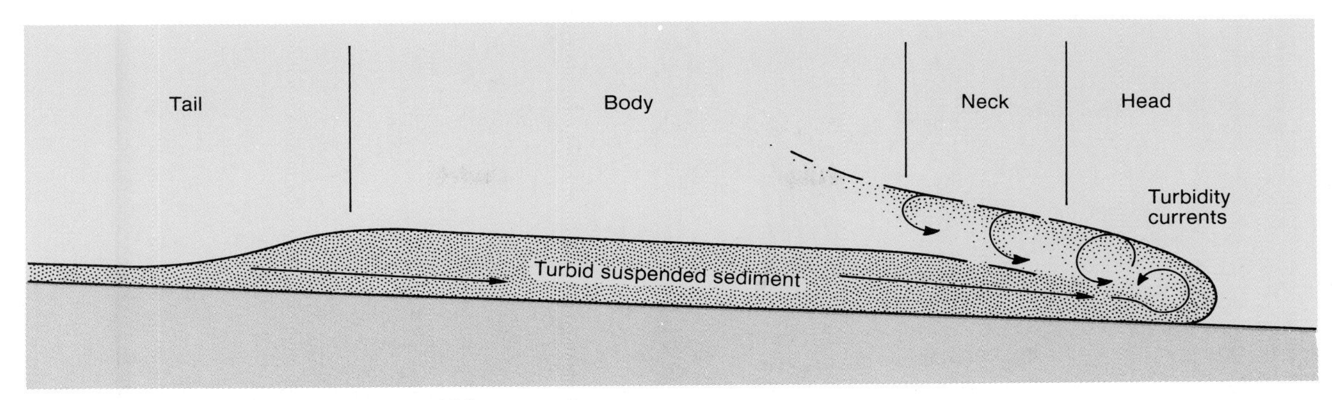

Figure 17.4 Schematic diagram of a turbidity current.

Turbidity Currents and Turbidites

There are multiple mechanisms for transporting terrigenous sediment to the outer continental margin. The most important are turbidity currents, which are a special type of density current caused by suspended sediment. In our earlier discussion of density currents (chapter 5) the density gradient was produced primarily by temperature differences with adjacent water masses generating slow currents among the deep water masses. Turbidity currents are density currents based on the same principles, except that the density gradient is due to suspended sediment that is present in one water mass and not in another. The water containing the suspended sediment has a greater density than the ambient water and thus the gradient is created that produces the current. Turbidity currents are local and move rapidly in comparison to the thermohaline currents of deep water masses.

A good example of a turbidity current is displayed when you drop a rock into a mud puddle or small pond. As the rock hits the bottom a cloud of mud is suspended, generally with well-defined boundaries. During the few seconds following the formation of this mud cloud, it spreads away from the place where the rock hit bottom and formed the cloud. This is a turbidity current. The muddy water has a greater density than the clear, ambient water and this density gradient produces the current. As the current spreads away from its origin, it loses its energy due to gravity and friction, and the suspended sediment settles to the bottom.

The sediment sequences that accumulate from turbidity currents are called turbidites, which are thin sequences that fine upward in grain size. Each of these sequences represents the deposition of a single turbidity current event and produces a predictable sequence of sediment layers. The originally well-mixed sediment-fluid combination settles to the substrate with the coarsest particles on the bottom and a gradual fining of particle size to the top. This grading is combined with particular types of layering to produce the **Bouma sequence**, named after the marine geologist Arnold H. Bouma, who first recognized the turbidite sequence. Each complete sequence contains five recognizable layers (fig. 17.5). 17.5).

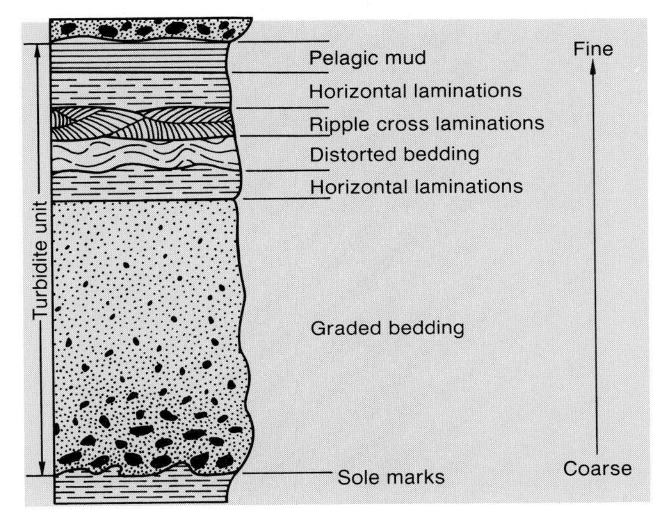

Figure 17.5 Diagram of a turbidite sequence, also called a Bouma sequence. The thickness of this sequence may be from about a centimeter to several meters.

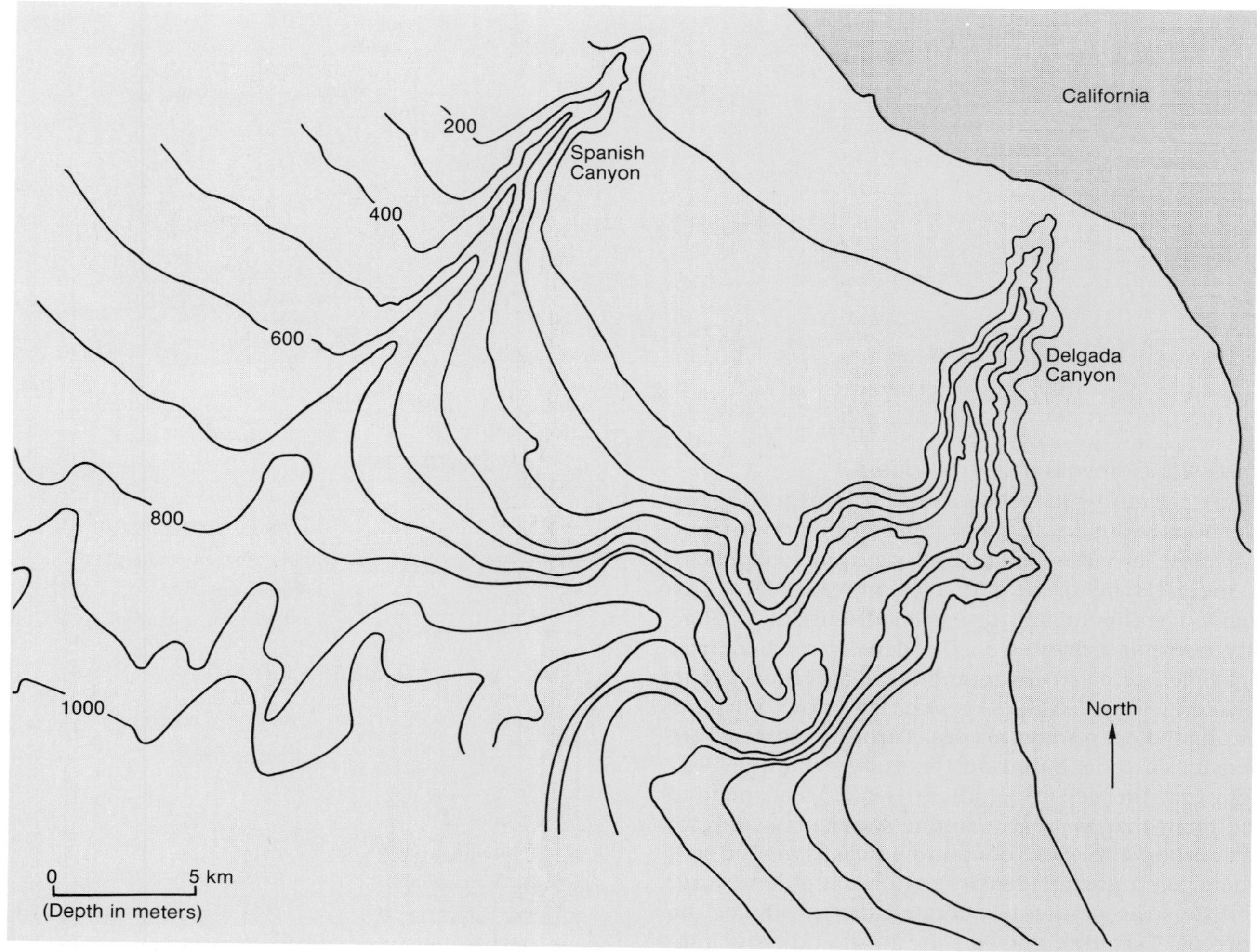

Figure 17.6 Diagram of a submarine canyon on the continental slope. Notice that although the canyon crosses the entire slope, it also extends across part of the shelf and continues into the valley on the continental rise.

The Bouma sequences range widely in thickness, from about a centimeter up to a meter or more. Most are toward the thin end of this range. There is a general relationship between grain size, thickness of the sequence, and position on the submarine fan, with the units fining and thinning toward the distal or far end of the fan. There may also be pelagic sediment incorporated into the Bouma sequence, especially in the top layer.

There is evidence that turbidity currents have been more common in the past than at the present time. For example, when sea level was lower during the glacial advances of the most recent ice age, the shoreline was near the present shelf-slope break. As a result, many large rivers dumped their sediment load at the top of the continental slope. This influx of sediment created density gradients that resulted in turbidity currents, which may have been nearly continuous at some locations.

Submarine Canyons

The greatest relief on the continental margins is at the sites of **submarine canyons**, the large valleys that are cut primarily into the slope, but also extend onto the shelf and the rise (fig. 17.6). These features

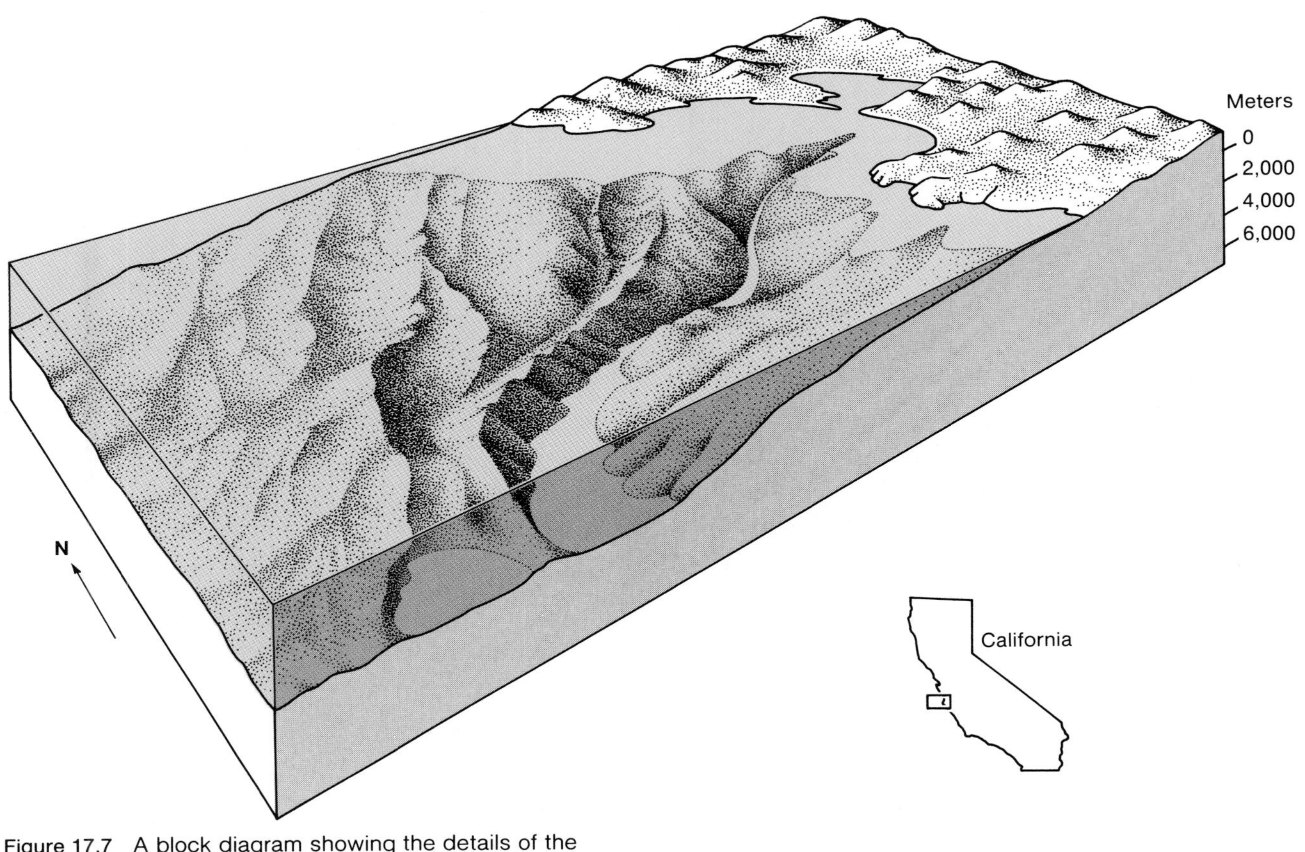

Figure 17.7 A block diagram showing the details of the bathymetry associated with the Monterey Canyon off the central California coast.

are present on outer continental margins around the world. Profiles perpendicular to the canyon axes are somewhat V-shaped with steep valley sides. The course of a submarine canyon may be straight or meandering. Tributaries are common (fig. 17.7). Submarine canyons are the major pathway for sediments carried from the continents to the deep ocean environment. The gradient of the canyon floor is steep, an average of nearly 60 m/km in short canyons and about 10–15 m/km in the long ones. While this may not seem very steep, it is from 5 to 30 times the gradient of the continental shelf. The canyons are cut into all rock types, including granite.

Many submarine canyons head near the mouths of rivers and some are in line with shelf valleys, which are themselves connected to river valleys on the continents. This arrangement is related to the origin of the canyons. The upper part of many canyons was excavated by streams during the low stands of sea level, and the lower part was formed primarily through erosion by turbidity currents and other sediment gravity transport phenomena. Data indicate that during low sea level stands of the past 5–10 million years, there were tremendous quantities of sediment introduced to the marine environment at or near the present shelf-slope break, which was the position of the shoreline during much of that time. This sediment generated nearly continuous, powerful turbidity currents that cut the deep canyons. Some years ago there was great skepticism about the ability of turbidity currents to erode these large quantities of sediment, especially over a rather short duration of geologic time. Data from both field observations and laboratory experiments have shown that both conditions are possible. In fact, the largest canyon yet discovered, the Zhemchug Canyon in the Bering Sea, was entirely cut during the past two million years.

Figure 17.8 Photograph of sand cascading down the slope of a submarine canyon off the coast of California.

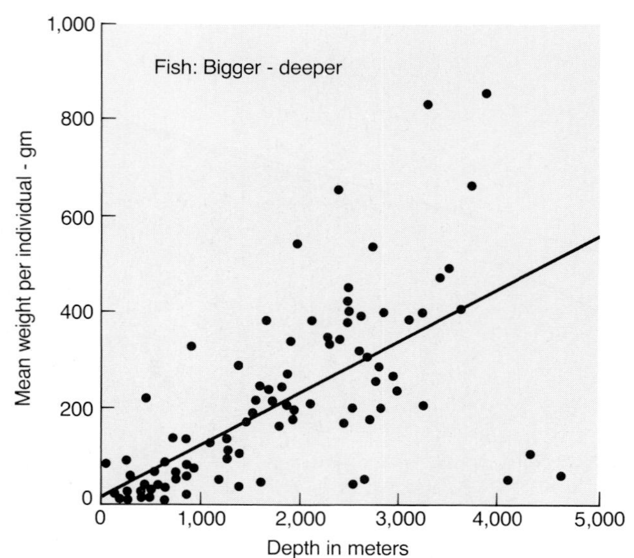

Figure 17.9 Graph showing the relationship of fish size with depth on the continental slope.

Although the submarine canyons serve as sediment conduits (fig. 17.8), there is also a considerable amount of sediment that accumulates on the canyon floors. This is indicated by the flat nature of canyon floors. Coring and seismic profiling confirm this characteristic. Downslope transport is demonstrated by the presence of ripples and scour marks on the canyon floors.

Biota of the Slope

The continental slope can be considered a separate and distinct biological province, although there are species that are shared with both the shelf and the rise. There is however, a dearth of knowledge about slope organisms, even in comparison with the abyssal floor of the ocean. This is partly due to the general absence of scientific investigation of the continental slope until about the 1970s.

Environmental factors such as oxygen, phosphate concentration, organic matter, and sediment texture affect the faunal zonation of the slope: oxygen and phosphate are necessary for all organisms, organic matter may limit some organisms because it is a primary food source, and certain benthic organisms are restricted to specific sediment types. As a result, there is a narrow depth range over which most species are distributed. Because of the relative homogeneity of the deep ocean, there is widespread distribution of the species within their depth ranges.

The three primary megafaunal groups on the continental slope are echinoderms, decapod crustaceans, and fish, with the last being the most diverse. The total biomass of the continental slope is greatly reduced from that of the shelf because slope benthos are far removed from the zone of photosynthesis and food supply is limited. There is a general trend of decreasing biomass downslope, but there are many exceptions, probably related to local areas of abundant organic matter due to downslope transport. Strangely enough, however, bottom-dwelling fish show an increase in size of individuals with depth down the slope, but the abundance of individuals diminishes (fig. 17.9).

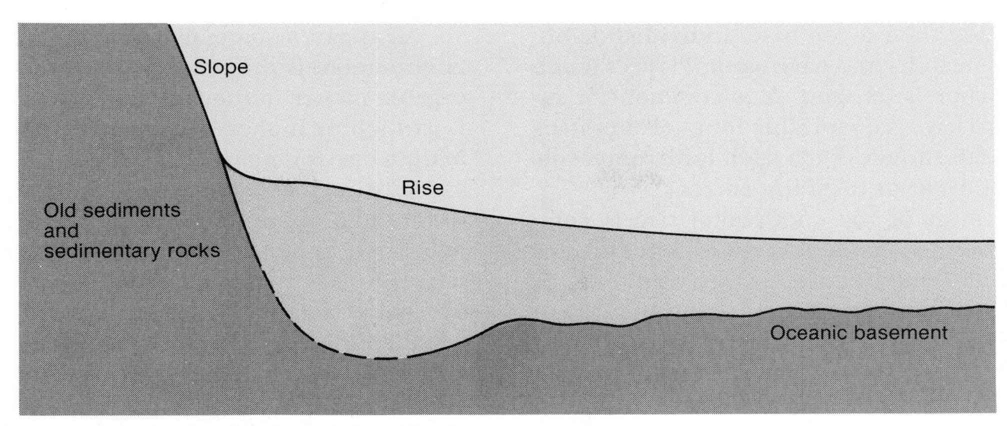

Figure 17.10 General diagram showing the relationship of the continental rise prism of sediments to the underlying crust.

Continental Rise

The most seaward province in the continental margin is the continental rise, which lies at the base of the continental slope. Although the rise is comprised largely of sediment derived from the continent, it rests on the oceanic part of the earth's crust (fig. 17.10). The boundaries of the rise are difficult to recognize because of its subtle bathymetric definition. Generally, there is a distinct break in gradient at the slope-rise break, but only a gradual transition from the rise to the abyssal plain. It has been suggested that the gradient of 1:1,000 be used for the rise, with the slope greater and the ocean floor less than that value. There is a wide range in the gradient of the continental rise, but the average is near 1:150.

Origin and Structure of the Rise

The continental rise is essentially an amalgamation of adjacent submarine fans, modified in shape by deep currents. These fans have accumulated at the base of the continental slope as the result of sediment gravity phenomena transporting sediments across the relatively steep slope. The wedge-shaped prism of sediments that comprise the rise is very thick, typically 2,000 m or more. This sediment laps onto the base of the slope and grades imperceptibly onto the abyssal floor of the ocean basin (fig. 17.10). Data from the rise indicate that these strata are not very old, with most of the accumulation being during less than 10 million years.

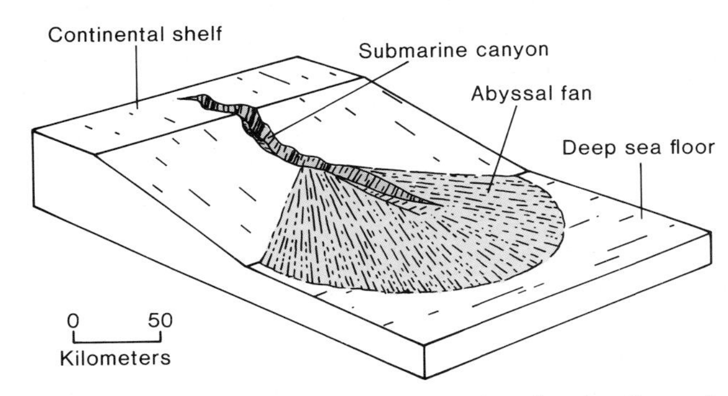

Figure 17.11 Diagram of a submarine fan showing the main valley and its relationship to the adjacent submarine canyon.

There is an obvious relationship in location between the rise and the submarine canyons that feed the rise. Fan-shaped accumulations of rise sediments have their apex at the mouths of submarine canyons. The individual submarine fans that comprise the rise typically have an incised valley at the head with multiple distributaries branching from this area. The valley emanates from the base of a submarine canyon (fig. 17.11). The fan itself is shaped like a broad cone with the point at the canyon base. Individual tributary channels meander and have natural levees much like fluvial channels on land. It is common for individual fans to have recognizable lobes or suprafans on the upper fan surface. Each such lobe represents a depositional event or turbidity current.

Although most of the continental rise is composed of submarine fans, sediments of other origins are also present. Contour currents (see chapter 5 for review) are significant contributors to the fans. These currents rework turbidite sediments and help to shape the rise. There is also a contribution of pelagic sediment to the rise, both biogenic tests and clay-sized mineral particles.

The internal nature of the continental rise as revealed by seismic reflection profiles is dominated by progradation or outbuilding. Sediment layers are extensive and uniform except where they are interrupted by slumping. This is an expected feature due to the combination of rapid accumulation of fine sediment on a sloping surface.

Physical and Chemical Environment of the Rise

Because the depth of even the shallow part of the continental rise is about 2,000 m, this environment is similar in many respects to the deep ocean floor. It is homogenous in terms of salinity and temperature. Salinity is at the normal marine concentration of 35‰ and temperature is about 0°C, but shows some slight variation.

Overall, the chemistry of ocean water in the rise environment is uniform. Oxygen is typically abundant and is not a limiting factor in controlling the biota, although there is some regular variation in oxygen content at these depths. The highest concentrations are in the high northern latitudes, with a regular (but modest) decrease to the high southern latitudes. Nutrient elements are also relatively abundant in waters over the continental rise. The upwelling of these deep margin waters with their contained nutrients and oxygen produces major plankton blooms and contributes greatly to the productivity of shallow outer margin waters at the shelf-slope break (see chapters 5 and 9).

An apparent anomaly in the chemistry of the rise environment is the relatively high concentration of organic matter in the fine-grained rise sediments. It is as high or higher here than in other continental margin environments (fig. 17.12). There is a logical explanation for this distribution and its relationship to the high oxygen content of rise waters. Organic debris is trapped in the rapidly accumulating sediment and is not in contact with the well-oxygenated overlying waters. Therefore, oxidation of the organic debris does not occur and the rise waters are not depleted of their oxygen content.

Physical processes, especially gravity-generated phenomena, dominate the dynamics of the rise, although biological processes are also important. The physical processes include sediment gravity phenomena, such as turbidity currents, debris flow, and slumping. Sluggish deep-water currents of the thermohaline variety are also present and are due to density gradients, primarily from temperature variation. Internal waves impact on the rise and in doing so may cause substrate disturbance, which can generate turbidity currents.

Biological processes that affect the ocean bottom may be divided into two types, much like other marine environments: (1) bioturbation due to sediment reworking by benthic organisms, and (2) biologic erosion where physical and chemical activities of the organisms cause breakdown of rock and mineral material. Benthic organisms of the rise carry on similar activities as do their shallow water counterparts. The rise substrate is covered with evidence of grazing and burrowing, and with abundant pellets (fig. 17.13). These activities in the soft muddy substrate rearrange and redistribute rise sediment. Biologic erosion caused by organisms boring into rocks, scraping rocks, and through reactions between secretions of organisms and rocks is restricted to hard substrates. It is therefore most significant in submarine canyons, although large particles or shells that occasionally reach the rise may host boring organisms.

Fauna of the Continental Rise

The continental rise has a relatively abundant and diverse benthic fauna, considering the depth and the rigor of such a deep, dark environment. There has, however, been only modest attention paid to the organisms that occupy this zone. Data show that the

Figure 17.12 Bottom sediments of the continental rise showing evidence of bioturbation.

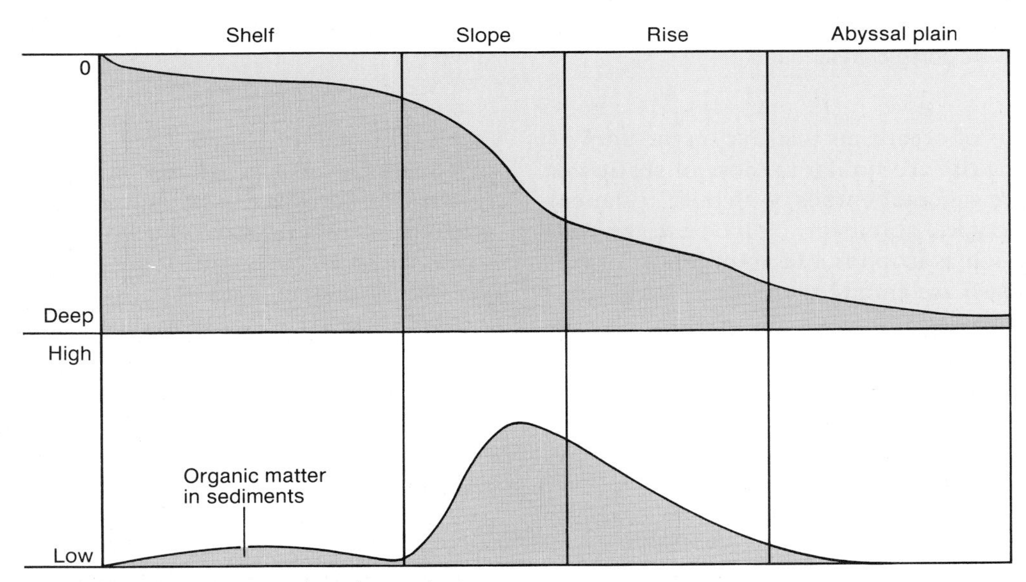

Figure 17.13 Diagram showing concentration of organic matter in sediments of the continental rise.

species diversity is higher on the rise than any other open marine environment including the continental shelf. This is attributed, in part, to the limited amount of competition among species at intermediate depths relative to the shallow and deep environments. Another contributing factor is the fairly high percentage of organic debris that accumulates in continental rise sediments (fig. 17.13). This detritus is carried to the rise from shallower areas of the shelf and slope and is incorporated in the rapidly accumulating turbidite sediments. Because of this type of accumulation, the organic debris is buried and is not exposed to the sediment surface, where oxidation occurs. Consequently, organic debris is available for the numerous deposit-feeding benthic organisms that forage over the continental rise.

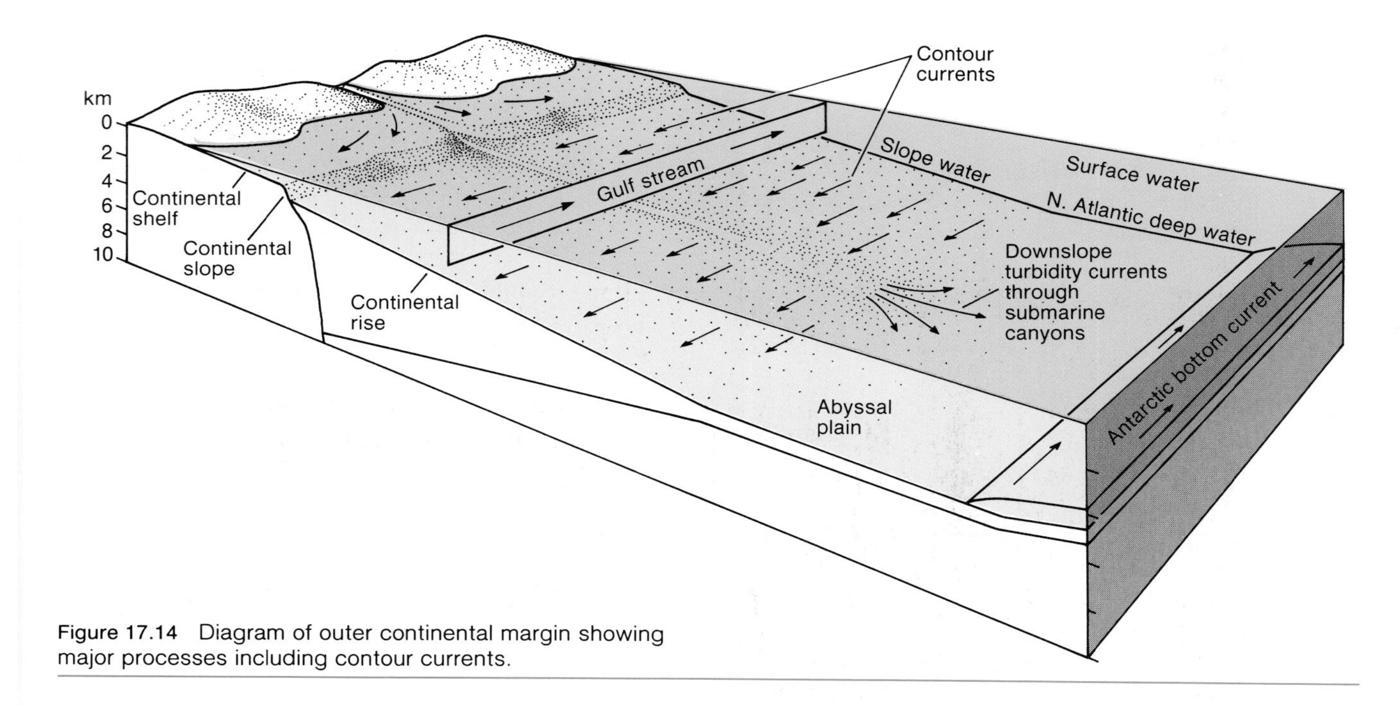

Figure 17.14 Diagram of outer continental margin showing major processes including contour currents.

The groups of organisms that live on the floor of the continental rise are similar to those of shallower zones. Most are vagrant benthos, with both epifaunal and infaunal types abundant. Polychaete worms, crustaceans such as isopods and amphipods, brittle stars, and starfish are among the more abundant varieties present. Some sponges may also be found and are one of the few sessile types present.

Evidence of bioturbation is not as common in the rise area as might be expected given the abundance of vagrant benthic organisms. The reason is the fairly rapid rate of sediment transport and accumulation due to the turbidity currents. These currents excavate previously deposited sediment and redeposit it along with new sediment from shallower environments. The result is a reworking and redeposition of sediment, which tends to destroy the bioturbation activities of the organisms. A modest diversity of small fish is also present.

Sediments of the Continental Rise

There are two types of sediments that characterize the continental margin: terrigenous and pelagic. **Terrigenous sediments** are those derived from the land and transported across the continental shelf to the outer margin, where they come to rest. These sediments may have almost any composition and grain size; however, mud and fine sand-sized particles of clay minerals and quartz are the most common. Pelagic sediments are composed primarily of tests or skeletal material of planktonic organisms (see chapter 10) that live in the water column above the outer continental margin. As these phytoplankton and zooplankton expire, the hard parts of the organism slowly settle to the bottom and become incorporated in the sediment. Some very fine mineral particles that are not biogenic in origin may also settle to the bottom and comprise a part of the pelagic sediment. Most of the outer margin sediments are terrigenous in origin, but this changes toward the deeper ocean floor where pelagic sediments are dominant (see chapter 10).

There are multiple mechanisms for transporting the terrigenous sediment to the outer continental margin. The most important is by turbidity currents, which were discussed in detail in an earlier section of this chapter. The other is via contour currents. Having discussed turbidites, it is also appropriate to consider **contourites**. Contour currents have been mentioned previously (chapter 5) but their products have not. The contour current is also an important process on the continental rise (fig. 17.14).

These geostrophic currents flow along the depth contours at speeds of up to 70 cm/sec, which is about equivalent to 1.5 mph, but commonly the speeds are in the range of 10–30 cm/sec. These currents can transport grains as coarse as fine sand. Structures such as ripples and sediment dispersal patterns on the outer margin demonstrate that much sediment is moved by contour currents. The currents tend to move with the ocean bottom water. The earth's rotation and the Coriolis effect tend to keep these currents on the upper portion of the rise.

The nature and location of the contour currents produce contourites. These are thin sediment layers that occupy, in part, the same site as turbidites but can be distinguished from them. The contourite sediment is composed of silt and clay that is thinly layered, well sorted, and like the turbidites, shows grading of particle sizes. Bed thickness tends to be less than turbidites. Some investigators have attributed much of the upper rise sediment sequences to a contour current origin. At one of the DSDP sites off North Carolina there is a sequence of nearly 1,000 m of contourites.

Summary of Main Points

The outer continental margin, composed of the continental slope and continental rise, represents the most seaward of the continental environments. The slope displays a great range in depth, from a few hundred meters to almost 3,000 m, whereas the rise has ocean basin depths.

The slope is largely a region of sediment transport to the deeper part of the ocean basin, although there is also some accumulation on this surface. The major path of the sediment transport is by means of submarine canyons, which are the highest relief features of the slope. Most sediment that makes its way through the canyons accumulates at the base of the slope in the form of submarine fans. The coalescing of adjacent fans produces the continental rise.

In addition to the dominating downslope processes of the slope and rise, there are also processes that move parallel to the slope and rise. These contour currents are also capable of transporting sediment; they redistribute much of the sediment that is deposited on the slope and rise.

The great change in depth across the slope and rise results in a major change in the organisms that occupy these environments. The decreased temperature, increased depth, and general decrease in food supply impact on the fauna. There is a truly deep sea character to the community with overall decrease in diversity and numbers as compared to the continental shelf. No producers are present, detritus feeders dominate the benthos, and carnivores are essentially restricted to deep-sea varieties of fish. Bottom sediments contain a fair amount of organic debris derived from the shelf and brought to the slope and rise by turbidity currents. This, plus the debris generated on the slope and rise, is the primary source of food.

Suggestions for Further Reading

Emery, K.O. 1960. **The Sea off Southern California, A Modern Habitat for Petroleum**. New York: John Wiley and Sons.

Hill, M.N., ed. 1963. **The Sea**, vol. 3. New York: Wiley Interscience.

Kennett, J.P. 1981. **Marine Geology**. Englewood Cliffs, New Jersey: Prentice-Hall, Inc.

Pilkey, O.H., ed. 1968. "Marine geology of the Atlantic continental margin of the United States," *Southeastern Geology,* v. 9.

Shepard, F.P. and R.F. Dill. 1966. **Submarine Canyons and Other Sea Valleys.** Chicago: Rand McNally.

18
The Pelagic Environment

Chapter Outline

Pelagic Zones
- Epipelagic Zone
  - Physical and Chemical Characteristics
  - Circulation
- Mesopelagic Zone
  - Physical and Chemical Characteristics
- Bathypelagic Zone
  - Physical and Chemical Characteristics

Deep-sea Circulation
- Atlantic Circulation
- Pacific Circulation
- Indian Ocean

Pelagic Communities
- Epipelagic Community
  - Productivity
  - Phytoplankton
  - Zooplankton
  - Nekton
- Mesopelagic Community
  - Zooplankton
  - Nekton
- Deep-water Community

Summary of Main Points

Suggestions for Further Reading

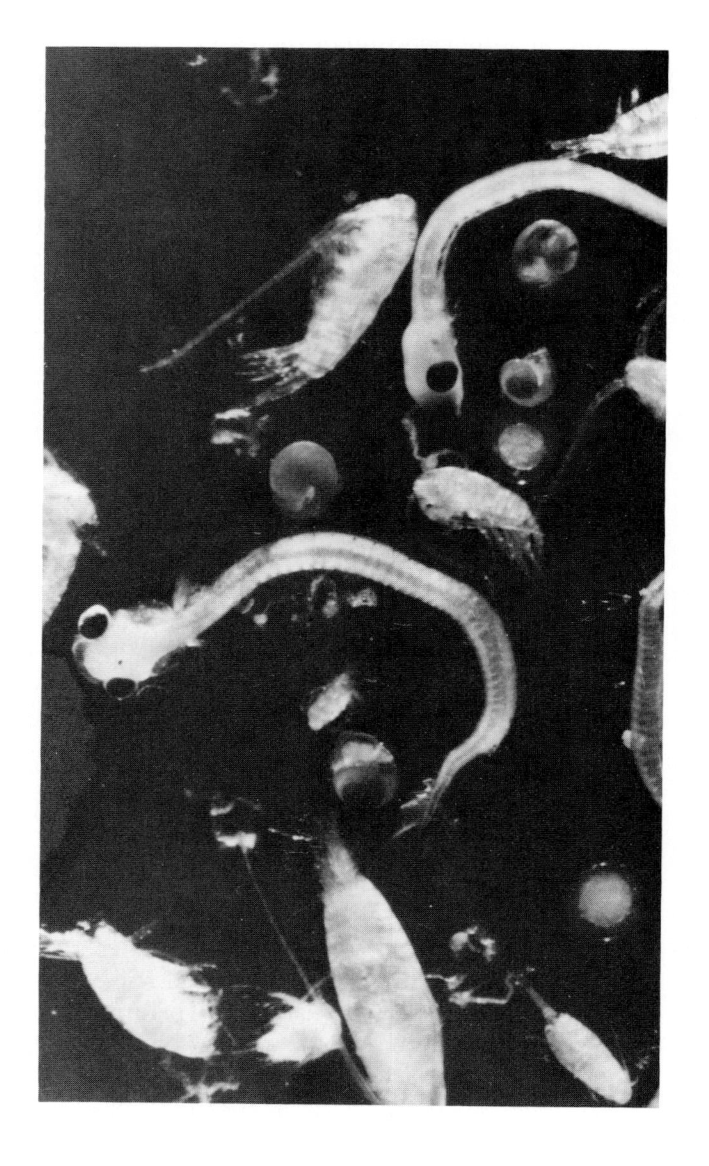

By far the largest of the marine environments is that encompassed by the enormous water volume beyond the coast and above the sea floor: the pelagic environment. As described in chapter 8, this includes two subenvironments: (1) the neritic, which is the water mass above the continental shelf, and (2) the oceanic, which is the remainder of the pelagic environment (fig. 8.1). The pelagic environment is subdivided by water depth into four zones, epipelagic, mesopelagic, bathypelagic, and abyssopelagic. Each has a rationale for its boundaries based on a combination of physicochemical parameters and the organisms that occupy the zone.

Pelagic Zones

The epipelagic zone is comprised of the upper 200 m including the neritic, or shelf waters, as well as the upper oceanic layer. The boundary is not absolute because the depth at the shelf-slope break ranges geographically. In the oceanic realm, this boundary is based partly on the distribution of organisms. The mesopelagic zone extends from the base of the epipelagic zone to depths of about 1,000 m. It represents a layer within which there is substantial change in life and water properties. From 1,000 m to a depth of around 4,000 m is the bathypelagic zone, which is the uppermost uniform zone in the ocean. The base of this zone corresponds closely to the depth of the abyssal floor. The abyssopelagic zone is essentially that of the deep-sea trenches and is therefore restricted to those geographic areas where trenches occur.

Epipelagic Zone

The uppermost zone of the pelagic environment is the most variable in space and time and the most important from the standpoint of marine life. It is here that photosynthesis takes place and biomass is the greatest. It is also the zone where surface circulation (horizontal circulation) dominates and where the temperature changes significantly with the seasons and with latitude.

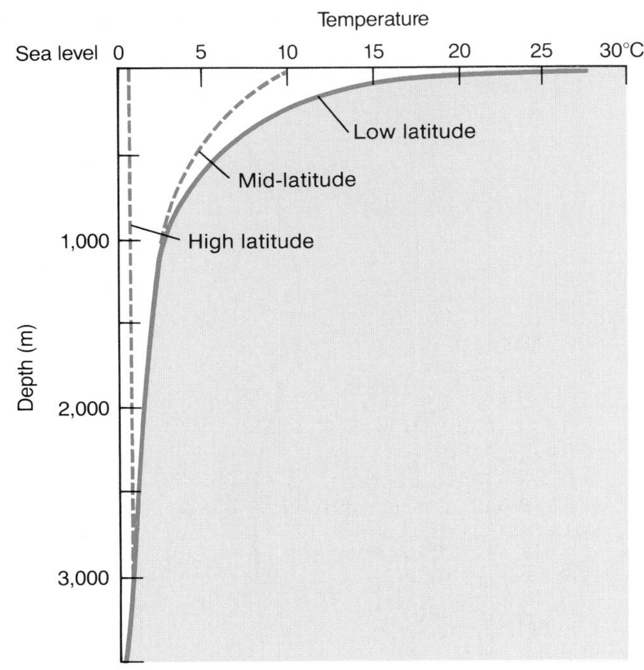

Figure 18.1 Temperature variation with depth in oceanic waters.

Physical and Chemical Characteristics

Not only is the epipelagic zone warm due to surface heating from the sun, but it also experiences latitudinal and seasonal variations in temperature. This ranges from being covered with ice during much of the year in the extremely high latitudes to marked seasonal variation in the mid-latitudes. In the tropics temperatures are commonly 24–26°C with little seasonal change. The mid-latitudes experience surface water temperatures of about 10°C during winter and 15°C in the summer. High latitudes are near 0°C throughout the year (fig. 18.1).

Epipelagic waters also show latitudinal variation in salinity. Generally, the highest salinities are found in the central portion of major oceanic gyres, giving mid-latitude values of 36‰–37‰, whereas high-latitude values are close to 34‰ (fig. 18.2). These salinity patterns, along with the previously noted temperature patterns, give the expected low density

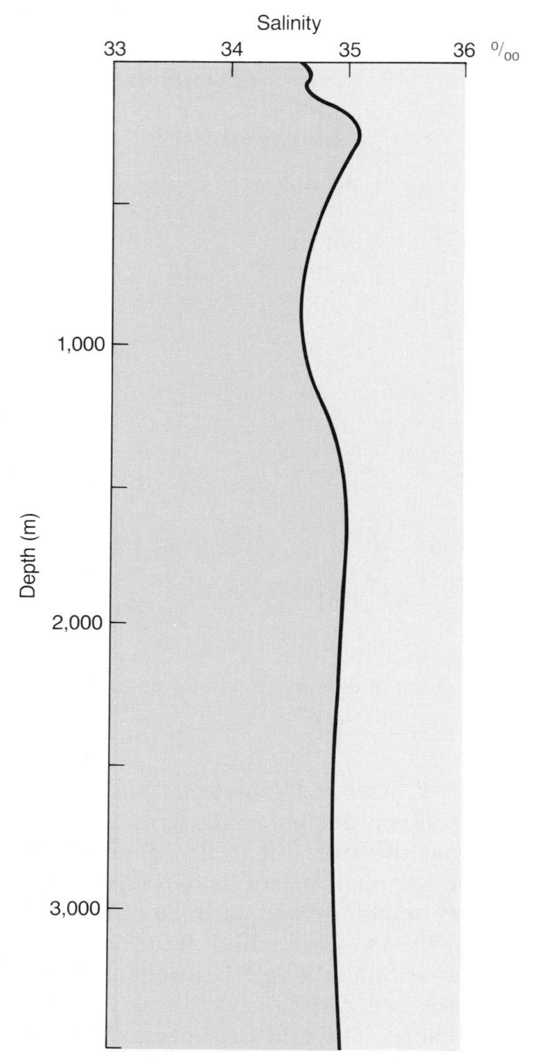

Figure 18.2 Profiles of salinity relative to depth for the oceans.

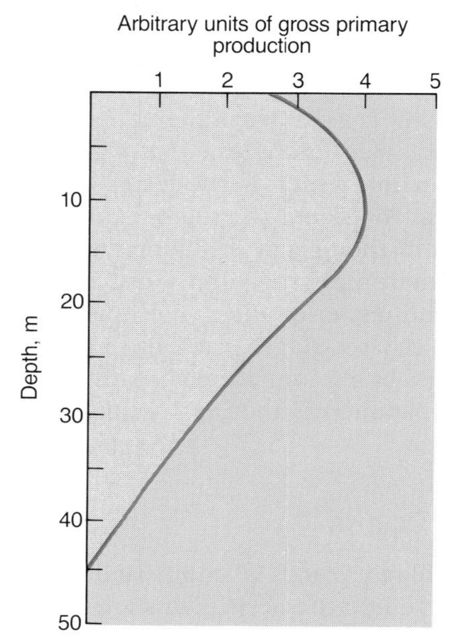

Figure 18.3 Profile of productivity relative to depth.

values typical of surface waters because the low density caused by high temperatures offsets the increase in density caused by high salinities.

The unique property of the epipelagic zone is the penetration of light throughout; it includes the photic zone, where light occurs. Typically this ranges to a depth of 100 m or so, but this varies due to suspended sediment, angle of incidence of light rays, water surface roughness, and other factors. The highest rate of productivity is about 10 m below the surface because very intense light tends to depress photosynthesis. The depth to which light penetrates varies depending on seasons and latitude because of the angle of the sun relative to the water surface.

The epipelagic zone is one where oxygen and carbon dioxide are abundant. Oxygen is provided by exchange with the atmosphere and through the photosynthetic process, whereas carbon dioxide comes from respiration of organisms and through interaction with the atmosphere. By contrast, the nutrient elements phosphorus and nitrogen may be limiting factors for production in epipelagic waters. They are used in the photosynthetic process and may be totally depleted, thus causing further production to cease. Relatively high concentrations of both phosphorus and nitrogen are present in deep waters. Upwelling of these nutrient-laden waters may result in blooms, which are water zones where there are extremely high rates of photosynthetic activity (fig. 18.3).

Circulation

Global prevailing wind patterns are the dominant factor in surface circulation in the open ocean environment. Currents driven by wind extend to a depth of about 100 m and serve as a primary basis for defining the epipelagic zone. The fundamental wind patterns that drive these currents are the trade winds and the westerlies in both the northern and southern hemispheres. These winds produce large,

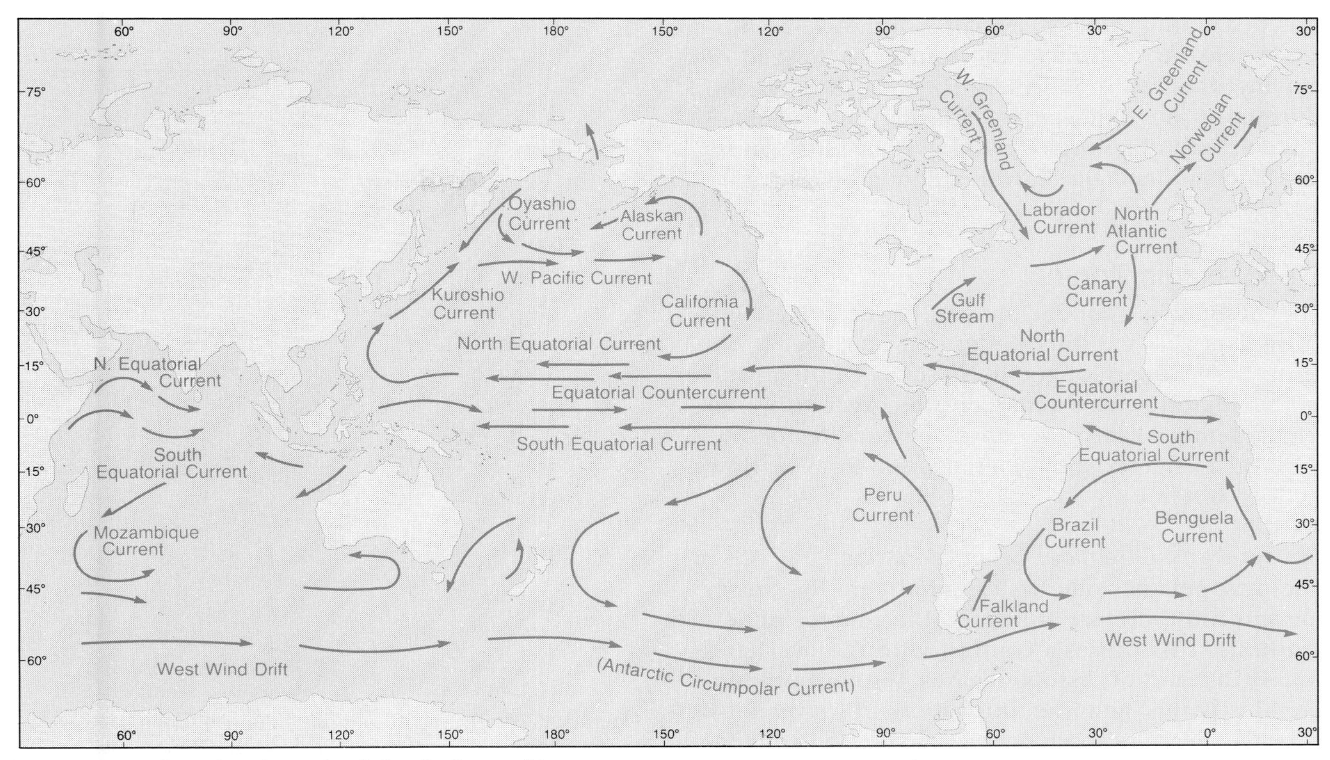

Figure 18.4 Map of surface circulation in the world showing the major current systems in each ocean basin.

circular circulation gyres, as discussed in chapter 5 and shown in figure 5.9. Subsidiary gyres and circulation are caused by the major gyres and other winds.

Although the general pattern of surface winds is simple and can be applied to all oceans, the details of surface currents are complex. Just a look at a world map of surface currents illustrates this point (fig. 18.4). Most of the complexities are caused by landmasses that interact with surface currents. An excellent example of this is seen in the high latitudes of the southern hemisphere. The Antarctic Circumpolar Current, or West Wind Drift as it is also called, flows from west to east around the Antarctic landmass (fig. 18.4). There is a countercurrent called the East Wind Drift that flows from east to west between the landmass and the Circumpolar Current. Both of these systems are wind driven.

Another important aspect of oceanic circulation, located in the same general vicinity of the currents mentioned in the preceding paragraph, is the convergence of water masses that occurs in the mid to high latitudes of the southern hemisphere. Here are the Antarctic Convergence, located between 50° and 60°S latitude, and the Subtropical Convergence, located near 40°S latitude. At the Antarctic Convergence, the cold surface water of the Antarctic meets the warmer water of the Subantarctic Zone and descends into the mesopelagic zone. A similar situation exists at the Subtropical Convergence with a relatively cold water mass converging with a relatively warm water mass.

Other circulation patterns show interactions of one gyre with an adjacent gyre. In the case of some of these, for example the tropical area of the Atlantic, there is circulation across what appear to be major water mass boundaries. In this area some water from the South Equatorial Current of the major gyre in the southern hemisphere is interactive with the major gyre in the northern hemisphere through the Caribbean Sea, where the mixing occurs (fig. 18.4).

The Aleutian Current in the Pacific Ocean and the currents around Greenland in the Atlantic Ocean are examples of circulation cells that are partly controlled by land (fig. 18.4). The Aleutian Current is part of a counterclockwise gyre, but the Aleutian Islands cause it to turn to the southwest instead of moving northward and merging with the Oyashio Current off the Kamchatka Peninsula. In Greenland the northern counterclockwise gyre is deflected by the landmass and the currents flow around it (fig. 18.4).

Mesopelagic Zone

The mesopelagic zone is the part of the pelagic environment that exhibits the greatest change. It extends from a depth of about 200 m to 1,000 m, and is the region where temperature, organisms, nutrient elements, and other environmental factors exhibit the characteristics of a transition from shallow to deep water.

Physical and Chemical Characteristics

The mesopelagic zone has essentially no light from the sun and therefore is beyond the zone of photosynthesis. This causes a contrast with the overlying waters in terms of dissolved gases. Without production by plants and the utilization of oxygen by animals, there is a sharp decrease in oxygen concentration within the mesopelagic zone. In fact, this portion of the mesopelagic zone is generally called the **oxygen minimum zone**. No significant mixing occurs with either the well-oxygenated surface waters or the deeper waters.

There is also a large decrease in temperature through the mesopelagic zone. This thermocline typically occupies the entire mesopelagic zone with temperatures decreasing from more than 20°C to only 2–4°C below the thermocline. This thermocline is best developed in low latitudes where surface temperatures are highest. Associated with the thermocline is the **pycnocline,** a zone of density change in seawater (fig. 18.5). Some salinity changes also contribute to the pycnocline because surface water salinities may be somewhat above that of normal seawater, as in the central part of the oceans, or below normal seawater, as in the high latitudes.

These changes in the density of water have an effect on the transmission of sound. The velocity of sound increases with an increase in temperature, salinity, and density. Near the surface, temperature is most prominent in controlling sound velocity; in deeper water, pressure becomes the primary factor. These variations produce a zone of minimum sound velocity that corresponds to the thermocline and therefore to the mesopelagic zone. The changes in sound velocity caused by temperature and pressure result in a sound channel due to refraction or bending of the sound waves toward the location of lesser temperature or pressure (fig. 18.6). The sound channel thus formed permits long-range signaling underwater and has been used for military communications, especially between submarines.

The concentration of nutrient elements is at its maximum near the base of the mesopelagic zone; there is a great increase in both phosphorus and nitrogen from the top of this zone to the base (see figs. 8.3 and 8.5). This increase is a response to the combination of organic material breakdown below the photic zone and the lack of nutrient uptake because photosynthesis is absent. The result is high concentrations of nutrient elements in all ocean basins.

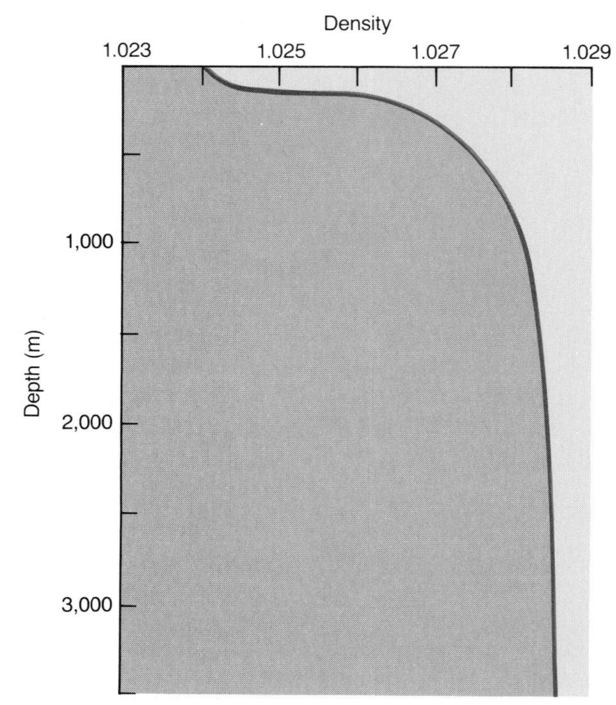

Figure 18.5 Vertical distribution of density in the oceans: the pycnocline.

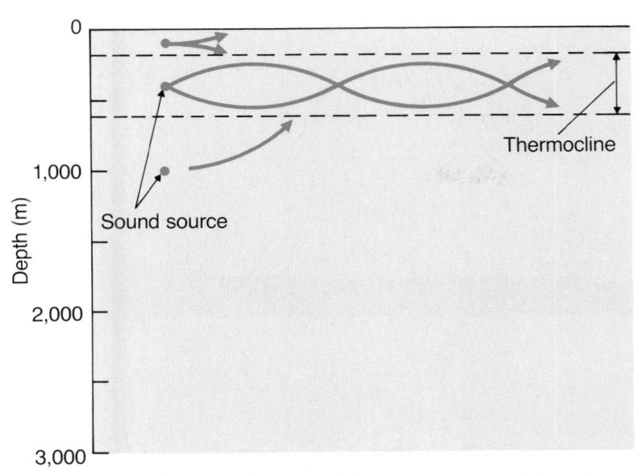

Figure 18.6 The combination of temperature and density variations in mesopelagic waters produces a sound channel in which sound tends to be trapped and move parallel to the surface.

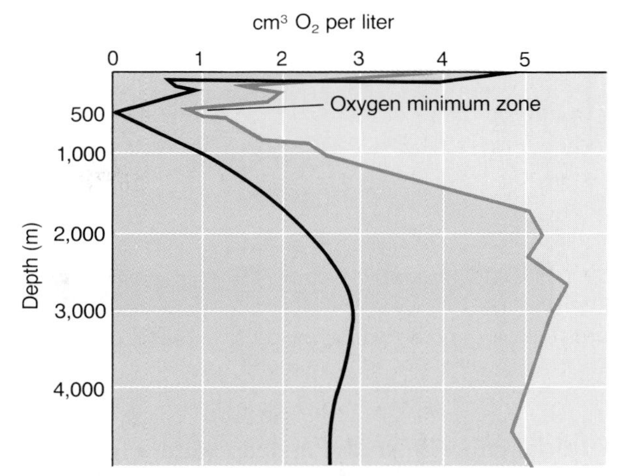

Figure 18.7 Distribution of oxygen and nutrients in the oceans.

Bathypelagic Zone

The bathypelagic zone is the true deep-sea pelagic environment, where there is a good amount of uniformity in the environment, especially in such factors as salinity and temperature. This zone extends from the base of the mesopelagic zone, at about 1,000 m, to the ocean floor. The discussion here will also include the abyssopelagic zone, which is the deep-sea trenches only. The greatest variations in characteristics of the deep pelagic environment occur just above the seafloor due to the concentration of certain dissolved constituents and the local presence of suspended sediment.

Physical and Chemical Characteristics

There is little variation in such parameters as temperature, salinity, and nutrient element concentration. The only marked change is that caused by pressure, which increases directly with depth. Temperature is nearly constant at about 1–2°C, but its subtle variation is responsible for much of the deep-sea circulation, discussed in a following section. Salinity is almost constant at about 34.7‰, the mean value for the ocean. These two parameters are dominant in determining the density of seawater and thus, sigma-t (see chapter 4) is constant.

Perhaps the greatest change in the environmental variables is displayed by oxygen, which shows a modest but steady increase with depth from the oxygen minimum of the mesopelagic zone (fig. 18.7). This phenomenon is caused by a combination of cold temperatures, which increase the solubility of oxygen, and by the descent of cold, dense high-latitude surface waters to the deep sea. These oxygen-rich waters provide much of this constituent to the bathypelagic zone. Nutrient concentration decreases slightly due to the continual decrease in biomass with depth.

Deep-sea Circulation

Circulation of the deep water of the ocean, also called thermohaline circulation (see chapter 5), affects all waters below the epipelagic zone or about 200 m. This accounts for about 98% of all waters in the ocean basins. Although much of this water is relatively homogenous, there is sufficient variation to result in distinct water masses, which move slowly and retain their separate identities in doing so. The following discussion will consider only the basic principles involved and the movement of major water masses. The reader should note that the actual situation is much more complicated than depicted here.

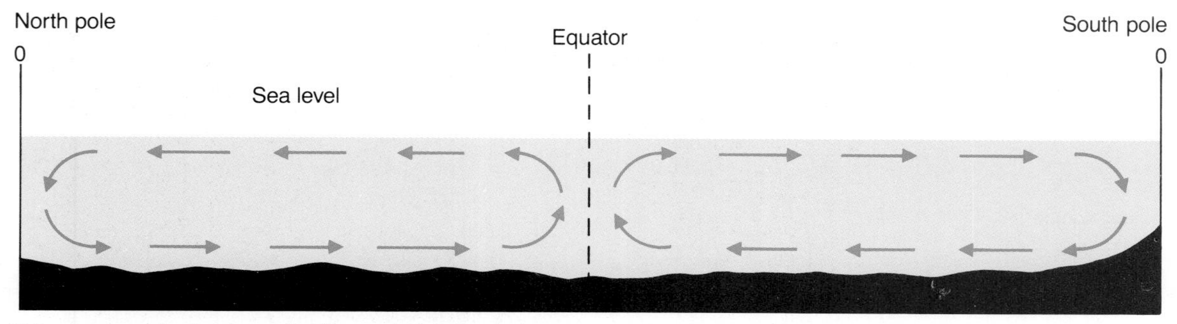

Figure 18.8 A general, simple model for oceanic circulation proposed by Alexander von Humboldt.

Initial efforts at studying deep-water circulation produced simplified models. A pioneer in the field was Alexander von Humboldt, after whom a current located off the coast of Peru is named. He believed that the entire ocean system circulated in a convective fashion (fig. 18.8). Von Humboldt deduced that the sun would heat ocean water near the equator more than near the poles, resulting in large, convection cell circulation. He did not consider the effects that heating has on salinity and therefore on density, and the scheme that he produced did not account for much of the observed variation. It was, however, a good first start at understanding deep-water circulation.

In the 1950s, Henry Stommel produced a general deep-water circulation model for the world that has proved to be a good representation. The combination of the cooling of surface waters in high latitudes and the increase in salinity from ice formation causes the water density to increase. The result is an unstable condition that makes the cold, dense water descend to near the bottom. Stommel showed that this phenomenon occurs primarily in a region near Greenland and in the Weddell Sea of the Antarctic (fig. 18.9), although it also takes place in other cold-climate areas.

Thermohaline circulation caused by density gradients flows generally from high latitudes toward lower latitudes, although some of the deep-water masses continue beyond the equator. The rotation of the earth results in a concentration of currents on the west side of the ocean basins (fig. 18.9). The deep circulation around the Antarctic landmass serves to connect the deep waters of the world ocean.

Atlantic Circulation

Thermohaline circulation in the Atlantic Ocean includes several large water masses that essentially converge. The source areas for the deep water mentioned previously produce Antarctic bottom water and North Atlantic deep and bottom water. These water masses all flow toward the equator, with the North Atlantic waters eventually overriding the Antarctic water mass (fig. 18.10). The result is a three-layered system of water masses in the Atlantic plus the tongue of Mediterranean water in the northern hemisphere. This Mediterranean water mass is high in salinity and temperature, relative to the Atlantic waters, and has an intermediate density. It moves westward or essentially perpendicular to the primary movement of Atlantic water masses. There is also a layer of shallow central water that originates in the Atlantic and does not extend below about 500 m depth.

The Antarctic bottom water and the North Atlantic deep and bottom water are the densest with sigma-t values of 27.86 and 27.85, respectively. Antarctic water is below 0°C and the NADBW is 2.5–3.1°C. In order to demonstrate how similar the various thermohaline masses are, the Antarctic intermediate water mass is 3–7°C and has a sigma-t value of 26.82–27.43 (fig. 18.10). These values are not markedly different, but they represent nearly the total spectrum of density for the deep waters of the Atlantic Ocean.

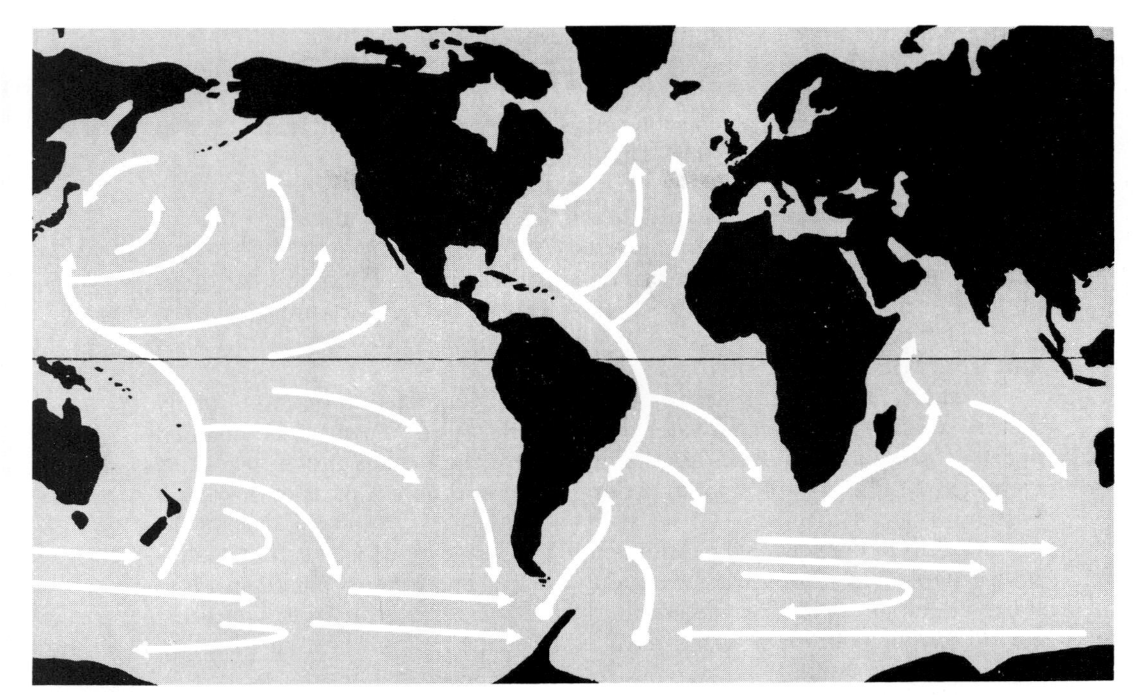

Figure 18.9 The deep-water circulation pattern as proposed by Stommel. This is accepted as the proper representation of deep circulation in the world ocean.

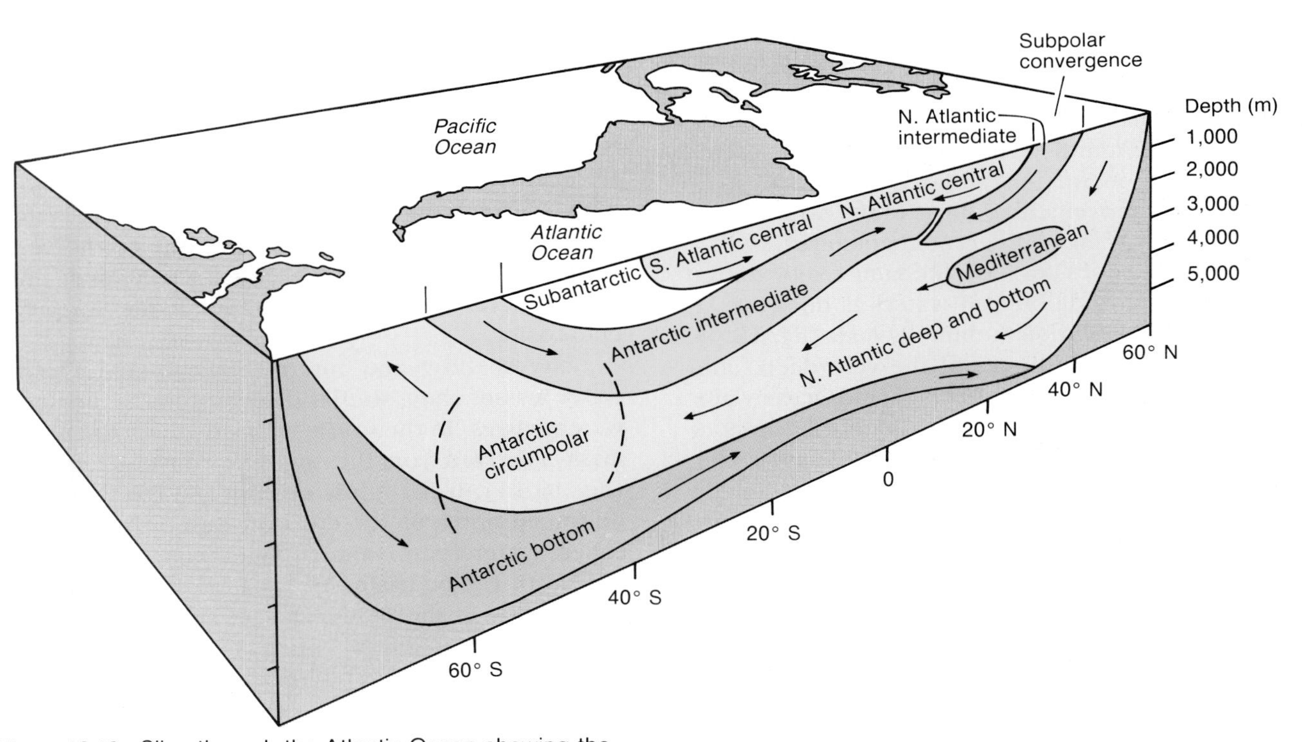

Figure 18.10 Slice through the Atlantic Ocean showing the distribution and movement of the major water masses.

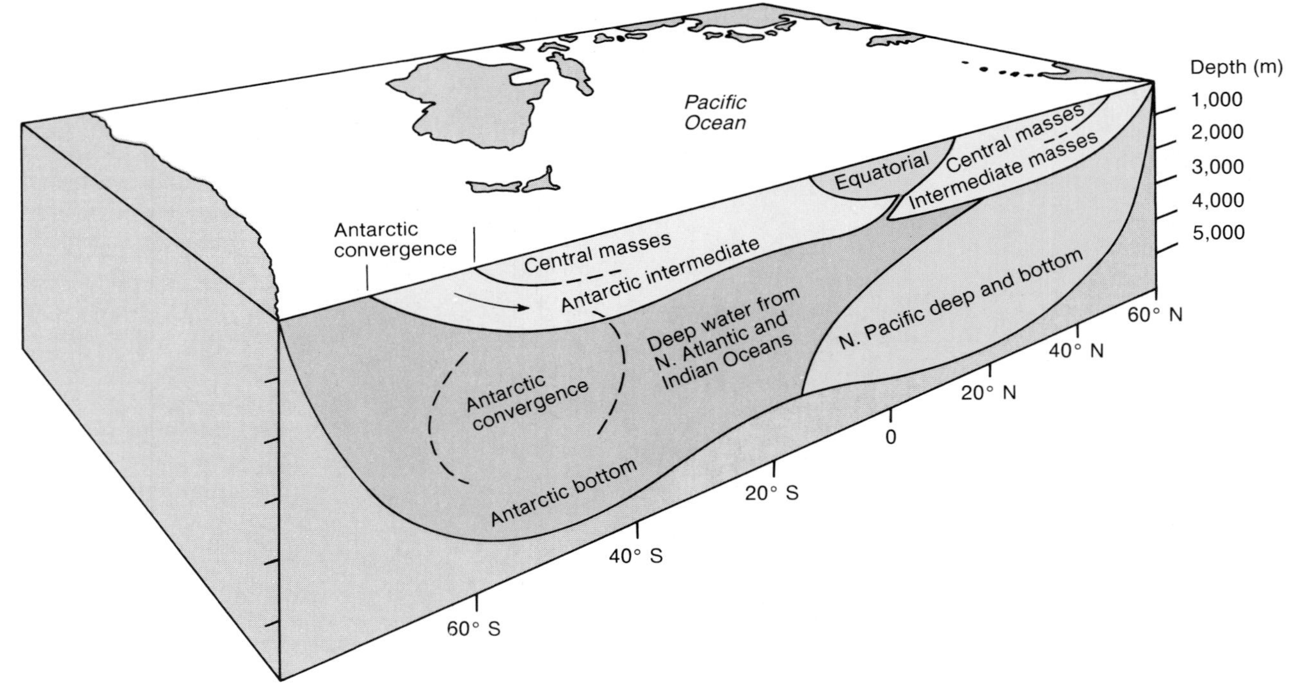

Figure 18.11 Slice through the Pacific Ocean showing the distribution and movement of the major water masses.

Pacific Circulation

Circulation in the Pacific Ocean is less well known than in the Atlantic due in part to its enormous size and the fact that it has not been studied in detail for as long as the Atlantic. The major water masses and their circulation in the Pacific Ocean display the same types of movement as those in the Atlantic, but the water masses have different characteristics.

The Pacific does not have the same major source of cold high-latitude bottom waters as does the Atlantic. This is shown on Stommel's map (fig. 18.9), which indicates no sources of the very deep, cold bottom water in this ocean. Deep Pacific Ocean water is derived from the westerly moving cold waters of the southern Atlantic and Indian Oceans (fig. 18.9). The result is that the Pacific Ocean does not contain water that is as cold or as dense as that in the Atlantic Ocean.

A cross section of the major water masses in the Pacific shows four major entities: two are deep and bottom water, and two are intermediate water (fig. 18.11). The deep and bottom water in the southern hemisphere extends to a depth of about 2,000 m and is derived from the cold waters that circulate westward from the other oceans. Its northern hemisphere counterparts, the North Pacific deep and bottom water, occupy a similar position (fig. 18.11). It does not exchange significant amounts of water with the South Pacific and it does not receive water from the extremely high latitudes due to the constriction at the Bering Strait. North Pacific deep and bottom water are effectively trapped in the North Pacific basin.

The shallow and intermediate Pacific Ocean water masses show similar distributions as the deep water masses. In the southern hemisphere the water mass is derived from the Antarctic convergence area (fig. 18.11). It has a low salinity (33.8‰) and can be traced north of the equator. Its northern hemisphere counterpart is similar, but is generated within the North Pacific basin. An equatorial water mass is also present in the Pacific Ocean. This water mass is characterized by low salinities as compared to other shallow water masses and has a marked thermocline at its base.

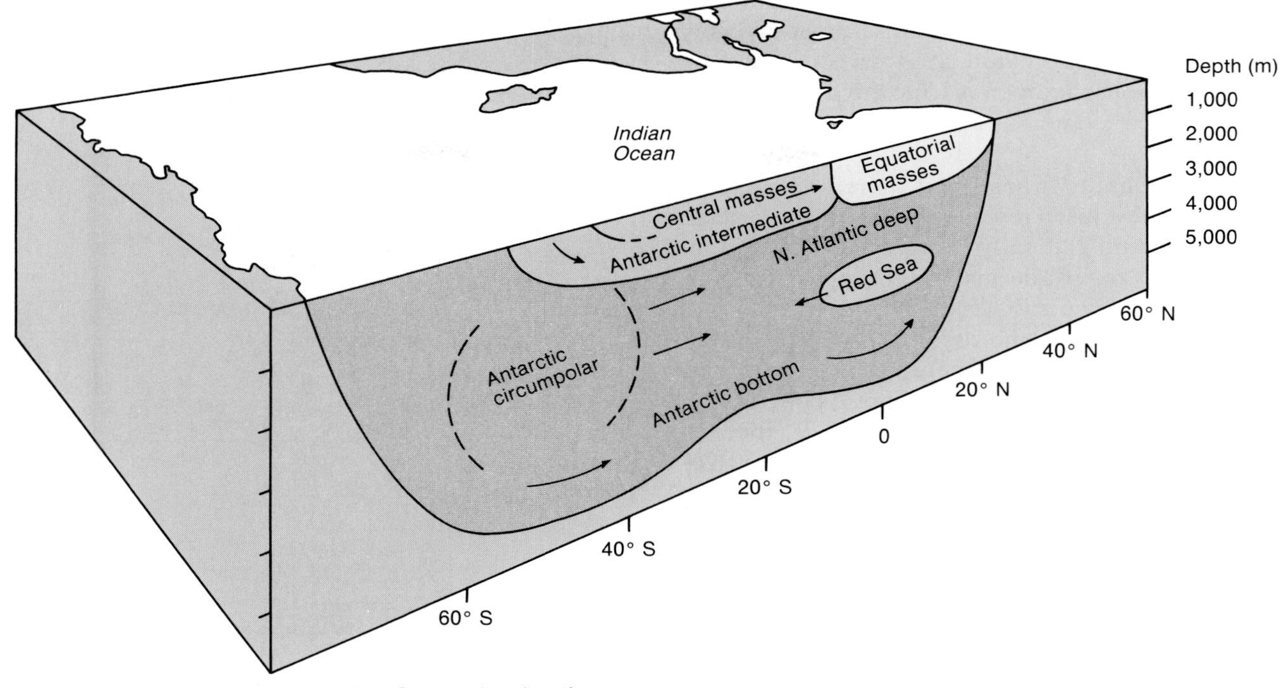

Figure 18.12 Slice through the Indian Ocean showing the distribution and movement of the major water masses.

Indian Ocean

The Indian Ocean differs markedly from the other oceans because of its limited extent in the northern hemisphere. Aside from that, its circulation and water masses are similar to the Pacific. Like the Pacific Ocean, the Indian Ocean does not have the great sources of deep cold water that are present in the Atlantic. The Indian Ocean receives its deep and bottom waters from the South Atlantic and from the Antarctic areas (fig. 18.9). The movement of these water masses is slow in the Indian Ocean due to the low density gradient that is developed within the water mass.

Much like the Pacific Ocean, the Indian Ocean has an Antarctic intermediate water mass and a shallow equatorial mass (fig. 18.12). A comparison of the south Pacific and the Indian oceans shows little significant difference. There is a relatively small but important water mass derived from the Red Sea that is also present (fig. 18.12). This mass is characterized by relatively high salinity and warm temperature. In many respects it is analogous to the Mediterranean water that is present in the Atlantic Ocean (fig. 18.10).

Pelagic Communities

The pelagic environment contains the bulk of the organisms in the world ocean. These organisms have great diversity as well as large numbers. There is geographic variation in the distribution of pelagic organisms due to the influence of landmasses and to climatic differences. There is also a major zonation in the distribution of pelagic organisms with depth of water.

Although the pelagic environment contains both a neritic division and an oceanic division, it is only the latter that will be discussed in this chapter. In a general way, the zones of the pelagic environment that were discussed at the beginning of the chapter can be applied to the communities of organisms that inhabit the ocean.

Epipelagic Community

The surface waters of the ocean contain the most diverse life in the marine environment and represent the most important zone in the sense that this is where productivity takes place. This upper 200 m includes not only the photic zone where photosynthesis takes place, but also contains a great diversity of zooplankton and nekton. It is without question the key to life in the open marine environment.

Productivity

The rate of photosynthesis and therefore, productivity, is greatest just a few meters below the surface of the ocean and gradually decreases to the bottom of the photic zone. The maximum depth at which photosynthesis occurs varies in both space and time, but in the open ocean photosynthesis may extend to about 200 m, the depth of the epipelagic zone. There is, however, a depth within the photic zone below which there is no net production. That is, there is more consumption of producers by herbivores than can be compensated for by photosynthesis (fig. 18.13). This is called the **compensation depth**.

Few things in the open ocean are important inhibitors of photosynthesis except perhaps for limited nutrient supply. The most prominent limiting factors in neritic waters are suspended sediment and turbidity, but they are not factors in oceanic pelagic productivity. Seasonal effects produced by the position of the sun and by waves, which cause reflection, are the primary inhibitors here. The more important limiting factor is the availability of the nutrient elements phosphorus and nitrogen. Because of their continual utilization in the photosynthetic process, they are taken up and then removed from surface waters as the organism expires and descends through the water column. It is important that epipelagic waters are replenished in these nutrients by upwelling and from the landmasses through surface circulation.

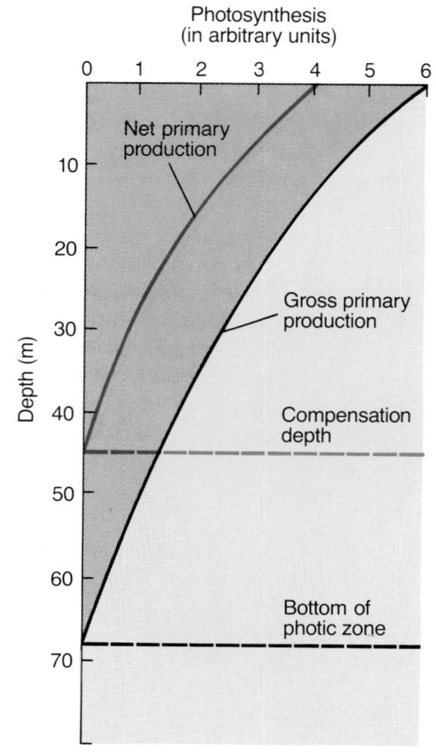

Figure 18.13 Relationship between gross and net productivity in the oceanic water column.

Phytoplankton

The pelagic phytoplankton are dominated by microscopic and submicroscopic organisms, including diatoms, coccolithophores, and dinoflagellates. A few macroscopic algae, such as the *Sargassum,* are also present, but they are insignificant by biomass as compared to the total phytoplankton mass.

The small size of most phytoplanktonic organisms is a distinct advantage in their environment. These small, unicellular organisms display intricate shapes and geometries (for example, fig. 9.2), which results in a tremendous ratio of surface area to volume. The large surface area facilitates the exchange of nutrients and waste with the surrounding waters. This efficiency has resulted in the domination of microscopic and submicroscopic organisms in the pelagic environment. The extreme surface-area-to-mass ratio also aids the organisms in floating. Additional aids for floating include secretion of certain low-density oils and gases.

Zooplankton

Planktonic animals are abundant and diverse in the epipelagic zone. Although both the phytoplankton and the zooplankton are abundant here, there is some difference in the distribution of these organisms relative to depth. In general the phytoplankton (producers) are nearer the surface than the zooplankton

BOX 18.1

Measurement of Productivity

The productivity of pelagic waters can be measured in a variety of ways. Some are simple and easily carried out where others are not. Regardless of the method employed, there are some drawbacks caused by the techniques used.

Productivity is the rate at which photosynthesis occurs and it may or may not be reflected in the biomass of photosynthetically produced organic tissue in a given environment. That is, there may be environments where the productivity is high, but due to a high rate of consumption, the biomass of the producers is low. Therefore we cannot determine an index of productivity by simply considering the amount of plant material in a given environment.

Measuring the rate of photosynthesis requires the measurement of some element involved in the photosynthetic process. The equation for this follows:

$$CO_2 + H_2O + \text{minerals} + \text{sunlight} \longrightarrow \text{organic matter} + O_2 + \text{heat}$$

The candidates are oxygen, carbon dioxide, and nutrients. Changes in the amount of these materials in seawater can be used as an index of productivity.

Oxygen is produced during the photosynthetic process in proportion to the organic matter produced and is an easily measured constituent of marine waters. One measurement method is to collect water samples over a diurnal sampling period so that data are available for both the daytime and the nighttime. The variation in oxygen content over the daylong period shows the rate of photosynthesis and the oxygen uptake as the result of respiration during the nighttime hours.

Another technique for measuring oxygen production is to collect a water sample and place half of it in a transparent bottle and the other half in an opaque bottle. Both contain the same microorganisms, but the sample in the transparent bottle will experience photosynthesis and respiration, mimicking the natural environment, whereas the sample in the opaque bottle will undergo only respiration. A comparison of the oxygen content at various time intervals will indicate the rate of productivity.

In a somewhat similar manner, the uptake of carbon dioxide in the photosynthetic reaction can also be used as an indicator of productivity. Like the technique for oxygen, this one also involves diurnal samples. The carbon dioxide concentration is related to pH and can be determined by very sensitive pH meters.

Uptake of the nutrient minerals phosphorus and nitrogen can also provide an index of productivity. Sequential samples are analyzed for either or both and compared much like the approach for oxygen and carbon dioxide.

There are several factors affecting the concentration of these materials in marine waters that are not related to the photosynthesis and therefore result in errors in productivity determinations. A very important factor is the diffusion of oxygen and carbon dioxide between the atmosphere and the water. This can be quite high, especially during rough water, but its impact is largely in the surface waters. The concentration of nutrient elements is affected by decay of organic tissues and precipitation of minerals in addition to photosynthesis. Carbon dioxide concentration is greatly affected by precipitation of minerals in shallow environments, especially reefs. There are, then, various methods for determining productivity, but there are problems that cause inaccuracies in all of them.

(primary consumers). Possible explanations for this include efficiency of producing and consuming, respectively. The zooplankton flourish just beneath the phytoplankton and consume the phytoplankton as they descend through the water column. It has also been suggested that the tremendous numbers of phytoplankton utilize so much space that they inhibit the zooplankton from occupying the shallow part of the epipelagic zone. The greatest concentrations of zooplankton tend to be at the margins of phytoplankton blooms, apparently as the result of feeding activities.

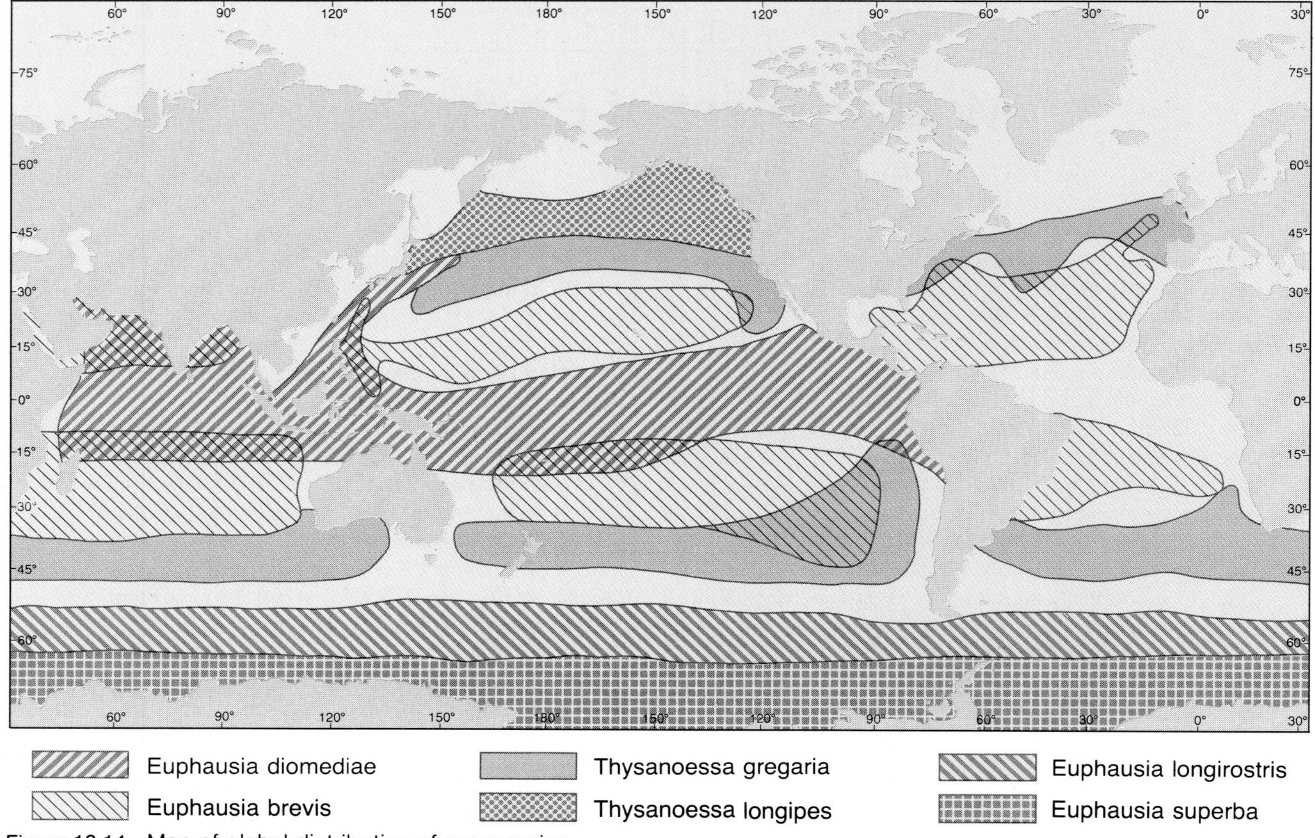

Figure 18.14 Map of global distribution of some major zooplankton species. These latitude-bound patterns are typical of zooplankton in general.

The geographic distribution of zooplankton shows distinct patterns that are controlled primarily by climate through water temperature. There are several geographic zones (fig. 18.14) that are climatic in nature: the tropical, subtropical, temperate, subpolar, and polar, going from the equator to the high latitudes. Although there is variation in the width of the zones and there is some overlap between zones, the basic pattern is one of latitudinal, and therefore climatic, control.

Nekton

The epipelagic zone also contains great diversity of swimming animals. These are dominantly carnivorous predators that feed upon the zooplankton or on other nekton. As a consequence, they are generally large, are good swimmers, and have various sensory capabilities to aid in their search for prey. There are also nekton that are herbivorous, such as anchovies and menhaden, which feed primarily on diatoms.

The fish that occupy the epipelagic environment display a drab coloration. They have dark green, blue, or gray dorsal sides and white or silver ventral sides. This affords camouflage from predators when viewed from above or below. If you look down on them, they will blend in with the darkness below and if you look at them from below toward the sky they will also blend in.

Mesopelagic Community

This zone also has a great abundance and diversity of life, but less so than the epipelagic zone. The primary difference is that the mesopelagic zone is below the zone of light penetration and therefore contains no producers, that is, no phytoplankton. Zooplankton are abundant but decrease with depth.

For the most part, the mesopelagic zone is beneath the thermocline and temperatures are only slightly affected by seasonal changes. They are typically about 10°C. This position below the thermocline results in an oxygen minimum layer and there is little interchange with the overlying waters. The oxygen minimum is due to the domination of animals (consumers) that consume oxygen and the absence of essentially all producers plus the lack of circulation with more oxygenated waters.

Zooplankton

There are large numbers of zooplankton in the mesopelagic zone. These organisms have various adaptations to living in this portion of the ocean. The advantage of living in the dark is that this condition makes it difficult for predators to locate prey. The cold water is denser and more viscous than surface water, thus slowing the descent of food and making it easier for primary consumers to obtain.

One of the particularly important attributes of mesopelagic zooplankton is their ability to migrate vertically through the water column, primarily for feeding. This is most commonly a diurnal migration cycle with the organisms ascending to the pelagic zone during the dark hours and descending to the aphotic zone during the day. The patterns of this migration are known from sampling at various depths over diurnal cycles and also from echo-sounders, which are generally used for bathymetry. The great quantities of zooplankton cause the sound pulses to be reflected and scattered, producing a pattern that is recorded on the ship's depth recorder (fig. 18.15). This **deep scattering layer,** as it is called, can be traced from the mesopelagic zone up to shallow water and then back again on a daily basis. The most likely organism that comprises this layer is the euphausiid shrimp (fig. 9.11), although small fish and planktonic coelenterates have also been suggested.

Nekton

The fish of the mesopelagic zone are markedly different from those of the epipelagic zone. They have some of the typical deep-water adaptations, such as small size, generally less than 10 cm, large mouths, numerous well-developed teeth, and lures to attract prey (fig. 18.16). All of these features aid the fish in being more efficient at gathering food and they help them to survive in this zone of limited food.

Another common adaptation of organisms in the mesopelagic zone is bioluminescence. Virtually all species in this region have some type of light-producing organ or structure. The glow emitted attracts prey and thereby reduces the energy expended in acquiring food. It also has the disadvantage of attracting some predators. Some of these luminescent structures may also be for species recognition in reproduction.

Deep-water Community

Beneath the mesopelagic zone the pelagic ocean is totally dark and essentially uniform in terms of temperature, salinity, and other factors except pressure, which increases with depth. Fish are small and dominate the environment, but are scattered. They must depend on other deep-sea creatures or descending organic debris for their food supply. The organisms of this cold water environment have very slow rates of metabolism, which is a distinct advantage in a zone of little food. They cannot migrate very far vertically because of their adaptation to high pressures. A migration upward would result in their bodies exploding due to the expansion of body gases in their air bladders. Their bones and muscles are poorly developed and their mouths are extremely large (fig. 18.16).

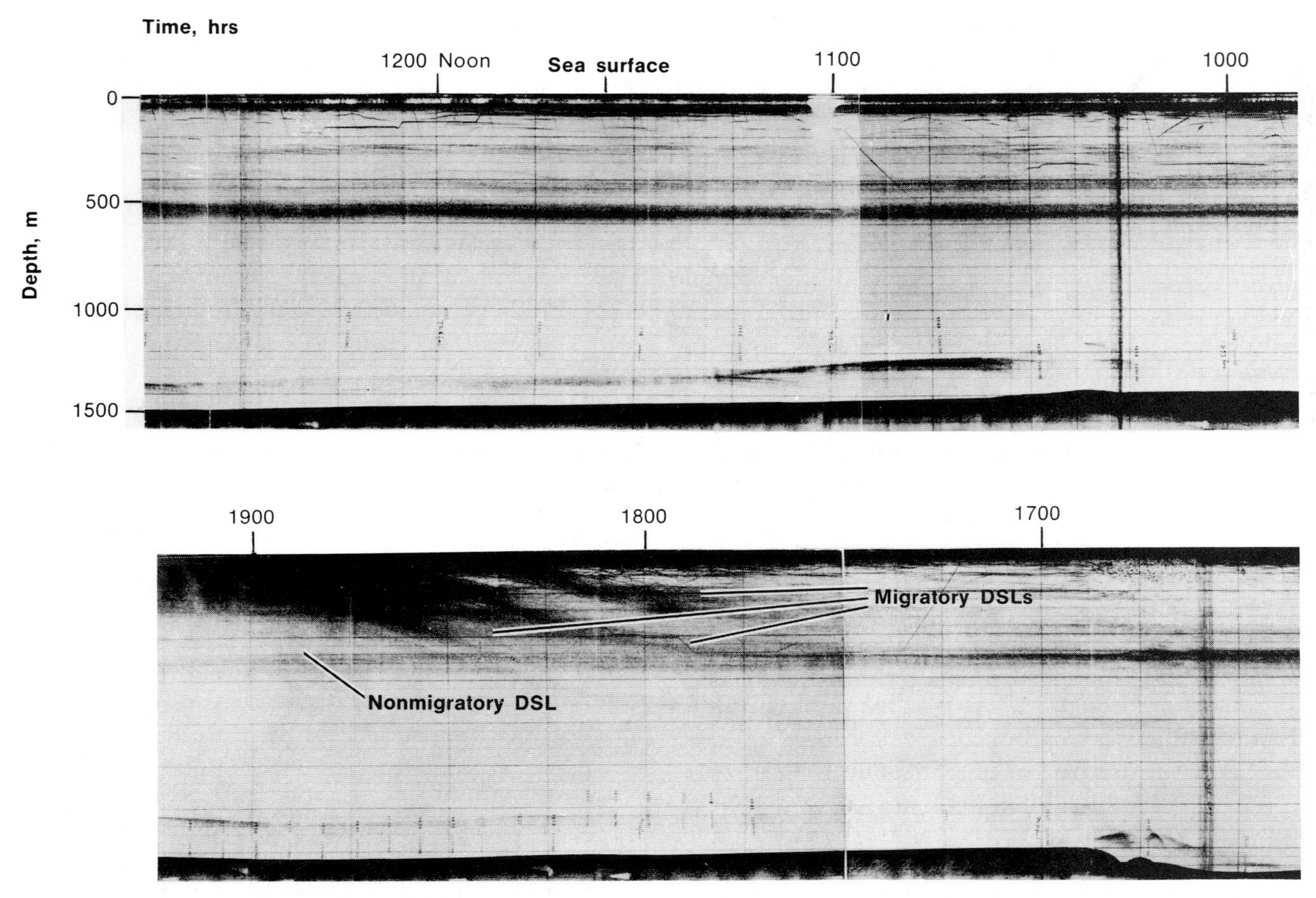

Figure 18.15 Precision depth recorder record of the diurnal migration of zooplankton for feeding.

Figure 18.16 Two examples of deep-sea fish with their typical large mouths and teeth.

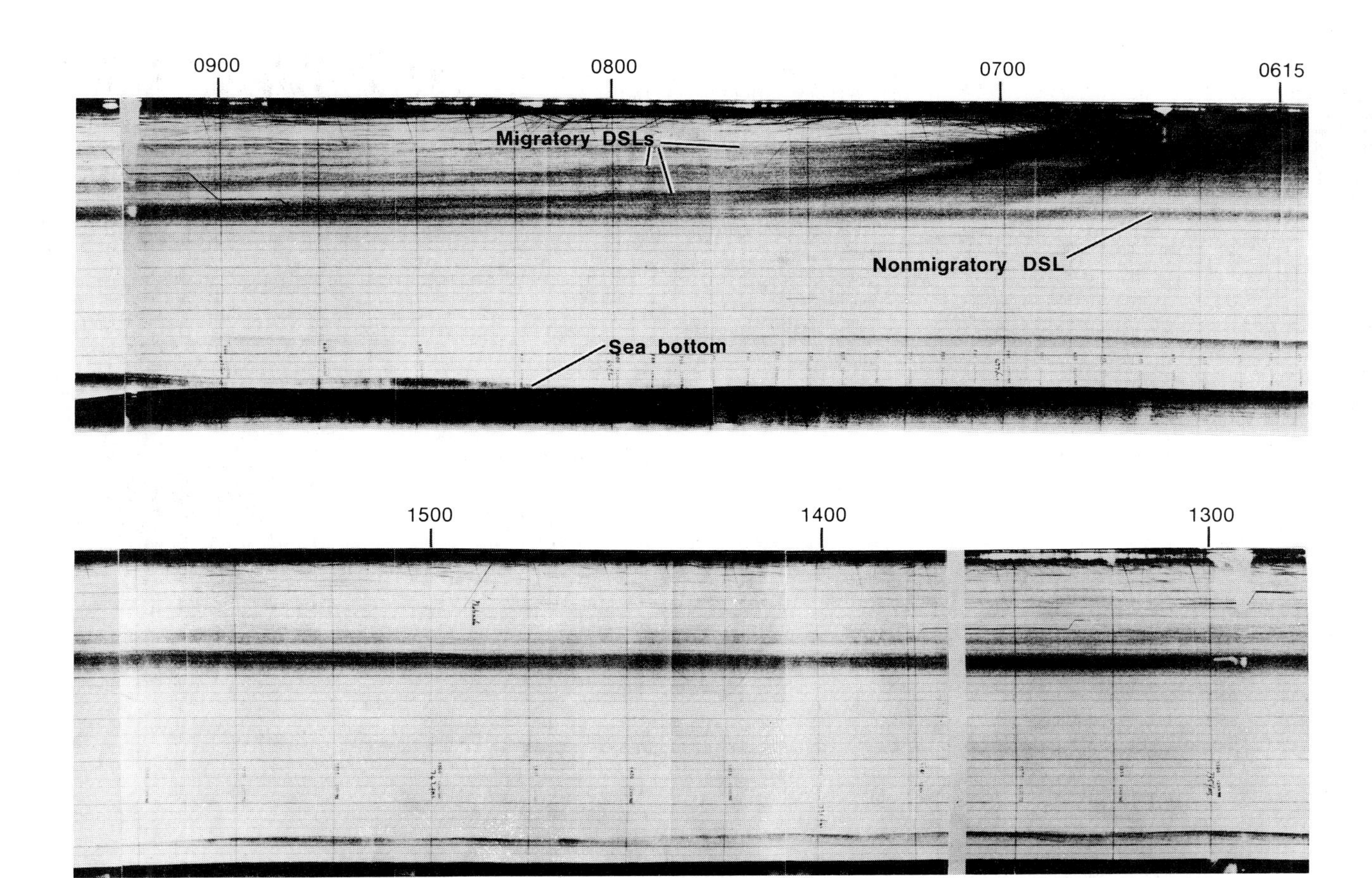

Summary of Main Points

The pelagic environment contains the vast majority of the water volume of the world ocean and also most of the life. Although there is great variation in the environment, its enormous size causes the pelagic environment to appear uniform relative to coastal and continental margin environments.

The thin surface water mass is controlled by global wind patterns, but most of the water in the world ocean circulates slowly as the result of density gradients. Although most of the density variation in the deep ocean is due to temperature differentials, salinity variation is also a factor. In general the thermohaline circulation moves from high latitudes where the water masses form to the lower latitudes. The north Pacific is an exception in that there are no deep-water masses formed in this area.

The pelagic water column shows significant changes from the surface to the ocean floor. These are in terms of the physical properties, chemistry, and the abundance and nature of the organisms present. Life is most abundant in the surface waters and decreases through the mesopelagic and bathypelagic zones. Plankton are concentrated in the epipelagic and upper mesopelagic zones. Fish are also abundant in upper waters, but show less decrease in deeper waters than plankton.

Suggestions for Further Reading

Coker, R.E. 1962. **This Great and Wide Sea: an Introduction to Oceanography and Marine Biology.** New York: Harper and Row.

Hardy, A. 1958. **The Open Sea: Its Natural History, part 1, The World of Plankton.** Boston: Houghton Mifflin.

Idyll, C.P. 1976. **Abyss—The Deep Sea and the Creatures that Live in It.** New York: T.Y. Crowell.

Munk, W. 1955. "The circulation of the oceans," *Scientific American,* v. 191:87–94.

Stommel, H. 1955. "The anatomy of the Atlantic," *Scientific American,* v. 190:30–35.

Thorson, G. 1971. **Life in the Sea.** New York: McGraw-Hill, Inc.

19
The Floor of the Ocean Basins

Chapter Outline

The ocean basins include the entire marine environment between the continental rises of the outer margins. This represents most of the world ocean both in terms of the area covered by the basins and the volume of water contained therein. It may seem that devoting only two chapters (18 and 19)to this enormous part of the oceans represents a lack of balance of treatment. That is really not true because of the great size and the homogeneity of the ocean basins with respect to water masses, circulation, life, sediments, and configuration relative to other marine environments.

There are three distinct physiographic provinces within the ocean basin proper. These are the abyssal floor, oceanic ridge, and deep-sea trenches. The latter represents only a small part of the total ocean basin, whereas the abyssal floor and oceanic ridge cover enormous areas (fig. 19.1).

Provinces

Abyssal Floor

The abyssal floor is a broad and relatively flat environment that is essentially the floor of the deep ocean. Included within the abyssal floor are the abyssal plain, which is the extremely flat part of this province, and the abyssal hills, which are gently rolling features that are scattered throughout the ocean floor.

Abyssal Plain

The gradient of the abyssal plain is defined as being less than 1:1,000, which serves to distinguish it from the adjacent continental rise. These plains are the most extensive flat environments on the surface of the earth. The water depth over the abyssal plain ranges from about 3,500 to 5,000 m. This seems like considerable variation for a flat environment, but it can be explained by the fact that the abyssal plains slope gradually for hundreds of kilometers.

The great, extensive abyssal plains occur in each of the three major ocean basins (fig. 19.1) and are also present in the large mediterraneans, such as the Gulf of Mexico and the Mediterranean Sea. Data from the Deep Sea Drilling Project as well as numerous seismic cross sections of the ocean basins show that the abyssal plains are underlain by thick sequences of sediments and sedimentary rocks. Although the thickest sections are up to a few kilometers thick, the sedimentary cover thins toward the oceanic ridges and thickens toward the continental margin.

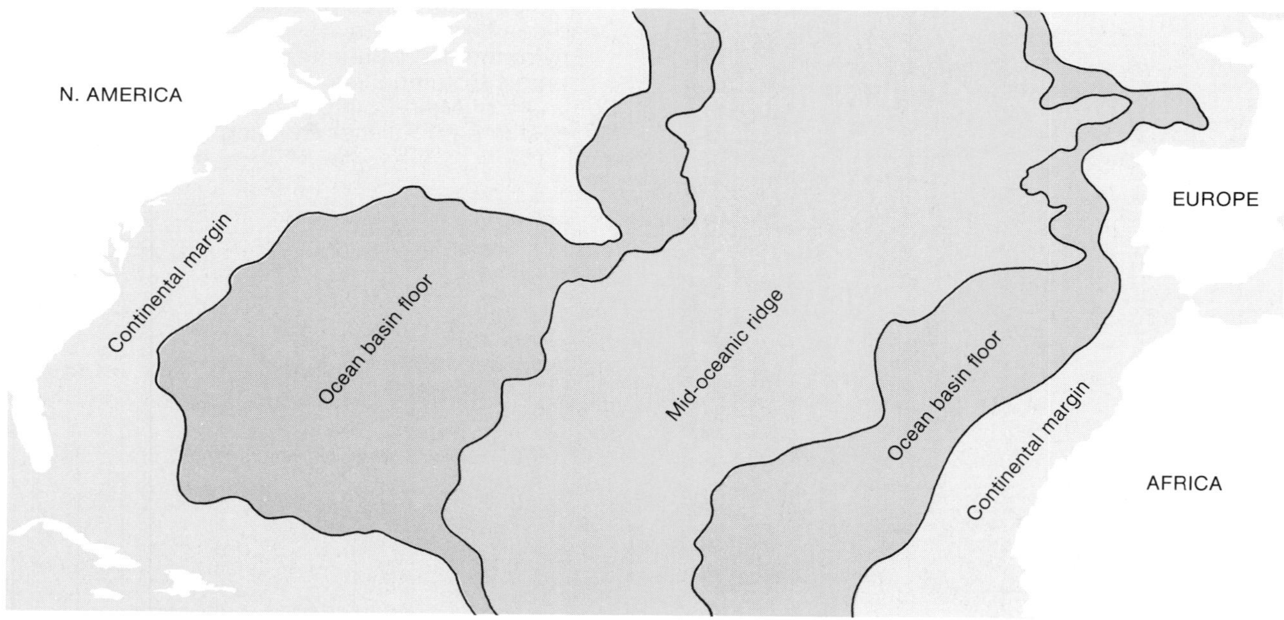

Figure 19.1 General physiographic map of the North Atlantic Ocean showing the major provinces. (From Heezen, B. C., et al, 1959, "The Floor of the Ocean-I, The North Atlantic," *Geological Society of America,* Special Paper 65.)

The seismic sections also demonstrate that the sediments of the abyssal plain cover a broadly irregular surface. This irregular surface is interpreted as being underlain by volcanic flows based on its configuration and the lithology of cores taken from the DSDP.

Abyssal Hills

These features range from 50–250 m in relief and may be up to several kilometers in diameter. They poke up through the sediment cover of the abyssal floor throughout the ocean basins. Abyssal hills are common in the Pacific Ocean and are scattered in the other two oceans. Their tendency to be more common toward the oceanic ridge system suggests something about their origin, namely, that the abyssal hills are old, partly covered piles of material with the same composition as the oceanic ridge, but that they have moved away from the spreading center (fig. 19.2).

Oceanic Ridge

The **oceanic ridge** is a mountainous feature that is continuous for 80,000 km over the ocean basins (see fig. 2.13). It has from a few hundred to 2,000 m of relief and its crest rises to an average depth of only 2,500 m below sea level. The ridge is about 1,000 km wide at its base, with a crest that rises 1–3 km above the abyssal floor of the ocean. Topography tends to be rugged, especially near the central area, where relief is greatest (fig. 19.3).

The first ridge discovered was in the Atlantic Ocean where the feature essentially bisects the ocean basin. The term **Mid-Atlantic Ridge** was applied and subsequently the term "mid-oceanic ridge" became used for this feature on a global scale, although the ridge is not in the middle of the Pacific or Indian Oceans. The oceanic ridge is characterized by high seismic activity and active volcanism. Locally, it rises above sea level and forms volcanic islands, the largest of which is Iceland.

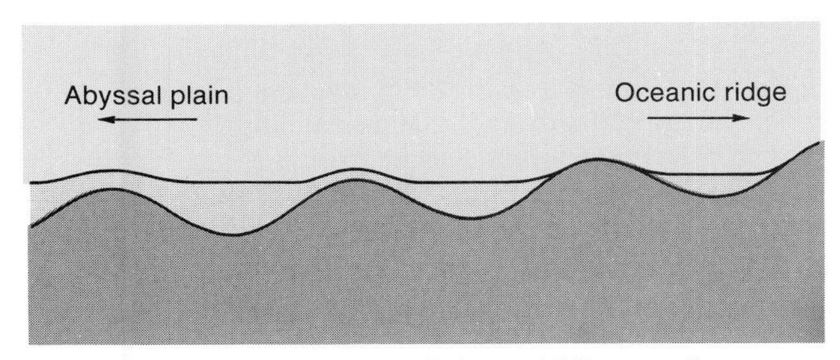

Figure 19.2 General sketch of abyssal hills near the oceanic ridge showing their relationship to cover of sediments.

The ridge, although continuous, is dislocated at hundreds of places by large fracture zones. These fractures or faults are the result of movement in the earth's crust. They are nearly parallel, cutting perpendicularly across the oceanic ridge (fig. 19.4). Fracture zones are tens of kilometers wide and some extend for over 3,000 km, with relief up to 4 km. Each fault may displace the ridge crest several hundred kilometers.

Mid-Atlantic Ridge

The oceanic ridge system in the Atlantic nearly bisects the basin and was one of the first lines of evidence used to support the idea of seafloor spreading. It also serves as a good example of the high-relief, rugged end of the morphologic spectrum within the oceanic ridge system. As seen in the topographic profile in figure 19.3, there is a general rise in the ocean floor from the abyssal floor to the crest of the oceanic ridge. Upon that gradient is superimposed the considerable relief of the Mid-Atlantic Ridge. It contains a central rift valley, which is up to 2 km deep and tens of kilometers wide. It is within this rift valley that the youngest rocks of the oceanic crust are found.

East Pacific Rise

The oceanic ridge system in the southeast Pacific has a different character than the Mid-Atlantic Ridge. In fact, when first discovered, this was not thought to be part of the same system as the Mid-Atlantic Ridge because it has such a different topography. The term "rise" was applied because of the similarities with other oceanic features, such as the Bermuda Rise and the Rio Grande Rise in the Atlantic Ocean.

The East Pacific Rise shares many features in common with the Mid-Atlantic Ridge, but has a

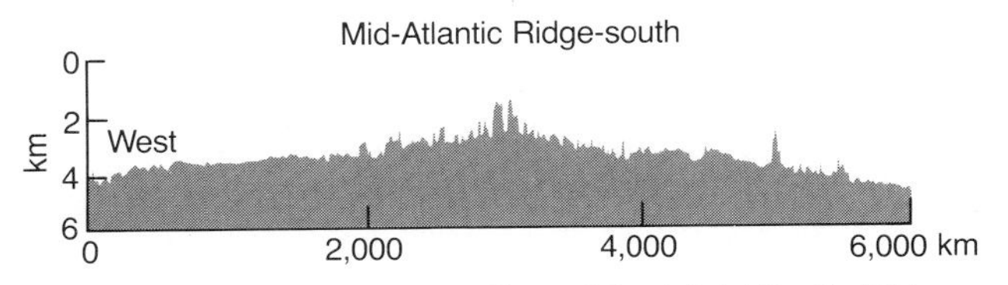

Figure 19.3 Topographic profiles of the Mid-Atlantic Ridge.

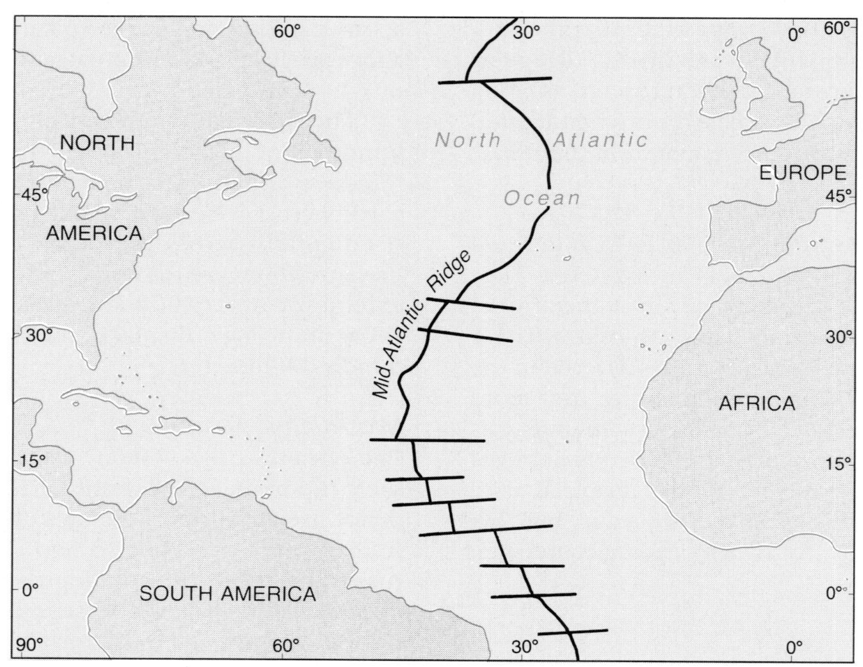

Figure 19.4 Map of part of the North Atlantic Ocean showing fracture zones and the dislocation of the ridge system.

markedly different profile: the relief on the East Pacific Rise is relatively low and it has no central rift valley. This difference is thought to be due to the rate of spreading, with the East Pacific Rise spreading more rapidly than the Mid-Atlantic Ridge.

Deep-sea Trenches

Trenches are more areally restricted than the other two provinces of the ocean basin, but they are spectacular in their configuration. A **deep-sea trench** is a long, linear, arcuate feature that is at least 6,000 m below sea level. Trenches tend to be thousands of kilometers long, hundreds of kilometers wide, and up to 11 km deep. The deepest part of the ocean is in the Challenger Deep in the Mariana Trench (11,035 m) in the western part of the Pacific Ocean.

The bathymetric profile of a trench is noticeably asymmetrical (fig. 19.5), with the seaward side of the trench being less steep than the landward side. Typically there is a volcanic island arc or mountain

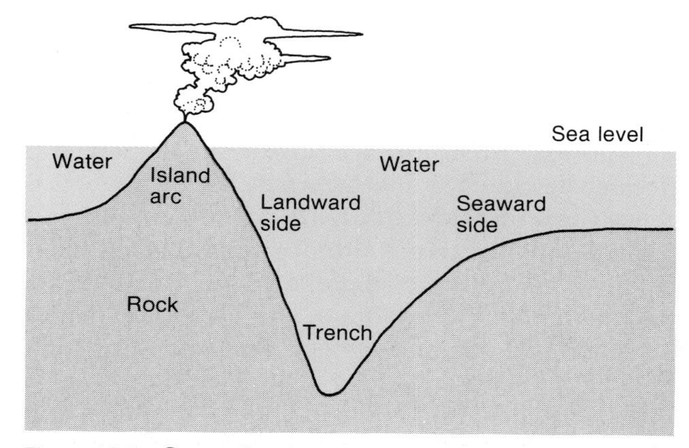

Figure 19.5 Generalized profile of a typical deep-sea trench, showing the characteristic asymmetry.

range on the landward side of the trench. For example, the Aleutian Trench is just seaward of the Aleutian Islands in Alaska, and the Middle-America Trench is adjacent to the volcanic ranges of Central America (fig. 19.6). This asymmetric profile and relationship of trenches to volcanic mountain ranges

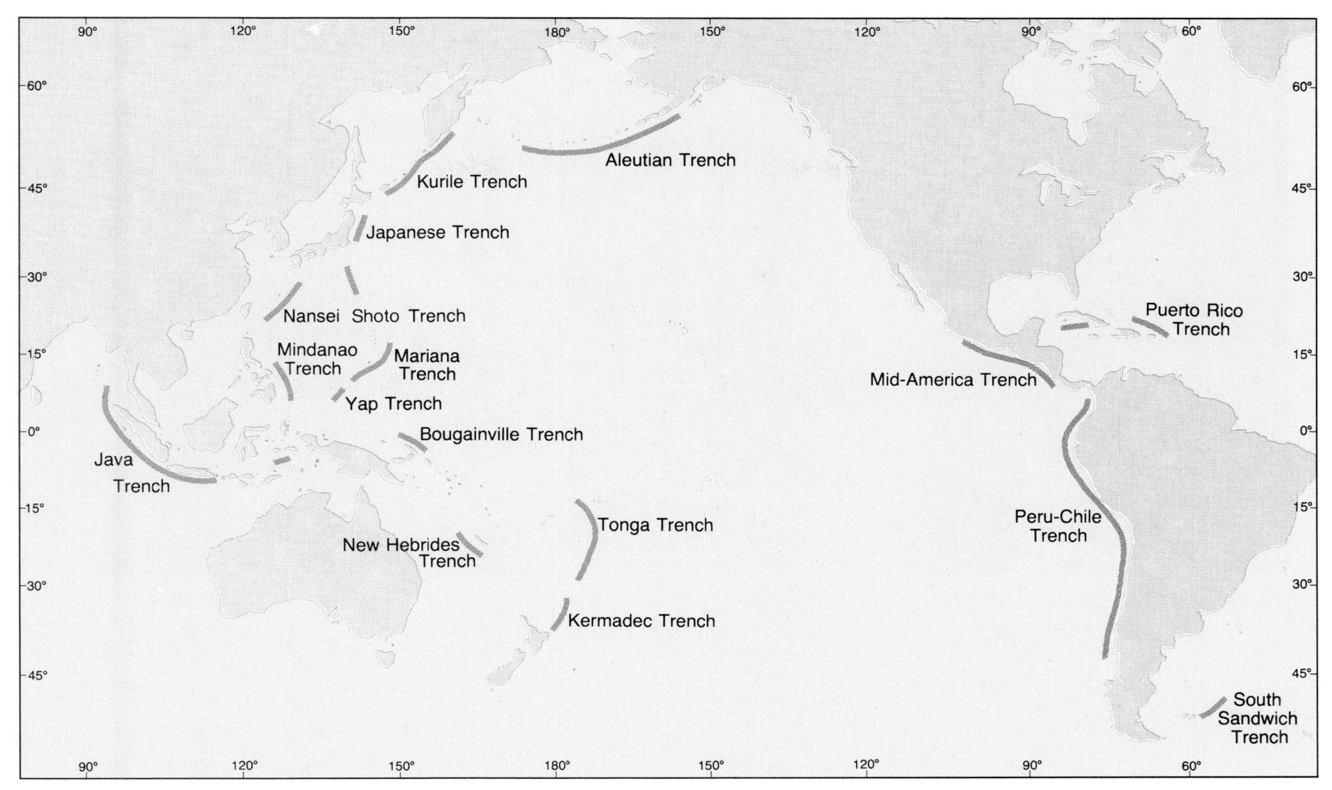

Figure 19.6 Map showing the global distribution of oceanic trenches. Notice that the vast majority is in the Pacific Ocean.

is associated with the concept of plate tectonics (see chapter 3). The trenches represent the subduction area where plate margins descend under other plates during collision and the volcanic mountains are the result of magmatic activity at these collision zones. The numerous earthquakes that occur in the areas surrounding the Pacific are also related to these collision zones and therefore to the trenches.

Distribution of the trenches shows that they are most common around the Pacific **ring of fire** where plate collision occurs. The term comes from the concentration of volcanoes around the Pacific Ocean where plates collide. Trenches also occur in the Caribbean Sea and South Atlantic, as well as off the coast of Java in the Indian Ocean. All are plate collision areas. Trenches also show a gently arcuate areal configuration with the convex side toward the ocean basin (fig. 19.6). Although the trenches cover only about 1% of the ocean floor, they are an important oceanic feature and are an integral element in the plate tectonics concept.

Processes

Typically one thinks of the deep-sea environment as a dark, cold, and still place where there is no activity except perhaps for a few benthic creatures. In fact, however, the deep sea is a dynamic environment with various processes that tend to mold the bottom. This discussion will consider three major process categories: physical, volcanic, and biologic. The physical processes are dominated by deep-ocean currents, which interact with bottom sediments causing their erosion, transportation, and deposition. There are many parts of the ocean floor, especially in the ridge areas, where volcanic activity plays a vital role in the development of the deep-ocean environment. The numerous benthic organisms interact with the substrate to cause various bioturbation features.

Figure 19.7 Stalked benthic organisms (crinoids) bending as the result of deep bottom currents.

Deep-sea Currents

The introductory discussion of deep oceanic currents (chapter 5) considered thermohaline density currents, which dominate in about 98% of the volume of marine water. These currents are sluggish, with average speeds of only 1 or 2 cm/sec (about 100 ft/hr). Photographs of the ocean floor show various types of evidence of swifter currents, including the bending of benthic stalked organisms (fig. 19.7), along with various ripples and other indications of sediment movement (fig. 19.8).

The oceanic scale pattern of deep currents is one of cold, dense water forming near the poles and descending to or near the ocean floor. These currents then flow toward the low latitudes and across the equator (fig. 18.9). Along the margins of the ocean basins, especially on the west sides of the basins, there are currents that move much faster than average because they are compressed by the rotating earth. These currents are responsible for the sediment transport shown in figure 19.8. Even deep currents are capable of moving sediment, although they are not considered swift in comparison with surface currents. Because the sediment on the ocean floor is very finely grained, rarely coarser than 20 microns (silt), and because it is also loosely packed on the ocean floor, it takes currents of only 10 or 15 cm/sec to move the particles. Local topography may cause rapid currents of up to twice the above speeds on the seafloor, for example, around a seamount or other feature that protrudes above the ocean bottom. Constrictions to flow, such as the Florida Straits, also produce elevated bottom current speeds.

Deep-ocean currents not only move sediment along the ocean floor, resulting in ripples and other bottom features, they also generate clouds of very fine sediment that may be transported long distances over the seafloor. These clouds are formed by resuspension of bottom sediment and they produce high concentrations of sediment in the lower 100–200 m of the water column. Some suspended sediment is present up to 1.5 km above the ocean floor (fig. 19.9).

Turbidity currents that originate on the outer continental margin may also make their way onto the ocean floor. These sediment-laden water masses produce currents that may persist for hundreds of kilometers. Consequently, turbidites may reach the abyssal plain environment where they fill in the low

Figure 19.8 Deep-sea photograph of ripples and other features indicative of sediment transport on the ocean floor.

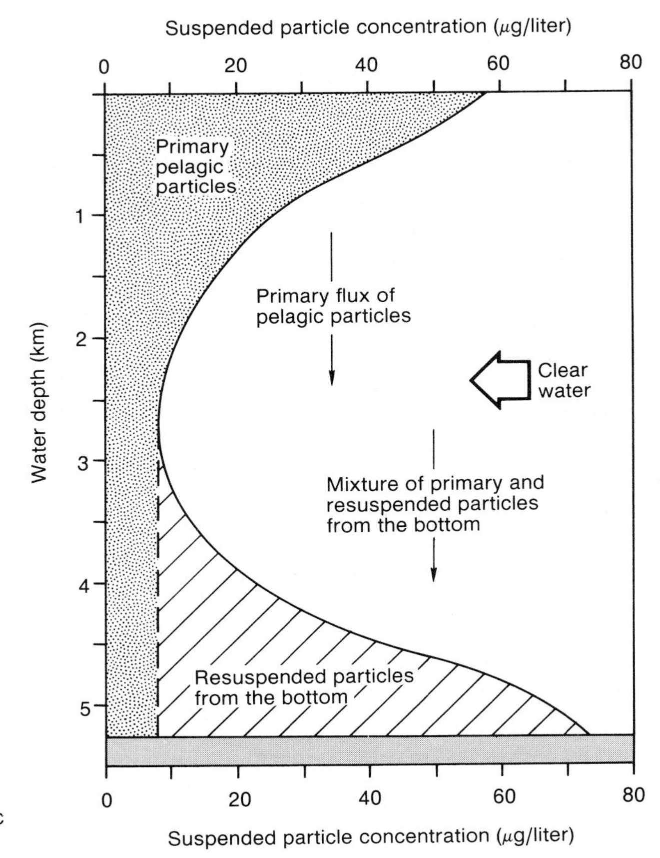

Figure 19.9 Vertical distribution of suspended sediment in the oceanic water column.

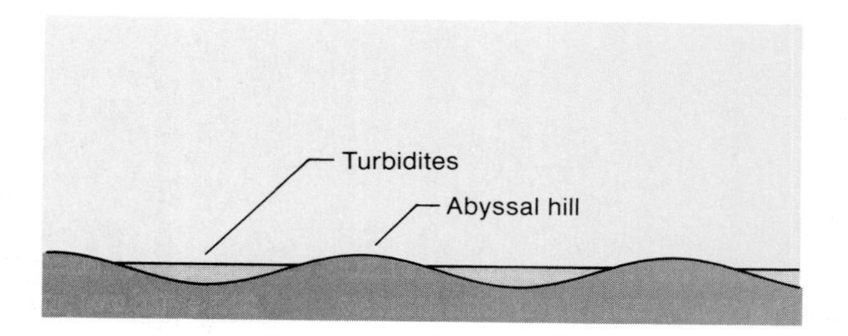

Figure 19.10 Turbidites filling in low areas among the abyssal hills on the ocean floor.

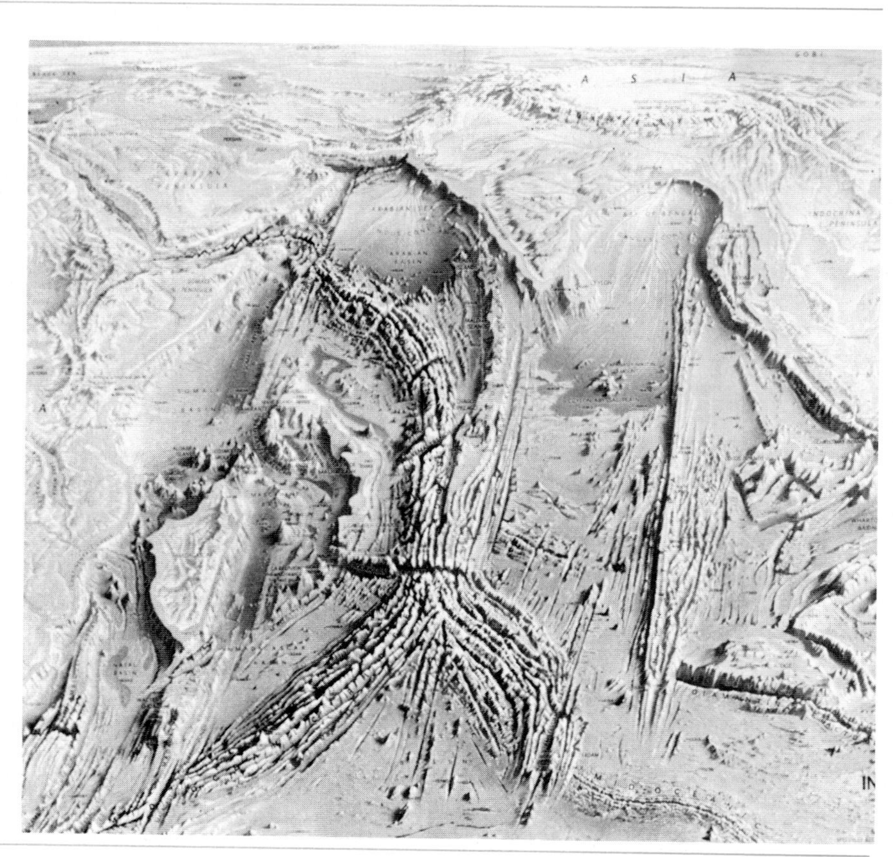

Figure 19.11 General topography of the Ninety East Ridge in the Indian Ocean.

areas between topographic high spots, such as abyssal hills (fig. 19.10). These abyssal turbidites tend to be thin and finely grained because they are deposited by the distal portion of the turbidity current.

Volcanic Activity

Volcanic eruptions are a fairly common phenomenon in the ocean basins. They occur in three primary associations: (1) those related to plate collision zones and the seismically active zones of the earth's crust, (2) those that are generated in spreading zones associated with oceanic ridges, and (3) those that are generated in various parts of the oceanic crust not related to plate boundaries or zones of seismic activity.

Oceanic ridges are one of the most extensive areas of volcanism and basaltic eruptions. Flows of lava are extruded in the area of the central rift valley and may extend for several kilometers from the center of the eruption. Piles of successive flows may be as high as 200 m. They tend to develop in linear masses along the central rift valley. As spreading from the rift occurs, these volcanic materials are carried

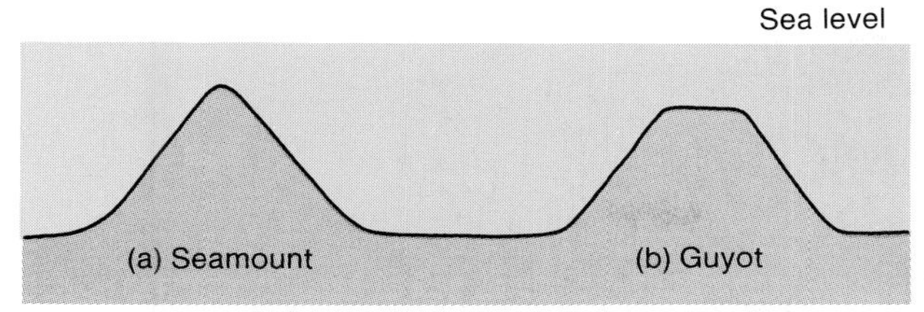

Figure 19.12 General profile sketch of (a) a seamount and (b) a guyot.

to more distant parts of the ocean floor. In some situations these flows are so thick and contain so much mass that the crust fails and breaks.

Abundant volcanism has also produced the large plateaus and aseismic ridges that are present on the ocean floor. These thick piles of volcanic rock may reach over a kilometer in thickness. They are scattered about the ocean basins, with older ones, such as the Blake Plateau, being covered with fine-grained deep-sea sediment. Younger versions include the Galapagos Platform in the southeastern Pacific, the Azores, and Iceland in the north Atlantic Ocean.

The aseismic ridges are similar in origin to the plateaus, but have a linear shape, usually straight. They may have originated as old fractures in the thin oceanic crust, with lava outpourings accumulating to great thicknesses. Some aseismic ridges may at one time have been in shallow water, based on the presence of certain carbonate sediments that form only under those conditions. The Ninety East Ridge in the eastern Indian Ocean is an example of such a feature (fig. 19.11).

Some of the most interesting and widespread of the volcanic features in the oceans are the seamounts and guyots. Individual or grouped volcanoes occur on the ocean floor with either of two distinct shapes. The **seamounts** are conically shaped structures that mimic their volcanic counterparts on land in size and shape. By contrast, guyots, or tablemounts as they are sometimes called, have a similar shape but with a flat upper surface (fig. 19.12). They are, in effect, decapitated seamounts. Guyots apparently were eroded near sea level by waves, producing the flat upper surface. They were then colonized by reefs, and they accumulated shallow-water sediments and organisms. The presence now of these sediments on the top of the guyot at depths in excess of a kilometer and the tilting of their upper surface indicate that subsidence has occurred since their formation.

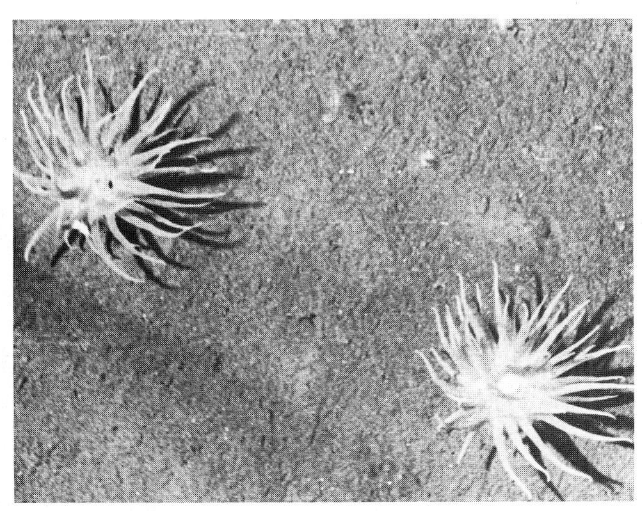

Figure 19.13 Photograph of the ocean floor showing deep sea benthic anemones.

Guyots and seamounts number in the thousands throughout the ocean basins of the world, with the Pacific having the most. Many are associated with volcanic island groups, such as the Hawaiian Islands, the Fiji Islands, and others. These groups of volcanic islands had their origins from fractures or hot spots on the seafloor. As the crust passed over these spots, volcanic structures developed, resulting in a chain of volcanic islands.

Volcanism is also commonly associated with the seismically active island arc systems along plate boundaries. These island arcs develop as the result of magmatic activity produced by the descending plate margin during collision (see fig. 3.11).

Biological Processes

Organisms on the floor of the ocean basins interact with the substrate in much the same fashion as those of the shallower environments. Among the main differences are that the organisms are fewer in numbers and they are typically much smaller than their shallow-water counterparts (fig. 19.13). Two types of biological processes are common in the benthic community of the deep ocean: (1) filter feeding by sedentary organisms and (2) grazing by active bioturbators. Essentially all of them are detritus feeders that produce large quantities of fecal pellets.

BOX 19.1

Project FAMOUS

The Mid-Atlantic Ridge was examined and sampled by scientists using the submersible *Alvin* (see chapter 1) during 1973 and 1974. This project, called FAMOUS (French-American Mid-Ocean Undersea Study), involved first-hand observation and sampling of the central rift valley of the ridge system southwest of the Azores (see fig. 2.13). Extensive preliminary work including side-scan sonar surveys, detailed bathymetric surveys, and bottom photography preceded the first submersible dive. The major objective was to determine the overall nature of the central rift valley of the oceanic ridge system including heat flow, water chemistry, the fauna, and the rocks and sediments that comprise the seafloor in this environment.

The entire project included fifty-one dives, seventeen of which were by *Alvin* with the others by French submersibles. The rift valley floor in this area is nearly 3,000 m deep and its walls rise about 1,500 m above the floor, giving a relief that is much like the Grand Canyon. The entire rift valley is 30 km wide, but the innermost valley is only a few kilometers wide and has steep walls that rise 300 m above the floor. The seafloor here is rugged with small-scale relief of meters to tens of meters. Volcanic flows are widespread (fig. B19.1) and thin veneers of sediment are locally present. Pelagic sediment accumulation takes place at a rate of about a centimeter per thousand years. Some surfaces are irregular and jagged due to viscous flows and explosions from escaping gases. In fact, some of the rock samples exploded after they reached the surface due to the great decrease in pressure on the gases trapped in them.

Thousands of photographs were taken and hundreds of samples were collected in order to systematically map the rift area. Dating of rock samples showed that the increase in age away from the central valley indicates a spreading rate of 2–3 cm/yr. Small earthquakes persist in the area of the central fracture zones.

A surprising abundance and diversity of life was observed by FAMOUS personnel, including mollusks, corals, sponges, echinoderms, and fish. Efficient filter-feeding organisms such as crinoids, sponges, and corals, seem to do well. Tracks and trails were also observed in the fine sediment, indicating the presence of a vagrant benthos fauna.

Figure B19.1 Photograph of volcanic flows in the oceanic rift valley.

Figure 19.14 Photograph of the deep-sea floor showing indications of bioturbation.

The sedentary varieties rely on the slow currents to carry detritus to them. Lack of selectivity causes much of their intake to be inorganic and is therefore converted to pellets after passage through the digestive system.

Most of the benthic organisms are bioturbators that forage through the substrate, ingesting large quantities of sediment. Because of the overall shortage of organic debris they also pass much inorganic sediment through their systems and produce fecal pellets. Although organisms are not large or abundant on the abyssal floor relative to shallower environments, the slow rate of sediment accumulation and the active feeding habits of the organisms tend to produce a markedly bioturbated substrate (fig. 19.14).

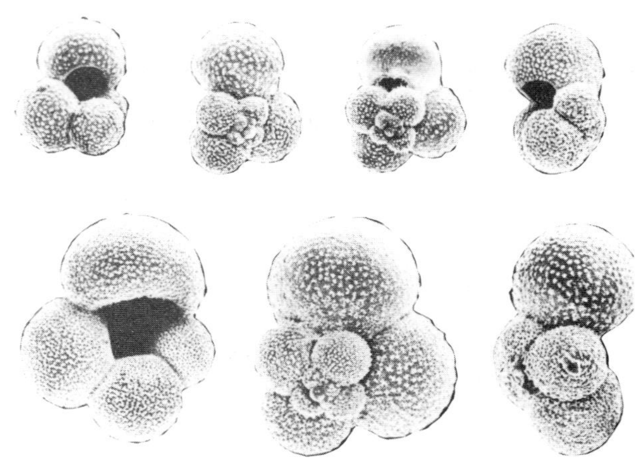

Figure 19.15 *Globigerina*, one of the dominant planktonic foraminifera that comprise calcareous ooze.

Figure 19.16 Pteropds, planktonic gastropods that are an important constituent of some calcareous oozes.

Deep-sea Sediments

Like most of the ocean basin characteristics, sediments also tend to be relatively homogenous as compared to continental margin environments. There are important differences and distribution patterns. The classification of deep-sea sediments can be treated in fairly simple terms and yet produce a good representation of the nature of their composition and distribution.

There are two broad categories of deep-sea sediments, terrigenous and biogenic. Terrigenous sediments are those that are derived from the land, that is, from previously formed rocks and sediments. Biogenic sediments are those that originate as skeletal material primarily from organisms that live in the pelagic environment. Both of these sediment types are abundant and widespread throughout the world ocean. There are two additional categories that each contribute a small amount to the total volume of ocean basin sediment. These are **cosmogenic sediments** (extraterrestrial particles), which fall into the ocean from beyond the earth's atmosphere, and direct chemical precipitates, including such things as manganese nodules.

Biogenic Sediments

On a global basis, biogenic sediments are more widespread than those of terrigenous origin. Most are composed of microscopic tests or skeletons of both plant and animal varieties. They include organisms with both calcium carbonate ($CaCO_3$) and siliceous (SiO_2) tests.

Calcareous Ooze

Calcium carbonate skeletal material is produced primarily by three types of organisms: coccolithophores, planktonic foraminifera, and pteropods.

Coccolithophores are submicroscopic, photosynthetic organisms that range from about 4 to 50 microns in diameter. They live in shallow open waters and have numerous calcareous plates held together by organic tissue. When the organism expires, the tissues decay and the plates disarticulate and fall to the ocean floor. **Planktonic foraminifera** comprise the most abundant of the calcareous oozes on the ocean floor. These bulbous microscopic organisms live in shallow water and fall to the bottom in the same fashion as coccoliths. The foraminifera are dominated by the genus *Globigerina* and other close relatives (fig. 19.15). They are extremely widespread and certain of the species are quite temperature sensitive. Distribution of these temperature-sensitive species has enabled scientists to use planktonic foraminifera as indicators of paleoclimates, the ancient climatic conditions on the earth. **Pteropods** are small, conically shaped gastropods (snails) that float in shallow open ocean waters. They are about a millimeter or so in length and have a fleshy structure that extends from the calcareous shell (fig. 19.16). Pteropods are the least abundant of the calcareous oozes.

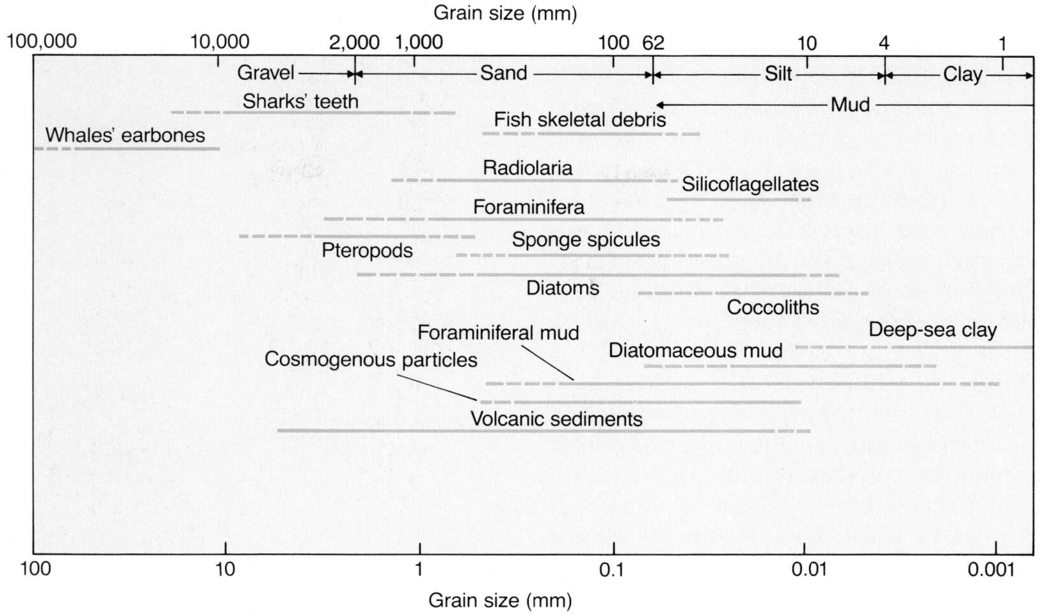

Figure 19.17 Chart showing the size of the important constituents of deep-sea sediments. (M. Grant Gross, *Oceanography: A View of the Earth,* 3d ed., © 1982, p. 88. Adapted by permission of Prentice-Hall, Inc., Englewood Cliffs, New Jersey.)

These combine to produce **calcareous ooze,** the term applied to fine-grained sediment containing at least 30% biogenic calcareous particles. The remaining percentage may be a variety of compositions, but is generally dominated by fine terrigenous clay.

Siliceous Ooze

There are also three primary varieties of organisms that contribute to deep-sea siliceous oozes. These are, in order of decreasing abundance: diatoms, radiolarians, and silicoflagellates. All are microscopic.

Diatoms are the most abundant group of organisms in the sea and as a consequence they provide abundant particles of siliceous sediment when they expire and descend to the seafloor. The diatom skeletons exhibit a great variety of shapes and range in size from about 4 to near 100 microns.

Radiolarians are microscopic, planktonic, single-celled animals that have a radial symmetry and range from about 50 microns to over a millimeter in diameter (about the size of a sharp pencil point). They have numerous delicate spines that assist in their floating mode of life.

Silicoflagellates are also primary producers. They occupy the same general niche as the diatoms and are similar in size. Although this group is not as diverse or abundant as diatoms, they are widely distributed in the world ocean.

Although the organisms that comprise deep-sea ooze are all small, there is variation depending on the group (fig. 19.17). Refer to chapter 9 for a review of these planktonic organisms.

Terrigenous Sediments

Sediment particles that are derived from previously existing sediments or rocks comprise the bulk by volume of the sediment in the deep ocean. Although many of the deep-ocean sediments are termed biogenic oozes, because only 30% of these oozes must be biogenic material to be so classified, they may also contain large amounts of terrigenous particles, typically clay minerals.

The terrigenous deep-sea sediments have two primary modes of origin: (1) they may reach the ocean floor by settling through the water column and are therefore pelagic, or (2) they may travel on or near the bottom. The latter are most abundant near land, with turbidites and glacial marine sediments the dominant types.

Turbidites are deep-sea terrigenous sediments whose origin was discussed in the previous chapter. These sediments commonly flow out onto the abyssal floor and fill in topographic low areas. **Glacial marine sediments** are those derived from glaciers. Sediment-laden ice sheets flow out over the ocean basin and melt or partially melt, thereby providing sediment to the ocean floor. Because glaciers contain a broad spectrum of composition and grain size, glacial marine sediments reflect this variety. Particles falling from melting ice sheets may be up to boulder size.

Most terrigenous sediment is of pelagic origin. The numerous rivers of the world provide great quantities of fine sediment particles, most of which eventually come to rest on the ocean floor. Wind also provides some of this sediment as do volcanic eruptions. Particle size of terrigenous sediments ranges from a few microns to several centimeters, with the glacial sediments being the coarsest. Pelagic clays are the most abundant and are less than 10 microns in diameter (fig. 19.17).

Miscellaneous Deep-sea Sediments

Manganese nodules comprise the bulk of direct chemical precipitates on the seafloor although some clay minerals and phosphorites are also precipitated in this environment. These nodules were first discovered in the south Pacific Ocean during the *Challenger* expedition. They have since been found throughout the world in deep waters. Manganese nodules are amorphous masses of manganese and iron oxides with a fair amount of SiO_2. They range in size from spherical nodules a few millimeters in diameter to platy encrustations over a meter in diameter. Most are a few centimeters in diameter. They occur scattered over the ocean floor (fig. 19.18), but are highly concentrated in some areas. The economic potential of these nodules is great and is discussed in chapter 20.

Figure 19.18 Photograph of manganese nodules on the seafloor.

Cosmogenic sediments are those derived from extraterrestrial sources. Meteorite showers produce particles that fall to the seafloor and become incorporated in the sediment. The total volume is small in comparison with other deep-sea sediments, although it is estimated to total an influx of a few thousand tons per year to the world ocean. The typical occurrence of these particles is in small spherules ranging from one to hundreds of microns in diameter. The composition varies, but nickel, iron, and silicon are the most abundant elements.

Distribution of Deep-sea Sediments

Because of the relative homogeneity of the ocean floor conditions and the domination of a few sediment types, the ocean basins of the world can be characterized by only five deep-sea sediment types: calcareous ooze, siliceous ooze, deep-sea clay, turbidites, and glacial marine sediments. Other varieties are minor constituents of seafloor sediments although they may be abundant locally.

The coarser terrigenous sediments of the ocean floor show the distribution patterns expected from sediment-gravity processes. Turbidites are generally proximal to land and glacial marine sediments are confined to high latitudes (fig. 19.19). In the case of glacial sediments, it is particularly noteworthy that the entire continent of Antarctica is surrounded by this type, whereas in the northern hemisphere, the

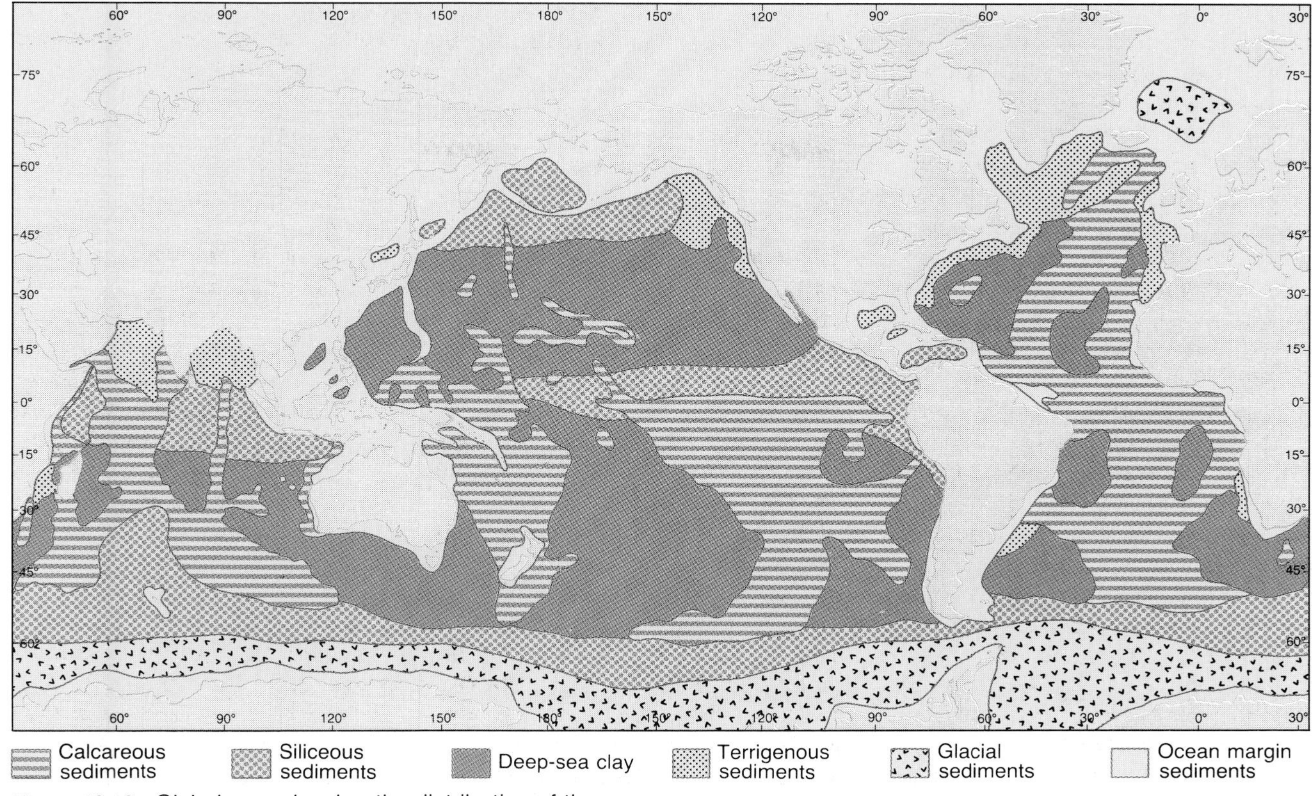

Figure 19.19 Global map showing the distribution of the major types of deep-sea sediments.

only major area covered is near Greenland, although small ice sheets are also present in the northern parts of North America, Europe, and Asia. This demonstrates the size of the Antarctic ice sheet and its contained volume of sediment as contrasted to the glaciers of the northern hemisphere.

Pelagic deep-sea sediments are dominated by clays and occur throughout the world, but are most widespread in the Pacific, particularly in the northern hemisphere (fig. 19.19). This concentration in the North Pacific is thought to be the result of windblown sediment from the Gobi Desert in Asia and also from the volcanism around the Pacific Ocean.

Diatomaceous oozes are the dominant deep-sea sediment in the high latitude zones of high productivity, discussed in chapter 8. This area has a combination of cold water, high nutrients, and good light penetration, thus generating an enormous diatom population. There is also a belt of siliceous ooze near the equator in both the Pacific and Indian Oceans. This is dominantly a radiolarian ooze and is due to the zone of high productivity that characterizes the area because of diverging surface currents and resulting upwelling.

Distribution of calcareous ooze is complicated by a phenomenon known as the **carbonate compensation depth**. Calcium carbonate is relatively soluble in seawater and, as discussed previously, its solubility is increased as pressure increases and temperature decreases, and carbon dioxide concentration increases. In other words, as a planktonic calcium carbonate organism expires and falls through the water column, it becomes unstable and begins to dissolve at a depth where the combination of temperature and pressure are such that calcium carbonate cannot exist in the solid state. This depth is the carbonate compensation depth (CCD) and it varies from place to place in the ocean. The mean depth of the CCD is about 4.5 km, but there is a range

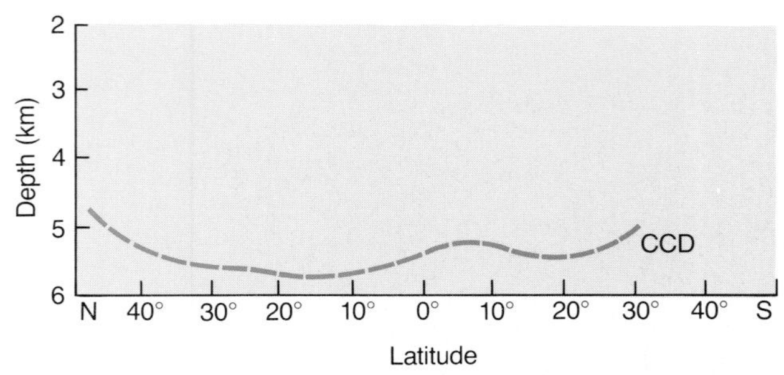

Figure 19.20 Location of the carbonate compensation depth (CCD) in the oceanic water column.

of about a kilometer (fig. 19.20). This range results from some variation in the controlling factors in the stability of calcium carbonate, especially carbon dioxide.

The smaller carbonate skeletal material such as coccolithophores are confined to shallow water because of carbonate solubility. Many of the larger foraminiferal tests are poorly preserved because of partial dissolution. It is this phenomenon of solution at depth that results in the complicated patterns for carbonate oozes as shown on the global distribution map (fig. 19.19). A comparison of this map with the general bathymetry of the ocean basins shows, for example, that carbonate oozes are dominant on and near the oceanic ridge systems and on plateaus that rise above the abyssal floor. These features are above the CCD.

Deep-sea Benthic Life

In the discussion on the early history of oceanography in chapter one it was pointed out that Edward Forbes believed that there was a depth below which no life existed in the ocean, which he called the azoic zone. This notion was dispelled not long thereafter, especially by the *Challenger* expedition, which recovered numerous species of organisms from the deep ocean floor. It was not until the middle of the present century, however, that the true nature of the abundance and diversity of life on the deep seafloor was known.

Environmental Characteristics of the Deep Sea

Organisms that occupy the floor of the deep sea have numerous problems or obstacles to overcome in order to survive this extremely rigorous environment. First of all, depths of 3,000 m or more are characterized by very cold water, with temperatures below 4°C throughout. These temperatures permit the water to contain high levels of oxygen because of the inverse relationship between temperature and gas solubility. As a result, oxygen is not a limiting factor in the distribution and abundance of life on the seafloor.

The hydrostatic pressure on the seafloor is tremendous: it is hundreds of times that at sea level and, in the deepest trenches, up to 1,000 times that at sea level. These pressures and the extremely cold water are thought to slow biochemical reactions that take place in this environment. The pressure/temperature conditions result in slowed metabolism, which in turn causes slow growth and reproductive rates, but results in relatively long life spans.

Salinity is about that of the mean value for the ocean (34.3‰) and shows little variation in this homogenous part of the ocean. Light is absent except for that generated chemically by certain organisms.

The key limiting factor in this environment is the relative scarcity of food. Organic debris must be generated by organisms that occupy the seafloor or settle from the overlying water column. The slow rate of descent, numerous pelagic organisms present, and the well-oxygenated water make it difficult for organic detritus to survive the trip to the ocean floor.

Studies have shown that there is about 1–2 gm/m^2 of organic matter on the ocean floor. While this seems like a low concentration, it is not much less than that for the continental shelf. This has led to speculation that bacteria play a more important role in the food chain of the deep sea than was previously believed. Many of the deep-sea organisms depend on bacteria as a major food source. Nevertheless, the abundance of benthic organisms on the continental shelf is over $10{,}000/m^2$, whereas it is less than $100/m^2$ on the deep ocean floor. Some of the reasons for this major difference include extreme pressures, very cold temperatures, and the limited number of living organisms that serve as a food source for carnivores.

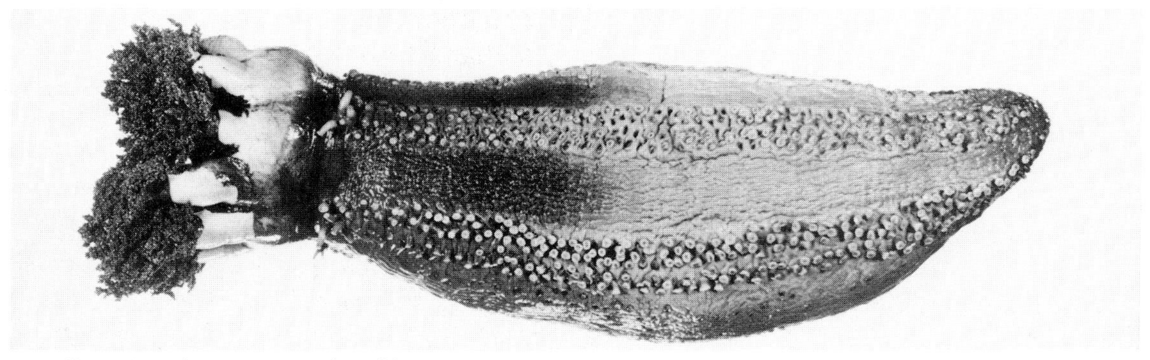

Figure 19.21 Sea cucumber, a vagrant benthic echinoderm.

Figure 19.22 Various sessile benthic organisms on the deep seafloor.

Benthic Communities

The abundant groups of benthic invertebrate animals on the seafloor differ markedly from the continental margin environments. Echinoderms such as brittle stars, some crinoids, and especially sea cucumbers (fig. 19.21) are abundant, along with polychaete worms and various crustaceans. Pogonophorans (long, tube-building worms) seem to be restricted to deep waters.

Many deep-sea species are scavengers that feed on organic detritus in the sediment. The sea cucumbers and some mollusks are among these. Sessile, filter-feeding benthic organisms like crinoids, polychaete worms, and pogonophorans are also relatively common (fig. 19.22). These organisms are preyed upon by carnivorous invertebrates as well as deep-sea fish.

Although the deep-sea floor is a dark, remote, and harsh environment, there is a community of some diversity and abundance. Organisms must be both hardy and efficient in order to survive.

BOX 19.2

Vent Community of the Galapagos Rift

In 1979 there was a research expedition to the southeastern Pacific Ocean to study the central rift valley near the Galapagos Islands. This was intended to be another project in the series of expeditions to study the oceanic ridge system, particularly the spreading center in the central valley. What the scientists found ranks as one of the major oceanic discoveries of all time. They discovered a tremendous community of bizarre organisms that are unlike any previously known. The scientists also discovered the plumes of extremely hot water and sulfide minerals that are associated with these organisms.

Previous ventures into rift valleys in 1976–1978 found warm waters rich in hydrogen sulfide and bacteria, large but dead clams, and inactive vents. The major discoveries in the Galapagos area took place in early 1979 using the *Alvin*. Although the geological discoveries are nearly as unusual as the biological ones, the discussion here will focus on the living community of the rift valley.

The physical and chemical environment in this rift valley is extremely unusual in that at a depth of 2.5 km the ambient water temperature is about 11°C, which is much higher than expected. Local plumes of superheated water reach temperatures of 350°C. This water contains dissolved ions that precipitate as sulfide minerals of iron, copper, and zinc. This mineral-laden water jets out of the hydrothermal vents as "black smokers," which are clouds of mineral particles.

Figure B19.2 Photograph of the huge tube worms that live in the hydrothermal vent environment of the Galapagos Rift Valley.

The organisms are truly amazing because they live in a harsh environment that lacks light and therefore lacks photosynthesis. The suspended organic material is hundreds of times greater at the vents than in the adjacent waters, and is much more productive than even the surface waters, where photosynthesis is at its maximum. The basis of the community in this environment is **chemosynthesis,** which is the formation of organic compounds from inorganic ones, using chemical energy. Tremendous numbers of chemosynthetic bacteria serve as the "producers" in this environment. Other organisms include crabs, very large tube worms, and huge clams.

The worms are extremely numerous (fig. B19.2) and reach 3–4 m in length, with the girth of a person's wrist. They are filter feeders that thrive near the hydrothermal vents. One of the most unusual aspects of these worms is that they have no mouth, intestinal tract, or anus. They feed by absorption of food and oxygen through small and abundant tentacles. The animal itself is housed in a long tube and has red flesh (fig. B19.2), which is red due to hemoglobin much like that of mammals. This blood carries food and oxygen throughout the living tissues of the organism.

Large clams, about 30 cm long and with similar red tissues, are also abundant in the vent community. Dating of the shells shows that they grow about 4 cm/yr, which is much faster than the shallow-water varieties. Crabs collected from this environment have been kept alive for several months in pressurized chambers that emulate their normal habitat.

Similar organisms with a chemosynthetic basis have been found at the base of the continental slope off west Florida. These organisms also are in sulfide-rich waters, but there is not any elevation of temperature above that expected at the ocean floor. Both of these unusual communities indicate that not only can marine life flourish in unexpected conditions, but that the overall productivity of the deep ocean may be much more than we thought just a few years ago.

Summary of Main Points

The deep ocean environment consists of the abyssal floor, the oceanic ridge system, and the oceanic trenches. The latter are areally insignificant, but are an important element in the dynamics of the earth's crust. These deep ocean benthic environments are relatively homogenous and static as compared to shelf and coastal areas, but there is constant change; it just takes place at a slow rate.

Ocean floor areas are flat to gently rolling, depending on their proximity to the oceanic ridge and on the amount of sediment that has accumulated since the basaltic crust formed. Sediment originates both from turbidity currents that extend to this environment and also from pelagic particles, both biogenic and otherwise.

Deep-sea sediments are dominated by biogenic oozes of both siliceous and calcareous composition. They originate from planktonic plants and animals. The calcareous particles may be dissolved before reaching the ocean floor or they may go into solution soon thereafter due to the combination of low temperatures and high pressures with the solubility of the mineral. Generally, high latitude areas and the deepest part of the ocean floor are dominated by siliceous oozes, with the remainder dominated by calcareous ooze. Only the North Pacific is dominated by non-biogenic sediments in the form of brown clay.

Although not abundant by coastal or shelf standards, there is a fair diversity and abundance of life on the ocean floor. These detritus feeders are both scavenging vagrant types and filter-feeding types. The sluggish but persistent currents serve to transport food to these opportunistic creatures.

Suggestions for Further Reading

Heezen, B.C., and C.D. Hollister. 1971. **The Face of the Deep.** London: Oxford University Press.

Menard, H.W. 1969. "The deep ocean floor," *Scientific American,* v. 221:53–63.

National Science Foundation. 1968–1984. **Initial Reports of the Deep Sea Drilling Project,** numerous volumes. Washington, D.C.: National Science Foundation.

Seibold, E., and W.H. Berger. 1982. **The Sea Floor: An Introduction to Marine Geology.** New York: Springer-Verlag, Inc.

PART FIVE

Applied Oceanography

20
Resources of the Sea

Chapter Outline

- Food
  - Plants
  - Fish
    - Commercial Varieties
  - Shellfish
    - Shellfish Mariculture
  - Mammals
- Mineral Resources from the Ocean Bottom
  - Minerals Taken from Seawater
  - Minerals on the Ocean Floor
    - Minerals of the Continental Margin
    - Deep-sea Minerals
- Mineral Resources from Beneath the Ocean Bottom
  - Petroleum
    - Offshore Exploration and Drilling
  - Subsurface Mining Beneath the Seabed
- Summary of Main Points
- Suggestions for Further Reading

The marine environment is one of the great natural resources of the earth because it contains a broad spectrum of things that humans collect, harvest, and mine. These include enormous quantities of food, a variety of minerals, and petroleum products from various sources. Even though the dollar amount taken from the ocean is huge, hundreds of billions of dollars per year, we have not really begun to utilize some of the potential resources that the oceans contain.

Going back to the early civilizations, the sea has provided us with three primary benefits: transportation, military power, and food. Since the industrial revolution, this base has been expanded to include petroleum, minerals, and energy. This chapter will focus upon three major resource categories: food, minerals, and petroleum.

Food

The oceans and adjacent coastal waters provide a tremendous variety of food for the entire world. Even many of the nations that do not border the world ocean consume some of its food products. Although there have been serious problems with overconsumption of some organisms, there is still a great potential to develop industries centered around various types of plants and animals that are as yet not harvested except locally. As world population increases and traditional food production, both on land and sea, falls short of providing the needed nutrition for third world countries, it will be necessary to expand our harvest of the sea to include some of these as yet underutilized sources of food.

The discussion here will consider the plant or photosynthetic food sources separately from the animal sources. Both are important commercially in the United States, however, the specific organisms harvested and their uses vary greatly on a worldwide basis. Additionally, although most organisms are simply harvested as they occur naturally, there are others that are actually farmed in the sense of **aquaculture,** the aquatic analog to agriculture. The term **mariculture** is used for marine aquaculture.

Plants

Various types of algae and sea grasses are harvested and used for a variety of purposes, including food. This is one area that still has significant potential for expansion, but some hurdles, especially those of taste, must be overcome for this to happen. In the United States there is no significant market for these products as they occur naturally. Only in certain health food and specialty food stores are such plants available. By contrast, there are many countries, especially in southeast Asia, where various types of seaweed are a major part of the diet.

Mariculture of plants is being carried out in many of these oriental countries, with the most successful and most sophisticated activities being those in Japan. During the past few years China has developed new techniques and is expanding mariculture

Figure 20.1 Photograph of a kelp harvester, which moves backward, cutting the kelp below the water surface and carrying it into the barge over the stern.

operations. As might be expected, this type of activity is restricted to shallow waters, most commonly in estuaries. Open waters make dispersal of nutrients and spores difficult, as well as predator or disease control.

At the present time, the largest use of algae and other marine plants is for their by-products. An important group of economic plants is the brown algae, especially the kelp. These large seaweeds are abundant and are harvested regularly along the west coast of North America. The harvesting technique utilizes large floating machines, similar to threshing machines that harvest wheat and oats (fig. 20.1). These algae are an important source of iodine, bromine, and various potassium compounds (potash). Algin, a complicated polysaccharide, is also produced from kelp. This compound is a binding or thickening additive in various types of food, such as salad dressings, and as a coating in papers and pills.

The most important economic group of seaweeds in terms of dollar amount of annual harvest is the red algae. The current annual harvest approaches a billion dollars. These algae are used as food for human consumption in southeast Asia and also as both fertilizer and animal food. Probably the most important use of the red algae in North America or Europe is as a source of two polysaccharides: **agar** and **carrageenan.** Agar is a chemical with a jelly-like property that has made it the primary medium in microbiological culturing. It is also used as an additive in gelatin capsules in the pharmaceutical industry and in foods that must be kept moist, such as prepared cake frosting. Carrageenan has similar properties and is an additive in ice cream, toothpaste, paint, cosmetics, and many other products that we use every day.

Some green algae (*Chlorella*) is used in Japan as a dietary supplement for protein, fat, and vitamins.

Fish

Various types of fish have been economically important since the first civilizations occupied the marine coast. This industry has expanded and diversified greatly over the past couple of centuries and still has room for more. Nevertheless, the fishes account for the greatest dollar amount of commercial food harvested from the sea.

The highly populated Asian countries are the biggest harvesters of finfish. China accounts for about half of the world total and has an annual harvest that is about four times that of India, the second place country. They are followed in order by the USSR, Japan, Indonesia, and the Philippines. The United States is not even in the top ten with an annual production of about 2.5 million metric tons.

Commercial Varieties

The group of fishes that includes the anchovies, sardines, herring, and menhaden represents about one-third of the world's catch and is the largest commercial type of finfish in terms of tons of catch. These fishes are individually small, generally less than 25–30 cm in length (fig. 20.2), but they occur in

Figure 20.2 Menhaden in the hold of a fishing vessel out of Beaufort, North Carolina.

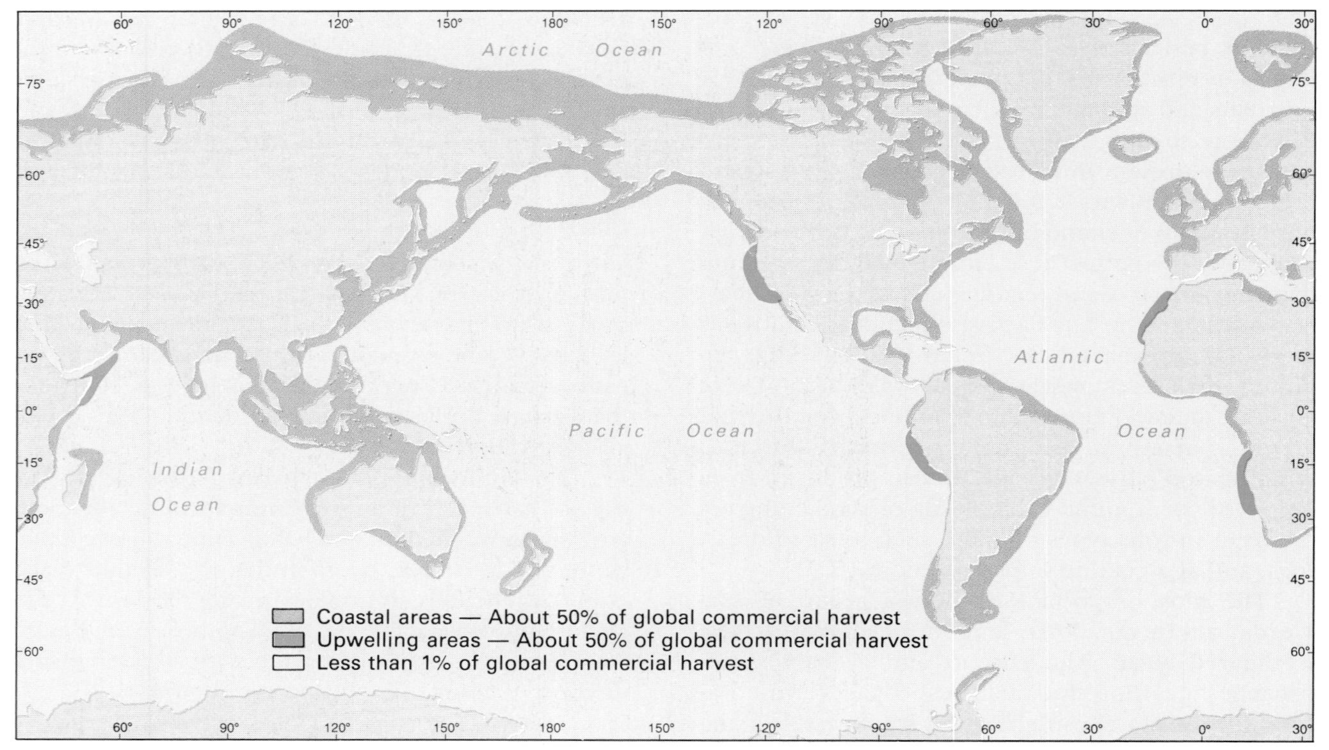

Figure 20.3 World map showing the most important fishing areas, which are concentrated on shallow shelves and in areas of upwelling.

enormous schools. This group of fishes feeds on phytoplankton and, like most important commercial groups, they are most abundant over the continental shelves of the world, especially in the areas of upwelling on the east sides of the oceans (fig. 20.3). They are harvested primarily by large nets, such as purse seines (fig. 20.4). These fish are used primarily for fertilizers, fish meal, and fish oil. The fish meal is typically used as a protein supplement in animal food. This meal has great potential as a supplement in food for human consumption because of its high protein value. For example, it takes only 6.6 kilocalories of energy of herring to produce a gram of protein whereas it takes 263 kilocalories of milk and about 700 of beef to produce a gram of protein. Obviously, the herring and other similar fish offer great potential for expanding their use for human consumption.

The second largest group of commercial fish is that of the demersal fish, which includes the cod, haddock, and hakes. These fish live in the shelf waters and are commonly caught using the otter trawl. The individuals are much larger than the herring, with a typical size of 50–60 cm. Some are used for fish meal and some are used for table food. Rockfish include sea bass, grouper, redfish, and perch. They account for about 10% of the total commercial fishery and are used primarily for table food. Much of the commercial fishing for these fish types is by long line (fig. 20.5) and by trawling.

The flatfish are demersal varieties that live in shallow shelf waters and are also popular table food. Included are the sole, flounder, halibut, and plaice. They are not only tasty but they have few bones and cleaning is easy. These fish are caught primarily by otter trawl but also by long line. In very shallow coastal waters fish traps are used (fig. 20.6). The fish are herded into the trap, much like capturing horses in a blind canyon.

Figure 20.4 Large fishing vessel setting a huge purse seine with the aid of a small launch.

Figure 20.5 Fish sorted in piles according to species on long-line vessel. The line is wound in and payed out of the large empty spool in the upper left of the photo.

Figure 20.6 Fish traps into which fish swim with the tides in estuaries of New England.

BOX 20.1

Peruvian Anchovy Industry

There are situations where a combination of natural environmental conditions and the fishing activities of humans result in serious problems within the fishing industry. One of the most striking examples is that of the anchovy (anchoveta) industry in Peru. During the late 1960s and early 1970s this was the largest fishing industry in the world based on tonnage. It is now less than 10% of what it was at the peak harvest.

The waters off the coast of Peru are characterized by upwelling. Circulation in this area is dominantly toward the north and this, in combination with the Coriolis effect of the southern hemisphere, yields an offshore component of the surface waters, resulting in upwelling of deep, nutrient- and oxygen-rich waters (see chapter 5). Such upwelling conditions result in great phytoplankton blooms and in turn, in a tremendous population of the herbivorous anchovy.

Starting about 1950 the harvesting of anchovies began to expand as this fish began to be used in fish meal. Initially a few thousand tons were taken per year but that amount grew to 2 million tons by 1960 and to 9 million tons by 1965. It was during the early 1960s when the anchovy industry was growing rapidly that it became apparent that nature had a mechanism for limiting the anchovy population.

There is an atmospheric condition called **El Niño,** meaning "the child," a name adopted because this condition is prevalent at Christmas time. El Niño changes the winds that blow along the Peruvian coast so that they come from the north, which causes warm surface waters and inhibits upwelling. This results in a paucity of phytoplankton and a natural depletion of the anchovy population. Prior to the extensive fishing of this anchovy population, there was no knowledge of this effect. Once the harvest had reached several million tons per year, the impact of an El Niño condition became obvious.

The El Niño year of 1965 resulted in a reduction of about 2 million tons of anchovies harvested (fig. B20.1). In following years the catch increased beyond previous highs and beyond the maximum sustained yield. That is the maximum amount of harvest that could be expected to sustain itself over the years based on known information about the anchovy population and life cycle. At the time about 95% of all anchovies in the area were being harvested each year.

A severe El Niño condition took place in 1970 and, in combination with the overfishing, caused a massive decline of the anchovy population, from which it has yet to recover. There was a minor rebound in the middle 1970s but this was followed by another decrease so that now the industry harvests less than it did in 1960 (fig. B20.1). This provides an excellent example of how human use of marine resources must be kept in line with nature's way of controlling or limiting a population. Proper management of this industry would have kept it healthy continuously, but overfishing has resulted in its demise.

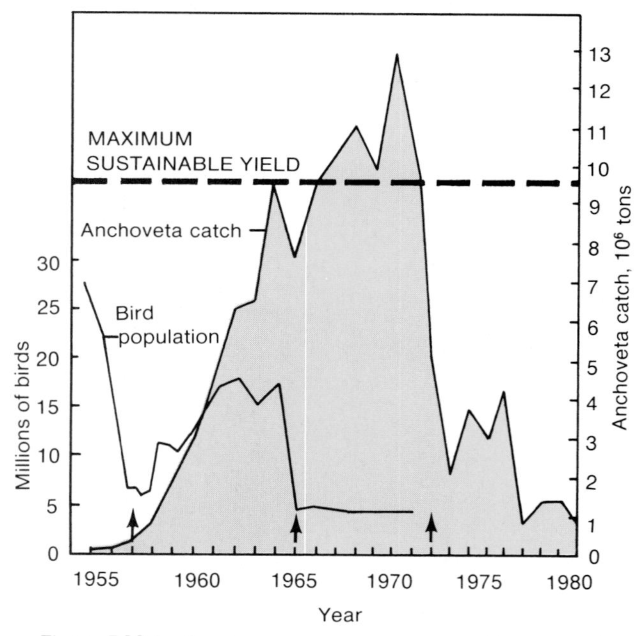

Figure B20.1 Graph of the anchovy harvest off the coast of Peru showing the influence of the El Niño, as indicated by the arrows.

Figure 20.7 Photo of a large school of tuna, each of which is about a meter long.

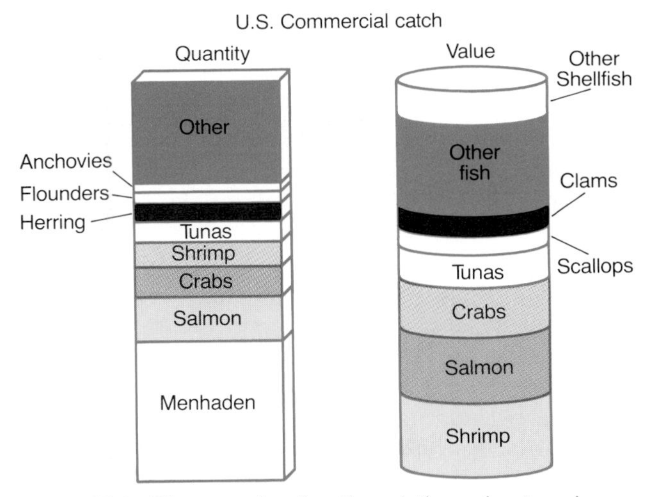

Figure 20.8 Diagram showing the relative value to volume of various types of important commercial fish. For example, shrimp comprise a small volume, but have a high value, whereas menhaden are by far the largest in terms of volume, but are not even listed separately on the value graph.

The large, fast-swimming carnivores, such as the mackerel and tuna (fig. 20.7), are caught in a variety of ways, including by a pole and line. Tuna are among the largest of the bony fishes, weighing up to 200 kg, and they are the only important finfish that is harvested from deep water.

Shellfish

The various types of shellfish are among the most important of all commercial foods taken from the marine environment because they are more valuable per unit weight than the finfish (fig. 20.8). Shellfish include mollusks, such as oysters, scallops, and clams, and crustaceans, such as the shrimp, crabs, and lobsters.

Commercial shellfish come primarily from shallow shelf and coastal waters where nutrient levels are high and food is abundant. Oysters and many commercial varieties of crabs and clams are concentrated in estuaries. The oysters especially prefer the sheltered, brackish waters of this environment. Some crabs, scallops, and lobsters prefer the shallow open marine environment, but most scallops prefer sandy and grassy bottoms, and most crabs and lobsters live on the rocky ocean floor, where they can hide from predators and await prey. The shrimp prefer the shallow shelf, although they are also taken from coastal bays.

All of the shellfish industries have experienced some severe problems during the past few decades. Most of these difficulties have occurred as the result of two situations: (1) overharvesting and (2) pollution. Although the problems are probably most significant in North America, they have taken place elsewhere as well, especially in Europe and Japan.

Estuaries are environments of high nutrient levels, which is a prime reason that they contain great concentrations of oysters, clams, and crabs. The nature of the estuarine environment is also conducive to pollution problems. The combination of runoff from the drainage basin that serves the estuary and the tendency for population centers to border estuaries has produced problems for the shellfish industries. Runoff from agricultural and populated areas produces concentrations of nutrients that may be too high for the health of the estuary. These extreme levels of nutrients cause phytoplankton blooms; eventually the great abundance of organic material results in oxygen demands that deplete the estuary of its natural oxygen levels and organisms are suffocated. A more common problem is the influx of toxic wastes from urban areas into the estuarine system. Storm sewers carry various petroleum chemicals from pavements, along with industrial and domestic wastes, both organic and inorganic, all of which cause serious problems for shellfish. Typically the most seriously affected commercial organisms are the oysters and clams. Both are filter-feeding sessile benthos that pass large quantities of water and suspended material through their systems. In some instances toxic compounds may cause large-scale destruction of the population, but the most common situation is the concentration of toxic substances in the tissues of shellfish, thus rendering them unfit for human consumption. The oyster industry of Chesapeake Bay has been severely damaged by chemical wastes, as has that of the Carolinas. Soft-shelled clams in New England and the clams in various areas of the southeastern United States have been declared unfit for consumption at various times during the 1970s and 1980s.

Overfishing has been a serious problem for the lobster and shrimp industry. In some situations, however, it is difficult to determine if the decrease in abundance of a particular commercial fish or shellfish is due to overharvesting or to the deterioration of the environment. This has been a long-standing argument between the fisherman and the agencies that legislate the fisheries industry. There are also problems that exist between the commercial and sportfishing industries. The commercial fishermen feel that sportfishermen interfere with their activities and vice versa. There is also strife between domestic fisheries and foreign counterparts that come into United States waters to fish. They do not obey the policies set up for United States fishermen and there is at present no mechanism for controlling this type of activity. The international agencies charged with regulation of fishing activities have been working extensively on solving this and related problems. A discussion of such activities is included in the next chapter.

The four leading shrimp harvesting countries are India, the United States, Thailand, and Mexico, which collectively account for one-half of the world harvest. Although the United States harvests over 80% of its available crop of shrimp, the worldwide percentage is quite low and there is much room for expansion of the shrimping industry, outside of North America. The United States also ranks among the top three countries in harvest of clams, oysters, lobsters, and scallops and thus is also at or near peak production of these marine products. The United States is not, however, even among the top four countries in several types of finfish harvesting, and that is where the expansion of effort can take place.

Figure 20.9 Cultivating oyster beds at low tide near Chincoteague, Virginia, where some of the largest oysters in the United States are harvested.

Shellfish Mariculture

Because there is little potential for increasing the natural availability of shellfish and because the demand for shellfish for human consumption is increasing, mariculture appears to be a means of increasing the production of these organisms. There is already, in some countries, broadly based mariculture activity that is producing oysters, clams, and mussels and that has been in operation for centuries. Most of the early mariculture efforts were aimed simply at cultivating natural areas where the shellfish grow well, such as oyster beds in estuaries (fig. 20.9). These methods are still used in some places. Most of this activity centers around providing a proper substrate on which the planktonic larvae will settle and mature. Generally this is a bed of old shells. Another more controlled technique utilizes shells or some other type of hard surface attached to strings suspended over large tanks. This provides a proper surface for the larvae of oysters or mussels to attach and as a result, for the growth of the desirable shellfish.

Lobsters are grown in ponds where the temperature is controlled and food is provided. Such conditions protect the lobsters from pollution or predators. Growth rates in these controlled conditions are 4–5 times those in nature. Shrimp have also become the object of mariculture activities and offer great possibilities for the future. Because of the intense pressure of harvesting on the natural population of shrimp, it is important to expand mariculture activities in this industry in order to meet the ever-increasing demand for the product.

Mammals

There is a large industry that harvests marine mammals in various ways for various reasons. Mammals harvested are primarily two varieties, the whales and the seals. The industry has come under great pressure to decrease the catch and has been the target of attack from various public interest groups.

Figure 20.10 Smaller whaling ship in the foreground showing the stern ramp through which the whales are winched from the sea. The large ship in the background is a factory ship where the whale is processed and canned so that the products are ready for shipping when the vessel docks.

Seals have been harvested primarily for their fur, which is used for clothing, although natives of high latitudes also hunt them for food, implements, and oil as well. The slow mobility and docile nature of seals make them easy prey for the hunter. These characteristics, plus their desirable fur, resulted in near-annihilation of the world's seal population in the nineteenth and early twentieth centuries. There are now protective regulations that control the number taken each year and, although there is argument about that number, it does appear that the population of most species is stabilizing or even increasing.

As an example of the near-extinction and subsequent repopulation of seals, consider the northern elephant seal. They were hunted vigorously during the nineteenth century until they were thought to be extinct. Then in 1892 a group of eight seals was discovered on a remote island. This small population was nurtured and protected so that there is now a population of elephant seals that numbers in excess of 50,000.

The near-demise of the whales is a classic example of overharvesting. The plight of this group of mammals has been followed by numerous conservation groups throughout the world. Whaling has been an important large fisheries industry for centuries, although it ceased operations in the United States in the 1920s. Throughout the early history of North America, whaling was a critical industry because whale oil was a necessary product, both here and in Europe. After the discovery of petroleum in 1859 there was a steady decline until the last United States whaling ship was wrecked off the Massachusetts coast in 1924.

At the time of the decline in this country, there was great expansion and modernization in Russia, Japan, Norway, and Great Britain. Technology improved the harpoons, catch boats, and other types of apparatus, and the whaling ships themselves became larger and accommodated more comprehensive processing operations. Whaling vessels were organized in fleets that included factory ships, where processing took place (fig. 20.10).

This increased efficiency of operation and expanded activity took its toll by the beginning of World War II. It became apparent that some type of world policy on whaling was needed and in 1946 the International Whaling Commission was established. The policies of this body were not well-founded scientifically and the quotas imposed were based largely on the capacity of the industry to harvest, not on the ability of the whale population to survive. Its regulations are not enforceable and consequently there has been little positive impact on the dwindling whale population. This dilemma is as yet unsolved; however, conservation groups have caused great public awareness and hopefully a solution will be forthcoming.

Mineral Resources from the Ocean Bottom

The oceans contain a great abundance and diversity of minerals, some of which are being utilized, but many that are as yet not economical to recover. Although all natural elements are dissolved in seawater, only a few occur in concentrations high enough to be extracted profitably. The ocean floor contains various types of minerals and compounds that have considerable value although many are widely dispersed in the sediments and others are at water depths that make recovery impossible with present technology.

Minerals Taken from Seawater

The water of the world ocean provides us with a variety of important and valuable minerals and elements. Probably the first mineral to be extracted from ocean water was halite or common salt (NaCl), whose production goes back to the ancient civilizations. Extraction is accomplished through evaporite ponds along coastal areas, such as lagoons, where salt crusts form and are scraped into large stockpiles for processing. A large salt industry exists along the margin of San Francisco Bay in California.

Some of the dissolved elements in seawater are extracted for use in various manufacturing industries. Magnesium, for instance, has been commercially extracted from the ocean for about a century. Bromine is the other major element that is produced from seawater, although the demand for this element is decreasing. It is used as an additive in gasoline, but the reduction in use of leaded gasoline may end bromine production from the ocean.

Perhaps the most important resource of the ocean water is the water itself. The future need for fresh water will likely exceed the present supply. As a consequence, it will be necessary to desalinize more and more salt water. There are hundreds of desalinization plants currently in operation, producing about 2 billion gallons of fresh water each day. The cost of this process is still high and only in places where water is very scarce is it cost-effective to desalinize seawater. As technology improves, it is likely that more plants will be installed, with larger capacities than those presently in operation.

Minerals on the Ocean Floor

The sediments and rocks of the seabed contain a wealth of valuable minerals. Some are localized, some are widely but thinly scattered, and others occur in abundance over large areas. The value of these materials ranges from diamonds to sand, but all are important.

Minerals of the Continental Margin

The minerals that are mined from the continental margin are concentrated in shallow marine environments. They are partially derived from the erosion of land areas and are partly formed in the marine environment itself. The discussion of these minerals will be based on location, from the coast to the outer margin.

Although not obvious to many people, the sediments that comprise the beach and adjacent nearshore area are valuable resources in themselves. Their value generally is in proportion to the scarcity of the available sediment. Along most coastal plain areas with broad barrier beaches, the value of sand as a commodity is low; however gravel may command a high price. An example is Florida, which has large volumes of sand, but essentially no gravel except for some coarse shell deposits. By contrast, rocky coastal areas commonly have a shortage of sand and its value increases accordingly. An extreme example is Puerto Rico, where sand is in very short supply. There, sand "rustlers" sneak to the beach with trucks at night to remove loads of sand under darkness. Authorities must patrol these areas to prevent the theft of this valuable commodity.

An important use for sand from shallow, nearshore environments is beach nourishment. The sand is dredged or pumped onto an eroded beach to repair the damage caused by erosion. In order to accomplish this task it is necessary to have a nearby source. In some situations this is not available and costs rise greatly.

A similar type of mineral product is calcium carbonate, which is pure and used in the cement and agricultural lime industries. Calcium carbonate occurs as sand-size spherical grains called **ooids** in the shallow warm waters of low latitudes. The Bahama

Figure 20.11 Large stockpile of ooids in the Bahamas where ooid shoals are mined as a source of pure calcium carbonate for agricultural lime and for the cement industry.

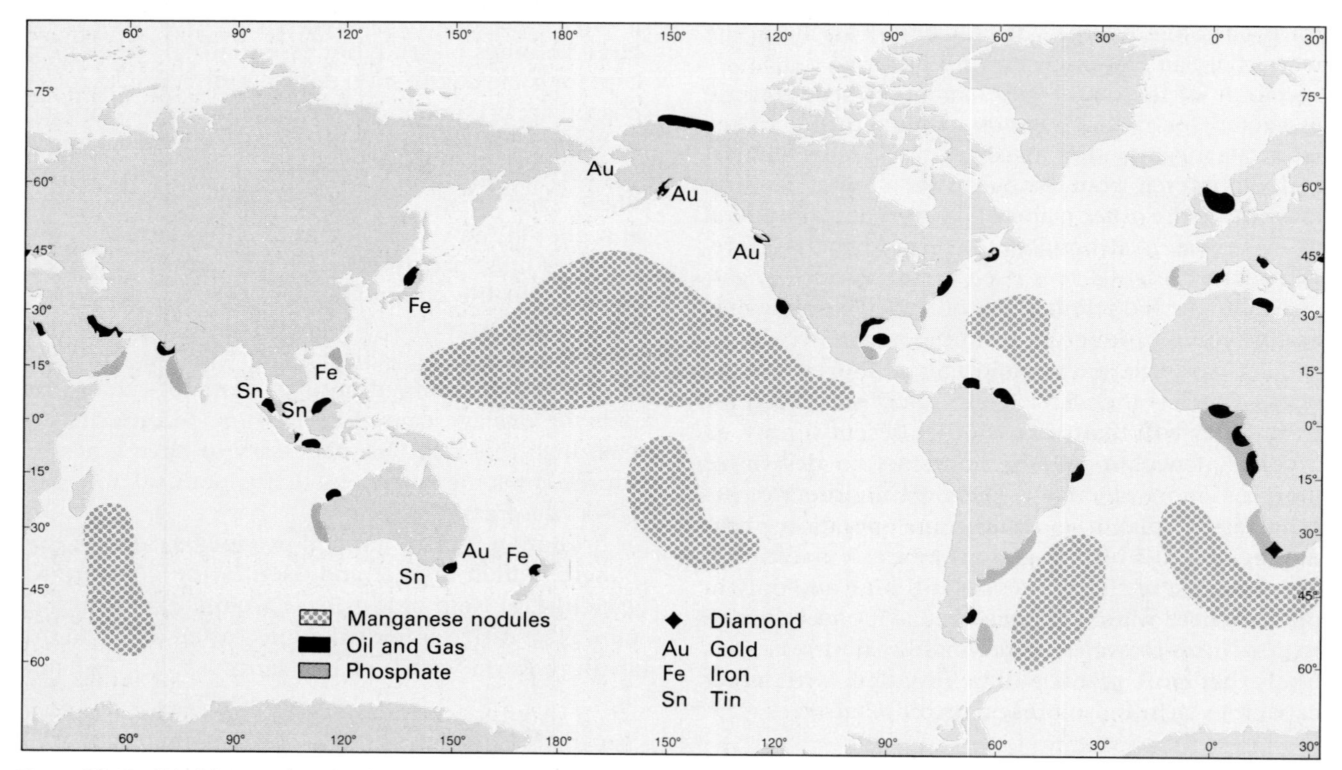

Figure 20.12 World map showing locations of primary marine mineral and petroleum resources.

Platform east of Florida is an excellent location for extensive ooid accumulations. Here the carbonate sand is mined by large dredges and draglines, placed on barges, and transported to the United States for processing (fig. 20.11).

Placer deposits are also important economic deposits that are commonly found in beach and nearshore sediments. These are concentrations of dense minerals, commonly valuable ones, by wave and current processes. The occurrence of placers along the coast is analogous to the placer gold deposits that occur in streams, where they are concentrated by current activity. Marine placer deposits may contain magnetite (iron oxide), zircon (a source of zirconium), ilmenite (a source of titanium), platinum, or gold. These deposits typically are local and tend to reflect the nearby geology in the source area from which the placer minerals are derived. Gold is present along the Pacific coast of Alaska and some of the most important ilmenite sources in the world are the placer deposits in southeast Asia (fig. 20.12).

Diamonds also occur in coastal deposits, but are scattered. They are found in beach deposits in South Africa, where they are derived from the source rocks that produce most of the world's diamond supply.

One of the potentially most valuable offshore mineral resources is phosphorite, a mineral assemblage of various calcium phosphate compounds. Phosphate compounds are used as fertilizers, in soaps, detergents, explosives, and many other important products. The availability of fertilizers is especially critical to expansion and improvement of agricultural activities in third world countries. Although new land-based phosphate mining operations have recently been developed in South America, Morocco, and Australia, there is still a need to further explore and recover new phosphate reserves.

Phosphorite occurs on the continental shelf and is especially abundant near the outer edge of the shelf. Deposition of the phosphorite is concentrated near areas of upwelling, where nutrient-laden, phosphate-rich waters rise near the surface. The warming of this water and the increase in pH results in precipitation of phosphate minerals. The cost of recovering these deposits is prohibitive at present, but it is likely that increased need and improved technology will bring about commercial phosphorite production from the shelf before the end of the century.

Deep-sea Minerals

The best known and most widespread mineral deposit of the deep-sea environment is that of manganese nodules. This iron and manganese oxide occurs as spheres up to 20 cm in diameter and also as encrustations on a variety of rocks, shells, and other objects.

Manganese nodules have been known since the *Challenger* expedition of the nineteenth century (see fig. 10.9). They occur in all ocean basins and cover millions of square kilometers, or nearly 25% of the deep-sea floor. Concentrations are very high in some areas, especially in the equatorial Pacific Ocean, where values of 100 kg/m^2 (about 25 lbs/ft^2) are commonly found. Unfortunately this abundant and valuable resource is on the floor of the ocean, about 4 km below sea level, which is not a place where recovery can be achieved economically. Here again, increased need and technology will likely come together and make the mining of manganese nodules an economic venture in the not too distant future. For now, the elements of secondary abundance, such as copper, nickel, and cobalt, have greater economic potential than the manganese.

The spreading centers of the ocean floor, where plates diverge and magma comes to the surface of the crust, are also places where valuable elements are present in high concentrations. The rocks that form here contain considerable zinc, copper, lead, and precious metals, such as silver and gold. The location where there is the best chance for economic recovery of these materials is in the Red Sea, where they occur at depths of about 2,000 m. Here the metals are associated with hot brine environments. Estimates of the worth of these deposits is in excess of 2 billion dollars; there are companies considering a mining operation in this area.

Mineral Resources from Beneath the Ocean Bottom

The strata that lie within the earth's crust beneath the bottom of the ocean contain abundant and diverse resources, the most valuable of which are petroleum and sulfur, but also include coal, iron, and other metals. Essentially all of these products are taken from beneath the shallow waters of the continental margin due to costs, although the future will probably include expansion to deeper waters.

Petroleum

Some might argue that it is inappropriate to discuss petroleum in this chapter because it does not actually come from the sea or even the seafloor, but is taken from within the earth's crust, below the seafloor. Nevertheless, the exploration and production of petroleum must consider the ocean environment and there is considerable potential for these activities to severely alter the marine environment. For these reasons, the topic will be considered.

The petroleum industry did not move to the marine environment until 1938, when a field was discovered about a mile (1.6 km) offshore of Louisiana. Only nominal activity took place offshore until after World War II. Then there was a great deal of drilling in shallow coastal waters, especially in the vicinity of the Mississippi Delta. Offshore exploration followed in California and Texas. At first the total production of oil and gas from offshore wells was a small part of the total, but this has now expanded to about 25% of the world's total. The development of large fields in the North Sea and on the north slope of Alaska has been largely responsible for this increase, but continental shelf areas all over the world are now producing oil and gas.

In the early 1970s, when it became apparent that petroleum reserves were in short supply, especially in the western world, there was a flurry of offshore exploration in many areas of the world. Prominent among them were the eastern Gulf of Mexico, the Atlantic shelf and the north slope of Alaska. To date, only the last has proven to be a major producing region.

Figure 20.13 Jack-up drilling platform is floated on its barge-like platform. When in place the legs are used to lift the platform off the water surface and stabilize the rig for drilling operations.

Offshore Exploration and Drilling

Finding and producing oil and gas offshore is quite different than on land. Some aspects of the task are easier offshore, but most are both more difficult and very expensive. Most of the primary exploration in water over the continental margin is done through use of seismic reflection surveys (see chapter 10 for review). Instruments are towed behind a ship over areas that may have some promise for detailed exploration activity. This type of activity is somewhat easier over the water than on land because it can be accomplished relatively quickly and the ship can proceed essentially anywhere. On land there are problems with terrain, access, buildings, and so on, which are not present at sea.

The drilling process at sea, however, is extremely expensive and complicated as compared to that on land. Drilling platforms may be permanently installed and then used for production after the wells have been completed (fig. 20.14). This is an expensive system, however, and commonly a jack-up drilling rig is used. Such a rig is on a large floating platform with large, long legs that rise above the surface during transport and are then lowered onto the shelf floor for drilling. The platform, its drilling rig, and all of the support facilities are jacked up above the water surface and become a stable platform (fig. 20.13).

Figure 20.14 Drilling platform in the Gulf of Mexico near the Mississippi Delta area.

There are also drilling rigs mounted on floating platforms that can be stabilized with a combination of ballast and positioning motors. This system works well except in extremely rough seas, but there are practical limitations to the depth of water in which drilling can take place. Most such activities are in water less than 200 m deep, but wells have been drilled in depths of up to 700 m. Some of the problems of water depth are related to limitations for the divers who are typically utilized for various installation and servicing activities of drilling equipment.

The expense of drilling offshore is enormous. Equipment, labor, insurance, logistics, and so on all are far more costly offshore than on land. A single drilling platform may cost tens to hundreds of millions of dollars. The cost of drilling offshore may be ten times that on land. Consequently, offshore wells must be large producers, compared to those on land, in order to make a profit. When the price of oil fluctuates substantially, there is a major impact on production. For example, in 1986 oil prices fell several dollars per barrel in a few months. Because of this decrease much offshore oil drilling was halted, especially on the north slope of Alaska, where expenses are high and transportation costs to refineries and the market are extreme.

At the present time offshore production is mainly on the continental shelf, but there is activity into the outer margin. Especially promising is the continental rise area, where thick, organic-rich sediments have been accumulating for tens of millions of years. Many people believe that the rise contains vast quantities of petroleum.

Subsurface Mining Beneath the Seabed

Sulfur is an extremely important industrial mineral. In the United States, there are large reserves of sulfur in the caps of salt domes that abound in the northern Gulf of Mexico. It is extracted beneath the sea in much the same fashion as it is on land: high-pressure, high-temperature water is pumped into the sulfur zone on the top of the salt dome, creating a solution. This sulfur-rich solution is pumped up to the surface, where cooling brings crystallization and the mineral sulfur is formed.

Some of the most interesting underground mining takes place beneath the continental shelves in various parts of the world, especially Japan and Great Britain. Both of these countries have limited size, high population density, and a highly developed industrial base. Such countries require certain minerals to sustain their activities and importing them can be extremely expensive. Consequently, shaft mining that extends nearly 50 km underground seaward of the coast line has been developed. In Japan, a large volume of coal is mined in this fashion and in Great Britain, coal and iron are mined in undersea shafts.

Summary of Main Points

The ocean and its floor provide considerable wealth in the form of both food and mineral resources. Some of these, such as the fish and shellfish table food industries, are obvious; however, there are also great quantities of fish harvested for fertilizers, protein supplements, and pet food. Oil and gas from the strata beneath the ocean are great revenue producers, but coal, iron, and sulfur are also produced from these rocks. There are localized concentrations of valuable minerals, such as zircon, magnetite, and other placer minerals, even gold. On the opposite end of the spectrum are such valuable materials as sand and gravel, which are abundant and cheap in many places, but scarce and expensive in others. As water deepens there is a general decrease in mining activity due to costs.

Even though we are taking great quantities of both renewable and nonrenewable resources from the sea, much is yet to be developed in this vast resource. Mariculture can increase finfish and shellfish production to provide food for many areas where it is not otherwise available, and there are potential food sources that haven't yet been utilized. The same can be said for mineral resources. In most cases the location and abundance of the minerals is known, but the costs of mining are presently prohibitive. This will change with time as needs increase and technology improves.

Suggestions for Further Reading

Bell, F.W. 1978. **Food from the Sea: the Economics and Politics of Ocean Fisheries.** Boulder, Colorado: Westview Press.

Cuyvers, Luc. 1984. **Ocean Uses and Their Regulation.** New York: John Wiley and Sons.

Ross, D.A. 1980. **Opportunities and Uses of the Ocean.** New York: Springer-Verlag, Inc.

21
Impact of Humans on the Marine Environment

Chapter Outline

The vastness of the world ocean has been emphasized throughout this book. The marine environment covers over two-thirds of the earth's surface and it extends to depths greater than the highest point on the earth. Nevertheless, humans have the capability of reaching virtually every part of the ocean—but in doing so, they can have a large impact on the environment. In most situations, the impact is adverse, but there are also circumstances where humans have benefited the marine environment.

The greatest impact of human activities is associated with dense populations; thus coastal environments and their adjacent shelves feel this impact more than the deeper areas. Some of these densely populated coastal areas have experienced severe problems in the forms of pollution, coastal erosion, and the general deterioration of the environment. This chapter will consider problems of the coast including pollution, coastal structures, and the impact of human activities on marine coastal communities. Problems in the open pelagic environment will also be treated, along with a discussion of marine policy as it is being developed worldwide.

People in the Coastal Zone

Because people live near the coast, they build ports and harbors on it, they depend on it for food and for recreation, and they greatly influence the coastal environments. Most of the world population lives on or within a short distance of the coast. All types of marine activities, including construction, collecting food, transportation, and military, take place in this fragile area between land and sea. Many of these activities are not compatible, although all are important. The following discussion looks at some of the problems that have arisen because of the overuse or improper use of coastal environments by people.

Construction in the Coastal Zone

Various types and scales of construction have taken place in coastal environments since the days of the ancient civilizations. There are two types of activities toward which most of this construction is directed: for the transport and protection of vessels, such things as shipping channels, harbors, and related support structures are built, and to protect property on land, structures such as dikes and seawalls are constructed. Although both are important and in many cases necessary, they also cause problems to the coastal environment.

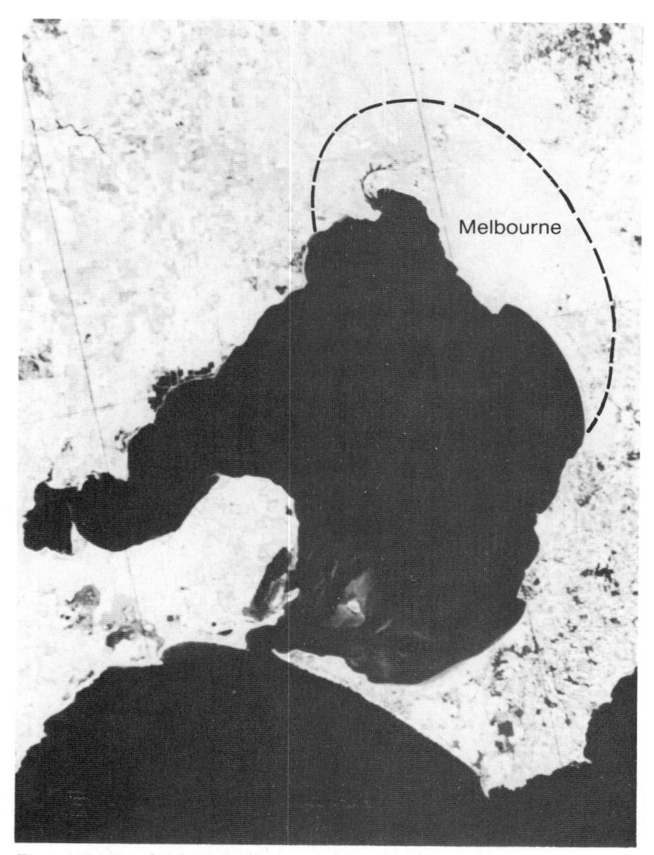

Figure 21.1 Aerial photograph of Melbourne, Australia, a port city situated in a natural harbor.

Construction for Transportation

All vessels regardless of their use require some type of harbor for shelter, repair, onloading and offloading materials, and so on. During the early days of seafaring activities, harbors were sought and developed around the terrain where embayments with natural openings provided the proper shelter and access. A look at a map of the United States shows that most of the major coastal cities border natural harbors. Good examples include Boston, New York, Baltimore, Miami, San Francisco, Seattle and Melbourne, Australia (fig. 21.1). As time has passed and the natural harbors have all been taken, it has become necessary to modify various coastal settings to provide either the embayment itself or access to it via a navigable channel.

Figure 21.2 Large dredge operating in a major shipping channel. The large bit rotates in the bottom and a slurry of water and sediment is pumped up, then placed either into barges and hauled away for disposal, or pumped through large pipes to a nearby disposal area.

These modifications typically involve large-scale construction, such as dredging a harbor, erecting breakwaters or docks, and perhaps deepening channels. Almost all of these harbors border estuaries with the associated marshes and tidal flats. Because construction activities are not compatible with the estuarine environment in many ways, it must be determined whether the disruption perhaps even destruction of the environment is justifiable on a cost-benefit basis; that is, is the cost of severely damaging the environment outweighed by the benefit that results from the completion of the project? Dredging (fig. 21.2) disturbs or destroys benthic communities. The disposal of dredged material also poses problems in that it may not be good-quality material and improper disposal could harm additional environments. Another problem commonly associated with dredging is that currents on the floor of the estuary tend to redistribute sediment and as a result, maintenance dredging is typically necessary to keep shipping channels open to the proper depth. This adds greatly to the cost of a harbor facility. During recent years the problems associated with dredging have grown tremendously. Larger and larger ships require deeper channels, which typically means that extensive dredging operations must be carried out in all major ports. This activity and the huge amounts of dredge spoil that it generates, combined with stricter federal regulations about its disposal, result in great expense. It is not unusual for channels to be dredged to depths of at least 20 m and for the volume of dredge spoil to reach 1,000,000 m^3 for major projects.

Construction for Protection of Property

The nature of coastal environments, especially those on the open coast like barrier islands, is that they are dynamic and under continuous change. Prominent types of changes are erosion of the beach, washover during storms, and migration of inlets. When these environments are developed and occupied by people, then there is potential for conflict between the natural dynamic system and human attempts to modify it.

In the earlier discussion of inlets (chapter 13) it was noted that under many situations the inlet tends to migrate along the coast due to longshore transport of sediment. Inlets may be used as passageways for vessels or they may have permanent buildings along their sides. In either situation, human activities do not accommodate the migration of the inlet (fig. 21.3) and structures called **jetties** are built in order to stabilize the location of the inlet. Generally the need for the jetties in the first place means that there will be problems with them once they are constructed. The longshore transport of sediment

Figure 21.3 Photo of Blind Pass, St. Petersburg Beach, Florida, which has migrated to the right. Even though the left side has been stabilized, longshore transport of sediment tends to fill the inlet.

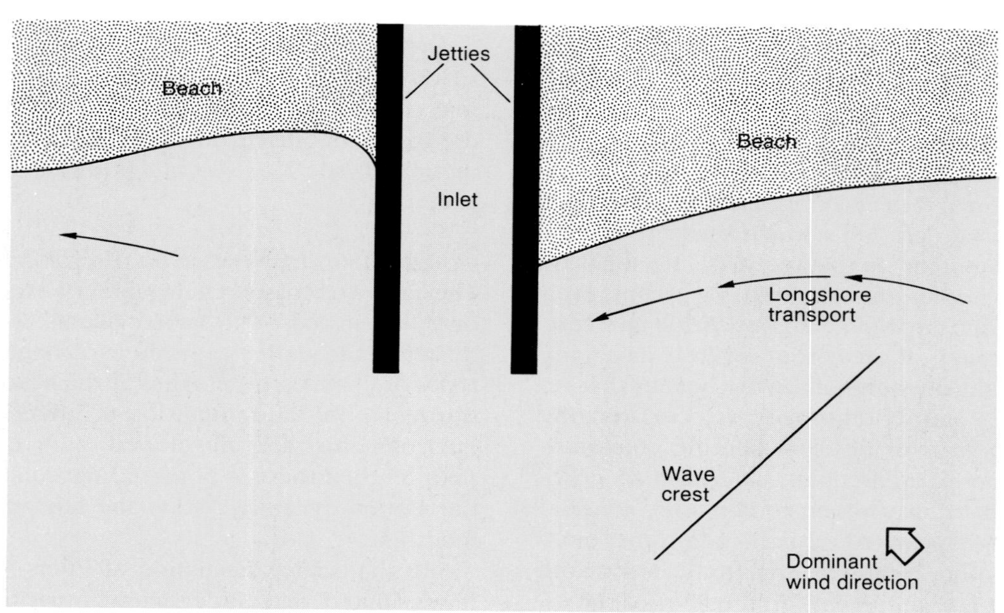

Figure 21.4 A diagram of a jettied inlet showing the accumulation of sediment that has resulted from the longshore transport of sediment.

that causes the natural inlet to migrate will also cause large quantities of sediment to accumulate on the updrift jetty. There will probably be an accompanying dearth of sediment on the downdrift side because the jetty acts like a dam and thus prevents sediment from passing the inlet (fig. 21.4). Some jetties may have sediment accumulated to the end of the jetty just a few years after construction (fig. 21.5). This renders it nearly useless and makes it necessary to either dredge the inlet or establish a system of sand bypassing, or both. Either of these is a great expense, one that may not be avoidable.

Figure 21.5 A new jetty at Clearwater Pass in Florida. This jetty had been in place less than 10 years at the time of the photo, yet a considerable amount of sediment has accumulated because of the large volume of littoral drift.

Figure 21.6 Homes at Indian Rocks Beach, Florida, that were undercut by beach erosion during Hurricane Elena in 1985.

Erosion of beaches is a widespread problem due in part to the global rise in sea level, but also due to development of waterfront property. The natural beach or barrier island system may erode or wash over during storms, with the net result being a rearrangement of the sediment within the system. Little or no sediment is lost and we have no major problem. As soon as there is development, whether in the form of roads, buildings, bridges, or other permanent structures that interfere with the natural coastal processes, then problems may arise (fig. 21.6). Waves

(a)

(b)

Figure 21.7 Examples of seawalls (a) that are poured concrete and vertical, and (b) that have rip-rap in front to dissipate some of the wave energy.

Figure 21.8 Photograph showing a groin complex that is well designed for its location as evidenced by the accumulation of sand on both sides and the nearly complete burial of the groins.

and currents carry sediment over the road or perhaps scour under the road and damage or destroy it. Similar processes may remove some of the shorefront land or damage buildings. These and other types of situations cause the construction of "protective" structures that are designed to alleviate these problems. The most common types of structures are **seawalls** and **groins.** Each of these comes in a variety of sizes and shapes, and may be constructed from a variety of materials. Seawalls are barrier structures built between the ocean and the structure that is to be protected. Generally they are parallel to shore at the landward margin of the beach. They may be poured concrete, timbers, steel sheeting, or boulders (fig. 21.7). Groins are made from similar materials and are perpendicular to the shoreline. The seawalls are intended to protect property from wave attack and groins are designed to trap sediment that is moving along the shore and thereby to build up the beach.

The success or failure of these structures is largely dependent upon the design and placement of them. There are problems inherent in both that must be overcome if the structure is to be considered successful. First, neither is nor can be aesthetically pleasing; they look bad and detract from the natural beauty of the coastal landscape. Second, the improper design and location of these structures can cause more damage than would occur if they were not present. Many seawalls are vertical and impermeable, thereby reflecting all wave energy. This often results in severe scour at the base of the seawall and accelerates beach erosion (fig. 21.7a). Reflection also transmits the wave energy to another location and may cause problems there. The most effective type of seawall is one that dissipates wave energy. This can be accomplished by a sloping surface or by a very irregular one, such as a boulder structure (fig. 21.7b). Unfortunately, this type of structure also has the worst appearance of the types available.

Groins are supposed to act as small temporary dams, trapping sediment moving along the beach due to littoral drift, and thereby building up the beach. In many cases this does not take place because of poor design or location of the structures. The elevation, spacing, and length of the groins must be designed so they trap some sediment, but still allow some to bypass the system. A properly designed and located groin system will only be visible for a short time, then it will become buried by sand as the beach accretes (fig. 21.8). A poor groin field will result in some of the same problems that occur with jetties (fig. 21.9).

Figure 21.9 A poorly designed groin, as evidenced by the difference in sediment accumulation on either side of the structure, which protrudes considerably above the beach surface.

Ineffective coastal structures are widespread throughout the marine coasts of the United States. Not only do they look bad, but they are at best temporary solutions to the problem. They also are expensive and, too often, they cause as many problems as they solve.

Pollution in the Coastal Zone

Since the late 1960s the term pollution has been widely used by the public media, by people in everyday conversation, and in legal and scholarly documents. Rarely, however, is there any definition of the term in these contexts. It is important in any type of discussion about a term and its implications to include a proper working definition. In this case a United Nations document of 1971 defines pollution as "the introduction by man, directly or indirectly, of substances or energy into the marine environment resulting in such deleterious effects as harm to living resources, hazards to human health, hindrance to marine activities including fishing, impairment of quality for use of seawater and reduction of amenities." The gist of this lengthy and comprehensive definition is that anything humans introduce into the marine environment that upsets the natural environment is considered pollution.

Pollution comes in a wide variety of materials, introduced in a broad spectrum of fashions. Some types of pollution are obvious because of the severity of their effect on the environment. This category includes oil spills, garbage dumps, and toxic chemicals. Other forms are very subtle and the results are not noticeable until their cumulative effects are seen, perhaps after several years. An example is the concentration of a toxic substance in the tissues of marine organisms, which then renders them unfit for consumption.

Pollution in the coastal zone may emanate from two general places: (1) human activities along the shore that lead to disposal, leaks, spills, or other introductions from land-based operations generally of an industrial nature, or (2) human activities in the marine environment and subsequent transport to the coastal zone. The most common sites of coastal pollution are in coastal bays, especially estuaries, on beaches, and on river deltas.

Figure 21.10 The oil tanker *Argo Merchant* that was wrecked in a storm near Cape Cod in December, 1976, and caused the tremendous oil spill seen in the photo.

Oil Spills

Most people think that any oil or petroleum product present in marine waters is pollution. This is not the case because there are many places in the world where natural seeps release oil, asphalt, and tar into the sea. The coastal areas of southern California are among the places where this seepage is common. Here it is because of the many fractures present in the California Borderland (see chapter 17), which provide pathways for hydrocarbons. The decay of organic tissues also produces hydrocarbons, such as methane, which is present in marine waters.

The vast majority of hydrocarbons in coastal marine environments is the result of pollution from offshore drilling, faulty pipelines, and ship accidents. Any of these release large quantities of oil into the shallow coastal waters with the typical result being the shoreward transport and eventual landfall of the petroleum. Heavily loaded oil tankers have run aground and broken up in rough seas causing thousands of barrels of oil to be released into the marine environment. This has happened in most parts of the world, with the most notable accidents being in the North Sea and off the northeast coast (fig. 21.10) of the United States, areas that have oil exporting facilities and refineries, respectively. Spills of similar magnitude can occur if a well blows out offshore or when a pipeline breaks.

The presence of large quantities of oil causes several problems for the environment. First, oil is toxic to most marine organisms. It can coat feathers of birds, clog the filtering mechanisms of benthic invertebrates, or inhibit locomotion of organisms. Some oil is dissolved into the ocean water and some sinks to the bottom where it is incorporated into sediments. The combination of winds and currents can disperse the oil rapidly and minimize the effects if the source is stopped. Often the landfall causes large amounts of oil to concentrate on beaches or in estuaries, thus affecting tourism and organisms.

Figure 21.11 Volunteers cleaning up an oil spill on the beach in California. Here straw is used to absorb the oil.

The negative impact of oil varies greatly depending on the specific coastal environment. For example, on exposed rocky coasts oil does not penetrate the substrate and is rapidly cleaned off due to high wave energy. In contrast, the marsh is one of the most adversely affected environments. The combination of low wave and current energy with extensive vegetation traps oil and results in the petroleum substances persisting in this environment for up to several years. Fine-grained, densely packed sediments tend to restrict penetration of oil, whereas coarse sand or gravel permits oil to penetrate up to a meter of sediment. In some situations, such as tidal flats, oil removal is difficult because it may be more harmful to the environment and its contained organisms than simply letting it degrade.

Various techniques have been used to disperse and clean up oil spills. Detergents have been tried as dispersants, but they are also pollutants. Fire can be effective, if it is used quickly when the spill is concentrated, but it also causes control problems and results in atmospheric pollution. Straw, Styrofoam, or other absorbent material commonly is used to collect the oil and thereby remove it from the environment (fig. 21.11).

Petroleum pollution also reaches coastal waters from industrial machinery and automobiles. Usually this type of oil pollution is in a dispersed state except for spills or illegal disposal from industrial complexes. The typical cause is storm runoff from roads, parking lots, and other areas where automobiles spread small quantities of oil as a function of their operation. The concentration of this oil is low, but the total volume is high and there is nearly continuous input via storm drainage. In fact, more petroleum reaches the marine environment by this route than any other and it is concentrated in the coastal waters. Proper maintenance of automobiles can reduce this input. Treatment of storm runoff could also help, but would be extremely expensive.

Dumping of Garbage and Sewage

There are many countries that routinely dump garbage and sewage into shallow coastal waters under the assumption that it will dilute and decompose in the marine environment. The United States is one of the primary contributors in this activity. In New York City, for example, there are huge barges that haul garbage on a daily basis to a site just a few kilometers offshore of the harbor. This site has been designated for garbage disposal and as such has caused severe deterioration of the shallow shelf environment throughout the New York bight. Most of the material dumped on these shallow shelf sites is biodegradable. It was initially thought that this garbage would simply oxidize, benefiting the environment with an additional food supply for organisms. Unfortunately the rate of decomposition in this environment is much slower than burial on land or exposure to the atmosphere.

Solid sewage waste has also been dumped in shallow shelf waters adjacent to densely populated metropolitan areas, such as New York and New Jersey. The result has been an adverse impact on the environment. The organic material utilizes great quantities of oxygen in the decay process and as a result causes serious problems for organisms living on or near the bottom. This condition has prevailed not only in the actual dumping area, but has spread over thousands of square kilometers.

These types of large-scale pollution are slowly being stopped by various types of legislation and larger sewage treatment facilities with more sophisticated treatment methods.

A widespread type of garbage pollution that is a tremendous eyesore, but is volumetrically not significant, is the nonbiodegradable garbage that washes onto the beaches from the coastal and shelf fishing industry. It has been common practice for most fishing vessels, especially the trawlers of the shrimping industry, to throw bottles, jars, and so on overboard with the tops on to prevent them from sinking and interfering with the nets. These floating objects wash onto the shore and pollute the coast. The practice has been particularly widespread in the Gulf of Mexico, but legislation and better policing has helped to control the problem.

Toxic Chemicals from Industry

Industrial complexes along the coast have contributed tremendous quantities of toxic chemicals to the marine environment, causing harm and death to organisms living in these waters. Particularly problematical are the heavy metals, synthetic organic compounds, and acids.

Two of the most common and problematical of the heavy metals in coastal waters are mercury and lead, although copper, nickel, zinc, and chromium are also toxic and have been introduced from industrial wastes and runoff. Mercury is used in a wide variety of industries, and although the rate of influx into marine waters per year may not be alarming, it is an element that builds up over time, both in the sediments and in organisms. Tissues of fish and shellfish concentrate mercury such that the organisms present a health hazard for humans if consumed. Several areas in the world have had to quarantine certain fish with accumulated levels of mercury that exceeded tolerable limits. Several coastal states and the Great Lakes in the United States, along with Canada, Finland, Norway, and Sweden have had these quarantines. The environment and its organisms have remarkable powers of recovery, however, and stopping the sources of mercury can reverse the trend in just a few years.

Lead is another hazardous heavy metal that enters the aquatic environment, both freshwater and marine. Most of the lead has been in the form of small quantities from gasoline, spread over most of the automated world largely from exhaust. Storm sewer runoff carries the lead into coastal environments where it is consumed by organisms and retained in their tissues. Like mercury, lead is a toxic substance that cannot be tolerated by living things. The laws against use of leaded gasoline are making a significant dent in the rate at which lead is introduced into the environment, but this substance tends to remain in marine waters for longer than originally thought.

There are many varieties of synthetic organic compounds that are extremely toxic and that have been introduced into marine waters, some in large volumes, including pesticides, various chemicals used in plastics, solvents, and other organic fluids. Probably the best known of these pollutants is DDT, the prevalent pesticide of the 1950s and 1960s. This organic liquid is carried in the atmosphere, in freshwater runoff, and also is introduced directly into the marine environment. It is not biodegradable and is concentrated by organisms. As a consequence, each trophic level in the food chain receives an increasingly concentrated dose of DDT. The top consumers are typically fish-eating birds or people. In most situations the concentration has not been a serious problem for humans as yet, but that is not the case for birds. The most important problem has been the reduction of calcium carbonate in organisms with high levels of DDT in their tissues. This results in eggs with very thin shells that break when incubated by the adult birds, thus threatening their populations. Fortunately DDT has been strictly regulated in

Figure 21.12 Low-tide photograph showing crusts of chemical sediment pollutants in the bay adjacent to a large phosphate plant near Tampa, Florida.

the United States and the trend in aquatic bird populations has turned around. There are other countries, however, where similar restrictions have not been adopted and serious problems still exist.

Acids have also been introduced into coastal waters as waste from industry. These may come from a variety of chemical plants, battery companies, or acid plants themselves. In most situations the pollution is the result of sloppy disposal operations or the occasional accident. An excellent example is a plant where sulfuric acid is produced for use in the processing of phosphate minerals for the fertilizer industry in Florida. Waste from this plant had been entering the northern part of Tampa Bay for several years and the pH of waters near the plant was as low as 3. There was essentially nothing alive in this part of the bay and unusual mineral precipitates were forming in the intertidal zone near the wastewater effluent (fig. 21.12). Environmental regulatory agencies enforced the ordinances and in a matter of about four years the bay had been returned to essentially normal conditions in which organisms flourished. This is but one indication of how responsive marine communities are to change and how rapid repopulation is once the environment becomes habitable again.

Radioactive Pollutants

Various industrial activities use radioactive materials, with nuclear power facilities being the most prevalent. As in other pollution situations, both disposal and accidents are the source of radioactive pollution. Up until the 1960s radioactive wastes were dumped into the ocean. Although most of the intentional disposal of radioactive wastes is in the deep ocean, accidents and poor quality control can cause it to enter coastal waters. Large numbers of nuclear reactors are located along coastal waters for cooling purposes. The potential for tremendous disasters is present, but quality control is generally good and regulations are strict.

Some industrial activities involve rather low concentrations of radioactive materials that occur naturally. The phosphate industry is among these and the introduction of radioactive materials is a serious problem. The phosphate minerals contain small quantities of uranium. In some situations the uranium is reclaimed during the beneficiation process because it can provide income for the industry. Some facilities, however, are not equipped to do this and at times the uranium price is depressed below the level necessary to make reclamation profitable. The overall consequence is that small quantities of uranium are introduced into coastal marine waters.

These radioactive substances tend to concentrate in the skeletons of organisms; the concentration levels may become high enough to prohibit harvesting some fish or shellfish. The uranium is primarily concentrated in the skeletal portions of organisms so this does not pose a major problem for most commercial organisms. One industry that does feel the impact of radioactive concentrations is the soft-shelled crab industry. The entire organism including the shell is consumed, thereby passing the radioactive material on to the human consumer.

Nutrient Pollution

A common problem in some coastal environments, especially the shallow, low-energy estuaries, is the influx of excess nutrient material and the eventual deterioration of the environment as a result. There are two primary sources for this high rate of nutrient influx: agricultural fertilizer runoff and human waste.

Agricultural fertilizers and those used on lawns contain high levels of phosphorus, a nutrient element. Although the intent is for the plants to take up this phosphorus, much of it is carried into coastal waters via runoff. Many detergents have similar effects, however, regulations have been implemented to control phosphate levels in these products. Sewage, both treated and untreated, may also be introduced into these estuarine environments, intentionally or unintentionally. It may emanate from septic tanks, spillage, or from primary and secondary treatment facilities that pump effluent directly into the coastal waters.

Such introduction of phosphorus can be advantageous if it occurs in low to moderate levels and if it is controlled. Unfortunately this is not typically the case. Large amounts of nutrient material dumped into an estuarine system that cannot disperse and dilute the nutrients cause tremendous blooms of producers: phytoplankton, benthic algae, and grasses. Most people have probably seen a water body literally clogged with algae and other vegetation, especially in the spring. Such an occurrence is the result of a high influx of nutrient material. The problem created by this type of extreme productivity comes with the very high volume of organic material produced by these blooms. There is commonly much more plant material than can be consumed by the herbivorous animals in the estuarine system. This results in an abundance of plant material that eventually dies off and decomposes (oxidizes). The oxidation process uses much oxygen and this typically results in an oxygen deficient environment that causes the death of many animals, especially fish. It is important, therefore, to control influx of nutrient material. This requires regulation on the composition and use of fertilizers and also stringent regulation in the disposal of sewage in this type of coastal environment.

Pollution in the Open Ocean

We commonly think of the open ocean waters as clear, blue, and pollution free. This is not at all the case although the open ocean is typically much less polluted than many coastal zone environments. The enormous volume of the ocean certainly does dilute pollution, but there are serious problems in this environment that must be addressed. Some of the primary causes for oceanic pollution are ships and the deep-sea disposal of radioactive and toxic chemicals.

Pollution from Ships

Ships of all types and sizes from small trawlers to huge super tankers are traversing nearly all parts of the world ocean on a continual basis. Merchant ships alone number nearly 25,000 and fishing vessels are even more numerous. The result of this great volume of machinery and activity is a significant amount of pollution. Many large ships are like small towns in terms of size, population, and complexity. They have problems of pollution of that magnitude as well.

Without a doubt, the most noticeable pollution problem associated with shipping is a major oil spill accident. Although most of these accidents have taken place in rather shallow water due to collisions or to bad weather and grounding or breaking up of the ship, there is also potential for spills in the open sea. As alarming and apparently catastrophic as these accidents may be, they are really a minor factor in the pollution of the ocean.

By far the most important type of pollution in terms of volume is the intentional disposal of chemical wastes in the open ocean. Chemicals that are disposed of in this manner include by-products of plastics and chemical industries, most of which are toxic and carcinogenic organic compounds. Most of the operations that involve such disposal take place on the continental shelf or close beyond. The impact of this type of chemical pollution is quite severe, especially to the fishing industry. The long-term impact is not yet known; however, it is potentially quite serious. Every year several million tons of chemical waste are dumped in this manner. Typically the dumping activities are cloaked in a veil of secrecy because of the obvious problems. On many occasions one government will learn of dumping activities and bring international pressure to halt such activities. An example took place in 1971 when a Dutch freighter carrying wastes from a plastics industry was prevented from dumping its cargo west of Norway. The ship then steamed to a location south of Iceland but the British government intervened and eventually the ship returned to port with its cargo of chemical wastes.

The preceding incident was a primary factor that led to the development of large incinerator facilities on ships for the purpose of large-scale disposal at sea. Now there is considerable disposal of chemical wastes using this technique, especially by European countries. There are toxic gases emitted by this process but they are far from population areas and the volume is estimated to be insignificant on a global basis. In any case, this method of disposal has been demonstrated to be much safer than disposal at sea or burning at the industrial site.

Ocean Floor Dumping

Since the dumping of munitions and poisonous gases during and shortly after World War II, there has been considerable disposal of toxic and radioactive wastes on the floor of the ocean. It is an "out of sight, out of mind" syndrome that initiated and has perpetuated this activity. It is painfully obvious that it is very bad practice. There are various restrictions and laws now that control this type of activity (see section on marine law).

Intentional and accidental disposal of munitions on the ocean floor was widespread and caused serious incidents that resulted in injury and death. Some munitions were simply units that did not explode during battle, some were dumped by planes on the way back after bombing missions during which all units were not deployed, and others were large-scale disposal of obsolete munitions and gases. It was common practice, especially by the United States, to load a ship with ammunition or with mustard gas, take it out to sea and sink it. Numerous problems resulted from this type of disposal. Fishermen retrieved grenades and shells in their nets and many empty canisters were recovered indicating that the contents had escaped into the marine environment.

At the present time it is not known to what extent this type of activity has impacted the marine environment. Certainly the overall volume of dumped material is small as compared to the overall volume of the world ocean. Just as certainly, any introduction of toxic materials will cause a reduction in the quality of the environment. The largest potential problem is the effect on benthic fauna. Dispersal rates of pollutants on the ocean floor are not great and it is likely that these compounds will kill or seriously impact the life cycles of benthic creatures. We know that such chemicals cause disease or genetic defects in humans and higher animals, so it is probable that marine organisms are similarly affected.

There are situations where intentional dumping or disposal has been designed to benefit the marine environment. This is the sinking of ships, automobiles, or other similar material to form an artificial reef (fig. 21.13) in order to stimulate the development of a reef community and provide ecological niches and food for various open marine plants and animals. Typically this is done on the continental shelf in mid- to low-latitudes and it has proven quite successful for both sport and commercial fishing.

Marine Law

There is probably no topic that is more complex than international marine law. The jurisdiction covers more than two-thirds of the globe and involves hundreds of governments. Considerations must include not only fishing but also the military, minerals, petroleum, waste disposal, and pollution. A systematic international approach to the problem began only in the 1950s, but there has been much progress and increasing attention and effort promise

Figure 21.13 Underwater photograph of an artificial reef composed of old tire casings held together by cables. These provide a good substrate for reef-type organisms.

that this may be one area where worldwide cooperation may indeed be achieved.

Marine Law Prior to 1958

Some type of global marine law has been in effect since the time of the ancient civilizations surrounding the Mediterranean Sea. The Greek and Roman empires recognized the need for freedom of the seas for trading and established what was known as Rhodian sea law named after the Greek island of Rhodes, an important trading center where ocean laws were practiced. Although various of the governments of the Mediterranean did exercise some types of jurisdiction over their adjacent waters, freedom of travel was permitted. This lasted until Venice became a maritime power and achieved control of the entire Adriatic Sea in the 14th century. This led to a tendency of governments to control adjacent sovereign waters in the Mediterranean and the Scandinavian countries. Finally in 1493 the Pope issued a papal bull that gave split control of the oceans between Spain and Portugal primarily in an effort to give these maritime powers exclusive rights to establishing trade with the East and West Indies.

During the two centuries following, there was conflict between those countries that preferred a free and open sea (*mare liberum*) and those that advocated their right to claim areas of the sea (*mare clausum*). Eventually the Spanish Armada was defeated in 1588 by the British and Dutch. Not long thereafter the sea became open to all. The practiced philosophy was that all nations were free to use the sea for any purpose as long as that use did not interfere with any other nation. It did become apparent, however, that the coastal nations should have some control over their adjacent waters. This led to the well-known three-mile limit, or territorial sea, over which the adjacent nation had sovereignty. Although there was never a formal treaty to this effect, it was commonly accepted and practiced up to the middle of the present century. Beyond the three-mile limit there was complete freedom for navigation, fishing, laying of communication cables, and other marine activities.

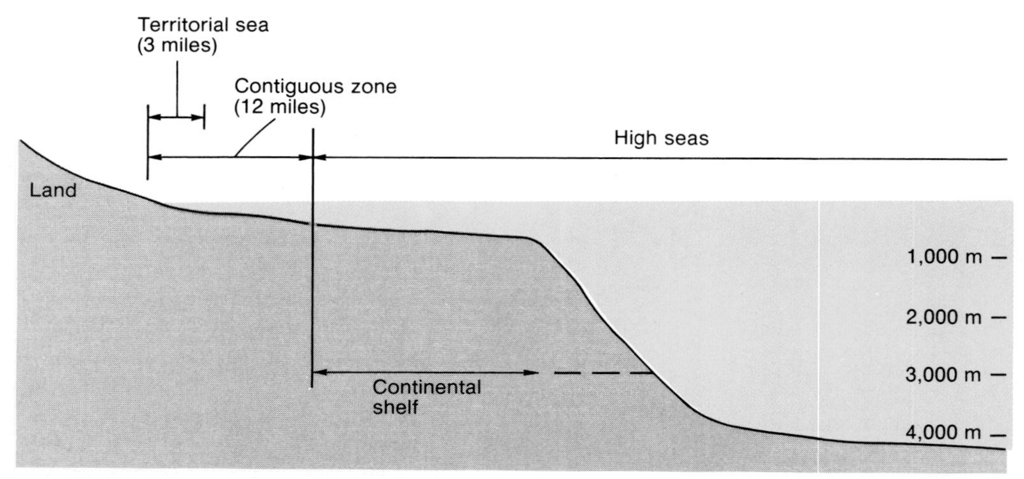

Figure 21.14 Zones designated by the first Law of the Sea Conference.

During World War II it became apparent that some countries had certain advantages over others in terms of using the seas and also that the resources of the ocean, especially fish, were not inexhaustible. In 1945 President Harry Truman proclaimed that the United States had jurisdiction and control over all of the contiguous continental shelf including its fisheries. Mexico followed with a similar statement and a year later Argentina did the same. In 1947 Peru and Chile claimed a 200-mile jurisdiction in order to permit them to control the rich fisheries in the waters beyond their narrow shelves.

These actions and the snowballing effect they had on other countries made it apparent that something must be done on an international level and the United Nations became involved. The International Law Commission of this body prepared draft articles covering the law of the sea, and after several preliminary meetings and revisions of the articles, the first United Nations Conference on the Law of the Sea was held in Geneva, Switzerland, in 1958.

1958 Law of the Sea Conference

There were 86 nations that participated in the first Law of the Sea Conference and much was accomplished to formalize international use of the world ocean. This conference adopted four major conventions: (1) the convention on the territorial sea and the contiguous zone, (2) the convention on the high seas, (3) the convention on the continental shelf, and (4) the convention on fishing and the conservation of the living resources of the high seas.

Important results included the agreement on the subdivisions of marine waters into three main categories: (1) the territorial seas, (2) contiguous zone and exclusive fishing zone, and (3) the high seas (fig. 21.14). The territorial seas extended three miles from the low-water line and nations had sovereignty over these waters as well as the air space above them. The contiguous zone to the 12-mile limit was part of the high sea, but the adjacent nation had control over fishing and fishing research in these waters. All other waters were considered the high seas in which total freedom existed.

The convention of the continental shelf gave control of all non-living resources to a depth of 200 meters to the adjacent nation but did not prohibit fishing or laying of communication networks by other nations. Marine research in these waters required the consent of the adjacent nation. The convention of fishing and conservation of living resources gave all nations the privilege of fishing on the high seas and stated that cooperation among nations should be instituted as necessary to conserve these living resources.

A second Law of the Sea Conference was held shortly thereafter in 1960. It served primarily to clarify some of the decisions made during the 1958 conference. The outcome of these two meetings caused some serious problems due to lack of definition and inadequate rationale for some of the decisions. There were three main points of contention: (1) the lack of definition of the width of the territorial sea, which was especially a problem to nations

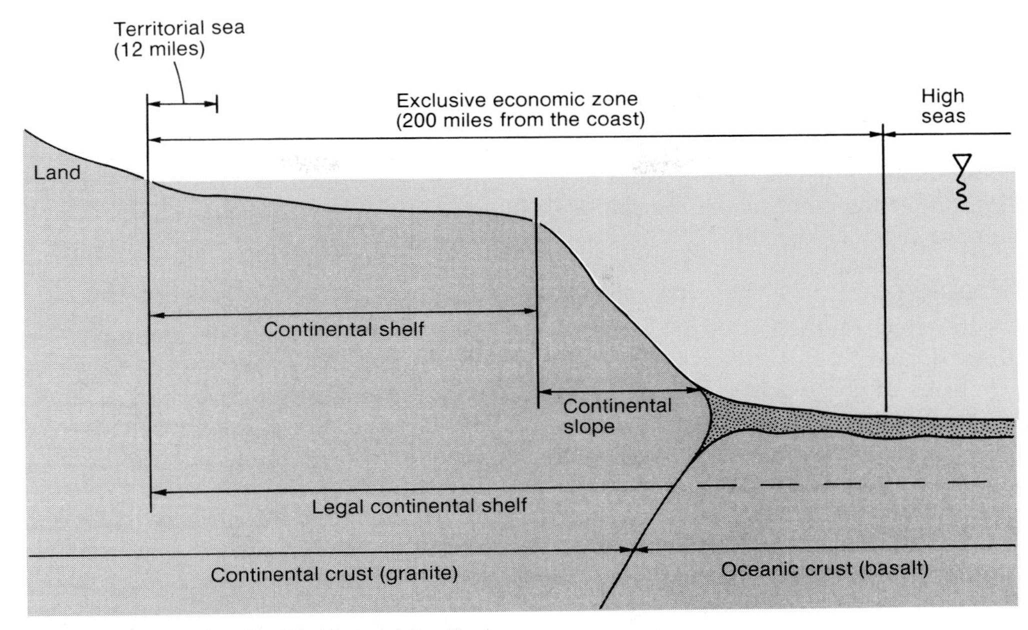

Figure 21.15 Zones designated by the third Law of the Sea Conference.

with very irregular coastlines or that faced each other closely, (2) the 200 m depth as a definition of the outer limit of the shelf, and (3) the potential for exploitation of marine minerals.

Third Law of the Sea Conference

After over a decade of discussion and dissatisfaction with the conventions drawn up at the first and second conferences, it became apparent that further efforts were necessary. Another reason for this conference was the pleas of the lesser developed nations who feared that the nations with advanced technologies would rape the world ocean of its mineral resources.

Because of the lack of clarity in the first conventions and the dissatisfaction with some of the articles, a number of countries began to act individually. Several declared territorial seas that extended for more than 12 miles. There were conflicts over fishing rights and even scientific research. The third conference began in 1973 in New York and continued at various locations until 1982 when the Law of the Sea Convention was adopted.

The main points that were addressed during the third conference were: (1) the width of the territorial sea and the degree to which the adjacent coastal government controlled it, (2) ownership of the resources both within the territorial sea and beyond, (3) navigation through and over narrow straits in international waters, (4) management of the living resources of the sea with special emphasis on migratory species, (5) pollution problems, (6) scientific research in the world ocean, and (7) a system for management of the high seas.

The third conference was very long not only because of the many important topics to be covered, but because nearly 150 countries participated, introducing more than 400 draft articles. Various alliances and interest groups consisting of countries banding together for a common purpose also furthered the complexity of the tasks at hand.

The third conference adopted a 12-mile territorial sea, with the right of free passage through straits that are less than this width. There are more than 100 such areas, and restrictions or control on passage through these locations would cause serious military and economic problems for many nations. The zone of exclusive economic control extends 200 miles from the shore regardless of where this falls relative to the continental shelf-slope break (fig. 21.15). The high seas area remains open to all. The third conference considered the deep seabed and established an international agency, the International Seabed Authority, to manage the mineral resources of this environment. Considerable time and effort were also devoted to problems of marine pollution. Most of the results speak to that caused by ships and give jurisdiction for enforcement of the regulations to the flag state of registry of the ship.

Remaining Problems

Although the conferences on the law of the sea have accomplished much, there is still much to be done in regulating the uses of the world ocean. Despite the designation of the adjacent coastal state to control the fishing industry in its economic zone, it is necessary to adopt some type of international scheme for management of fisheries on the high seas. Many species migrate great distances and populations do not adhere to artificial political boundaries. This situation is unlike the mineral resources which can effectively be managed by the coastal government.

Pollution remains a big problem, but waste disposal is a legitimate use of the ocean. This makes it all the more difficult to regulate and control. One of the major problems is knowing how the ocean deals with the various types and amounts of pollution and this requires large-scale research by consortiums of nations. High-level and far-reaching cooperation among all nations with marine interests is essential to the proper management of the high seas.

We must also look to the oceans for additional uses beyond fisheries, minerals, and waste disposal. Harnessing the energy of the ocean must be given increased attention. Military uses are changing and expanding and this must also be considered. Because of the enormity of the world ocean and the huge number of countries involved, it may be more effective to consider regional efforts to achieve some of these goals. Some areas lend themselves to this approach, such as the Gulf of Mexico, the Mediterranean Sea and the Persian Gulf, but even in the large ocean basins this may be an effective way to achieve more faster.

Summary of Main Points

The world ocean offers tremendous potential for natural resources and other uses. These include food, military and transportation activities, waste disposal, mineral resources including petroleum, and energy. At the same time the ocean presents tremendous problems of regulation and management.

Fishing and shellfish industries have greatly impacted some species of desirable animals but there have been great strides made in proper management of these species. Mariculture offers potential for greatly increasing our harvest of desirable species, and high protein food additives may also lead to expanded fisheries industries and help the global starvation problem.

Human activities must be better controlled if the oceans are to remain a viable environment for fisheries. This includes pollution of both coastal and open waters as well as overfishing. Much regulation of disposal techniques and locations has improved this trend but there is still more to do.

Because of the desirability of coastal property, there has been much abuse and poorly designed and managed development of this type of property. Until the late 1970s humans have given little consideration to developing coastal properties so that natural processes will not be greatly impaired by construction. There is still great pressure to develop these properties, with economics as the prime consideration, but that trend is being overcome by a more conservation-minded citizenry and government.

Governance of the world ocean is extremely complicated and we are just in the infant stages of seeing the effects of international marine law. As time passes and these laws are refined to suit the needs of the world community, the benefits of this type of regulation and management will become apparent.

Suggestions for Further Reading

Center for Ocean Management Studies, University of Rhode Island. 1981. **Comparative Marine Policy.** New York: Praeger Publishing.

Cuyvers, Luc. 1984. **Ocean Uses and Their Regulation.** New York: John Wiley and Sons.

Gerlach, S. A. 1981. **Marine Pollution.** Berlin: Springer-Verlag (translated from 1976 German version).

Ross, D. A. 1980. **Opportunities and Uses of the Ocean.** New York: Springer-Verlag.

APPENDIX

A

Conversions and Constants

Length

1 micrometer (μ) = 0.0000394 inch; 0.0001 cm; 0.001 mm

1 millimeter (mm) = 0.0394 inch; 0.001 m; 0.1 cm

1 centimeter (cm) = 0.394 inch; 0.01 m

1 meter (m) = 39.4 inches; 3.28 feet; 0.001 km

1 kilometer (km) = 0.621 mile; 0.540 nautical mile; 1,000 meters

1 fathom = 6 feet; 1.83 meters

1 nautical mile = 1.85 kilometers; 1.15 statute miles; 1 minute of latitude; 1,853 meters

Area

1 square kilometer = 0.39 square mile; 247.1 acres

Volume

1 milliliter = 0.001 liter; 1.0 cubic centimeter; 0.06 cubic inch

1 liter = 1,000 milliliters; 1.06 U.S. quarts; 33.8 fluid ounces

Speed

1 centimeter per second = 0.036 kilometer per hour; 0.02 mile per hour; 1.97 feet per minute

1 kilometer per hour = 27.8 centimeters per second; 0.621 mile per hour

1 knot = 1 nautical mile per hour; 51.5 centimeters per second

1 meter per second = 2.23 miles per hour; 3.6 kilometers per hour

Pressure

1 atmosphere = 14.7 pounds per square inch; 760 millimeters of mercury at 0°C; 1.013 bar

Time

1 lunar month = 29 days, 12 hours, 44 minutes

Appendix B

General Classification of Marine Organisms

Kingdom—Monera

- Phylum—Schizophyta (bacteria)
- Phylum—Cyanophyta (blue-green algae)

Kingdom—Protista

- Phylum—Chrysophyta (diatoms or golden-brown algae)
- Phylum—Pyrrophyta (dinoflagellates)
- Phylum—Protozoans, all environments
  - Class—Sarcodina, locomotion with pseudopodia
    - Order—Foraminifera
    - Order—Radiolaria
  - Class—Ciliata, locomotion with cilia

Kingdom—Plantae

- Division—Rhogophyta (red algae)
- Division—Phaeophyta (brown algae)
- Division—Chlorophyta (green algae)
- Division—Tracheophyta (vascular plants)
  - Class—Angiospermae (flowering plants)
    - Subclass—Monocotyledoneae (sea grasses, marsh grasses, etc.)

Kingdom—Fungi

Kingdom—Anamalia

- Phylum—Porifera, sponges, mostly marine, sessile benthos on solid substrate
- Phylum—Cnidaria (also called Coelenterata), mostly marine, benthic and pelagic
  - Class—Hydrozoa, colonial, planktonic and sessile benthos
  - Class—Scyphozoa, essentially all planktonic (jellyfish)
  - Class—Anthozoa, many colonial, mostly sessile (corals, anemones)
- Phylum—Ctenophora, marine, planktonic
- Phylum—Platyhelminthes, all environments, planktonic and vagrant benthos, infauna and epifauna (flat worms)
- Phylum—Nemertina, mostly marine, planktonic, benthic, some commensal
- Phylum—Gastrotricha, mostly marine, benthos, shallow, sandy and muddy environments
- Phylum—Nematoda, all environments, mostly vagrant benthic, some parasitic
- Phylum—Entoprocta, mostly marine, shallow benthic, encrusting sessile
- Phylum—Endoprocta, marine and fresh water, shallow benthic, encrusting colonial
- Phylum—Brachiopoda, mostly marine, sessile benthos (lamp shells)
- Phylum—Mollusca, mostly marine, all environments, most with external shell
  - Class—Amphineura, shallow marine, vagrant benthos (chitons)
  - Class—Gastropoda, mostly vagrant benthos, few planktonic (snails, limpets, slugs)
  - Class—Scaphopoda, benthic infauna (tusk shells)
  - Class—Cephalopoda, benthic and pelagic (nekton), most without shell
  - Class—Pelecypoda, benthic bivalves, vagrant and sessile, infaunal and epifaunal (clams, oysters, scallops)
- Phylum—Sipunculida, all marine, benthic, intertidal to abyssal (peanut worms)
- Phylum—Pogonophora, marine, deep water, sessile benthic, tube dwelling worms.
- Phylum—Hemichordata, marine, sessile benthos (acorn worms)
- Phylum—Chaetognatha, marine, pelagic, weakly swimming (arrow worms)

Phylum—Annelida, all environments, benthic sessile and vagrant (segmented worms)
  Class—Polychaeta, mostly marine, benthic, some with tubes
  Class—Hirudinea, leeches
Phylum—Arthropoda, all environments, segmented bodies with exoskeleton
  Class—Merostomata, benthic nearshore (horseshoe crabs)
  Class—Pycnogonida, benthic nearshore with 4 pairs of legs (sea spiders)
  Class—Crustacea, mostly marine, dominant class of marine arthropods, pelagic and benthic
    Subclass—Branchiopoda (brine shrimp)
    Subclass—Ostracoda, pelagic and generally microscopic
    Subclass—Copepoda, pelagic and benthic, most are microscopic
    Subclass—Cirripodia, sessile benthos (barnacles)
    Subclass—Malacostraca
      Order—Mysidacea, benthic and pelagic, small
      Order—Cumacea, infaunal, shallow to deep
      Order—Isopoda, benthic, shallow (sand hoppers)
      Order—Amphipoda, benthic and pelagic
      Order—Stomatopoda, benthic (mantis shrimp)
      Order—Euphausiid, pelagic (krill)
      Order—Decapoda, benthic and pelagic (crabs, shrimp, lobsters)
Phylum—Echinodermata, marine, mostly benthic, spiny skinned animals, pentameral symmetry
  Class—Echinoidea, generally with long spines (sea urchins and sand dollars)
  Class—Asteroidea, 5 arms or a multiple of 5 (starfish)
  Class—Ophiuroidea, rapidly moving (brittle stars)
  Class—Crinoidea, flexible and delicate arms (sea lillies)
  Class—Holothuroidea, no arms, fleshy bodies (sea cucumbers)
Phylum—Chordata
  Class—Agnatha, jawless fishes (sea lampreys and hagfishes)
  Class—Chondrichthyes, cartilaginous fishes (sharks and rays)
  Class—Osteichthyes, bony fishes
  Class—Reptilia, turtles, sea snakes and crocodiles
  Class—Aves, birds
  Class—Mammalia, animals born alive and that nurse young
    Order—Carnivora, seals, walruses and sea lions
    Order—Cetacea, whales
    Order—Sirenia, manatees

Glossary

a

abyssal Pertaining to the ocean floor environment that is deeper than 3,700 meters.

abyssal hills Gently undulating features of the deep ocean surface, produced by volcanic flows, and having relief of tens of meters.

abyssal plain The flat portion of the abyssal zone which together with the abyssal hills comprise the abyssal floor.

abyssopelagic zone The portion of the ocean that is below a depth of 3,700 meters.

acid A generally sour-tasting substance that is a proton donor which dissociates to produce to free hydrogen ions.

agar A gelatinous substance from red algae that is used in various products, such as bacterial medium, and as a milk thickener.

alcyonarian corals Soft, colonial corals that include the sea fans, sea whips, and encrusting varieties.

algal mats Thin layers of blue-green filamentous algae that commonly develop in the intertidal zone and that may trap sediment and form stromatolites.

algal ridge A ridge of coralline algae (red) that forms on the windward crest of a coralgal reef.

algal stromatolites Accumulations of layered sediment, generally carbonate, that is commonly formed in the intertidal zone by blue-green filamentous algae.

algin A complex substance produced by brown algae and used in various manufactured products, such as ice cream, paint, and toothpaste.

amphidromic point The nodal point on a cotidal chart from which the cotidal lines radiate. The tidal waves rotate around this no-tidal point in a single tidal cycle.

aquaculture The cultivation of various aquatic organisms under controlled conditions.

asthenosphere A somewhat-plastic layer in the upper mantle that extends from the base of the rigid crust (lithosphere) to depths of a few hundred kilometers.

atoll A somewhat circular reef that is neither attached to nor contains a landmass and that has a central lagoon.

b

backshore That portion of the beach that is above high tide and is approximately horizontal. It is the supratidal part of the beach.

baleen whales Whales that have a filtering structure on their upper jaw that traps the plankton they feed upon.

barrier reef A reef that is separated from the adjacent land mass by a lagoon.

base A soapy and bitter-tasting substance that accepts protons in a water solution.

bathyal zone That portion of the ocean bottom between depths of 200 meters and 3,700 meters; essentially, the continental slope and rise.

bathythermograph An instrument shaped like a small torpedo which records depth and temperature in waters less than about 300 meters deep.

beach The zone of unconsolidated sediment between low tide and the landward marked change in material or physiography and which is subjected to waves.

beachrock Lithified beach sediment that forms in low latitudes where sediment is dominantly carbonate.

beam trawl A type of net for collecting marine organisms that has a frame to keep the net open. It is typically towed along the ocean bottom.

benthic Pertaining to the bottom environment.

benthos Organisms that live on or in the bottom.

bioerosion Erosion that is caused by organisms and that may be due to a combination of physical or chemical breakdown of hard substrates.

biogenic sediments Sediments that are comprised of some type of skeletal material derived from organisms.

biomass The total mass of organic matter in the ecosystem. Typically expressed as the dried weight.

bioturbation The disturbance of the substrate by organisms' burrowing, boring, grazing, and so on.

borers Organisms that bore into hard substrates by a combination of physical or chemical means.

Bouma sequence The characteristic five-part sequence produced by the deposition of sediments from a turbidity current; a turbidite sequence.

brackish Water of less than normal salinity, generally between 0.5‰ and 17‰.

burrowers Organisms that physically burrow into the unconsolidated substrate.

c

calcareous ooze Deep-sea ooze that is characterized by the tests of planktonic organisms composed of calcium carbonate. Examples are foraminifera, coccolithophores, and pteropods.

carbonate compensation depth (CCD) The depth in the ocean below which calcium carbonate is dissolved more rapidly than it accumulates from above. It ranges over a few hundred meters, but averages a depth of 4,500 m.

carrageenan A carbohydrate that is produced in brown algae and is similar to algin. It is used commercially as a thickener in paints, ice cream, and toothpaste.

centripetal force Force that tends to move objects in toward the center around which they are moving.

chemosynthesis Formation of organic compounds using energy derived from the oxidation of certain compounds, such as ammonia or methane, and elements, such as iron, sulfur and hydrogen.

chlorinity The concentration of chloride content in seawater and the value from which salinity measurements are determined.

chondrites Stony meteorites comprised of mafic materials.

coccolithophores A microscopic planktonic algae encased by very small disks of calcium carbonate that settle to the ocean bottom and become an important constituent of oceanic sediment.

compensation depth The depth below which oxygen consumption is greater than that which is produced. Plant populations cannot be sustained below this depth.

continental borderland A type of continental margin that exhibits great relief because of its complex faulting, which produces a basin-and-range type of topography.

continental crust That portion of the earth's crust that has a granitic composition, a density of about 2.7 grams per cubic centimeter, and reaches a thickness of about 35 km.

continental drift The term first applied to the theory that the earth's crust was mobile and that continents moved great distances over the earth's surface.

continental margin The physiographic province of the earth's surface that extends from the shoreline to the base of the continental rise. It is geologically associated with the continents and includes the shelf, slope, and rise.

continental rise The most oceanward part of the continental margin, which is between the continental slope and the ocean floor.

continental shelf The shallowest part of the continental margin, which extends from the shoreline to the seaward change in gradient where the slope begins.

continental slope The steepest portion of the continental margin, located between the shelf and the rise.

continental terrace A term sometimes applied to the continental shelf and slope.

contour current Oceanic currents that are driven by density gradients and move parallel to bathymetric contours over the outer part of the continental margin.

contourites Sediments that are reworked and deposited by contour currents.

core The central dense zone of the earth, which is divided into an inner solid core and an outer liquid core. The core extends to a depth of 2,900 km below the surface of the earth.

Coriolis effect An effect on moving particles traveling over a rotating sphere (earth) as viewed by an observer. The path is characterized by a deflection to the right in the northern hemisphere and to the left in the southern hemisphere.

cosmogenic sediment Sediment that originates in outer space.

cotidal line Line on a tidal map that passes through all points where high tide occurs at the same time. Cotidal lines emanate from an amphidromic point.

crevasse splay A thin fan-shaped accumulation of sediment that is deposited during flooding on a delta, when the natural levee is breached.

crust The outer portion of the earth, which ranges up to about 35 km in thickness, and the base of which is the density change referred to as the Mohorovicic discontinuity.

d

debris flow A flowing mixture of mud and water that also supports and transports large particles.

deep scattering layer A layer of planktonic or feeding swimming organisms that is commonly recorded on echo sounders and migrates vertically through a few hundred meters on a diurnal basis for feeding in the photic zone.

deep sea General term for the open ocean environments including the abyssal floor and the trenches. Essentially that part of the ocean below about 3,500 m.

deep-sea trench Narrow and deep ocean environment that is below 6,000 m depth.

deep water waves Oceanic gravity waves that travel in water that is deeper than one-half of their wave length and are therefore unaffected by the bottom.

delta An accumulation of sediment at the mouth of a river where it empties into a basin.

delta front The portion of the river delta that is subjected to regular wave activity and is therefore sandy. It is just seaward of the subaerial part of the delta.

delta plain The upper surface of a river delta that is comprised of several environments and is partially subaerial and partially subaqueous. The most landward part of the delta.

demersal Living on or near the bottom of the ocean; refers to a mode of life of many types of fishes.

diatoms Microscopic planktonic photosynthetic organisms that are the most abundant primary producers in the ocean. They have siliceous tests and are important contributors to deep-sea sediments.

diffraction The spreading of wave energy due to obstructions.

distributaries An outflowing branch of a river on a delta; the opposite of a tributary.

diurnal Daily; pertains to phenomena that occur in a twenty-four hour cycle or a daily cycle.

downdrift offset The offset of the shoreline on opposite sides of a tidal inlet whereby the downdrift side relative to littoral sediment transport is offset seaward of that on the updrift side.

downwelling The descent of warm surface waters due to the interaction of a landmass and the Coriolis effect.

dunes Large, mobile accumulations of sand that commonly form landward of the beach. They may become stabilized due to vegetation.

e

ebb tidal delta A tidal delta associated with a tidal inlet that is formed by ebb tides and is located on the seaward side of the inlet.

Ekman spiral The representation of the relationship between the direction of the surface wind and upper layer of water which shows that surface waters move at an angle of 45° to the wind and that this angle increases with depth. The direction is related to the Coriolis effect and is to the right in the northern hemisphere and to the left in the southern hemisphere.

Ekman transport The net transport of water as shown by the Ekman spiral. In the northern hemisphere it is 90° to the right of the wind direction and in the southern hemisphere it is to the left.

El Niño Unusually warm current that flows to the south along the northwest coast of South America due to climatic changes near Christmastime.

epifauna Those animals that live on the ocean bottom as contrasted with the infauna that live in the substrate.

epipelagic zone The upper portion of the pelagic zone. It extends from the surface to a depth of about 200 meters.

estuary A coastal embayment that receives appreciable freshwater runoff from land and that experiences open tidal circulation with the ocean.

Eulerian current measurement Measurement of currents past a fixed point.

euryhaline Adaptable to a wide range of salinities.

eutrophic Containing abundant to excess organic matter.

f

fjords A special type of estuary that is narrow and deep, and was excavated by glaciers.

flood tidal delta A tidal delta associated with a tidal inlet, formed by flood tides, and located on the landward side of the inlet.

forced wave A wave that is formed and maintained by a continuous force, for example, tides.

foredunes Coastal dunes that are located immediately landward of the beach.

foreshore That portion of the beach that is seaward-sloping and is intertidal.

fringing reef A reef that is immediately adjacent to a landmass without the presence of a lagoon.

front The boundary between adjacent weather systems. They may be abrupt as is typical for cold fronts or they may be somewhat diffuse.

fry Young fish or hatchlings. They are not able to swim with sufficient strength to control their location and are at the mercy of oceanic currents.

g

geostrophic flow Horizontal fluid transport that results from a balance of gravitational flow and the Coriolis effect.

gill net A type of commercial fishing net that has rather large mesh so that a fish will swim into the net and be caught by the gills in the mesh.

glacial marine sediments Sediments that are derived from glaciers and deposited in the marine environment.

gravimeter An instrument that measures the pull of the earth's gravity field.

gravimetrics That part of the discipline of geophysics that measures the earth's gravity field.

gravity anomaly The variation in the earth's gravity field that deviates from the value based on a uniform earth. These anomalies may be positive or negative.

gravity waves Waves whose velocity is controlled primarily by gravity. The typical wind waves on the water's surface are gravity waves.

greenhouse effect The warming effect on the earth, similar to that of a greenhouse, that is caused by the emission of carbon dioxide, dust, and other materials into the atmosphere, thus trapping much of the solar radiation within the atmosphere.

groin A type of coastal structure that is placed on the beach perpendicular to the shoreline in an effort to trap sediment and stabilize the beach.

guyot A submarine volcano that has had its crest eroded and is not below the ocean surface.

gyres A somewhat circular form taken by surface ocean current systems as the result of the combination of surface winds, the Coriolis effect, and the rotation of the earth.

h

Hadal zone The deepest zone of the ocean; that within the oceanic trenches.

hermatypic corals Those that are reef builders and that have symbiotic algae.

hertz The term used to measure the frequency of sound impulses. One hertz is one cycle per second.

Holocene The period of geologic time since the end of the last glacial advance; approximately 18,000 years.

Holocene transgression The rise of sea level that has taken place since the glaciers began melting about 18,000 years ago.

holoplankton Those organisms that are planktonic throughout their life cycle.

hot spots A location on the surface of the earth where there is a persistent rise of molten magma; usually volcanoes are associated with these locations.

hurricane pass A tidal pass or inlet that is opened as the result of intense waves and high storm surge associated with a hurricane.

hypersaline A salinity level that is above normal concentrations, typically due to evaporation and lack of circulation.

hypsographic curve A graphic representation of the respective elevations and depths on the surface of the earth relative to sea level.

i

infauna Those organisms that burrow or bore into the substrate.

internal waves Waves that develop on the interface that separates two water masses.

island arc A group of volcanic islands, typically arranged in a curving linear pattern and associated with a deep-sea trench.

isobars Contours of equal barometric pressure on a weather map.

isohaline Contours of equal salinity within a water mass.

isostacy A condition analogous to an object floating in a fluid where by the lighter portion of the earth's crust is an equilibrium condition with the denser material below.

j

jetty A coastal structure used to stabilize inlets. It is parallel to the inlet margins and perpendicular to the shoreline.

k

kelp An economically important type of large benthic brown algae that lives in the cold waters of the rocky continental shelf and is harvested for its algin.

krill Small planktonic shrimp-like organisms that are among the primary food of baleen whales.

l

lagoon A type of coastal bay that lacks appreciable influx of fresh water and is restricted from open tidal circulation with the ocean. Commonly the term lagoon is used in a broader sense for all somewhat shore-parallel and shallow coastal bays.

Lagrangian current measurement The type of current measurement that involves tracing the movement of the water.

land breeze The breeze that moves from the land to the sea during the evening as the result of the cooling of the land mass.

lithosphere The outer rigid portion of the earth that includes the crust and outer mantle and extends to a depth of about 100 km.

littoral zone The intertidal benthic zone.

long line A type of commercial fishing apparatus that consists of a line up to several kilometers in length with hooks attached at regular intervals of a few meters.

longshore bar A shore-parallel sandbar that is within the surf zone or just beyond. There may be multiple longshore bars at a given location and they are typically interrupted by rip channels.

longshore current Nearshore currents generated by refracting waves that move parallel to the shoreline.

m

macrotidal The tidal range that is greater than 4 m.

magnetometer An instrument that measures the magnetic field of the earth.

manganese nodules Roughly spherical masses of primarily iron and manganese oxides that are common throughout the deep-ocean floor. They range from only a few millimeters to up to nearly a meter in diameter.

mantle That portion of the earth between the crust and the core.

marginal plateau A plateau-like feature in the ocean that is adjacent to the continental margin and is deeper than the margin, but shallower than the ocean floor.

mariculture The aquaculture that takes place in the marine environment.

marsh The vegetated upper intertidal zone that typically rims an estuary.

mean grain size The statistical average grain size of a sediment population.

mediterranean A large body of salt water that is surrounded by land, but has an opening or openings to the ocean.

meroplankton Organisms that spend only a part of their life cycle as plankton.

mesopelagic zone The part of the oceanic province of the pelagic environment that is between 200 and 1,000 meters in depth.

mesotidal The tidal range between 2 and 4 meters.

microplankton Planktonic organisms that are microscopic, but are easily recovered by a fine mesh net; also called net plankton.

microtidal The tidal range that is less than 2 m.

Mohorovicic discontinuity The rather abrupt discontinuity or boundary between the earth's crust and mantle; also called the Moho.

mud Sediment composed of the subequal mixture of silt and clay particles.

n

nannoplankton Planktonic organisms that are submicroscopic in size and are too small (generally less than 50 micrometers) to be trapped in nets. Also called centrifuge plankton.

natural levee A generally thin, narrow accumulation of sediment along the bank of a river or distributary that is deposited as the stream floods over its banks.

neap tide The minimum tidal range in the lunar tidal cycle, which occurs every two weeks, at first quarter and third quarter.

nearshore The zone just seaward of the beach that includes the longshore bars; essentially equivalent to the zone of breaking waves.

nekton Those organisms that can swim.

neritic province The pelagic part of the ocean that is above the continental shelf.

net plankton Those planktonic organisms that are large enough to be retained by plankton nets of the standard mesh size of 60 microns.

o

oceanic crust The basaltic portion of the earth's crust that underlies the ocean basins and is typically about 5 km thick.

oceanic province The portion of the pelagic environment that is beyond the continental shelf.

oceanic reefs Reefs associated with volcanic islands that rise from the ocean floor.

oceanic ridge The general term for the continuous ridge that extends throughout the world and separates major plates of the lithosphere.

oceanic rise A broad and gently sloping elevation of the ocean floor.

oceanic trench The deep, narrow part of the ocean floor that extends below 6,000 meters.

oligotrophic The condition of aquatic environments when there is little nutrient material.

ooids Small, spherical grains of sand that are formed by concentric precipitation of calcium carbonate around a detrital nucleus.

oozes The general term given to deep-sea sediments that contain at least 30% of biogenic skeletal material.

orthogonals Lines of equal energy on a refraction diagram that are drawn perpendicular to the wave crests.

orthophosphates Compounds or minerals of phosphate that are stable and naturally occuring, as compared to detergents and other synthesized phosphate compounds.

osmoregulation Maintenance of the proper saltwater concentration in a cell or organism through osmosis.

otter trawl A type of net trawl, kept open during operation by two otter boards, which are spread by the passing current.

oxygen minimum zone The zone below the photic zone where oxygen content is low due to the abundance of animals and the absence of producers.

p

paleomagnetism The ancient magnetism preserved in strata that shows the orientation of the earth's magnetic field at the time the minerals formed or were deposited. It is used extensively to support the concept of plate tectonics.

Pangaea The single supercontinent that existed about 200 million years ago, prior to the breakup and drifting of continents toward their present positions on the earth's surface.

partially mixed estuary An estuary in which there is some mixing between the fresh water provided by runoff from land and the marine waters that enter the estuary via tidal currents.

peat An unconsolidated sediment deposit of semicarbonized plant remains with a high water content. Deposits typically accumulate in marshes or swamps.

pelagic sediments Very fine-grained sediments that settle from suspension in the open ocean without benefit of currents.

pelagic zone The entire water environment of the world ocean, including the neritic and oceanic provinces.

period Term used to describe the length of waves: the time it takes for a complete wave length to pass a given point.

pH The negative logarithm of the hydrogen ion activity; commonly considered as the negative log of the hydrogen ion concentration.

phase velocity The rate at which waves are propagated.

phosphorite nodules Generally amorphous calcium phosphate masses that precipitate near the shelf-slope break and are related to upwelling areas.

photic zone The upper layer of the pelagic environment where light penetrates and photosynthesis takes place.

photosynthesis The organic process whereby organisms containing chlorophyll produce carbohydrates from carbon dioxide and water in the presence of sunlight.

phytoplankton Planktonic organisms that have the ability to photosynthesize.

placer deposits Accumulations of dense mineral concentrations following the removal of accompanying relatively light minerals, such as on the beach, in dunes, or tidal channels.

plankton Organisms that float or swim feebly and are not able to control their position in the water column.

planktonic foraminifera Single-celled animals that are floaters and that contribute to deep-sea sediments through their calcium carbonate tests.

plate tectonics The theory of mobility within the earth's crust that accounts for mountain building, earthquakes, volcanism, and the distribution of landmasses.

plunging breakers The type of breaking wave in which the crest steepens and breaks instantaneously.

polar easterlies High-latitude winds that move from the polar regions to lower latitudes, where they converge with the westerlies at the polar front.

pollution The introduction by man of materials into the environment that results in a negative impact on the environment.

polyhaline bay A type of coastal bay that is similar to a lagoon, but has a seasonal salinity that ranges from brackish during the wet season to hypersaline during the dry season.

primary productivity The rate at which photosynthesis takes place.

prodelta The most seaward portion of a delta. It is commonly the thickest of the three major parts and it contains the finest sediment.

pteropods Small planktonic gastropods that have a conical-shaped test that contributes to the calcareous pelagic sediment of the deep sea.

purse seine A type of commercial fishing net that encircles an area and is closed by a drawline similar to that of a lady's purse.

pycnocline The layer in which water density changes most rapidly with respect to depth.

r

radiolaria Microscopic single-celled animals that are planktonic and contribute their siliceous tests to deep-sea sediments.

radiometric age dating The absolute dating of materials or events in the history of the earth using radioactive isotopes of various commonly occuring elements, such as carbon, potassium, uranium, and thorium.

reef An organic framework that is wave-resistant.

reef flat The upper surface of a reef environment, typically controlled by a combination of waves and tidal level.

reflected waves Waves that strike a vertical and impermeable surface have essentially all of their energy reflected, according to Snell's law. Most waves that strike the coast have some of their energy reflected, with the amount controlled by the slope and permeability of the coast.

reflection profiling A type of seismic profiling of the sub-bottom layers of the ocean floor by acoustic signals that reflect off interfaces between layers.

refraction The bending of waves caused by interference of the bottom with wave propagation.

relict sediments Sediments that were deposited in an environment different than the one they now occupy. They are out of equilibrium with their present environment.

residence time The length of time a given element would be expected to remain in solution in the ocean assuming a steady flow and equilibrium conditions. Each element has its own residence time, ranging from just a few years to hundreds of millions of years.

respiration The process by which organisms use organic materials as a source of energy. Oxygen is used and carbon dioxide is produced.

reversing thermometer A thermometer that fixes the temperature recorded by reversing its orientation, thereby enabling determination of temperatures at great depth.

ridge and runnel An intertidal bar and trough produced by the erosion of a beach, and which slowly migrates back toward the beach between storms to repair the erosion.

rift valley A narrow trough formed by faulting in a zone of plate separation.

ring of fire The term given to the zone of volcanic activity and earthquake activity around the Pacific Ocean.

rubble zone The zone of reef debris that occurs on the upper surface of the windward reef.

s

salinity A measure of the dissolved solids in seawater. Specifically, it is defined as the total amount of dissolved solids in parts per thousand by weight when all carbonate is converted to oxide, the bromide and iodide are converted to chloride, and organic matter is oxidized.

salinometer An instrument for determining salinity. It may be based upon electrical conductivity or refractive index.

sea Waves under the direct influence of the wind.

sea breeze A breeze that flows from the sea toward land due to the warming effect of the land during the day.

seafloor spreading The process that produces lithospheric material by the upwelling of magma at the oceanic ridge and movement away from the ridge by the plates.

seamounts Submarine volcanoes that rise at least 1,000 m above the ocean floor.

seawall A type of protective coastal structure that is typically shore-parallel and may be constructed with a variety of materials.

sediment sink An environment where sediment tends to collect and be deposited, for example, estuaries or deltas.

seiches A standing wave that occurs in an enclosed body of water. The wave may have a period of minutes to hours and will continue to move back and forth across the basin after the initiating force has been discontinued.

seismic reflection Measurement and recording of acoustical energy that is reflected from surfaces that separate rock or sediment layers.

seismic refraction The bending or deflection of seismic waves due to differing densities as they pass from one layer to another.

seismology The study of earthquakes or earth vibrations.

sessile A mode of life whereby the organisms live on the bottom and are attached to it without benefit of any mobility.

shallow water wave A surface gravity wave that is being interfered with by the ocean bottom because the wave length is at least 20 times the water depth.

shelf reef Reefs that rise from the continental shelf; contrast with oceanic reefs.

significant wave height The average height of the highest one-third of the waves in a particular series of waves.

silicoflagellates Planktonic photosynthesizers that provide siliceous tests for incorporation in deep sea sediments.

slump The downslope movement of a mass of unconsolidated material which may occur in submarine canyons and may be caused by earthquakes.

sorting value The measure of the spread of distribution or uniformity of particle size in a given sediment population; essentially the standard deviation.

spilling breaker The shallow water breaking wave that is characterized by slow breaking over a modest distance as the wave moves shoreward.

spring tide The maximum tidal range at a given location caused by the alignment of the sun, moon, and earth during the new and full moons every two weeks.

spur and groove Structure commonly formed on the upper slope of the windward reef, composed of buttress-shaped coral ridges or spurs and narrow channels or grooves.

standing crop The amount of living tissue per unit area or volume at a given point in time.

standing waves Waves in which the water surfaces oscillate back and forth between fixed points called nodes; there is no propagation.

stenohaline Pertains to organisms that have little or no tolerance for salinity variation.

storm-dominated shelf The continental shelf environment that has its sediment distribution controlled by infrequent, but intense, storms.

stratified estuary The type of estuary that is characterized by a distinct horizontal layering between the overlying fresh water and the underlying salt water.

stratigraphy The study of the layered sediments and rocks of the earth's crust.

subduction zone The inclined plane where a lithospheric plate descends into the mantle as the result of collision with another plate. It is characterized by high seismic activity.

sublittoral zone The portion of the benthic environment that extends from low tide to the edge of the continental shelf.

submarine canyon One of several types of submarine valleys in the ocean. It is typically V-shaped, steep-sided, deep, and is most prominent on the continental slope.

supralittoral The zone just above high tide; equivalent to supratidal.

supratidal flat The part of the coast that is just above high tide and is inundated only during storms.

surf Shallow-water waves that break due to oversteepening.

surging breaker A type of breaker that steepens and surges onto the beach as contrasted to the spilling and plunging breakers.

swamp An intertidal environment characterized by trees, as contrasted to a marsh, which is vegetated by grasses.

swell Waves that have moved beyond the direct influence of wind.

t

terrigenous sediments Oceanic sediment derived from land and transported by bottom currents.

test The hard mineral exoskeleton of microscopic organisms.

thermistor An instrument that senses the temperature of a medium and is used in temperature-measuring devices.

thermocline The zone in the water column in which there is great temperature change relative to depth.

thermohaline Type of density circulation where the density gradient is caused by variation in temperature and/or salinity.

tidal bore A rapid rise of the tide which takes the form of a progressive wave and which may be meters high in some narrow estuaries having high tidal ranges.

tidal constituents The harmonic elements that collectively comprise the mathematical expression for the tides. Each one is based on some predictable change in the earth-moon-sun system.

tidal deltas Accumulations of sediment at both ends of a tidal inlet as the result of loss of capacity by tidal currents.

tidal flat The unvegetated part of the intertidal zone that is protected from direct wave attack.

tidal inlet A pass or break in a barrier island through which tidal currents pass.

tidal prism The amount of water exchanged in a given area during the tidal cycle; a tidal water budget.

tidal species The individual astronomical frequencies that combine to give the tidal range.

tide The regular and predictable change in water elevation caused by the mass attraction of the earth, moon, and sun.

tide-dominated shelf A continental shelf environment where the physical processes are dominated by tidal currents.

toothed whales The smaller fish-eating whales, which include porpoises, dolphins, killer whales, and sperm whales.

totally mixed estuary An estuary in which the water has been essentially homogenized, with no separation of fresh water and seawater.

trade winds A low-latitude wind system that blows toward the equator from the northeast in the northern hemisphere and from the southeast in the southern hemisphere.

transform faults Faults associated with and perpendicular to the oceanic ridge system, and along which crustal plates move relative to each other.

transgression Landward migration of the shoreline, typically associated with a rising sea level.

transitional wave A wave that is beginning to feel the effects of the bottom and has a wave length between 2 and 20 times the water depth.

transmissivity The ability of a fluid to transmit light; commonly a measurement of the turbidity of a water mass.

triple junction The site where three lithospheric plates meet.

trophic level A successive stage of nourishment as represented by links of the food chain. Primary producers are the lowest and the top carnivore is the highest.

tsunamis A very long-period and high-velocity sea wave produced by some disturbance on the floor of the ocean, such as an earthquake, volcanic eruption, or landslide. Upon reaching shallow water, these waves steepen greatly and may be more than 10 m high.

turbidite The sequence of sediment that is deposited by a turbidity current.

turbidity current A type of sediment gravity flow whereby the density gradient caused by suspended sediment results in flow down the gradient.

U

ultraplankton Plankton smaller than 5 micrometers.

upwelling The rising of deep, cold water caused by diverging oceanic currents or offshore movement of coastal currents influenced by the Coriolis effect.

V

vagrant The mobile mode of benthic life, as contrasted to sessile.

volatile compounds Compounds that readily change or dissociate.

W

washover fan A thin, fan-shaped sediment accumulation caused by storm activity and located on the landward side of a barrier island.

water mass A body of water characterized by a particular set of physical and/or chemical parameters.

wave amplitude One-half of the vertical distance between the wave crest and the adjacent trough.

wave-cut platform A nearly horizontal surface cut in a rocky coast near sea level by wave action.

wave-dominated shelf A continental shelf where physical processes are dominated by waves; synonymous with a storm-dominated shelf.

wave height The vertical distance between the wave crest and the adjacent trough.

wave length The horizontal distance between two corresponding points on successive waves, such as crest to crest.

wave period The time that is takes for two successive corresponding points on a wave to pass a given point. An indirect way to measure wave length.

wave steepness The ratio of the wave height to the wave length.

westerlies A belt of prevailing winds in the mid-latitudes where air moves away from the subtropical belt toward the higher latitudes.

wind-driven circulation Surface oceanic circulation that is driven by prevailing atmospheric circulation patterns.

wind tidal flats Tidal flats that are inundated by water only under onshore wind conditions; supratidal flats.

Z

zooplankton The animals that occupy the planktonic mode of life.

zooxanthellae Microscopic brown algae that live symbiotically with hermatypic corals.

Credits

ILLUSTRATIONS

Chapter 1

fig. 1.1: From Groen, P., *The Waters of the Sea.* © 1967 Van Nostrand Reinhold, New York, N.Y. **fig. 1.3:** Chart courtesy of Philip L. Richardson, Woods Hole Oceanographic Institute, Woods Hole, MA. **fig. 1.5:** W. J. Cromie, *Exploring the Secrets of the Sea,* © 1962, p. 38. Adapted by permission of Prentice-Hall, Englewood Cliffs, New Jersey. **fig. 1.6:** From Hill, M. N., (editor), *The Sea,* Volume 3. Copyright © 1963 John Wiley & Sons, Inc., New York, NY. Reprinted by permission of John Wiley & Sons, Inc.

Chapter 2

fig. 2.1: Duxbury & Duxbury, *The World's Oceans,* © 1984, Addison-Wesley Publishing Company, Inc., Reading, Massachusetts. Figs. 1.20, 2.6, 3.12, 4.12, 7.16, 8.15, and 14.15. Reprinted with permission. **figs. 2.4 and 2.15:** From Plummer, Charles C. and David McGeary, *Physical Geology,* 3d ed., © 1979, 1982, 1985 Wm. C. Brown Publishers, Dubuque, Iowa. All Rights Reserved. Reprinted by permission. **fig. 2.8:** From Guilcher, A., "Continental Shelf and Slope," in *The Sea,* M. N. Hill (editor), vol. 3. Copyright © 1963 John Wiley & Sons, Inc., New York, NY. Reprinted by permission of John Wiley & Sons, Inc. **fig. 2.9:** Adapted from Figure 150 (p. 322) in *Submarine Geology,* 2nd edition by Francis P. Shepard. Copyright 1948, 1963 by Francis P. Shepard. Reprinted by permission of Harper & Row, Publishers, Inc. **fig. 2.10:** From C. D. Hollister and B. C. Heezen, *Studies in Physical Oceanography,* A. Gordon (editor), vol. 2, p. 59. Copyright © 1972 Gordon & Breach Publishers, Inc., New York, NY. Reprinted by permission. **fig. 2.12:** Duxbury & Duxbury, *The World's Oceans,* © 1984, Addison-Wesley Publishing Company, Inc., Reading, Massachusetts. Pg. 43, Fig. 2.8. Reprinted with permission. **fig. 2.14:** From Heezen, B. C., and M. Ewing, *The Sea,* M. N. Hill, (editor), vol. 3, p. 405. Copyright © 1963 John Wiley & Sons, Inc., New York, NY. Reprinted by permission of John Wiley & Sons, Inc.

Chapter 3

figs. 3.1 and 3.5: Drawings from *Continental Drift* by Don and Maureen Tarling. Copyright © 1971, 1975 by G. Bell & Sons Ltd. Reprinted by permission of Doubleday & Company, Inc. G. Bell & Sons, Ltd., London, England. **fig. 3.2:** After A. Wegener, 1924 (reprinted 1966), *The Origin of the Continents and Oceans,* Dover, Mineola, New York. **fig. 3.3:** From Bullard, Sir E., J. E. Everett, and A. G. Smith, "The Fit of the Continent Around the Atlantic," *Philosophical Transactions,* vol. 258. © 1965 The Royal Society, London, England. Reprinted by permission of the publisher and the authors. **fig. 3.8:** From Baranzangi and Dorman, *Bulletin of Seismological Society of America,* 1969. Reprinted by permission. **fig. 3.10:** After Holmes, 1965, *Principles of Physical Geology,* Van Nostrand Reinhold (UK) Wokingham UK. **fig. 3.12a:** From Plummer, Charles C., and David McGeary, *Physical Geology,* 3d ed. © 1979, 1982, 1985 Wm. C. Brown Publishers, Dubuque, Iowa. All Rights Reserved. Reprinted by permission. **fig. 3.13:** From a map by W. C. Pitman, III, R. L. Larson, and E. M. Herron, 1974, Geological Society of America. Reprinted by permission. **fig. 3.15:** Duxbury & Duxbury, *The World's Oceans,* © 1984, Addison-Wesley Publishing Company, Inc., Reading, Massachusetts. Figs. 1.20, 2.6, 3.12, 4.12, 7.16, 8.15, and 14.15. Reprinted with permission.

Chapter 4

fig. 4.2: From Kennet, James, *Marine Geology.* © 1982 Prentice-Hall, Inc., Englewood Cliffs, NJ, adapted from Pipkin, B. W., D. S. Gorsline, R. E. Casey, & D. E. Hammond, *Laboratory Exercises in Oceanography.* Copyright © 1977 W. H. Freeman Company and Publishers. Reprinted by permission. **figs. 4.3, 4.8, and 4.9:** Reprinted with permission from Pickard, G. L., and W. J. Emery, *Descriptive Physical Oceanography,* Copyright 1982, Pergamon Books Ltd. **fig. 4.5:** Sverdrup/Johnson/Fleming, *The Oceans,* © 1942, p. 105. Adapted by permission of Prentice-Hall, Inc., Englewood Cliffs, New Jersey. **fig. 4.6:** Copyright, 1969, Bell Telephone Laboratories, Incorporated, reprinted by permission. **fig. 4.7:** Duxbury & Duxbury, *The World's Oceans,* © 1984, Addison-Wesley Publishing Company, Inc., Reading, Massachusetts. Figs. 1.20, 2.6, 3.12, 4.12, 7.16, 8.15, and 14.15. Reprinted with permission.

Chapter 5

fig. 5.3: From Plummer, Charles C., and David McGeary, *Physical Geology,* 3d ed. © 1979, 1982, 1985 Wm. C. Brown Publishers, Dubuque, Iowa. All Rights Reserved. Reprinted by permission; and after Bowditch, N., 1977, *American Practical Navigator,* Volume 1, U.S. Defense Mapping Agency, Hydrographic Center, Washington, DC. **fig. 5.4:** Base map by permission of Rand McNally and Company. **fig. 5.5:** From W. L. Donn, 1951, *Meteorology with Marine Applications,* second edition, McGraw-Hill Co., New York. Reprinted by permission. **fig. 5.9:** After J. Williams, 1962, *Oceanography, An Introduction to the Marine Sciences,* Little, Brown & Co., Inc., Boston, p. 136. Reprinted by permission of the author. **fig. 5.10:** From Stow, K. S., *Ocean Science.* Copyright © 1979 John Wiley & Sons, Inc., New York, NY. All Rights Reserved. Reprinted by permission of John Wiley & Sons, Inc. **fig. 5.12:** From von Arx, W. S., *An Introduction to Physical Oceanography.* Copyright © 1962 Addison-Wesley, Reading, MA. Reprinted by permission of The Benjamin/Cummings Publishing Company, Menlo Park, CA. **fig. 5.13:** From Stommel, H., "Summary Charts of the Mean Dynamic Topography and Current Field at the Surface of the Ocean, and Related Functions of Mean Wind-Stress," *Studies in Oceanography.* Copyright © 1964 University of Washington Press, Seattle, WA. All Rights Reserved. Reprinted by permission. **fig. 5.14:** Duxbury & Duxbury, *The World's Oceans,* © 1984, Addison-Wesley Publishing Company, Inc., Reading, Massachusetts. Figs. 1.20, 2.6, 3.12, 4.12, 7.16, 8.15, and 14.15. Reprinted by permission. **fig. 5.16:** From Ross, David A., *Introduction to Oceanography.* © 1982 Prentice-Hall, Inc., Englewood Cliffs, NJ, adapted from Hartline, B. K., "Coastal Upwelling: Physical Factors Feed Fish," *Science,* Vol. 208, pp. 38–40, Fig. 4, April 1980. Copyright 1980 by the American Association for the Advancement of Science. Reprinted by permission. **fig. 5.17:** Reprinted with permission from H. Stommel, "A Survey of Ocean Current Theory," *Deep Sea Research,* Vol. 5, p. 82, copyright 1958, Pergamon Press. **fig. 5.18:** Sverdrup/Johnson/Fleming, *The Oceans,* © 1942, © renewed 1970, p. 164. Adapted by permission of Prentice-Hall, Englewood Cliffs, New Jersey. **fig. 5.19a:** John A. Knauss, *Introduction to Physical Oceanography,* © 1978, p. 168. Reprinted by permission of Prentice-Hall, Englewood Cliffs, New Jersey. **fig. 5.19b:** From Worthington and Wright, *North Atlantic Ocean Atlas, Woods Hole Oceanographic Institution Atlas Series,* Vol. II. © 1970 Woods Hole Oceanographic Institute, Woods Hole, MA. Reprinted by permission.

Chapter 6

fig. 6.7: From: Kinsman, Blair: *Wind Waves,* Dover Publications, Inc., New York, 1984. Reprinted by permission. **fig. 6.9:** Richard A. Davis, Jr., *Principles of Oceanography,* © 1977, Addison-Wesley Publishing Company, Inc., Reading, Massachusetts. Pg. 125, Fig. 7.10. Reprinted by permission. **fig. 6.11:** From Dietrich, G., *General Oceanography.* © 1963 John Wiley & Sons, Inc., New York, NY. Reprinted by permission of Gebruder Borntraeger, Stuttgart, FEDERAL REPUBLIC OF GERMANY. **fig. 6.13:** Duxbury & Duxbury, *The World's Oceans,* © 1984, Addison-Wesley Publishing Company, Inc., Reading Massachusetts. Figs. 1.20, 2.6, 3.12, 4.12, 7.16, 8.15, and 14.15. Reprinted with permission.

Chapter 7

figs. 7.7 and 7.9: From von Arx, W. S., *An Introduction to Physical Oceanography.* Copyright © 1962 Addison-Wesley, Reading, MA. Reprinted by permission of The Benjamin/Cummings Publishing Company, Menlo Park, CA. **fig. 7.10:** From Davies, J. L., *Geographic Variation in Coastal Development.* © 1980 Longman Group Limited, England. Reprinted by permission.

Chapter 9

fig. 9.13a: After Sir A. Hardy, 1959, *The Open Sea: Its Natural History.* Used by permission of Houghton Mifflin Company and Collins Publishers, London, England. **figs. 9.21 and 9.26a:** Duxbury & Duxbury, *The World's Oceans,* © 1984, Addison-Wesley Publishing Company, Inc., Reading, Massachusetts. Figs. 1.20, 2.6, 3.12, 4.12, 7.16, 8.15, and 14.15. Reprinted with permission.

Chapter 10

fig. 10.4: After F. Hjulstrom, 1939, *Recent Marine Sediments,* American Association of Petroleum Geologists, Tulsa, OK. Reprinted by permission. **fig. 10.10:** From Davies, T. A., and D. S. Gorsline, *Chemical Oceanography,* Part 5, 2d edition, edited by J. P. Riley and R. Chester. Copyright © 1976 Academic Press, Orlando, FL. Reprinted by permission of the publisher and the authors. **fig. 10.12:** From the National Science Foundation and Deep Sea Drilling Project news release, 1970. **fig. 10.14:** From Hersey, J. B., "Continuous Reflection Profiling," in *The Sea,* M. N. Hill, (editor), vol. 3, p. 47. Copyright © 1963 John Wiley & Sons, Inc., New York, NY. Reprinted by permission of John Wiley & Sons, Inc. **fig. 10.17:** From Hill, M. N., "Single Ship Seismic Refraction Shooting," *The Sea,* M. N. Hill, (editor), vol. 3, p. 40. Copyright © 1963 John Wiley & Sons, Inc., New York, NY. Reprinted by permission of John Wiley & Sons, Inc.

Chapter 11

figs. 11.7 and 11.14: From R. A. Davis, ed., 1985, *Coastal Sedimentary Environments,* Springer-Verlag, Inc., New York, NY. Reprinted by permission. **fig. 11.11:** From Sumich, James L., *An Introduction to the Biology of Marine Life,* 3d ed. © 1976, 1980, 1984 Wm. C. Brown Publishers, Dubuque, Iowa. All Rights Reserved. Reprinted by permission.

Chapter 12

figs. 12.1 and 12.2: From R. A. Davis, ed., 1985, *Coastal Sedimentary Environments,* Springer-Verlag, Inc., New York, NY. Reprinted by permission. **fig. 12.7:** From Galloway, W. E., *Deltas, Models for Exploration,* edited by M. L. Broussard. Copyright © 1975 Houston Geological Society, Houston, TX. Reprinted by permission.

Chapter 13

fig. 13.26: After Rusnak, G. A., 1960, in *Recent Sediments Northwest Gulf of Mexico,* Shepard, F. P. and Moore, D. G., editors, American Association of Petroleum Geologists, Tulsa, OK. Reprinted by permission. **fig. 13.34:** From Hill, G. W., and Hunter, R. E., 1976, in *Beach and Nearshore Sedimentation,* Davis, R. A., and Ethington, R. L., editors, S.E.P.M. Special Publication #24, Tulsa, OK. Reprinted by permission.

Chapter 15

fig. 15.7: From J. L. Wilson, 1975, *Carbonate Facies in Geologic History,* Springer-Verlag, Inc., New York, NY. Reprinted by permission. **fig. 15.8:** From Maxwell, W. G. H., *Atlas of the Great Barrier Reef.* Copyright © 1968 Elsevier Science Publishers, Amsterdam, The Netherlands. Reprinted by permission. **fig. 15.13a:** From E. A. Shinn, 1963, *Journal of Sedimentary Petrology,* v. 33, p. 295. Reprinted by permission of the Society of Economic Paleontologists and Mineralogists, Tulsa, OK. **fig. 15.23:** From C. M. Hoskin, 1963, *Recent Carbonate Sediments on Alacran Reef.* National Academy of Science, National Research Council, Publication #1089, Washington, DC. Reprinted by permission of the author. **box fig. 15.1:** From Sumich, James L., *An Introduction to the Biology of Marine Life,* 3d ed. © 1976, 1980, 1984 Wm. C. Brown Publishers, Dubuque, Iowa. All Rights Reserved. Reprinted by permission.

Chapter 16

fig. 16.1: From Emery, K. O., *The Sea off Southern California.* Copyright © 1960 John Wiley & Sons, Inc., New York, NY. Reprinted by permission of John Wiley & Sons, Inc. **fig. 16.4:** Figure 10.1 (p. 280) from *Submarine Geology,* 3rd edition by Francis P. Shepard. Copyright 1948, 1963, 1973 by Francis P. Shepard. Reprinted by permission of Harper & Row, Publishers, Inc. **fig. 16.5:** From Redfield, A. C., "The Influence of the Continental Shelf on the Tides of the Atlantic Coast of the United States," *Journal of Marine Research,* 17:432–448, 1958. Reprinted by permission of Woods Hole Oceanographic Institution, Woods Hole, MA. **fig. 16.6:** From Houbolt, J. J. H. C., 1968, *Geologie en Mijnbouw,* v. 47, pp. 245–273. Reprinted by permission. **fig. 16.8:** H. E. Wright, Jr. and David G. Frey, eds., *The Quaternary of the United States.* Copyright © 1965 by Princeton University Press. Fig. 2, p. 725, reprinted with permission of Princeton University Press. **fig. 16.9:** From Plummer, Charles C., and David McGeary, *Physical Geology,* 3d ed. © 1979, 1982, 1985 Wm. C. Brown Publishers, Dubuque, Iowa. All Rights Reserved. Reprinted by permission. **figs. 16.10 and 16.11:** After Emery, K. O., 1968, *Bulletin,* American Association of Petroleum Geologists, Tulsa, OK. Reprinted by permission. **fig. 16.12:** From Hayes, M. O., *Marine Geology,* vol. 5, p. 123. Copyright 1967 Elsevier Science Publishers, Amsterdam, The Netherlands. Reprinted by permission. **fig. 16.13:** From Milliman, J. M., et al., *Geological Society of America Bulletin,* vol. 83, pp. 1315–1333, copyright 1972. Reprinted by permission of the authors.

Chapter 17

figs. 17.2a–c and 17.11: From Plummer, Charles C., and David McGeary, *Physical Geology,* 3d ed. © 1979, 1982, 1985 Wm. C. Brown Publishers, Dubuque, Iowa. All Rights Reserved. Reprinted by permission. **fig. 17.3a:** After Gorsline, D. S., 1978, *Journal of Sedimentary Petrology,* vol. 48, p. 1056. Reprinted by permission of the Society of Economic Paleontologists and Mineralogists, Tulsa, OK. **fig. 17.4:** From Middleton, G. V., and M. A. Hampton, "Sediment Gravity Flows: Mechanics of Flow and Deposition," Part I, pp. 1–38 in Middleton, G. V., and A. H. Bouma, *Turbidites and Deep-Water Sedimentation.* © 1973 Society of Economic Paleontologists and Mineralogists, Tulsa, OK. Reprinted by permission. **fig. 17.9:** From G. T. Rowe and R. L. Haedrich, 1979, Society of Economic Paleontologists and Mineralogists, Special Publication No. 17, Tulsa, OK. Reprinted by permission. **fig. 17.14:** From Heezen, B. C., et al., "Shaping of the Continental Rise by Deep Geostropic Contour Currents," *Science,* Vol. 152, p. 507, Fig. 4. Copyright 1980 by the American Association for the Advancement of Science. Reprinted by permission.

Chapter 18

fig. 18.4: From Stowe, K. S., *Ocean Science.* Copyright © 1979 John Wiley & Sons, Inc., New York, NY. All Rights Reserved. Reprinted by permission of John Wiley & Sons, Inc. **fig. 18.9:** Reprinted with permission from H. Stommel, "A Survey of Ocean Current Theory," *Deep Sea Research,* Vol. 5, p. 82, copyright 1958, Pergamon Press. **figs. 18.10, 18.11, and 18.12:** From *Oceanography: An Introduction* by Ingmanson, D. E., and W. J. Wallace. © 1985 by Wadsworth, Inc. Used with permission of the publisher. **figs. 18.13 and 18.14:** From Sumich, James L., *An Introduction to the Biology of Marine Life,* 3d ed. © 1976, 1980, 1984 Wm. C. Brown Publishers, Dubuque, Iowa. All Rights Reserved. Reprinted by permission.

Chapter 19

fig. 19.9: From Kennet, James, *Marine Geology.* © 1982 Prentice-Hall, Inc., Englewood Cliffs, NJ, adapted from Biscaye, P. E., & S. L. Eittreim, "Suspended Particulate Loads & Transports in the Nepheloid Layer of the Abyssal Atlantic Ocean," *Mar. Geol.* 23:155–72, 1977. **fig. 19.19:** From Davies, T. A., and D. S. Gorsline, *Chemical Oceanography,* Part 5, 2d edition, edited by J. P. Riley and R. Chester. Copyright © 1976 Academic Press, Orlando, FL. Reprinted by permission of the publisher and the authors.

Chapter 20

fig. 20.12: From McCormick, J. M., and J. V. Thiruvathukal, *Elements of Oceanography,* 2d ed. © 1981 W. B. Saunders Company, Philadelphia, PA, adapted from Kesler, S. E., *Our Finite Mineral Resources.* © 1976 McGraw-Hill Book Company. Reprinted by permission. **box fig. 20.1:** From Sumich, James. L., *An Introduction to the Biology of Marine Life,* 3d ed. © 1976, 1980, 1984, Wm. C. Brown Publishers, Dubuque, Iowa. All Rights Reserved. Reprinted by permission. (Adapted from Schaefer 1970 and FAO catch and landing statistics).

PHOTOS

all part openers: © Marty Snyderman. **chapter openers 1, 4, 18:** NOAA. **3:** © Marty Snyderman. **15:** National Park Service.

Chapter 1

fig. 1.7a,b: NOAA, courtesy A. C. Newman. **fig. 1.8a,b:** Scripps Institution of Oceanography, courtesy Deep Sea Drilling Project. **fig. 1.9:** Courtesy Oceanic Drilling program. **fig. B1.1:** NOAA.

Chapter 5

fig. 5.7: NOAA.

Chapter 6

fig. 6.2: United States Coast Guard.

Chapter 9

fig. 9.3b: Courtesy W. W. Hay. **fig. 9.4:** Scripps Institution of Oceanography. **fig. 9.15b:** © William J. Weber/Visuals Unlimited. **fig. 9.24:** NOAA. **fig. 9.26b:** © C. R. Wyttenbach, University of Kansas/Biological Photo Service. **fig. 9.37b:** © James Sumich.

Chapter 10

fig. 10.8: Courtesy Professor Sherwood W. Wise, Jr. **fig. 10.15:** Courtesy O. H. L. Pilkey.

Chapter 11

fig. 11.1: National Parks Service. **fig. 11.12:** © H. W. Pratt/Biological Photo Service.

Chapter 12

fig. 12.14: Courtesy L. D. Wright.

Chapter 14

fig. 14.2: U.S. Geological Survey. **fig. 14.16:** © J. Robert Waaland/Biological Photo Service.

Chapter 16

fig. B16.3: Courtesy A. D. Short.

Chapter 17

fig. 17.8: Scripps Institution of Oceanography.

Chapter 18

fig. 18.15: Scripps Institution of Oceanography. **fig. 18.16:** © James Sumich.

Chapter 20

fig. 20.1: Courtesy Kelco Industries. **figs. 20.2, 20.4, 20.5, 20.6, 20.7, 20.10:** NOAA.

Chapter 21

figs. 21.10, 21.11, 21.13: NOAA.

ILLUSTRATORS

Martha Blake

Figures 8.1, 8.2, 8.4, 8.7, 9.2, 9.3a, 9.4, 9.6a & b, 9.7a, 9.8, 9.9, 9.10, 9.11, 9.12, 9.13a & b, 9.14a, 9.15a, 9.16a, 9.18, 9.19, 9.21, 9.26a, 9.29a, 9.30, 9.31a, 9.32a, 9.36, 9.37a, 9.38a, 9.40a & b, 9.43a, Box 9.2, Box 9.3, 14.18

Bowring Cartographics

Figures 1.5, 1.6, 2.8, 2.12, 2.15, 3.1, 3.2a–c, 3.3, 3.4, 3.5, 3.6, 3.14, 4.3, 4.8, 5.4, 5.10, 5.13, 5.16, Box 5.1, 7.7, 7.8, 7.10, 8.9, 10.10, 10.12, 11.8, 13.1, 13.26, 13.41, 15.7, 16.1, 16.6, 16.13, 18.4, 18.14, 19.4, 19.6, 19.19, 20.3, 20.12

Fine line

Figures 1.2, 2.2, 2.3, 2.9, 2.14, Box 2.1a & b, 4.1, 4.2, 4.4, 4.5, 4.6, 4.7a & b, 4.9, 4.10, 4.11, 4.12, Box 4.1a, 5.9, 6.7, 7.1a & b, 7.2, 7.3, 7.9, 8.3, 8.5, 8.6, 8.8, 8.10, 10.4, 10.11, 10.14, 10.17, 11.10, 12.7, 13.10c, 13.34, 15.23, 16.5, 16.8, 16.10, 16.11, Box 16.1, 17.9, 18.1, 18.2, 18.3, 18.5, 18.6, 18.7, 18.8, 18.13, 19.3, 19.17, 19.20, 20.8

Norm Frisch

Figures 1.1, 2.1, 2.5, 2.6, 2.7, 2.10, 2.11, 3.7, 3.10, 3.11, 3.12a & b, 3.15, 5.1, 5.3, 5.5a & b, 5.6, 5.11, 5.12, 5.14, 5.15a & b, 5.18, 5.19a & b, 6.1, 6.8, 6.9, 6.10, 6.11, 6.12, 6.13, 6.15, 6.16, Box 7.1a & b, 10.2a & b, 10.3a & b, 10.13, 11.2a & b, 11.4a–c, 11.5a & b, 11.6a–c, 11.7a & b, 11.9, 11.11, 11.14, 11.25, 12.2, 12.5a, 12.6, 12.8, 12.10, 12.12, 13.1, 13.2, 13.7, 13.12a, 13.16, 13.17, 13.27, 13.29a & b, 13.30, 13.32, 13.33, 14.4, 14.8, 14.11, 15.5, 15.8, 15.13a, 16.2, 16.4a & b, 16.7, 16.9a & b, 16.12, 17.1, 17.3a & b, 17.4, 17.5, 17.6, 17.7, 17.10, 17.13, 17.14, 18.10, 18.11, 18.12, 19.1, 19.2, 19.5, 19.9, 19.10, 19.12, 21.4, 21.14, 21.15

Index